National Center for Construction Education and Research

AF834670

Pipeline Corrosion Control Level One

Annotated Instructor's Guide

Prentice Hall

Upper Saddle River, NJ
Columbus, OH

contren™ Learning Series

American Petroleum Institute

Pipeliner Training and Assessment Program (PTAP)

Prentice Hall
nccer
contren™ Learning Series

10 9 8 7
ISBN 0-13-046685-9

Preface

This volume was developed by the National Center for Construction Education and Research (NCCER) in response to the training needs of the construction, maintenance, and pipeline industries. It is one of many in NCCER's *Contren™ Learning Series*. The program, covering training for close to 40 construction and maintenance areas, and including skills assessments, safety training, and management education, was developed over a period of years by industry and education specialists.

NCCER also maintains a National Registry that provides transcripts, certificates, and wallet cards to individuals who have successfully completed modules of NCCER's *Contren™ Learning Series*, when the training program is delivered by an NCCER Accredited Training Sponsor.

The NCCER is a not-for-profit 501(c)(3) education foundation established in 1995 by the world's largest and most progressive construction companies and national construction associations. It was founded to address the severe workforce shortage facing the industry and to develop a standardized training process and curricula. Today, NCCER is supported by hundreds of leading construction and maintenance companies, manufacturers, and national associations, including the following partnering organizations:

PARTNERING ASSOCIATIONS

- American Fire Sprinkler Association
- American Petroleum Institute
- American Society for Training & Development
- American Welding Society
- Associated Builders & Contractors, Inc.
- Association for Career and Technical Education
- Associated General Contractors of America
- Carolinas AGC, Inc.
- Carolinas Electrical Contractors Association
- Citizens Democracy Corps
- Construction Industry Institute
- Construction Users Roundtable
- Design-Build Institute of America
- Merit Contractors Association of Canada
- Metal Building Manufacturers Association
- National Association of Minority Contractors
- National Association of State Supervisors for Trade and Industrial Education
- National Association of Women in Construction
- National Insulation Association
- National Ready Mixed Concrete Association
- National Utility Contractors Association
- National Vocational Technical Honor Society
- North American Crane Bureau
- Painting & Decorating Contractors of America
- Portland Cement Association
- SkillsUSA-VICA
- Steel Erectors Association of America
- Texas Gulf Coast Chapter ABC
- U.S. Army Corps of Engineers
- University of Florida
- Women Construction Owners & Executives, USA

Some features of NCCER's *Contren™ Learning Series* are:

- An industry-proven record of success
- Curricula developed by the industry for the industry
- National standardization providing portability of learned job skills and educational credits
- Credentials for individuals through NCCER's National Registry
- Compliance with Apprenticeship, Training, Employer, and Labor Services (ATELS) requirements for related classroom training (CFR 29:29)
- Well-illustrated, up-to-date, and practical information

Acknowledgments

This curriculum is a product of the farsightedness and leadership of the following sponsors:

Bass Engineering Company
Enhance Co., Inc.
Green Equipment Company
Marathon Ashland Pipeline
TEPPCO
Enbridge Energy Partners
Equistar Pipeline
Koch Pipeline Company
Phillips Pipeline Company
Valero Logistics Operations

This curriculum would not exist were it not for the dedication and unselfish energy of those volunteers who served on the Authoring Team. A sincere thanks is extended to:

Clay Brelsford
Roger Floyd
Wayne Guild
Kimberly Harris
David Selvage
Keith Wooten
Allen Carlile
Dan Franzese
Greg Guy
Tom Pickthall
Stephen Waguespack
Will Wright

A final note: This book is the result of a collaborative effort involving the production, editorial, and development staff at Prentice-Hall, Inc., and the National Center for Construction Education and Research. Thanks to all of the dedicated people involved in the many stages of this project.

Contents

Module 66101-02

Introduction to the Pipeline Industry

Introduction to the Pipeline Industry
Annotated Instructor's Guide

Module 66101-02

MODULE OVERVIEW

This module provides an overview of the pipeline industry. Topics covered include basic functions and purposes of pipelines and facilities, roles of control personnel, and operations procedures.

PREREQUISITES

Please refer to the Course Map in the Trainee Module. Prior to training with this module, it is recommended that the trainee shall have successfully completed the following:

Core Curriculum

OBJECTIVES

Upon completion of this module, the trainee will be able to:

1. Explain the basic functions and purposes of pipelines and facilities and identify the characteristics and hazards of common pipeline products.
2. Identify maps and drawings used to depict pipelines and facilities.
3. Explain the roles of control personnel and equipment in the overall operation of a pipeline.
4. Explain liquid pipeline hydraulics and gas pipeline pneumatics.
5. Explain the types and purposes of pipeline equipment.
6. Explain pipeline electrical power systems and corrosion control.
7. Review operations, maintenance, and emergency procedures and perform documentation required for pipeline operations.

PERFORMANCE TASKS

This is a knowledge-based module – there is no performance testing.

NCCER STANDARDIZED CRAFT TRAINING PROGRAM

The National Center for Construction Education and Research (NCCER) provides a standardized national program of accredited craft training. Key features of the program include instructor certification, competency-based training, and performance testing. The program provides trainees, instructors, and companies with a standard form of recognition through a National Craft Training Registry. The program is described in full in the *Guidelines for Accreditation*, published by the NCCER. For more information on standardized craft training, contact the NCCER by writing us at P.O. Box 141104, Gainesville, FL 32614-1104, calling 352-334-0911, or e-mailing info@nccer.org. More information may be found at our Web site, www.nccer.org.

HOW TO USE THIS ANNOTATED INSTRUCTOR'S GUIDE

Each page presents two sections of information. The larger section displays each page exactly as it appears in the Trainee Module. The narrow column ties suggested trainee and instructor actions to each page and provides icons to call your attention to material, safety, audiovisual, or testing requirements. The bottom of each page includes space for your notes.

Teaching Tip

If you see the Teaching Tip icon, that means there is a teaching tip associated with this section. Also refer to any suggested teaching tips at the end of the module.

SAFETY CONSIDERATIONS

Ensure that the trainees are equipped with appropriate personal protective equipment. Emphasize the importance of following all safety precautions and procedures when working with power tools.

PREPARATION

Before teaching this module, you should review the Module Outline, Objectives, Performance Tasks, and the Materials and Equipment List. Be sure to allow ample time to prepare your own training or lesson plan and gather all required equipment and materials.

MATERIALS AND EQUIPMENT LIST

Transparencies

Markers/chalk

Blank acetate sheets

Transparency pens

Pencils and scratch paper

Copies of Quick Quiz*

Module Examinations**

Overhead projector and screen

Whiteboard/chalkboard

Appropriate personal protective equipment

Copies of your company's policy and procedures manual

Samples of P&IDs, blueprints, and strip maps/alignment sheets

Posterboard or other large sheets of paper

Colored markers

Assorted valves

Copies of *API 650* and *API 653*

Copies of *API 510* and *2510*

Assorted fittings

Copies of *49 CFR Part 192* (Gas) and/or *49 CFR Part 195* (Liquid), including *Parts 192.327* and *195.248*

Assorted corrosion coupons

Various examples of documentation, including:
Operations logs
Work orders
Event logs

* Located at the end of this Annotated Instructor's Guide.
** Located in the Test Booklet packaged with this Annotated Instructor's Guide.

TEACHING TIME FOR THIS MODULE

An outline for use in developing your lesson plan is presented below. Note that each Roman numeral in the outline equates to one session of instruction. Each session has a suggested time period of 2½ hours. This includes 10 minutes at the beginning of each session for administrative tasks and one 10-minute break during the session. Approximately 15 hours are suggested to cover *Introduction to the Pipeline Industry*. You will need to adjust the time required for hands-on activity and testing based on your class size and resources.

Topic	Planned Time
Session I. Introduction	
A. Origination	________
B. Transport	________
1. Laboratory	________
C. Storage	________
1. Laboratory	________
D. Delivery	________
Session II. Pipeline Products and Map Drawings	
A. Natural Hydrocarbons	________
B. Refined Products	________
C. Special Products	________
D. Protective Measures	________
E. Piping and Instrumentation Diagrams	________
F. Blueprints	________
G. Strip Maps/Alignment Sheets	________
Session III. Operations, Fluids and System Hydraulics, and Gases and System Pneumatics	
A. Scheduling	________
B. Pipeline Operations	________
C. Control Systems	________
D. Communications	________
E. Fluids	________
F. Basic Hydraulics	________
G. The Ideal Gas Laws	________
H. Measuring Energy Gases	________
I. Commercial Natural Gas Combustible Limits and Specific Gravity	________
Session IV. Pipeline Equipment	
A. Pumps and Motors	________
B. Compressors	________
C. Valves	________
1. Laboratory	________
D. Actuators/Operators	________
E. Tanks	________
F. Pressure Vessels	________
G. Pipe	________
H. Fittings	________
I. Manifolds	________
J. Traps	________
K. Measuring/Sampling Equipment	________

Session V. Electrical Power Systems and Corrosion Control

A. Power Generation and Transmission ____________

B. Power Distribution ____________

C. DC Power Distribution ____________

D. Types and Causes of Corrosion ____________

   1. Laboratory ____________

E. Inspection and Testing ____________

F. Maintenance and Protection ____________

Session VI. Regulations, Procedures, and Documentation

A. Regulations ____________

B. Abnormal Operating Conditions ____________

C. Company Policies and Procedures ____________

D. Qualifications ____________

E. Documentation ____________

F. Summary

   1. Summarize module ____________
   2. Answer questions ____________

G. Module Examination ____________

   1. Trainees must score 70% or higher to receive recognition from the NCCER.
   2. Record the testing results on Craft Training Report Form 200 and submit the results to the Training Program Sponsor.

Module 66101-02

Introduction to the Pipeline Industry

Instructor's Notes:

Instructor's Notes:

Introduction to the Pipeline Operator Qualification Rule, API, and NCCER

It could be said that petroleum products keep America running. Petroleum products like oil and natural gas are the principal sources of energy in the United States and are used in many manufacturing and chemical processes. Petroleum products are primarily transported by pipeline. There are more than a million miles of liquid and natural gas pipeline in this country, ranging in diameter from two inches to four feet. These pipelines bring crude oil and natural gas from production fields to processing plants and refineries—and ultimately to the consumer.

Pipeline Operator Qualification

Petroleum products are considered hazardous, so pipeline personnel must take extra precautions and follow specific procedures when working with them. To help ensure the safe handling and transportation of petroleum products, the United States Department of Transportation (DOT) mandates safe practices as federal regulations.

One such regulation, the Operator Qualification (OQ) rule, requires pipeline operators to develop a written qualification plan to evaluate their personnel and their contract personnel (see *49 CFR Part 192 Subpart N—Gas,* and *49 CFR Part 195 Subpart G—Liquid*). The evaluation hinges on the operator's ability to perform certain covered tasks and to be able to recognize and respond to abnormal operating conditions that may be encountered while performing these covered tasks.

Covered Tasks

A pipeline performance task is deemed *covered* if the activity passes the following four-part test:

1. Is it performed on a pipeline facility?
2. Is it an operations or maintenance task?
3. Is it performed as a requirement of *49 CFR Part 195 (Liquid)* or *49 CFR Part 192 (Gas)*?
4. Does it affect the operation or integrity of the pipeline facility?

If the answer to all four questions is *yes*, then the activity is considered *covered* and must figure into the evaluation process of pipeline personnel.

American Petroleum Institute

The OQ rule and its related covered tasks provided the catalyst to develop a unified pipeline training and assessment program—so that pipeline operators and owners have ready access to the assessment and training tools they need. This effort was the direct result of the vision and leadership provided by The American Petroleum Institute (API) and API members.

Efforts to unify the industry are nothing new to API. Indeed, the impetus for forming the Institute in 1919 was the need to standardize engineering specifications for drilling and production equipment. Today, the most pressing issues revolve around public perceptions and government policies toward the industry. Speaking with one voice on these issues has become as essential as having interchangeable parts in the field. API's members determine the petroleum industry's positions on public policy as well as the standards that govern its day-to-day operations worldwide.

Consortium on Operator Qualification

API assembled a focus group whose primary purpose was to identify covered tasks and provide compliance recommendations to pipeline operators.

The Consortium on Operator Qualification (COOQ) became the guiding force for the identification and development of covered tasks for the liquid pipeline industry. Their work product, API publication 1161, *Guidance Document for the Qualification of Liquid Pipeline Personnel,* formed the basis for the resulting Pipeliner Training and Assessment Program.

The covered task list created in API publication 1161 has now evolved to include gas covered tasks and is maintained by NCCER. The most current covered task list can be downloaded at www.nccer.org.

National Center for Construction Education and Research

With the guidance and support of API, its members, and the Consortium on Operator Qualification, the National Center for Construction Education and Research (NCCER) was able to design and implement the Pipeliner Training and Assessment Program.

NCCER is a not-for-profit 501(c)(3) education foundation established in 1995 by the world's largest and most progressive construction and maintenance contractors and national contractor associations. NCCER was founded to address the

severe workforce shortage facing the construction and maintenance industries and to develop a standardized training process and curricula. Today, NCCER is supported by hundreds of leading construction and maintenance companies, manufacturers, and national associations and developed more than 35 construction and maintenance titles.

NCCER also maintains a National Registry, which contains the records of trainees and instructors who have received NCCER training. All completion information for every trainee is sent to the NCCER and filed within the Registry. The Registry can then confirm training and skills for workers as they move from company to company, state to state, or even within their own company.

Standardized Curriculum (NCCER's Contren™ Learning Series)

NCCER curricula are developed using the expertise of industry subject matter experts (SMEs). The use of SMEs as an integral part of every new development or revision ensures that the curriculum is current, industry-driven, and indicative of industry practices. The development and/or revision process is facilitated from start to finish by NCCER project managers, who work closely with the SMEs and a select team of technical writer(s) on the development and presentation of the curriculum.

Pipeline: Five Training Paths

The pipeline curriculum focuses on essential skills and competencies within five skill areas:

- Corrosion Control
- Electrical and Instrumentation
- Maintenance
- Mechanical
- Operations (Field and Control Center)

SMEs from the pipeline industry have determined the skills and competencies needed based on API's covered tasks and the consensus of the group/industry. Specific company information is generally not incorporated into the curriculum unless the SMEs determine that the information is the industry standard.

Training Modules and Covered Tasks

The smallest unit of instruction in the curriculum is a *module*. Each curriculum is composed of modules (the equivalent of stand-alone chapters) that are logically sequenced into complete *levels* of instruction. Every module is task-oriented and requires the successful completion of a written (knowledge-based) examination and a performance (hands-on) test. Modules may be individually selected and/or combined for efficient task training. Modules in the pipeline curricula, to the extent possible, will align with API-developed covered tasks. While not all modules will relate to covered tasks, all covered tasks will relate to at least one module.

Benefits of Training to the Pipeline Marketplace

Applying NCCER's standardized training process allows pipeline owners and contractors in the United States to establish the benchmark for acceptable skill and competency levels within the pipeline industry. This process will assist training organizations in developing and/or improving standardized pipeline training programs that effectively match their resources and operating conditions while responding to regulatory demands and to the needs of the industry. The process will also recognize training organizations that meet the minimum standards of competence, warranting regulatory, public, and professional confidence. This in turn will foster national unity within the pipeline industry for the mutual benefit of individual technicians, owners, contractors, and ultimately pipeline-users and society.

Additional Resources

To review the Pipeliner Training and Assessment Program Implementation and Resource Guide and the latest version of pipeline covered tasks, visit NCCER's Web site at www.nccer.org, or contact:

National Center for Construction Education and Research
PO Box 141104
Gainesville, FL 32614-1104
(888) 622-3720
Web site: www.nccer.org

For more information on API, please contact:

American Petroleum Institute
1220 L Street, Northwest
Washington, D.C. 20005
(202) 682-8000
Web site: www.api.org

To order *API Pub 1161*, contact Global Engineering Documents by calling 1-800-854-7149 or visiting their Web site at www.global.his.com.

Additional sources of information on the pipeline operator qualification regulation include:

DOT – Office of Pipeline Safety
www.ops.dot.gov

Transportation Safety Institute
www.tsi.dot.gov

Instructor's Notes:

COURSE MAP

This course map shows each curriculum path in the Pipeliner Training program. The suggested training order begins on the left with the Pipeline Core modules and proceeds to the right. Which curriculum path you follow depends on your areas of specialty and interest. On a given curriculum path, skills levels increase as you advance to the right on the course map, although some are single-level curricula. The local Training Program Sponsor may adjust the training order.

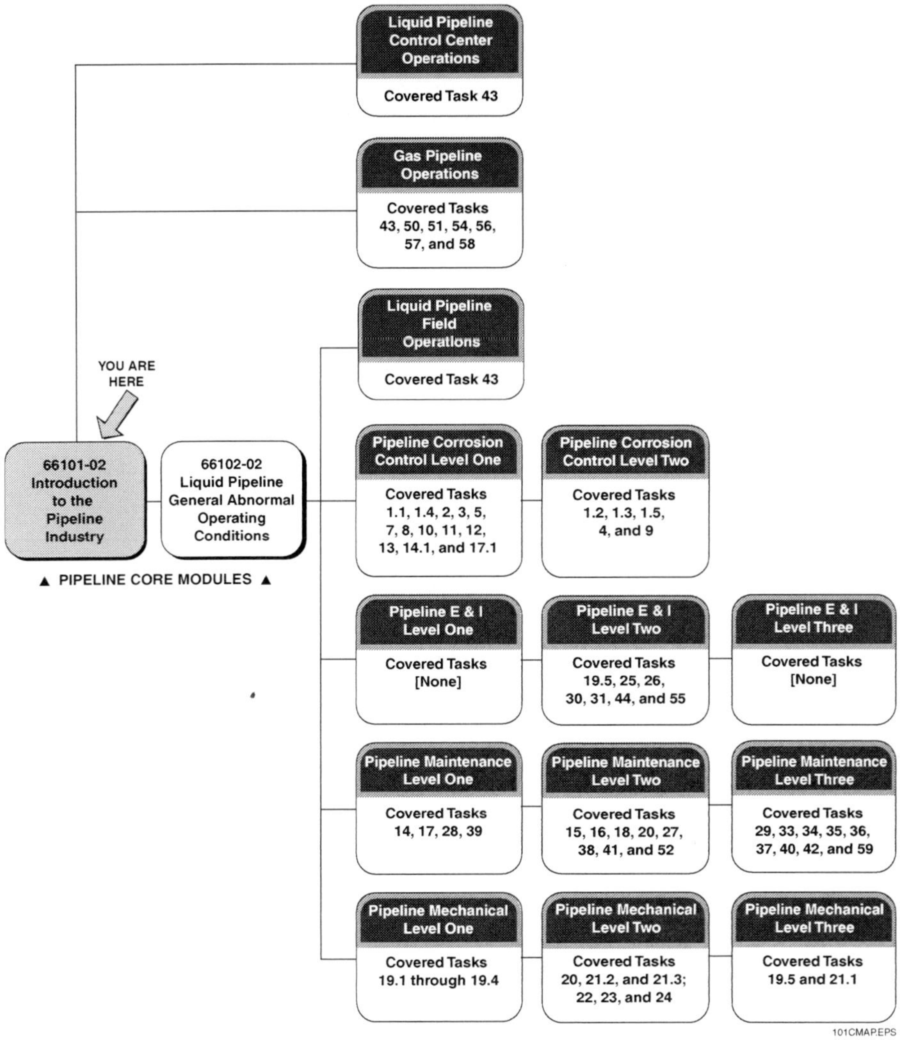

Homework

Assign reading of Section 1.0.0.

Instructor's Notes:

MODULE 66101-02 CONTENTS

MODULE 66101-02 CONTENTS (Continued)

Instructor's Notes:

Figures

Instructor's Notes:

Tables

Instructor's Notes:

Introduction to the Pipeline Industry

Materials

Ensure that you have everything required to teach the course. Check the Materials and Equipment List at the front of this module.

Classroom

Explain that terms shown in bold are defined in the Glossary at the back of this module.

Audiovisual

Show Transparency 1, Course Objectives.

Teaching Tip

See the general Teaching Tip at the end of this module.

Classroom

Review the history of pipeline systems. Emphasize the importance of protecting people and the environment.

Objectives

When you have completed this module, you will be able to do the following:

1. Explain the basic functions and purposes of pipelines and facilities and identify the characteristics and hazards of common pipeline products.
2. Identify maps and drawings used to depict pipelines and facilities.
3. Explain the roles of control personnel and equipment in the overall operation of a pipeline.
4. Explain liquid pipeline hydraulics and gas pipeline pneumatics.
5. Explain the types and purposes of pipeline equipment.
6. Explain pipeline electrical power systems and corrosion control.
7. Review operations, maintenance, and emergency procedures and perform documentation required for pipeline operations.

Prerequisites

Before you begin this module, it is recommended that you successfully complete the following: Core Curriculum.

1.0.0 ◆ INTRODUCTION

Pipelines are used to transport **petroleum**, petroleum **products**, liquid carbon dioxide, and other liquids and gases over great distances. In spite of the costs of constructing pipelines, they are a relatively economical means of transport. A pipeline system consists of pipes, pump stations, compressor stations, and other facilities to support the safe, continuous movement of liquids and gases. Pipelines carry petroleum from the source, the oil and gas fields, to the **refineries**, to and from gas processing plants, and eventually to consumers. Pipes are most often buried in the ground, but they are also laid under water and sometimes erected above the ground.

Today, a network of gas and liquid pipelines crisscrosses the United States. The first pipeline on a continental scale was completed in the 1930s, but the real boom in cross-country lines came during World War II when the Big Inch and the Little Big Inch were constructed to provide fuel for wartime industry and the military. The Big Inch began operating in 1943, was 24 inches in diameter, and stretched 1,341 miles from Longview, Texas, to Phoenixville, Pennsylvania, and Linden, New Jersey. This made it the largest **crude oil** pipeline in the world. The Little Inch, opened in 1944, was 20 inches in diameter and 1,475 miles long. It carried **refined products** from Beaumont, Texas, to Linden, New Jersey.

The most famous crude oil pipeline in the U.S. crosses 800 miles of permanently frozen ground in Alaska. The Trans-Alaska Pipeline, which began operations in 1977, is 48 inches in diameter and carries crude oil from the Prudhoe Bay field to the shipping terminal at Valdez, Alaska. Along the way, it crosses 34 major rivers and streams and three major mountain ranges, going from sea level to as high as 4,800 feet and back to sea level. Protection of the delicate tundra, permafrost, wilderness environment, and wildlife in its path were and still are critical concerns.

All pipeline operators are concerned with safety and environmental impact, particularly those carrying petroleum products that are hazardous to the environment as well as to the people who live near the pipeline. Good construction

Classroom

Ask a trainee to name the four functions of a petroleum pipeline system.

Review how oil and gas are formed. Refer to Figure 1. Explain the four types of traps.

Discuss drilling and completion. Ask trainees why a well may not be completed.

practices, good maintenance and operations practices, continuous monitoring, and rapid response are necessary to prevent or to minimize the effects of pipeline leaks. Operations of a pipeline must be conducted in a fashion that will ensure the safety of employees and the general public and protect the environment.

The petroleum pipeline system can be divided into the following functions:

- Origination
- Transport
- Storage
- Delivery

1.1.0 Origination

All petroleum products produced or shipped originate from the domestic and foreign crude oil or natural gas from the oil and gas fields. Exploration teams from petroleum companies use geological and geophysical methods to find possible deposits of oil and gas deep under the layers of the earth. Drilling confirms and accesses the find; the well is completed and production begins.

1.1.1 Exploration

Oil and gas are formed in the earth from decaying organic matter that was buried long ago. The combination of time, pressure, and heat causes the matter to undergo a chemical change into hydrocarbons. These fluids and gases then tend to move, or migrate, upward in the earth until they either escape at the surface or are trapped in the earth's layers. It is this trapped fluid and gas that is sought by exploration.

The oil/gas-bearing formation is basically the layer of earth in which the hydrocarbon deposit is trapped until it is drilled. The formation must have the porosity to hold the fluid and gas and have the structure to keep it from moving further upward. Traps can have a variety of structures, and some of the most typical are the anticline trap, stratigraphic trap, fault trap, and salt dome trap, as shown in *Figure 1*.

1.1.2 Drilling and Completion

When a possible deposit is indicated, a well must be drilled to try to find it. Many wells have been drilled without finding the sought-after oil and gas. There may not be enough to warrant production costs or it cannot be drawn out economically. However, when a workable deposit is found, the well is completed for production. Completion includes casing the well with concrete, perforating the casing with

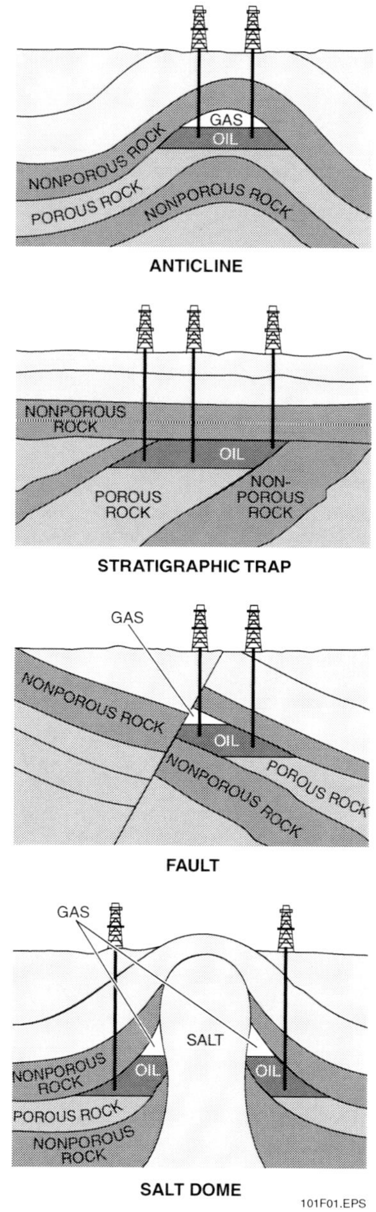

Figure 1 ◆ Types of traps.

Instructor's Notes:

holes for the product to flow through tubing within the casing, and installing pumps or valves on the top of the hole to control the flow. Tests of the producing well may show a large enough deposit to warrant other wells being drilled and completed in the area. All of these wells comprise a field. It is from these wells that the oil or gas is gathered to transport through the pipeline.

Many crude oil reservoirs have a gas cap. The oil produced from these reservoirs often contains various amounts of associated gas. Some formations, however, trap only gas. This trapped gas forms a gas or gas-condensate reservoir, which produces only gas or gas with varying amounts of liquid gas condensate. Drilling gas wells is very similar to drilling and completing crude oil wells. The primary difference between crude oil and gas wells is in the type of surface facility equipment needed to process the oil or gas after it is produced.

1.1.3 Production

Different types of pumps or valves, such as a Christmas tree and a beam pump, are used to cap the well depending on the properties of the product being withdrawn and the formation producing it (*Figure 2*). Oil and gas are often found in the same formation along with some amount of water. The amount of each may vary in relation to each other, and this may affect the production of the well. Since gas is lighter than oil, and oil is lighter than water, they are layered in that order: gas on the top, then oil, then water on the bottom.

Because the formation had to be closed off in some way for the product to accumulate, pressure is usually also trapped in the formation. When it is drilled, that pressure can push the oil, gas, and water into the hole with some force. When it is controlled, this pressure is good because it can push the oil and gas up to the surface with little or no pumping. Generally a newer well will have more pressure than an older one, so in time, pumping power may be needed to pull the product out. However, some wells start with only a little pressure and must be pumped from the beginning. Control of the flow is important since producing a well too fast can affect the life of the well. These conditions, along with other properties, will determine how the well is capped and produced. A Christmas tree of valves may be installed to control the flow, and a beam pump may be used to pull the product out. Gas wells are completed with only a Christmas tree.

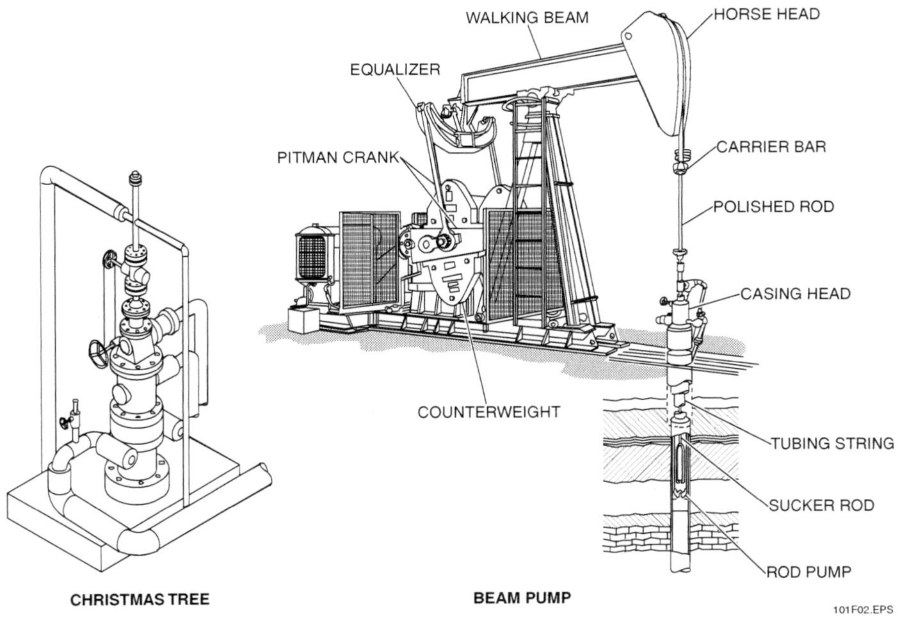

Figure 2 ◆ Christmas tree and beam pump.

Classroom

Ask trainees to explain the primary difference between crude oil and gas wells.

Discuss the effects of pressure on production and how to compensate for too little or too much pressure. Answer any questions trainees may have.

Laboratory

Divide trainees into pairs. Have each pair write down the steps in the process of moving oil and gas from the field to the consumer. Review and discuss the answers.

Classroom

On the whiteboard/ chalkboard, have trainees list the characteristics of the different types of pipe used in pipelines.

1.2.0 Transport

Lines from the oil and gas fields are called *gathering lines* or *flow lines*. These gathering lines bring the oil and gas together into a facility for separation and processing. The vapor gas must travel in a different pipeline from the crude oil, so the gas and oil are separated in huge high-pressure **tanks**, called *gas/oil separators*. *Figure 3* shows the high-pressure separators.

From the separation or gas processing facility, the crude oil and gas is directed into a large **main line**, or trunk line, for the longest part of the journey to a refinery or a shipping terminal. The gas may be stored in tanks or routed through gas pipelines to market. Gas main lines transport gas to a gas distribution center for distribution to consumers, industrial facilities, or storage caverns. From the refinery, a large main line, or trunk line, will carry products across country with distribution lines branching off the main line along the way. A large distribution line may carry the liquid to a central distribution point, such as a regional or city station, from which smaller lines or vehicles carry it to consumers. Pipelines are constructed to meet *ASME Piping Standard, B31.4-1998, Pipeline Transportation Systems for Liquid Hydrocarbons, and other Liquids*. *Figure 4* shows the transportation chain.

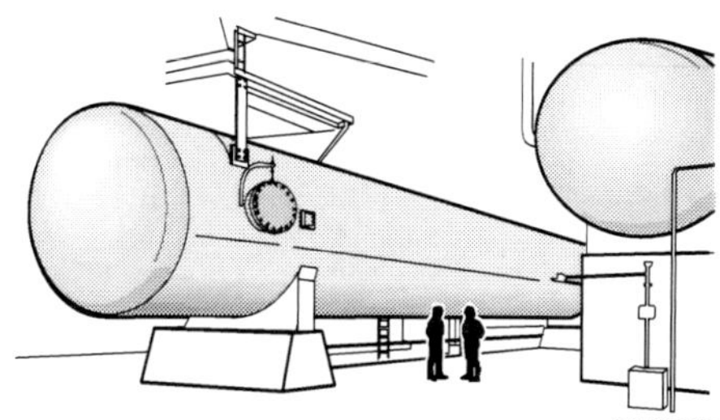

Figure 3 ◆ High-pressure separators. 101F03.EPS

1.2.1 Pipe

The size of pipe used in pipelines varies from 2-inch diameter pipe for gathering lines or small distribution lines, to 48-inch pipe for main lines or trunk lines. Most modern pipe is made of steel, either seamless or with a welded longitudinal seam, that is produced in lengths that are 40-60 feet long. These are often welded into 80-foot sections before being delivered to the construction site. Steel pipe is vulnerable to **corrosion**, which can cause damage to the pipe and reduce wall thickness, compromising the integrity of the pipe. Pipe coating and wrapping are the primary

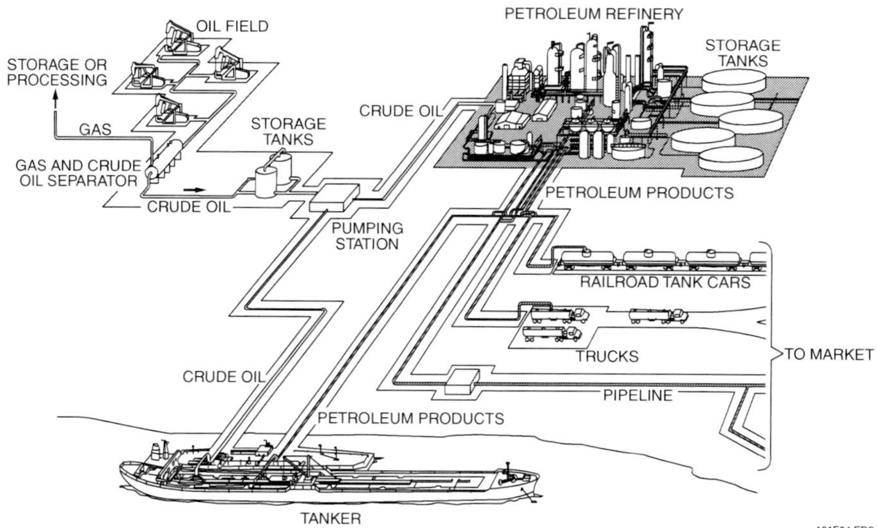

101F04.EPS

Figure 4 ◆ Transportation chain.

Instructor's Notes:

means of protecting the pipe. **Cathodic protection** is used after the pipe is buried to inhibit a corrosive environment.

1.2.2 Construction

At the time of construction, permits are obtained for a specified right-of-way for the pipeline. Earthmoving equipment clears the land, an access road is built along the route, and trenches are excavated for the pipe. Trenches require a minimum depth of coverage as stated in *Pipeline Safety Regulations, Part 195.248* for liquid and *Part 192.327* for gas.

Pipe is hauled to the right-of-way and welded in strings by teams of welders. After welding, radiographic inspection is used to check the welds. Sometimes, pipe must be bent by bending machines to fit the terrain or to go around obstacles. To inhibit corrosion, the pipe must be coated and wrapped before burial. This can be done at the mill or at the site.

The trench is dug and then prepared for the pipe with a layer of protective bedding. The pipe is strung or laid in the trench by sideboom tractors. Long sections are tied-in or welded together to form a continuous conduit. The trench is then filled with backfill to cover the pipe (see *Figure 5*). The pipe must be **hydrostatically tested** or checked for leaks. This involves filling the line with water and pressurizing it to see if the system will maintain the pressure intended.

When piping crosses lakes and rivers, it may be laid and weighted with concrete or steel anchors to keep it on the bottom. *Figure 6* shows an add-on weight installation. In most rivers, the crossings are drilled under the river bed without disturbing the bottom.

Pipelines are usually buried except at pump and compressor stations, some valve stations, and terminals. However, pipe is sometimes erected aboveground by using H-shaped supports, called bents, to hold the pipe. These bents are located at specified intervals along the line. The uprights on the bents are vertical support members (VSMs). In areas like Alaska, the VSMs have to be self-cooling to avoid heating the permafrost. This is not the normal installation, but it is necessary in frozen environments. *Figure 7* shows a bent with self-cooling VSMs.

Supports have sliding plates, or shoes, to allow the pipe to contract or expand with weather or to move with **seismic activity**. The shoes allow the pipe to move as much as 20 feet back and forth and 10 feet up and down. Aboveground lines are also constructed in a zigzag pattern instead of a straight line to allow for movement. *Figure 8* shows a common elevated pipeline section.

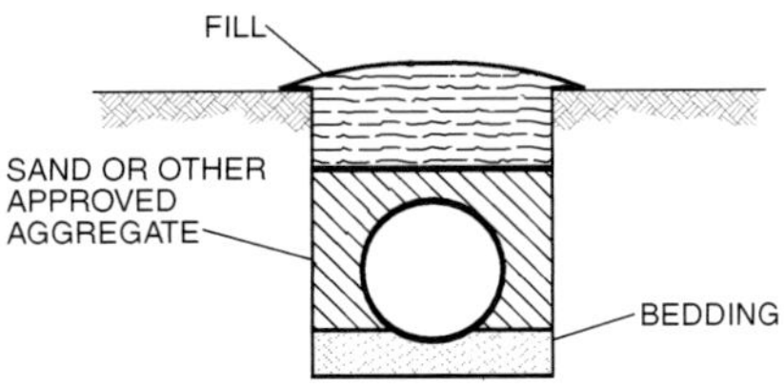

Figure 5 ◆ Pipe burial.

Figure 6 ◆ Add-on weight installation.

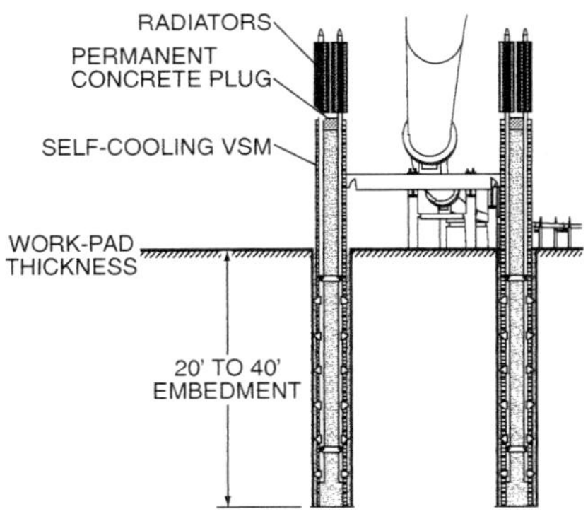

Figure 7 ◆ Bent with self-cooling VSMs.

Classroom

Hand out copies of *49 CFR Parts 195.248* and *192.327*. Ask trainees to identify the minimum depth requirements.

Review the process of preparing and laying pipelines.

Audiovisual

Show Transparency 2 (Figure 5). Discuss various types of approved aggregates.

Classroom

Referring trainees to Figure 9, trace the path of product through a liquid pipeline pump station.

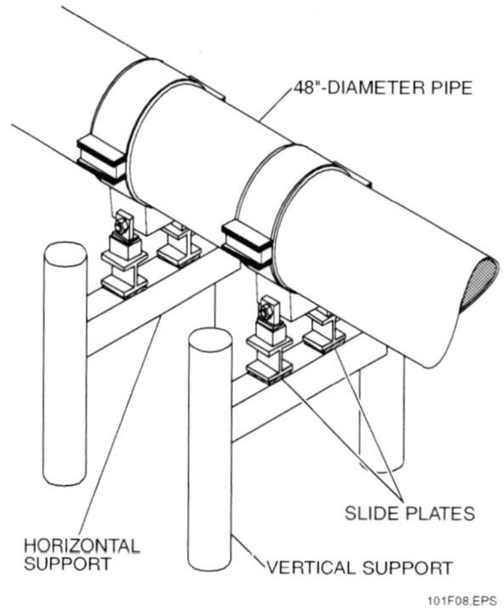

Figure 8 ◈ Common elevated pipeline section.

Pump and compressor stations are spaced strategically along the pipeline at intervals that are dependent on ground elevation, terrain, and other pipeline design factors to maintain the desired flow rate in the line. The pressure that causes the fluid or gas to flow is lost mostly through friction of the product but also by changes in elevation. This pressure must be changed by pumps and compressors to maintain flow. Drag-reducing agents (DRA) can also be added to the product in the line to reduce the friction loss.

A pump or compressor station facility includes pumps or compressors for moving great volumes of liquid or gas, methods of controlling the flow and pressure, means of measuring and accounting for the commodities being moved, storage facilities, and other devices for quality control. *Figure 9* shows an example of a liquid pipeline pump station layout. *Figure 10* shows an example of a gas pipeline compressor station layout.

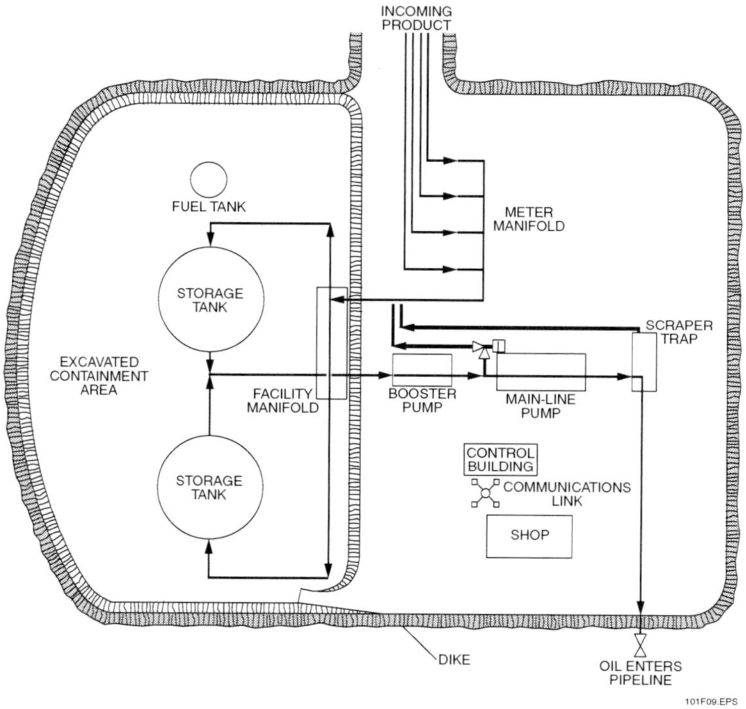

Figure 9 ◈ Example of a pump station layout.

1.6 PIPELINE CORE — TRAINEE MODULE 66101-02

Instructor's Notes:

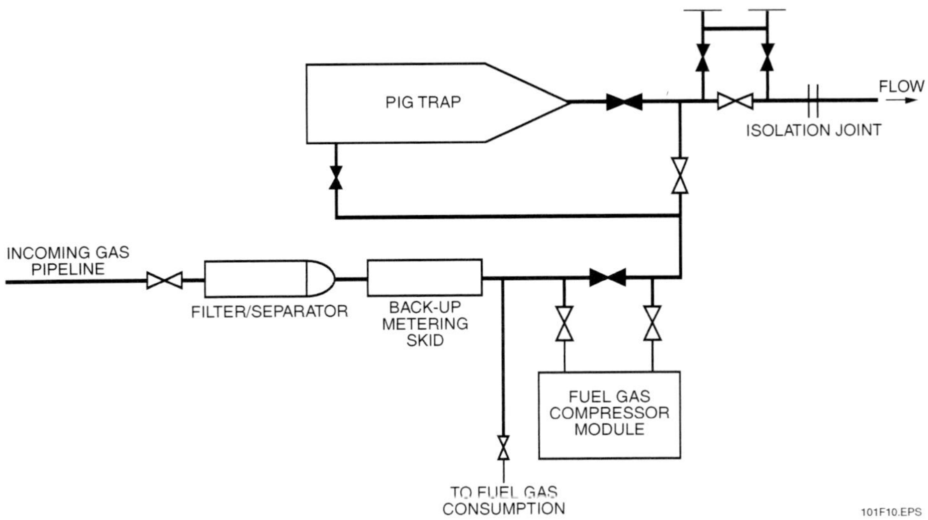

Figure 10 ◈ Example of a gas pipeline compressor station layout.

1.2.3 Operation and Maintenance of the Pipeline

The pipeline is monitored closely and inspected for potential equipment failure and line integrity. Crews identify problems and dig up the pipe to perform repairs. Small **anomalies** may be repaired temporarily while large anomalies may require replacing the pipe. In this case, a bypass line may be constructed so that flow continues while the damaged line is repaired.

The inside of the pipe must also be cleaned periodically to remove wax, scale, and debris. A scraper **pig** is inserted into the line at one point and flows with the liquid or gas to another point, where it is removed. The pig scrapes the inside walls using the hydraulic force of the flow behind it. An instrument pig, also called a smart pig, is inserted into the line periodically to check for anomalies.

1.2.4 Petroleum Products

Different petroleum products can be shipped consecutively from the point of origin in the same pipeline. This is called batching. To maintain quality, batches of each product are arranged in the line and pumped under pressure so that little or no mixing occurs during transit.

1.3.0 Storage

Storage facilities may be located along the pipeline route, at the origination sites, at pump stations, and at terminals. Storage is necessary to meet seasonal needs and other demand fluctuations. Storage is also necessary for the efficient operation of production equipment, pipelines, tankers, refineries, and for emergency use. Large volumes of natural gas are typically stored in underground storage fields throughout the United States.

The trend towards giant tankers created a need for large storage facilities at the loading and unloading ends of tanker runs. A large crude oil terminal typically has several tanks, each with capacities up to half a million **barrels**. A small pump station might have only one tank with a capacity of about 50,000 barrels. Stations along a product line may need separate tanks for storage of each product coming through. These are sometimes called relief tanks or surge tanks. *Figure 11* shows some storage tanks.

1.4.0 Delivery

A wide range of delivery methods is used for gas and petroleum products depending on their destination, intended use, and transport properties.

Audiovisual

Show Transparency 3 (Figure 10). Trace the path of product through a gas pipeline compressor station.

Classroom

Ask a trainee to explain how a pipeline is maintained. Answer any questions trainees may have.

Define batching. Ask a trainee how quality is maintained during batching.

Laboratory

Divide trainees into pairs. Have each pair list five things they know about storage. Then ask trainees to share their lists with the class.

Audiovisual

Show Transparency 4 (Table 1). Ask trainees which types of transport are used for ocean routes and which for rivers and canals.

Homework

Assign the reading of Sections 2.0.0–3.0.0 for the next session.

Materials

Ensure that you have everything required for demonstrations during this session.

Classroom

Have trainees write down the three basic stages involved in producing petroleum products from crude oil. Review and discuss the answers.

Referring to Figure 12, explain the refinery process.

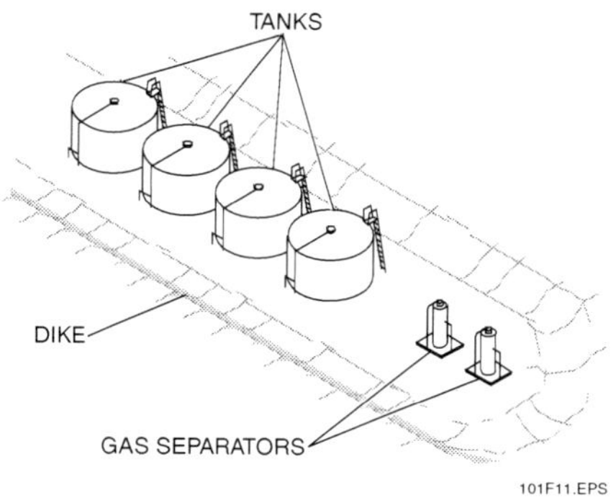

Figure 11 ◆ Storage tanks.

Petroleum is carried chiefly by pipeline, tanker, barge, tank truck, and railroad tank car. Nearly all product moves through pipelines for at least part of the route. Some products are delivered to city or regional stations by pipeline. From there they may be piped or delivered to local consumers or distribution sites. Gasoline, diesel, and kerosene are delivered by tank truck to gas stations for purchase by users. Industrial products may be shipped to companies by pipeline, truck, or railroad. Supertankers are used to move product on ocean routes, while barges are used on rivers and canals. While liquid petroleum products are typically measured in barrels, gas is typically measured using cubic feet. Depending on the volume being measured, gas quantities can be measured using billion cubic feet (BCF), thousand cubic feet (MCF), or hundred cubic feet (CCF). *Table 1* shows typical capacities that can be moved by different means of transport.

Table 1 Means of Transport

Transport	Capacity
Tanker Truck	200 barrels
Railroad Tanker	1,500 barrels
Barge	15,000 barrels
Supertanker	1 million barrels+
Large Liquid Pipeline	2 million barrels+/day
Large Gas Pipeline	1 BCF/day

101T01.EPS

2.0.0 ◆ PIPELINE PRODUCTS

A pipeline can carry a wide range of petroleum and gas products. Each of these products has its own characteristics and rules for handling. Each of these products also presents its own set of hazards for the pipeline workers, for the public, and for the environment. An understanding of the products in the pipeline, the hazards involved, and protective measures to be taken is essential to safe and secure operations.

Petroleum products are produced from crude oil in three basic stages. The first stage is separation and consists of separating the oil into its various fractions, or parts. The main process in this stage is fractional distillation, which separates light, medium, and heavy fractions. The second stage is conversion, in which less useful fractions are converted into more valuable ones. The third stage is treatment, which improves the quality and performance of petroleum products. *Figure 12* shows a diagram of the refinery process.

Products in a typical pipeline are discussed in terms of their characteristics and associated hazards. These products may include:

- Natural hydrocarbons
  - Crude oil
  - Liquefied petroleum gas
  - Wellhead natural gas
- Refined products
  - Gasoline
  - Diesel fuel
  - Fuel oil
  - Jet fuel
  - Consumer-ready natural gas
- Special products
  - Chlorine gas
  - Hydrogen chloride gas
  - Propylene
  - Acetone
  - Benzene
  - Ethane and ethylene
  - Anhydrous ammonia
  - Methane gas
  - Carbon dioxide

2.1.0 Natural Hydrocarbons

Natural hydrocarbons include crude oil, liquefied petroleum gas, and natural gas.

2.1.1 Crude Oil

Crude oil is unrefined petroleum—the source of the myriad of petroleum products. Its specific gravity ranges from 9 to 55 as compared with the specific gravity of water, which is 10. A light crude is one with a higher gravity, and a heavy crude has a lower gravity.

Because the petroleum may have different ingredients or fractions, its color can range from yellow

Instructor's Notes:

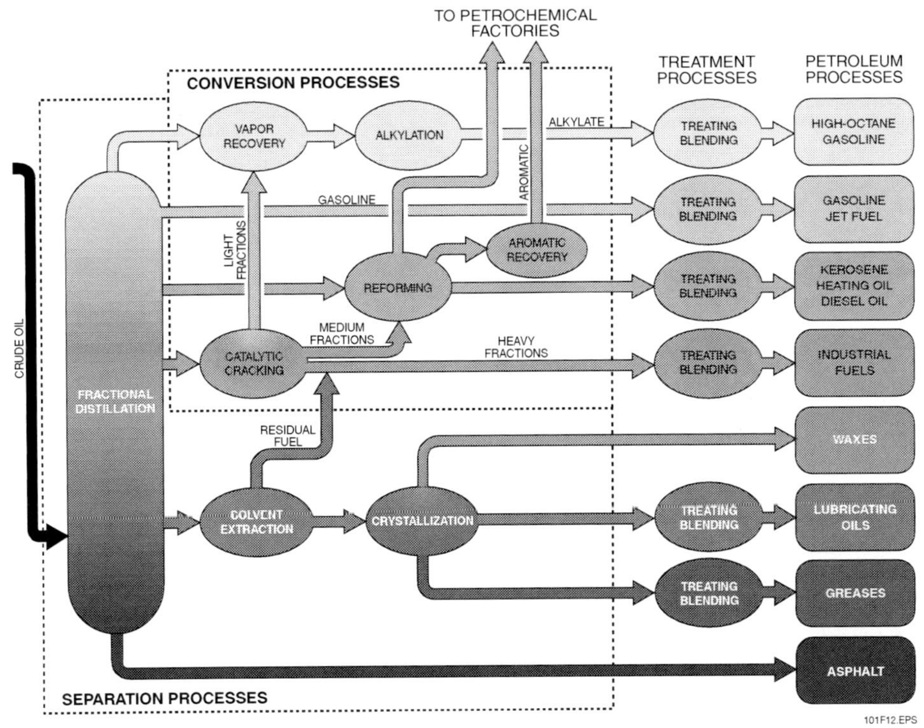

Figure 12 ◈ Diagram of the refinery process.

to black. It may have a paraffin, asphalt, or mixed base. Crude oil also contains substances other than hydrocarbons. These impurities include metallic compounds and sulfur and may make up as much as 10% of the oil. Oil containing a large amount of sulfur compound is called a sour crude. A sweet crude contains little or no sulfur.

2.1.2 Liquefied Petroleum Gas

Liquefied petroleum gas (LPG) is a mixture of heavier, gaseous, paraffinic hydrocarbons, principally butane and propane. These gases, easily liquefied at moderate pressures, may be transported as liquids and converted to gases on release of the pressure. Thus, LPG is a portable source of thermal energy that finds wide application in areas in which it is impractical to distribute natural gas. LPG is used in industry for cutting and welding metals and on farms for operating various kinds of machinery. The principal sources are natural and refinery gas, from which the liquefied petroleum gases are separated by fractionation.

Butane/butylene mix, a form of LPG, is extremely flammable. It releases vapors at well below ambient temperatures and readily forms flammable mixtures with air. If ignited, it will burn in open air or explode in confined spaces. Its vapors are heavier than air and may travel long distances to a point of ignition and then flash back. Alkane/chlorine gas mixtures have produced explosions. Closed systems will form a white frost at the point of the release. Liquid spills will vaporize, forming cold, dense vapor clouds that should be avoided even with respiratory equipment.

2.1.3 Wellhead Natural Gas

Natural gas is the name given to the naturally occurring hydrocarbon mixtures found in many underground wells. Natural gas is an unrefined

Classroom

Have a trainee explain the difference between light crude and heavy crude. Ask trainees to name the various ingredients and impurities crude oil may have.

Stress the dangers of various gas mixtures. From your own experience or readings, share a story of a major gas explosion and the damage it caused.

Safety

Stress the danger of exposure to petroleum products. Remind trainees to wear personal protective equipment and to follow safety procedures.

Teaching Tip

See the Teaching Tip at the end of this module.

product that is a naturally occurring gaseous mixture of hydrocarbons that may also contain nitrogen, helium, carbon dioxide, hydrogen sulfide, and water vapor. Because the content or makeup of natural gas can vary from well to well and region to region, it is normally referred to as wellhead natural gas. Natural gas is processed to recover the **natural gas liquids (NGLs)**, the heavier hydrocarbon components that include butane, propane, and natural gasoline.

> **WARNING!**
> Exposure to petroleum products can cause eye, skin, and respiratory tract irritation. Always wear proper personal protective equipment (PPE) and refer to company procedures and the product material safety data sheet (MSDS) for specific details.

Natural gas is highly compressible, highly expandable, and has a low specific gravity. Although gaseous at normal temperatures and pressures, the gases making up the mixture are variable in form and may be found as gases or as liquids under varying temperatures and pressures.

2.2.0 Refined Products

Refined petroleum and gas products include gasoline, jet fuel, diesel fuel, heating oil, and kerosene. Other refined products may include propane, butane, consumer-ready natural gas, and blending components.

2.2.1 Gasoline

Gasoline is a volatile, flammable liquid hydrocarbon refined from crude oil and used universally as a fuel for internal-combustion, spark-ignition engines. Natural gasoline, or casinghead gasoline, is the liquid hydrocarbon recovered from wet natural gas. Gasoline is classified as regular, premium, and aviation grades according to how smoothly it burns in an engine.

2.2.2 Diesel Fuel

Diesel fuel is a light hydrocarbon mixture for diesel engines, similar to furnace fuel oil. It has a boiling range just above that of kerosene. Diesel fuel requires less refining than gasoline.

2.2.3 Fuel Oil

Refined fuel oil comes in two grades. The specifications for both are defined by *American Society for Testing and Materials (ASTM) Specification D 396.*

No. 1 distillate, such as kerosene, is a light fuel. No. 2 fuel oil diesel is a distillate fuel oil prepared for use in atomizing-type burners or for smaller industrial burner units. Lesser grades of fuel oil, called *residual fuel oils,* are made from the topped crude of refinery operations. Residual fuel oil is used for the production of electric power, space heating, and various industrial purposes.

2.2.4 Jet Fuel

Jet fuel is a highly refined kerosene petroleum distillate. Jet fuel has good flow properties in low-temperature conditions, and it burns efficiently and cleanly. It must be delivered pure and free of rust, water, and sediment.

Military jet fuel is a highly refined kerosene distillate that meets stricter military specifications.

2.2.5 Consumer-Ready Natural Gas

Wellhead natural gas is normally run through a refining process to remove any undesired contents, remove contents that can be used for other commercial purposes, and achieve a more consistent safe product and heat content. The final product is known as commercial natural gas or consumer-ready natural gas.

Consumer-ready natural gas is still a mixture of different gases but is made up primarily of methane. Other gases, such as ethane, propane, nitrogen, and carbon dioxide can also be found in trace quantities. The composition of the consumer-ready natural gas remains fairly consistent over time. The heat value range of typical consumer-ready natural gas is 1,020 to 1,045 Btus per cubic foot.

2.3.0 Special Products

Special products include other by-products that are produced and carried through pipelines.

2.3.1 Chlorine Gas

Chlorine gas is a product used for water purification, chemical processes, and the manufacture of plastic. Chlorine gas is a strong oxidizing agent; it can react explosively with organic products and it is poisonous. Liquid and gas chlorine under pressure can be hazardous. Chlorine is nonflammable and nonexplosive, but like oxygen, it can support combustion of certain substances.

Instructor's Notes:

It can react violently or form explosive compounds with many chemicals, such as acetylene, turpentine, ether, ammonia gas, and hydrogen, and finely divided metals.

2.3.2 Hydrogen Chloride Gas

Hydrogen chloride gas is a fuming, poisonous gas used to make hydrochloric acid, an important industrial chemical used in many chemical compounds, in metallurgy, in food processing, and as a cleaning agent. Hydrogen chloride is nonflammable and noncombustible. However, if dissolved in water, it produces hydrochloric acid, which attacks most metals.

2.3.3 Hydrogen Sulfide

Hydrogen sulfide is a flammable, colorless, gaseous compound of hydrogen and sulfur, which in small amounts has the odor of rotten eggs. It is formed naturally by the decomposition of organic animal and vegetable matter by bacterial action. Hydrogen sulfide is found in crude oil, natural gas, sewers, and swamps and is created by a number of industrial processes. It is extremely corrosive and poisonous. Hydrogen sulfide causes damage to the eyes, skin, breathing passages, and lungs, attacks and paralyzes the central nervous system, especially the parts that control the lungs and heart. In large amounts, it deadens the sense of smell. Hydrogen sulfide is an extremely dangerous compound, and exposure is deadly.

2.3.4 Propylene

Propylene is used in plastics manufacturing. It is extremely flammable and will readily ignite at ambient temperatures. The flammable vapors, being heavier than air, may travel long distances along the ground to a source of ignition and then flash back. Propylene presents the danger of rocketing cylinders, ruptured containers, and explosion. Reignition or explosion can occur if the flames are extinguished without stopping the source flow. An electrostatic charge may build up during handling if the equipment is not grounded.

2.3.5 Acetone

Acetone is a clear, colorless liquid with a sweet, pungent odor that is used as a solvent. It is extremely flammable and may cause flash fire. It is incompatible with a variety of materials, including inorganic acids, caustic, amines, alkanolamines, oxygen, halogens, aldehydes, ammonia, oxidizing agents, and chlorinated compounds. Contact with these should be avoided during transport, handling, and storage.

2.3.6 Benzene

Benzene is an extremely flammable liquid, readily igniting at ambient temperatures. The vapor forms an explosive mixture with air. Flammable vapors, being heavier than air, may travel long distances to a point of ignition and flash back. Incomplete combustion can produce carbon monoxide.

2.3.7 Ethane and Ethylene

Ethane and ethylene are simple asphyxiants. They are dangerous when exposed to heat or flame and can react violently with oxidizing materials. Both are flammable over a wide range in air, with an even broader range in pure oxygen.

2.3.8 Anhydrous Ammonia

Anhydrous ammonia is the liquid form of pure ammonia gas. It is used as a fertilizer because of its high nitrogen content. It is also used as an industrial refrigerant in cold-storage and ice manufacturing plants, since it absorbs a large amount of heat when going from a liquid back to a gas. Anhydrous ammonia is made by compressing dry ammonia gas.

Ammonia is a colorless alkaline gas made up of one part nitrogen and three parts hydrogen. It is lighter than air and has a sharp, stinging odor. Ammonia can be safely inhaled if it is greatly diluted in air, but concentrated ammonia gas can cause suffocation and death. It does not burn in air, but it burns in oxygen with a weak yellow flame. Ammonia is made by combining hydrogen and nitrogen, and as a by-product in the production of coal and coke gas.

2.3.9 Methane Gas

Methane is a member of the alkaline or paraffin series of hydrocarbons. It is called marsh gas because it forms by anaerobic bacterial decomposition of vegetable matter in swampy land. Coal miners know it as *firedamp* because it is combustible when mixed with air. It is a major component of natural gas and of coal gas, and forms in large amounts in sewage disposal processes.

In addition to its use as a fuel, methane is important as a source of organic chemicals and of hydrogen. In the presence of steam and catalysts, it yields carbon monoxide and hydrogen, which can be converted to liquid hydrocarbons or to methanol and other alcohols. The incomplete combustion of methane with air produces carbon black, which is used in rubber and ink. Chlorination of methane yields various chlorides and chloroform.

Teaching Tip

See the Teaching Tip at the end of this module.

Classroom

Have trainees identify the special products that are extremely flammable (propylene, acetone, and benzene).

Classroom

Ask trainees to list the information contained on an MSDS.

Stress the importance of handling and containing spills carefully. Review Figure 13.

2.3.10 Carbon Dioxide

Carbon dioxide (CO_2) is a colorless, odorless, tasteless gas which is about 15 times heavier than air. Under normal conditions, it is stable, inert, and nontoxic. The decay of all organic matter and the burning of hydrocarbon fuels produces carbon dioxide. It is obtained commercially from four sources: naturally occurring underground CO_2 wells, fermentation, combustion of carbonaceous fuels, and as a by-product of chemical processing.

Carbon dioxide gas may be liquid or solid based on compression and lowered temperatures. It may be used as a refrigerant, inerting medium, chemical reactant, neutralizing agent for alkalies, and pressurizing agent. Liquid carbon dioxide is an effective fire-extinguishing agent because it rapidly reduces the temperature of burning materials below the ignition point. The gas, being heavy and inert, displaces air, which is the source of the fire's oxygen, and blankets the flames. Oil tankers, barges, storage vessels, and pipelines are quickly and safely made inert with carbon dioxide to allow for welding and other repairs.

2.4.0 Protective Measures

Material safety data sheets (MSDSs) are good sources of information on these products and must be made available to all employees of the company. There is an MSDS for each hazardous substance, listing its properties and hazards, including a hazard rating for health, fire, reactivity, and other hazards. MSDSs also detail safe handling, personal protective equipment requirements, and first aid/emergency procedures.

All products pose hazards to the environment if spilled or disposed of improperly. Areas of possible spills should have containment structures and/or procedures. These include dikes around storage and transfer facilities and containment booms around tankers being filled. The marine environment is particularly sensitive to oil spills because the tides and currents spread the oil and complex marine life is so vulnerable. *Figure 13* shows temporary containment of a spill from a tank.

Oil may be harmful on land, too. It can seep into the groundwater or streams where it contaminates plant and animal life and eventually human

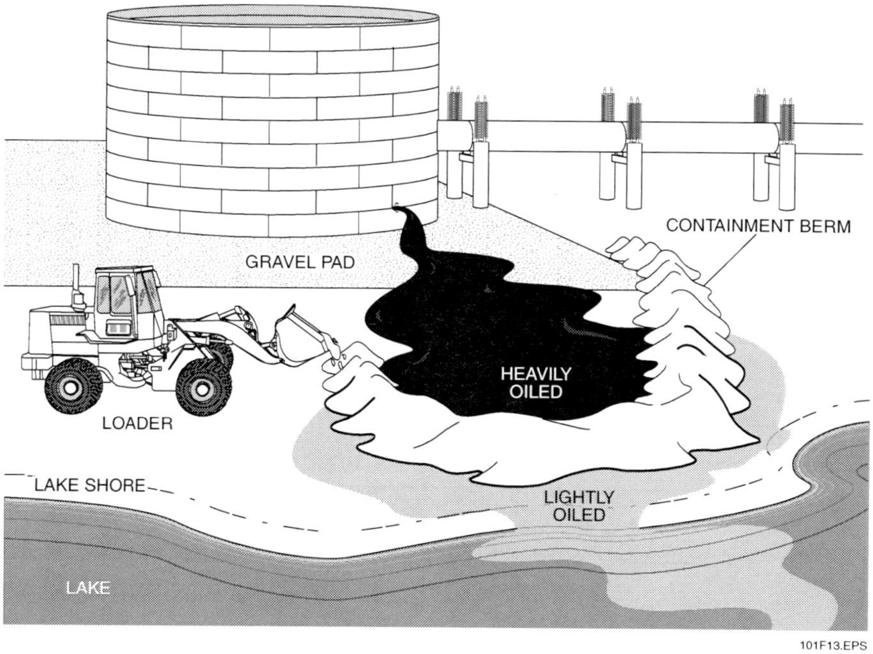

Figure 13 ◆ Temporary containment of a tank spill.

Instructor's Notes:

life. Handling and disposal of hazardous chemicals are addressed in company procedures and MSDSs. Detailed procedures exist for handling all petroleum products, and the observance of these procedures is strictly regulated and enforced by federal, state, and local agencies.

3.0.0 ◆ MAPS AND DRAWINGS

Maps and drawings are an important means of communicating information about pipeline equipment, flow, and processes. Some of these are piping and instrumentation diagrams (P&IDs), strip maps, blueprints, and alignment sheets.

3.1.0 Piping and Instrumentation Diagrams (P&IDs)

A P&ID depicts the functional layout of a piping system and its components, valves, and instrumentation as clearly and accurately as possible. It is accurate to the extent that all components are connected to each other as shown in relation to flow-path orientation. A P&ID does not represent the actual physical layout of equipment; for example, a valve that may appear to be right at the discharge of a pump can actually be located some distance away or even on another floor. Many times, a P&ID will use a broken line around groups of equipment to indicate that they are all in the same building (see *Figure 14*).

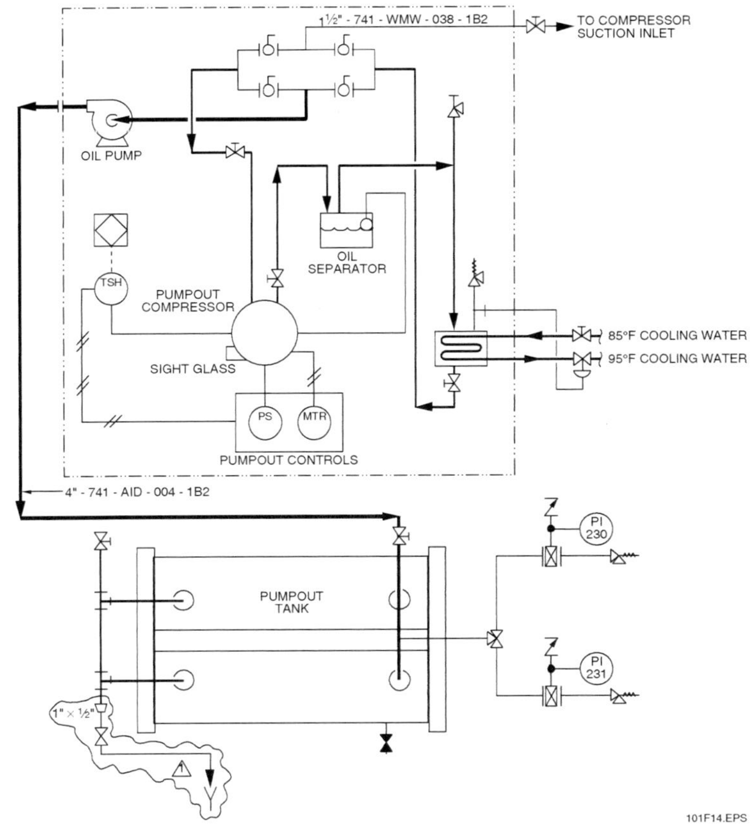

Figure 14 ◆ Sample P&ID.

Demonstration

Hand out samples of P&IDs, blueprints, and strip maps/alignment sheets. Have trainees familiarize themselves with each.

Classroom

Referring to Figure 14, point out the broken line and arrows. Ask trainees what these show.

Classroom

Ask trainees to name the information in a blueprint. Write their answers on the board.

Distinguish between strip maps and alignment sheets. Answer any questions trainees may have.

The P&ID depicts all components of a particular system, including pipe sizes, flange sizes, valve sizes, flow direction, and references to other diagrams. Standardized symbols are used for these components instead of pictures. Industry standards for these symbols have been developed and adopted by organizations, such as the American National Standard Institute (ANSI), the Instrument Society of America (ISA), and the American Society for Mechanical Engineers (ASME), to aid interpretation. The P&ID title block and legend should be examined closely for explanations of the diagram.

3.2.0 Blueprints

Blueprints are duplicates, or copies, of plans for buildings, assemblies, systems, and other constructions. These are working drawings, including specifications for a project that usually includes site plans, floor plans, elevations of buildings, cross sections, and detail drawings. They were originally made by a photographic process that produced white lines against a blue background, hence, the name. However, they are now produced by computer-aided design (CAD) software and generally have black or blue lines on a white background. The blueprint shows the size and location of all structures and equipment in the project to scale and how they are physically put together. *Figure 15* shows a site-plan blueprint.

3.3.0 Strip Maps/Alignment Sheets

Strip maps (see *Figure 16*) are combinations of different types of maps or drawings over the same area such as the pipeline route. The strips of each type of map or drawing are laid out side by side

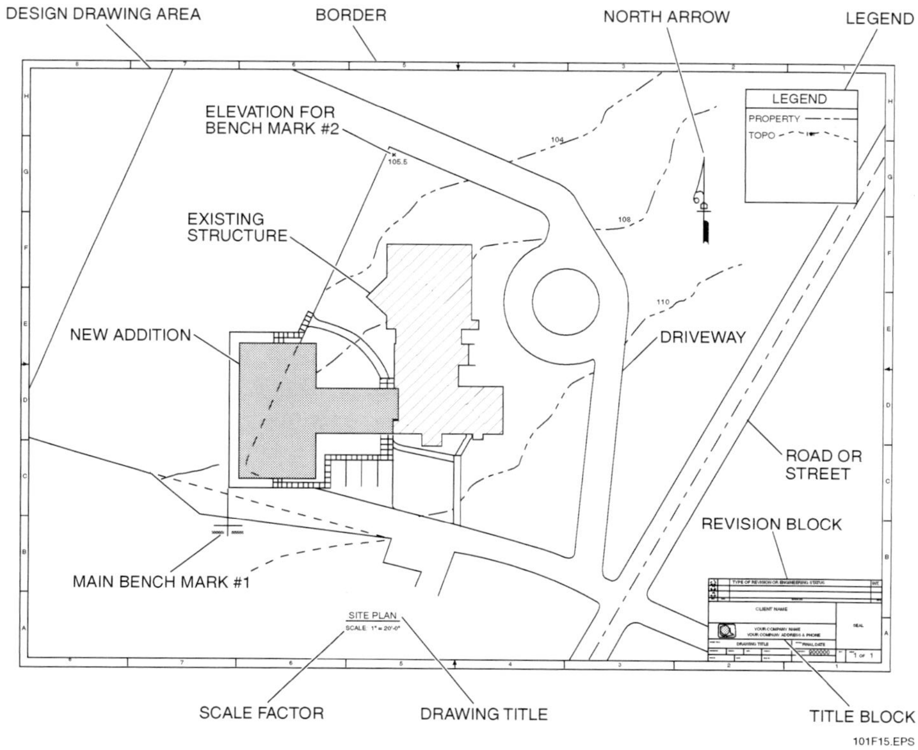

Figure 15 ◈ Site-plan blueprint.

Instructor's Notes:

101F16.EPS

Figure 16 ◆ Sample strip map.

Homework

Assign the reading of Sections 4.0.0–6.0.0 for the next session.

Materials

Ensure that you have everything required for demonstrations during this session.

Classroom

Have trainees list the tasks often performed by a scheduler. Write their answers on the board.

on the same sheet so that they line up along the route. There may be views shown of the ground from an aerial, or plan view, and side, or elevation view, which shows the ground profile along a specific line. The pipeline will be superimposed on all views. Strip maps help to coordinate surface features with topography as well as with other pipeline data at any point on the route.

Alignment sheets are various types of maps, drawings, and diagrams that show different features along the pipeline route. The alignment sheets will show the relationship of the pipeline to true north, geographic features, political boundaries, fences, rivers, property lines, and other pipelines. *Figure 17* shows one type of alignment sheet.

Strip maps and alignment sheets vary greatly from company to company. These names may be used interchangeably or may mean very different things.

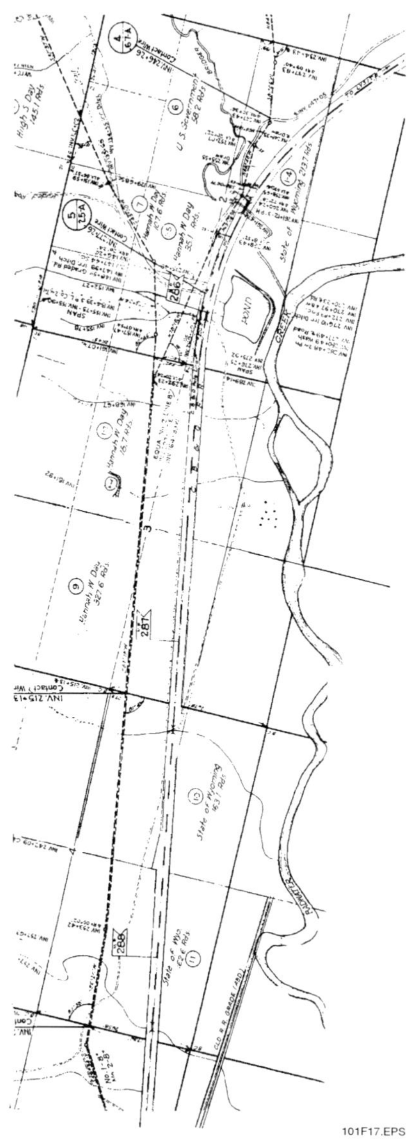

101F17.EPS

Figure 17 ◆ One type of alignment sheet.

4.0.0 ◆ OPERATIONS

Controlling the fast-paced, continuous, and complex operation of the pipeline is a critical job done largely by the scheduler, controller, and dispatcher with the help of other personnel and systems like Supervisory Control and Data Acquisition (SCADA), programmable logic controllers (PLCs), and remote terminal units (RTUs). All control functions are documented by filling out forms or printing out reports.

4.1.0 Scheduling

Before a shipment of product reaches the pipeline and comes under the control of the dispatching office, it has to be accumulated and scheduled with the pipeline scheduler. The scheduler at a pipeline facility typically performs the following functions:

- Takes nominations from shippers
- Creates pumping orders for pipelines to meet shippers' needs
- Validates and reviews operating information for schedule adjustments
- Receives delivery information in time for shipments
- Validates all shipping information for billing
- Utilizes power optimization to save costs and maximize throughput efficiency

Instructor's Notes:

4.2.0 Pipeline Operations

The pipeline is monitored closely and inspected for potential equipment failures and line integrity. These problems can arise from bad welds, corrosion spots, faulty valves, third party damage, or shifting of the pipe. When aboveground, the pipe may also be damaged by accidents or sabotage. In areas with faults or history of earthquakes, seismic activity must also be closely monitored. Flow can also be interrupted by obstructions in the pipe, such as stuck pigging devices, foreign objects, and debris.

Computers and automation are used to control and monitor the pipeline operation. Communication is maintained between all stations regarding the status of the operation. Controllers have the ability to shut down the pipeline flow if a problem is detected and to pinpoint the location of the problem. *Figure 18* shows a typical control center layout.

4.3.0 Control Systems

SCADA is the software used to monitor and control the flow of products through the pipeline every day. Wireless telemetry and SCADA systems provide remote alarm, control, and data collection for many types of pipeline operations.

SCADA is installed with the main computer system and hooked up through a server to RTUs and PLCs in the field. Sensors along the pipelines precisely measure operating conditions, flow rates, and the status of control devices, such as flow and pressure regulators and valves. The RTUs collect vital information in the field and transmit the data via telephone, microwave, or satellite circuits to the control center.

4.4.0 Communications

Communications are the systems that keep the pipeline operating. All along the pipeline, everyone must have open communications at all times because everything that occurs on the pipeline affects the entire system.

A variety of communications equipment and devices may be used in a pipeline company's operations. These are set up for continuity during normal or emergency conditions. It is important that communications be reliable and always available. Most companies have redundant, or backup, systems to ensure that this is true. Some of the equipment used includes:

- Dedicated satellites or microwave-based radio communications systems that provide linkage between the control room personnel and various pump stations and valve locations
- Hardwire communications that link the control room and all offices and all remotely monitored and operated facilities, such as pump stations and RTUs
- Portable cellular phones or vehicle phones provided to all key and supervisory personnel
- Locally activated pagers or nationwide pagers provided to key personnel for use in routine operating or emergency events

101F18.EPS

Figure 18 ◈ Typical control center layout.

Classroom

Have trainees name some problems that can arise with pipelines. Point out that computers and automation are used to prevent problems from occurring.

Discuss the setup of control systems. Ask trainees what SCADA and RTUs do.

Explain the equipment used to facilitate communications. Ask if trainees have any questions.

Classroom

Define hydraulics and fluid. Explain the different properties of fluids.

- Fixed repeaters installed at points within the system that provide internal long distance and local radio communications

In the event of a power loss, backup systems are provided, which may include standby generators for powering communications equipment, batteries, or uninterruptible power systems (UPS). Portable self-contained mobile repeaters may be used at an incident scene and utilized to augment existing radio systems or replace any that may be inoperable. A radio link (base station) may be used with portable units to allow for emergency communications between facilities in the event that telephone and/or microwave transmissions become inoperable. Redundant communications and power systems should be installed at key facilities to allow for control in the event of loss of primary systems.

5.0.0 ◈ FLUIDS AND SYSTEM HYDRAULICS

The operation of any liquid pipeline is built on the principles of hydraulics, or the science of the behavior of fluids at rest or in motion. These principles are in turn built on the properties and characteristics of fluids.

5.1.0 Fluids

A fluid is a substance that flows, and some flow more easily than others. A fluid is also a substance that changes form with a slight force or pressure; it is elastic, however—meaning it will try to return to its previous form when the pressure is removed. Some fluids are more elastic than others. Elasticity is a measure of the fluid's resistance to change in shape.

Fluids include all liquids and gases. Water at ordinary temperature is a liquid fluid; at high temperatures, a gas fluid. Air is a gas fluid. All fluids can be greatly affected by changes in temperature and pressure.

No fluid is perfect and frictionless, and different fluids have different levels of friction, or resistance to flow. This resistance to flow is called *viscosity*.

Some fluids are thicker than others. This is called *density*. Density can be a large factor in the viscosity of a fluid. It is a measure of how tightly packed the fluid molecules are or how much of the fluid's matter exists within a given unit of volume.

Measuring density in a liquid is readily done with a hydrometer because they are not usually compressible. Measuring density in a gas is more difficult because gases can easily be compressed. Compressibility is the ability to be squeezed, or forced into a smaller volume.

Density helps determine the compressibility of a fluid. When a fluid is compressed, it becomes denser. The specific gravity of a substance is derived from its density. Specific gravity is the ratio of the density or mass of a given volume of the substance to the density of the same volume of a standard substance, such as water. The water and the liquid substance must be at the same temperature. For gases, the gas must be at the same temperature and pressure.

American Petroleum Institute (API) gravity is a unit of measure developed specifically for liquid petroleum products. The heavier the substance, the lower the number of its API gravity. Temperature will affect API gravity and should always be corrected to 60°F.

The properties of fluids that are used to describe the fluids are density, viscosity, compressibility, inertia, and elasticity. *Table 2* shows some fluid properties.

Table 2 Fluid Properties

Fluid	API Gravity @ 60°F	Specific Gravity @ 60°F	Vapor Pressure @ 100°F	Absolute Viscosity	Compressibility @ 1000 psi @ 60°F
EP Mix 80/20	214	0.41	550 psi	0.11 cP	7.3%
NGLS	146	0.51	325 psi	0.15 cP	3.8%
Raw Feed Propane	146	0.51	150 psi	0.15 cP	3.8%
Butane	112	0.58	88 psi	0.18 cP	2.0%
Anhydrous Ammonia	97	0.62	200 psi	0.25 cP	1.6%
Gasoline	63	0.66	2 psi	0.70 cP	0.8%
Jet Fuel #1	42	0.82	2 psi	1.4 cP	0.8%
Diesel Fuel #2	33	0.86	1 psi	4.7 cP	0.5%
Crude Oil	50 and below	0.78–0.98	> 1 psi	6–400 cP	> 1.0%
Water	10	1		1.12 cP	0.3%

101T02.EPS

Instructor's Notes:

Other properties of fluids, particularly those found in pipelines, include volatility and flammability. Volatility is the tendency of a fluid to readily vaporize. Sudden vaporization can cause a violent reaction or explosion. Flammability is a measure of how readily a fluid catches fire when in contact with heat or sparks. The volatility and flammability of specific pipeline products is described in another section of this module.

5.2.0 Basic Hydraulics

Some laws of hydraulics apply to gases as well as liquids but only under certain conditions. In general, the following sections pertain only to liquid fluids.

Hydraulics is usually divided into two branches: hydrostatics for liquids at rest and hydrodynamics for liquids in motion.

5.2.1 Hydrostatics

Hydrostatic principles are fundamental to the design of many machines and instruments. The design of hydraulic presses and the operation of instruments, such as manometers and barometers, depend on these principles.

A simple hydraulic press consists of a liquid-filled cylinder that contains two pistons, one smaller than the other. When a force is applied to the smaller piston, the force travels through the liquid to the larger piston, producing a force larger than the initial one. The increase in force is proportional to the ratio of the larger piston to the smaller one. *Figure 19* shows a simple hydraulic press.

The laws of hydrostatics are the following:

1. The amount of pressure caused by the weight of a column of liquid is determined by the height of the column.

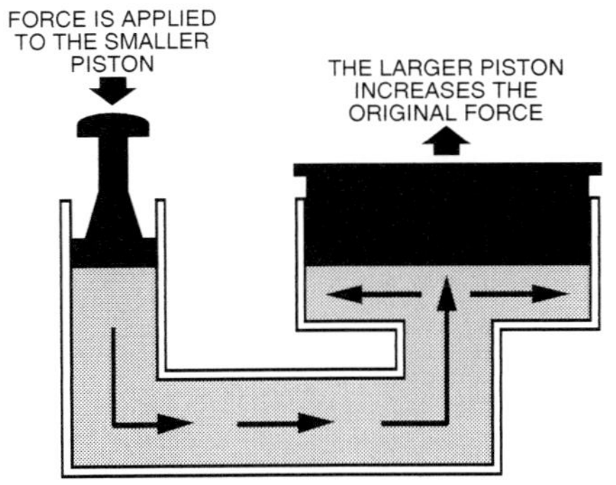

Figure 19 ◆ Simple hydraulic press.

2. An object placed in a liquid loses an amount of weight equal to the weight of the liquid it displaces.
3. A fluid in a container transmits pressure equally in all directions.

5.2.2 Hydrodynamics

Hydrodynamic principles are basic to hydraulic engineering, to the design of certain machines, and to pipeline engineering. Hydrodynamics define the plans for water systems, canals, and irrigation systems, as well as airplane design and hydraulic machines like water turbines.

Pipelines for water and other fluids have to take into account the pipe size and the fluid's friction on the pipe to determine fluid velocity through the pipe. Liquid flow can be steady or unsteady. An unsteady flow results from changes in the velocity, temperature, or pressure of a liquid or its movements around obstacles. The three basic laws of hydrodynamics apply only to steady-flowing liquids. The laws of hydrodynamics are the following:

1. The velocity of a liquid flowing through a pipe increases as the area of the pipe decreases and vice versa.
2. The pressure of a liquid increases as its velocity decreases and vice versa.
3. The velocity with which a liquid flows through an opening in a container equals the velocity of a body falling from the surface of the liquid to the opening.

Figure 20 shows liquid velocity and pressure relationships.

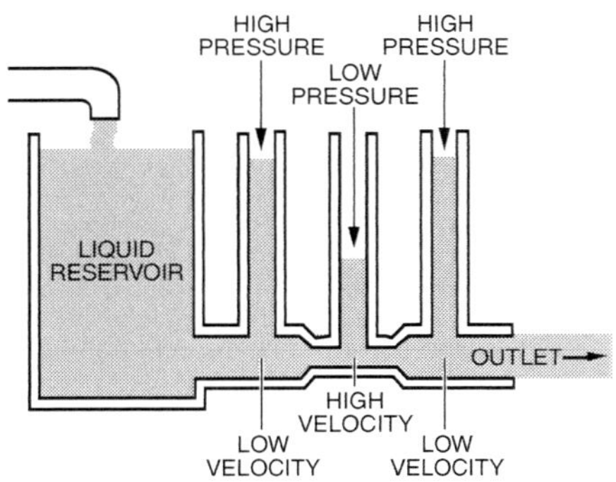

Figure 20 ◆ Liquid velocity and pressure relationships.

Audiovisual

Show Transparency 5 (Figure 19). Have a trainee explain how a hydraulic press works.

Classroom

Explain the three laws of hydrostatics.

List the three laws of hydrodynamics on the board and discuss each law. Note that the laws apply only to steady-flowing liquids.

Audiovisual

Show Transparency 6 (Figure 20). Explain the relationships of liquid velocity to area and of pressure to velocity.

Classroom

Ask trainees to list on the board the primary factors in pipeline design. Explain MOP and MAOP.

Define pneumatics. Ask trainees what the primary goal of gas compression is.

Ask trainees to name the three variables that have to be calculated and controlled when compressing gas.

Teaching Tip

See the Teaching Tip at the end of this module.

5.3.0 Pipeline Hydraulics

Fluid properties are major factors which affect the design of the pipeline. The primary factors in pipeline design are:

- Fluid characteristics
- Desired flow rate
- Maximum operating pressure (MOP)
- Maximum allowable operating pressure (MAOP) (for gas)
- Elevation profile/terrain
- Fittings/valves
- Installation costs
- Environmental concerns

Pressure is referenced in several ways in the pipeline industry. The maximum operating pressure (MOP) is defined as the maximum internal pressure allowed liquid during the operation of a pipeline. The term MOP is typically used more in the liquid pipeline industry and the regulations are referenced in *49 CFR 195*. The MAOP is defined as the maximum internal pressure permitted during the operation of a gas pipeline. The term MAOP is typically used more in the gas pipeline industry and the regulations can be referenced in *49 CFR 192*.

Although fluids are defined by their ability to flow, something has to start the flow and overcome their inertia and friction. A force, such as difference in pressure, gravity, or an external force physically pushing it, can do this. Usually, this is done by pumps. Stopping the flow once started will require another force, such as resistance (friction), or a stationary object, such as a closed valve.

The resistance or friction of the fluid against the pipe will be constantly trying to slow it down. Most pumps work by providing a difference in pressure and must continue providing that force all along the line to overcome the friction. There are drastic differences in the amount of pressure required to deliver the proper volume of different types of products. But the pressure in the system must be kept under the maximum limits.

Friction is also affected by the roughness of the pipe interior, the pipe diameter and length, the product density and viscosity, the flow rate, and obstacles, such as valves and bends. Pipeline design can help reduce some of these, but higher fluid viscosity and density will increase the friction.

When the fluid is traveling uphill, pumps must be used to overcome the fluid's inertia and resistance and the force of gravity. When the flow in the pipe is downhill, gravity may be all that is needed.

Another factor in the flow of the pipeline is the temperature. The higher the temperature, the lower the viscosity and vice versa. When the product is hot, it flows better. Sometimes heaters are included in the pipeline equipment to keep the product at the optimum temperature.

6.0.0 ◈ GASES AND SYSTEM PNEUMATICS

Pneumatics, as it relates to the pipeline industry, is the study of the storage, compression, and transmission of gases. The compression of gas has one primary goal, to deliver the gas at a pressure higher than the original pressure. The original pressure may vary from very low absolute pressures to several thousand pounds. The pressure rise may vary from a few ounces to many thousand pounds, and the volumes handled may vary from a few cubic feet per minute to hundreds or thousands of cubic feet.

There are three variables that have to be calculated and controlled when compressing and moving gas in a pipeline. These variables are pressure (p), volume (v), and temperature (k). These three characteristics of gases have a close relationship with one another and are explained in the ideal gas laws.

6.1.0 The Ideal Gas Laws

Pressure, volume, and temperature of gases have a close relationship with one another. By keeping one of them constant, the other two will have either a direct or inverse proportion. The relationships between pressure, volume, and temperature are explained using the ideal gas laws. These ideal gas laws are Boyle's law, Amonton's (Gay-Lussac's) law, and Charles' law.

6.1.1 Boyle's Law

Boyle's law states that when temperature is kept constant and volume is decreased, pressure will increase. His law gives the relationship between pressure and volume if temperature and amount are held constant.

- If the volume of a container is increased, the pressure decreases.
- If the volume of a container is decreased, the pressure increases.

If the volume is increased, the gas molecules have farther to travel and they will impact the container walls less often. This means the gas pressure will be less because there are fewer molecule impacts per unit of time.

If the volume is decreased, the gas molecules have a shorter distance to travel, thus striking the walls more often. This results in pressure being increased because there are more molecule impacts per unit of time.

Instructor's Notes:

6.1.2 Amonton's (Gay-Lussac's) Law

Amonton's law states that when volume is kept constant and temperature is increased, pressure also increases. Amonton's law was discovered by Joseph Louis Gay-Lussac in the early 1800s. His law gives the relationship between pressure and temperature when volume and amount are held constant.

- If the temperature of a container is increased, the pressure increases.
- If the temperature of a container is decreased, the pressure decreases.

If the temperature is increased, the gas molecules will move faster and they will impact the container walls more often. This means the gas pressure inside the container will increase. Gay-Lussac's law is a direct mathematical relationship. This means there are two connected values and when one goes up, the other also increases.

6.1.3 Charles' Law

Charles' law states that when pressure is kept constant and volume is increased, temperature also increases. The pressure that a gas exerts on the walls of its container is determined by the momentum of the atoms and molecules of the gas, which in turn is determined by the temperature. As the temperature increases, the atoms and molecules move faster, and so exert a greater pressure on the walls.

This law gives the relationship between volume and temperature if pressure and amount are held constant.

- If the volume of a container is increased, the temperature increases.
- If the volume of a container is decreased, the temperature decreases.

If the temperature is increased, the gas molecules will move faster and they will impact the container walls more often. This means the gas pressure inside the container will increase, but only for an instant. The greater pressure on the inside of the container walls will push them outward, thus increasing the volume. When this happens, the gas molecules will now have farther to go, thereby lowering the number of impacts and dropping the pressure back to its constant value.

6.2.0 Measuring Energy Gases

Quantities of natural gas are measured in volume units. A cubic foot of natural gas at a temperature of 60°F and an atmospheric pressure of 14.7 pounds per square inch gauge (psig) is the common unit of measure. Gas production from wells and supplies to power plants are measured in thousands or millions of cubic feet (MCF and MMCF). Resources and reserves are calculated in billions of cubic feet (Bcf) or trillions of cubic feet (Tcf). To put this in perspective, a trillion cubic feet is enough to fill a cube with sides 2 miles long.

The amount of energy that is obtained from the burning of a unit volume of natural gas is measured in British thermal units (Btus). One Btu is the amount of heat required to raise the temperature of 1 pound of water from 60° to 61°F at normal atmospheric pressure (14.7 psig). At sea level, it would take about 75 Btus to heat water enough for a cup of tea. A cubic foot of natural gas on the average gives off 1,000 Btus, but the range of values is 500 to 1,500 Btus. Therefore, 1 cubic foot of some natural gas may make only 7 cups of tea, while another would make as many as 20 cups of tea.

Energy content of natural gas varies because natural gas accumulations vary in the amount and types of energy gases they contain: the more noncombustible gases in a natural gas, the lower the Btu value. In addition, how much of any energy gas that is present in a natural gas accumulation—the mix of combustible gases—also influences the Btu value of natural gas. The more carbon atoms in a hydrocarbon gas, the higher its Btu value. To illustrate, methane typically represents more than 80% of energy gases. Methane contains one carbon atom per molecule; burning 1 cubic foot of methane gives off 1,012 Btus. Butane, possessing four carbon atoms, has a Btu value more than three times larger than that of methane. Molecular hydrogen, on the other hand, though combustible, contains no carbon atoms; its Btu value is three times smaller than that of methane.

6.3.0 Commercial Natural Gas Combustible Limits

Commercial natural gas is very flammable when sufficient amounts of oxygen (the oxygen present in air is more than adequate) are present. Mixtures of natural gas and air are explosive only when the mixture is less than 15% or more than 4% natural gas. If the natural gas is less than 4%, it will be too lean a mixture to burn, and over 15%, too rich a mixture to burn. The fact that only a small amount of gas is necessary to create an explosive mixture makes even small leaks in confined spaces very serious.

As mentioned earlier, an odorant is added to commercial natural gas so that it can be easily smelled if there should be a leak. However, it is possible for odorants to be filtered out or partially

Classroom

Explain the ways of measuring natural gas.

Explain in greater detail that the danger of explosion is greatest when the mixture of gas and air is between 4 and 15 percent. Ask a trainee why this is true.

Classroom

Define specific gravity. Ask a trainee whether a gas with a specific gravity of .5 will float or sink. What about a gas with a specific gravity of 1.5?

Homework

Have trainees read Section 7.0.0 for the next session.

Materials

Ensure that you have everything required for demonstrations and testing during this session.

Classroom

Have trainees take turns listing pieces of major pipeline equipment on the board.

Audiovisual

Show Transparency 7 (Figure 21). Explain how a centrifugal pump works. Discuss how and why pumps are used.

filtered out of commercial natural gas. This could happen when the natural gas passes through a filtering substance (soil or other). The presence of natural gas should be determined by testing the surrounding air with available instruments such as a combustible gas indicator or another suitable instrument.

6.4.0 Commercial Natural Gas Specific Gravity (Weight)

Specific gravity is a term used in the comparison of the weight of a gas compared to the weight of the same amount of air measured under the same temperature and pressure. Some gases may or may not be heavier than air. Air is judged to be the standard weight of one (1) with commercial natural gas weighing only .6 or 6/10 when compared to air. This means that natural gas weighs less than air so natural gas will rise up or actually float from the ground and fade away into the atmosphere when a leak occurs. Also, commercial natural gas will dissipate rapidly unless confined in a building, vault, or other structure. However, propane has a specific gravity of 1.5, is heavier than air, and will sink or fall to low places in buildings or in the ground. Butane is even heavier, weighing 2.0 or twice the weight of air, and will cling to the ground. It can be extremely dangerous when leaks occur.

Commercial natural gas is lighter than air and will normally rise and travel the path of least resistance. This means that a leak might travel uphill underground until it reaches a blockage and then travel back down the hill because that is the easiest path to follow.

7.0.0 ◆ PIPELINE EQUIPMENT

The pipeline system includes all the pipes, equipment, stations, terminals, and other facilities necessary for operation. The major pipeline equipment consists of:

- Pumps and motors
- Compressors
- Valves
- Actuators/operators
- Tanks
- Piping
- Manifolds
- Traps
- Measurement/sampling equipment

7.1.0 Pumps and Motors

Pumps provide the pressure and force to move product. Various types of pumps may be used in the pipeline, including centrifugal, positive displacement, hydraulic, and rotary pumps. Each is used in applications requiring different capacities or other pump characteristics. Power is supplied from many different sources, such as turbines, hydraulics, diesel engines, and electric motors. One pump station may use positive-displacement pumps driven by diesel engines, whereas another station may use centrifugal pumps driven by electric motors.

7.1.1 Centrifugal Pumps

Centrifugal pumps work by rotating and forcing product outward as product flows into the center. Revolving blades or vanes create the suction that pulls product into the pump. The product takes up the space between the vanes and rotates with them. It is then pushed outward until it leaves through the discharge, or exit, pipe. *Figure 21* shows how centrifugal pumps work.

7.1.2 Positive-Displacement Pumps

In a typical positive-displacement pump (see *Figure 22*), a piston slides back and forth in a cylinder. With each movement, it traps part of the product to be moved. This product is then moved toward the discharge end of the pump, where the pressure rises as the product is squeezed by the piston. The pressure forces the product out. There are several different types of positive-displacement pumps, including lift, or suction, pumps and force pumps.

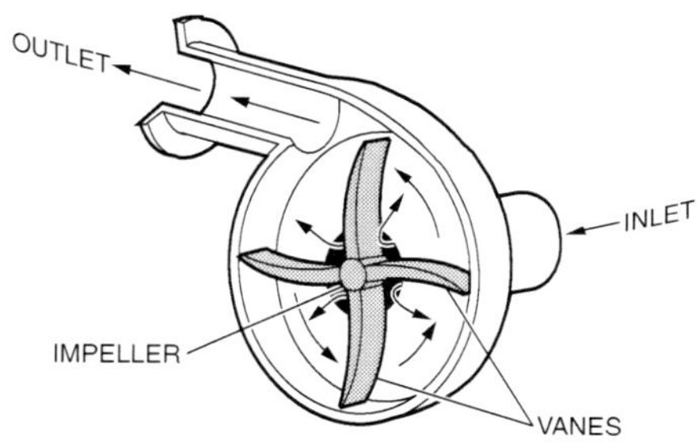

Figure 21 ◆ How the centrifugal pump works.

Instructor's Notes:

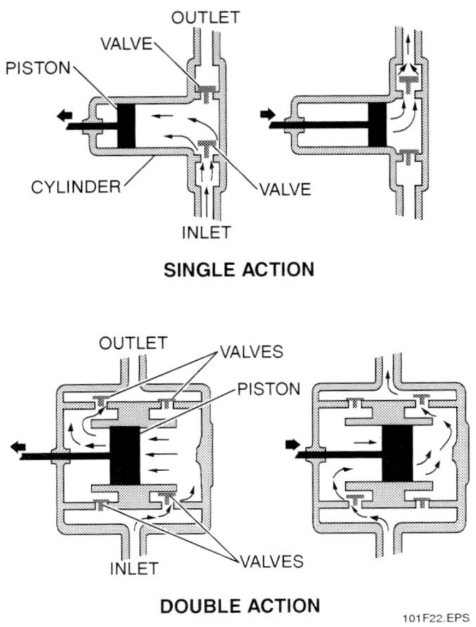

Figure 22 ◈ Two positive-displacement pumps.

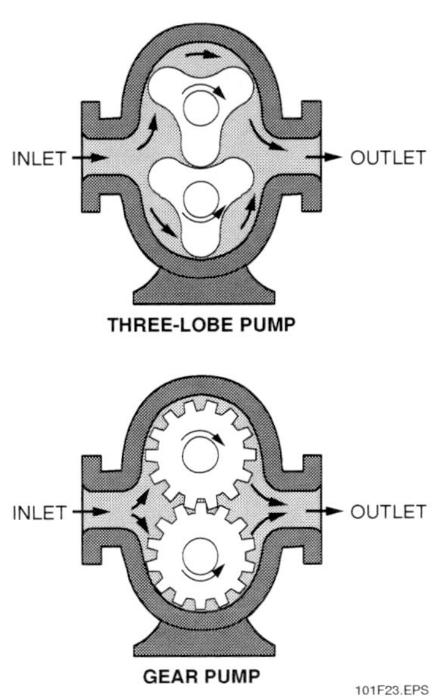

Figure 23 ◈ Two rotary pumps.

7.1.3 Rotary Pumps

Rotary pumps pump air as well as fluids. They have propeller-like blades that rotate on a single shaft. The rotation at the inlet side speeds up the fluid flow in a direction parallel to the shaft, building up pressure between the rows of blades. The pressure moves the fluid to the discharge side of the pump as the blades continue rotating.

A gear pump, which uses gear teeth instead of a piston, is used to pump thick products. The gears rotate in a housing, trapping product between the gear and the casing, moving it to the outlet side. The pressure in the outlet side builds up until it is high enough to discharge the product. Many types of drivers are used to drive these pumps. *Figure 23* shows two rotary pumps.

7.1.4 Drivers

Crude oil, refined liquid products, and gas are pushed through the pipeline grid by powerful pumps and compressors. The size of this machinery depends on the type of pipeline and the products and volumes being transported. Liquids are usually moved through the pipelines by huge pumps, while gases are moved by compressors. These compressors use pressure to push the gas through the pipeline. The pump and compressor drivers may be electric motors with fixed-speed or variable-frequency drive (VFD), diesel or gas engines, or natural-gas-fueled turbines. The choice of driver depends on the power sources available. The most common drive is a fixed-speed electric motor, although variable-frequency drives are becoming more popular. In remote locations, diesel-powered equipment is often used. Diesels are also used as backups during power failures.

7.2.0 Compressors

Compressor stations in a gas line act like pump stations in a fluid line. Compressors boost and maintain the pressure in the line, which keeps the gas flowing. In any compression process, the pressure of gas is increased from one level to another by the addition of energy to the gas being processed.

The centrifugal compressor is classified as dynamic, as opposed to a positive-displacement machine. In a reciprocating compressor, a quantity of gas is drawn into a cylinder, a piston moves, and the trapped gas is compressed by direct volume reduction or positive displacement. By this process, mechanical energy is transformed into pressure energy. *Figure 24* shows a gas compressor.

Audiovisual

Show Transparency 8 (Figure 22). Explain how each type of positive-displacement pump works.

Show Transparency 9 (Figure 23). Discuss how the two types of rotary pumps shown operate. Answer any questions trainees may have.

Classroom

Discuss the various types of drivers and how they work.

Referring trainees to Figure 24, trace the compression process.

Classroom

Ask a trainee how the compressor flow can be controlled. Ask another trainee to explain what the impeller does.

Laboratory

Place an assortment of valves on a table. Have trainees work in pairs to examine the valves and write down the basic types of valves and applications for each.

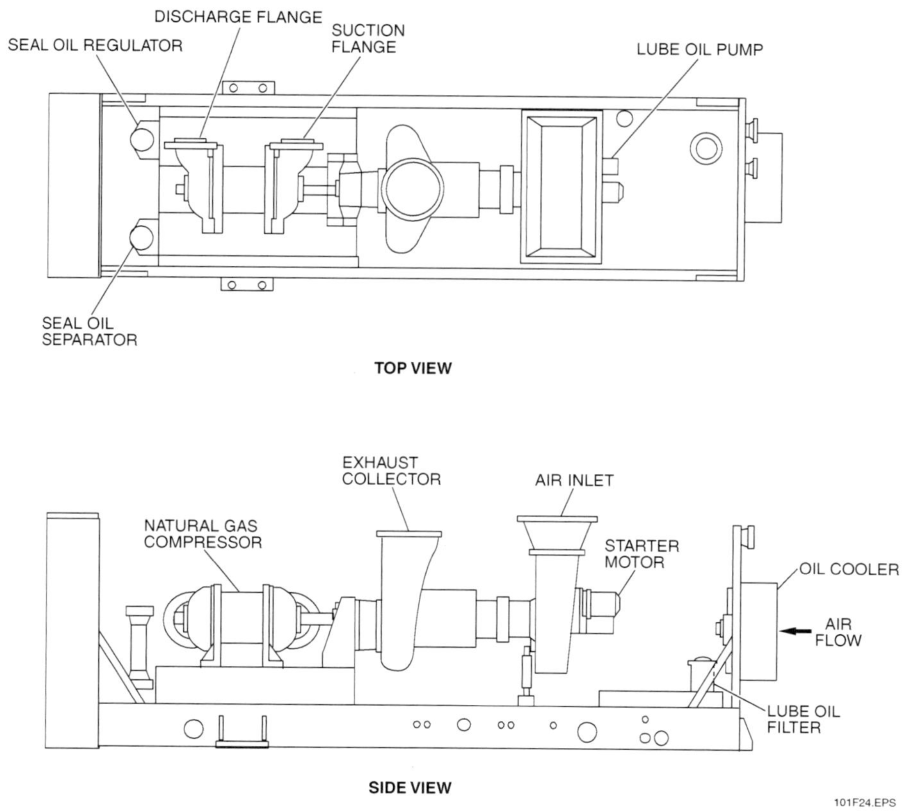

Figure 24 ◆ Gas compressor.

The combination of inlet vane, impeller, and diffuser or diaphragm is referred to as a stage. A single centrifugal stage has a relatively low pressure capability. When larger ratios are desired, additional stages are added in series. The compressor flow can be controlled and varied by changing the pressure differential across the compressor or by changing the power level. Centrifugal compressors apply inertial forces to the gas. The inertial forces are transmitted by an impeller which, by dynamic centrifugal motion, adds velocity energy through acceleration of the gas. The gas flows from the impeller into the diffuser, where the gas decelerates, and velocity energy is transformed to pressure energy.

7.3.0 Valves

Valves are a critical part of pipeline systems. They are used to control the rate of flow in a line, to open or shut off a line completely, or to serve as automatic safety devices. Valves may be identified by type or by function, and can be equipped with locking devices to prevent accidental operation of the valve. Valves may be equipped for manual operation, but most frequently operated valves in the pipeline are equipped with electric-, pneumatic-, or hydraulic-power-driven mechanical operators. Those used extensively include gate, plug, ball, butterfly, check, control, and relief valves.

Instructor's Notes:

7.3.1 Gate Valves

In a simple gate valve, a wedge-shaped disc or gate is pressed between inclined seats to close the valve. Operating on the same principle but with different configurations of gates and seats, there are other more complex gate valves. Conduit gate valves have an elongated gate and a port in the lower part for a fluid conduit. *Figure 25* shows a simple gate valve.

7.3.2 Plug Valves

Plug valves (see *Figure 26*) are similar in design to the stopcocks used to control flow from containers. The plug usually has an elongated port that may be smaller in cross-sectional area than the pipe. The valve is operated by giving the plug a quarter turn. Operation may be by a lever fitted to the stem, by a gear system, or by power-driven mechanical operators. Plug valves are used where reliability and fast operation are important. There is considerable sliding contact area between the plug and the valve body, and as size increases, the operating force required also increases. Many plugs have interior channels and grooves in which lubricants can be forced into the contact areas to improve turning and sealing. Plug valves are generally run in pipe in which scrapers are not run.

7.3.3 Ball Valves

Ball valves are an adaptation of a plug valve, using a ball for the plug. The sealing contact area of the ball valve is that of two seal rings. These seals are made practical by use of tough, low-friction, oil-resistant elastomers.

7.3.4 Butterfly Valves

Butterfly valves (see *Figure 27*) control product flow by use of a disc mounted on a stem that can be turned to block or allow flow. This simple lightweight valve is used in parts of the pipeline. Butterfly valves have two valuable features. They are quick-operating and take up minimal space.

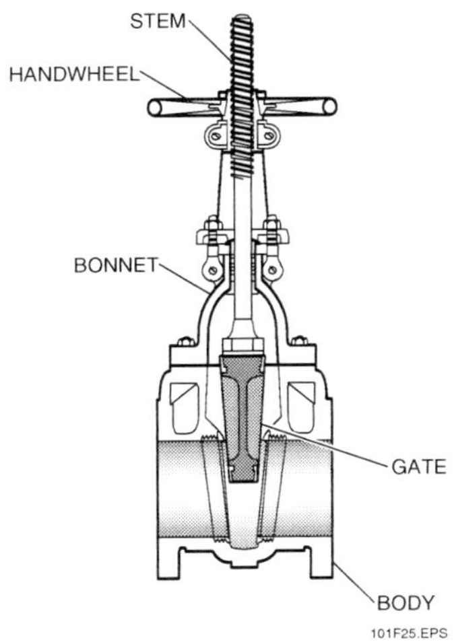

Figure 25 ◈ Simple gate valve.

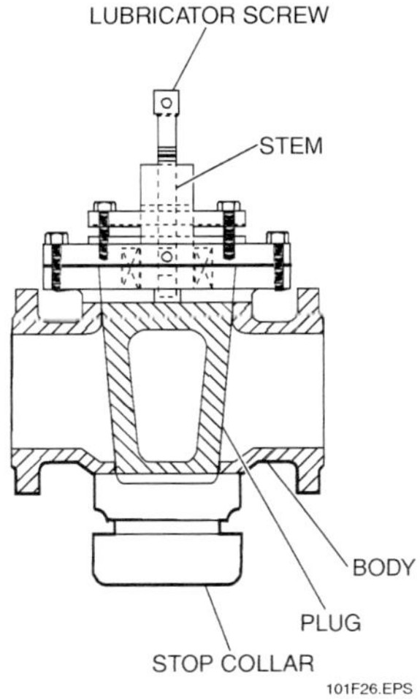

Figure 26 ◈ Plug valve.

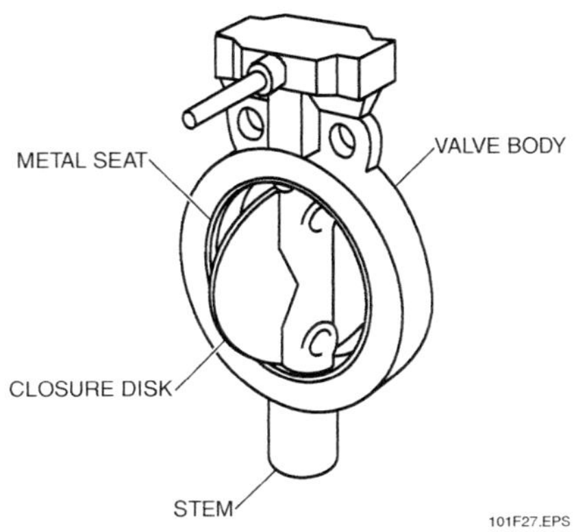

Figure 27 ◈ Butterfly valve.

Demonstration

When discussing each type of valve, show and pass around an assortment of actual valves. Have trainees compare them with the figures. Explain any differences.

Audiovisual

Show Transparency 10 (Figure 25). Point out the parts of the simple gate valve and explain how a simple gate valve works.

Show Transparency 11 (Figure 26). Explain how plug valves work.

Show Transparency 12 (Figure 27). Explain the operation of a butterfly valve. Note the disc mounted on the stem that can be turned to block or allow flow.

Audiovisual

Show Transparency 13 (Figure 28). Ask a trainee what makes a check valve close.

Show Transparency 14 (Figure 29). Have trainees explain how relief valves work.

Show Transparency 15 (Figure 30). Note when globe valves are used.

Classroom

Distinguish between how manual and automatic valves work.

7.3.5 Check Valves

Check valves (see *Figure 28*) are considered shutoff valves since they permit full flow or prevent flow in one direction. They close automatically when pressure in the flow direction is removed. Check valves can be designed to allow passage of scrapers and are most often used in gathering lines and in station piping. They are often used in conjunction with other shutoff valves.

7.3.6 Control Valves

Automatic control of flow rates and pressures is provided by control valves. They may or may not provide for complete shutoff because flow control is their major function. Port sizes and flow patterns through these valves must be continuously adjusted to provide control by throttling action, increasing their tendency toward damaging vibration, erosion, and corrosion. These valves are used principally at pump stations, although they may be needed at other points in the line.

7.3.7 Relief Valves

Relief valves (see *Figure 29*) may be called safety valves. Their function is to remain closed until pressures rise above design levels. Pressure is released to prevent pressure damage when released products are piped to safe discharge points.

When not opened at set pressure levels, the valves are kept closed by spring pressure, a nitrogen charge, a weight load, or other means. They normally close automatically when the pressure is relieved but may be designed to stay open until reset. They may be operated and monitored remotely.

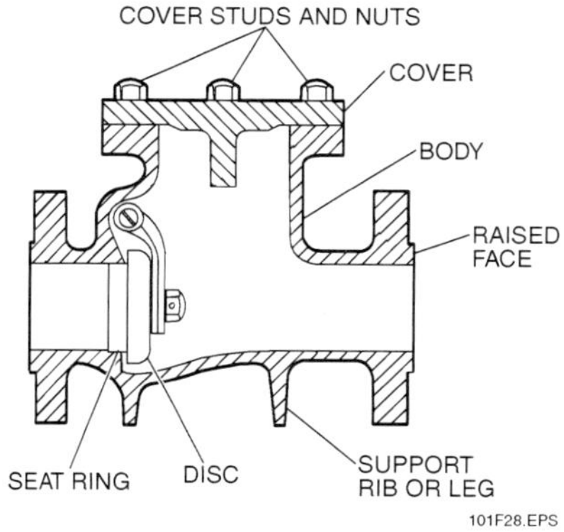

Figure 28 ◆ Check valve.

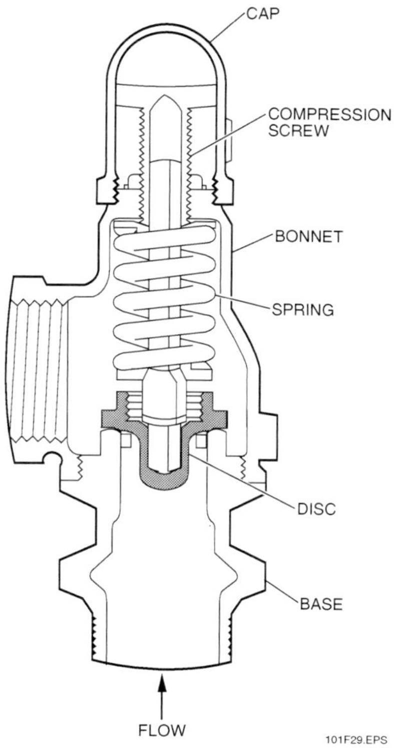

Figure 29 ◆ Relief valve.

7.3.8 Globe Valves

Other valves in the pipeline include globe valves (see *Figure 30*). Globe valves are used for shutoff and throttling, usually in auxiliary systems, such as steam and water systems.

7.3.9 Needle Valves

Needle valves are a type of small, simple, rugged device for throttling and shutting off extremely high pressures compared with other valves. They are used mostly in small auxiliary piping systems.

7.4.0 Actuators/Operators

A valve actuator, commonly called an operator, is a device that opens and closes a valve either manually or automatically. Manual actuators come in the form of handwheels, levers, and

Instructor's Notes:

wrenches that are attached to the stem. Larger valves that require great effort to turn must be equipped with some type of powered mechanism, including electric, pneumatic, or hydraulic units, mounted on the stem. Some valve actuators use a gearbox with wide reduction ratios. Some manual actuators employ high and low gear ratios to enable personnel to turn the valve.

Valves with automatic actuators are controlled from a console in a control center by remote control, sometimes from many miles away. Such automatic actuators are used when the valve is remote from the main work area, when the frequency of operation would require unreasonable human effort, or if rapid opening and closing of the valve is required. Maintenance crews often carry portable valve actuators to work in outlying installations.

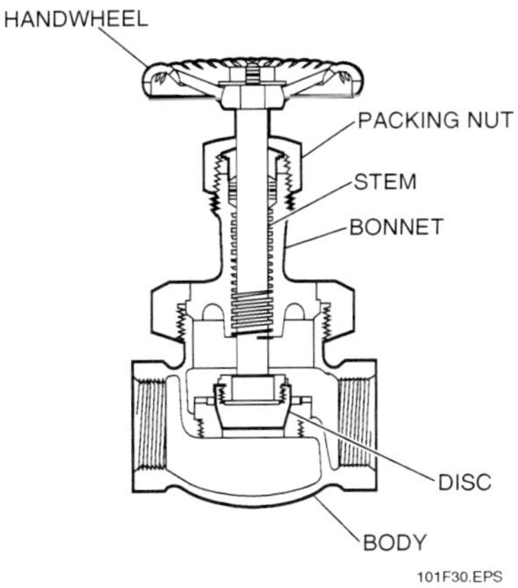

Figure 30 ◈ Globe valve.

7.5.0 Tanks

Tanks are containers used for liquid storage, relief, and breakout. Tanks may exist along the length of the pipeline, at the production site, at pump stations, at terminals, and at delivery points. Although the types and sizes of tanks at each of these sites may vary, they are usually round and made of steel. All atmospheric storage tanks are constructed per *API 650* and repaired per *API 653*.

Tanks must be calibrated for capacity. The dimensions are strapped, or measured, to be used in strapping tables. From these tables, the volume of liquid contained in the tank can be determined. Tank gauges show the tank number and gauge reading from which corresponding volume in barrels can be determined.

Cylindrical tanks may have flat floating roofs, fixed cone-shaped roofs, or both. The first can be used for gasolines and aviation-grade kerosenes, while the second can be used for fuel oils. Some tanks have fixed geodesic domes covering a floating pan. Spheroid tanks are used for volatile products because of their high vapor-pressure capacity.

The floating roof tank (see *Figure 31*) is designed to reduce losses through evaporation of volatile components of crude oil and to reduce the corrosion from sour crude by eliminating the vapor space between the oil and the roof. Seals around the perimeter of the roof prevent vapor loss and rain contamination. The wind girder on the outside rim of the tank adds strength to the shell.

Tanks must also be maintained and repaired per industry standards. Although a mixer helps to keep sediment in suspension in the product, sludge forms in the bottom and must be cleaned out periodically. The tanks are repainted inside and out at specific intervals. Structural inspections and preventive maintenance guard against internal and external damage.

7.6.0 Pressure Vessels

Just as tanks are used for the storage of liquid products, pressure vessels are used along the gas pipeline for the storage of gas products. Liquid vessels are typically cylindrical, but like tanks, can vary in size and shape from site to site, depending on application, capacity, and pressure ratings.

Cylinders and pressure vessels can be fabricated of steel, aluminum, or composite materials. They must be manufactured, inspected, marked, tested, and retested in accordance with United

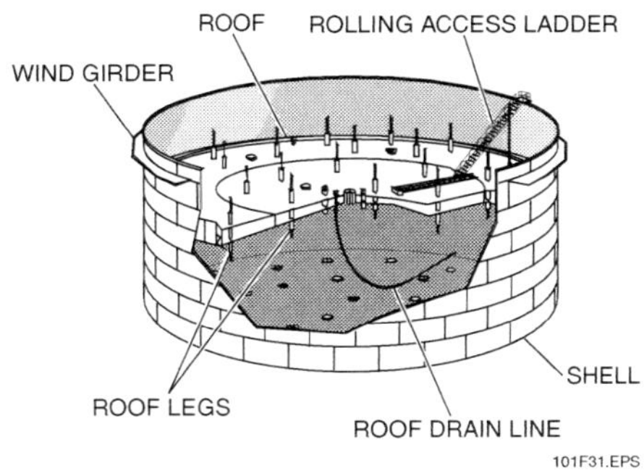

Figure 31 ◈ Open floating roof tank.

Classroom

Review *API 650* and *API 653*. Familiarize trainees with the construction and repair standards for atmospheric storage tanks.

On the whiteboard/ chalkboard, write *Cylindrical Tank with Flat-Floating Roof*, *Cylindrical Tank with Cone-Shaped Roof*, and *Spheroid Tank*. Have trainees contribute facts about each.

Review *API Standards 510* and *2510*. Discuss what the regulations cover. Ask trainees what materials can be used to construct pressure vessels.

Classroom

On the whiteboard/ chalkboard, write the four types of pipe mentioned in the text. Have trainees take turns writing a fact about each on the board.

Audiovisual

Show Transparency 16 (Figure 33). Review the types and identifying characteristics of the fittings.

Teaching Tip

See the Teaching Tip at the end of this module.

States **Department of Transportation (DOT)** regulations and exemptions for the type of service that they are used for. Pressure vessels must also be designed and repaired in compliance with *API Standards 510 and 2510*. These regulations cover the manufacture, marking, maintenance, repair, reconditioning, and testing of seamless pressure vessels, and they prescribe the requirements for transportation of hazardous materials within vessels. *Figure 32* shows a pressure vessel.

7.7.0 Pipe

Line pipe is made of steel through various methods. The main types of pipe used in the oil and gas industry are seamless, electrical resistance welded (ERW), and double submerged arc welded (DSAW). Other piping, less commonly used now, is welded steel pipe.

Seamless pipe is made from a solid length of metal pierced lengthwise by a mandrel with a rounded nose. The manufactured steel tubes and pipes are used in water, gas, and crude oil transmission and distribution lines and as casing pipes in the petroleum industry.

ERW is the most common form of manufacture for pipe in sizes from 2 to 22 inches outside diameter (OD). While seamless pipe is traditionally stronger and more expensive than comparable ERW pipe, ERW technology is improving and being used more and more frequently.

DSAW pipe is pipe that has longitudinal or spiral butt joints. The joints are produced by at least two passes, including at least one on the inside and one on the outside of the pipe. Pressure is not used, and filler metal for the inside and outside welds is obtained from the electrode or electrodes. This pipe is primarily intended for use in the gas and oil transmission industry.

Welded steel pipe is made by bending strips of steel into the form of a tube and welding the longitudinal seam either by electric resistance, by fusion welding, or by heating the tube and pressing the edges together. Two types of older piping that are still found on pipelines are furnace lap-welded and furnace butt-welded piping.

7.8.0 Fittings

Fittings include elbows, tees, reducers, flanges, plugs, and unions. Elbows are connections for directional changes. Tees are used to join three pipe ends. A reducer is used to connect pipes of different sizes. Flanges provide a means of opening and closing piping and attaching pipe ends to mating flanges on valves, tanks, pumps, and other equipment. Plugs and caps close the end of a pipe. A union provides a means of opening and closing piping and permits closing of circuitous piping systems. *Figure 33* shows types of fittings.

Fittings are made in a range of materials based on their service requirements. *Table 3* shows a partial list of materials and their pressure ranges.

Table 3 Materials and Pressures

Material	Pressure
Cast Iron	150 psig
Malleable Iron	250 psig
Cast Steel	150–2,500 psig
Forged Steel	1,000–6,000 psig

101T03.EPS

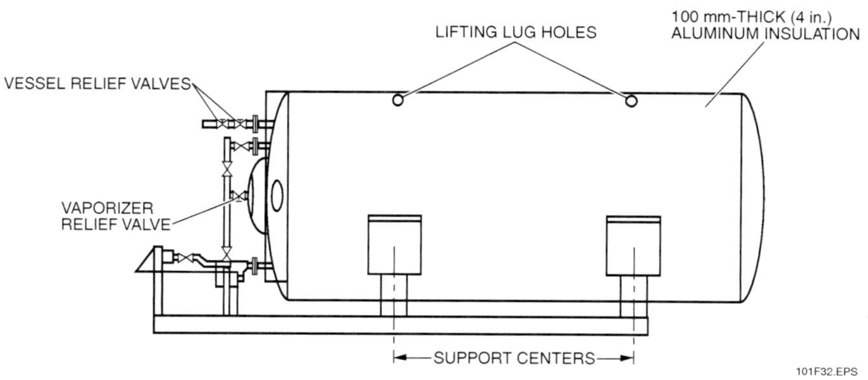

Figure 32 ◈ Pressure vessel.

Instructor's Notes:

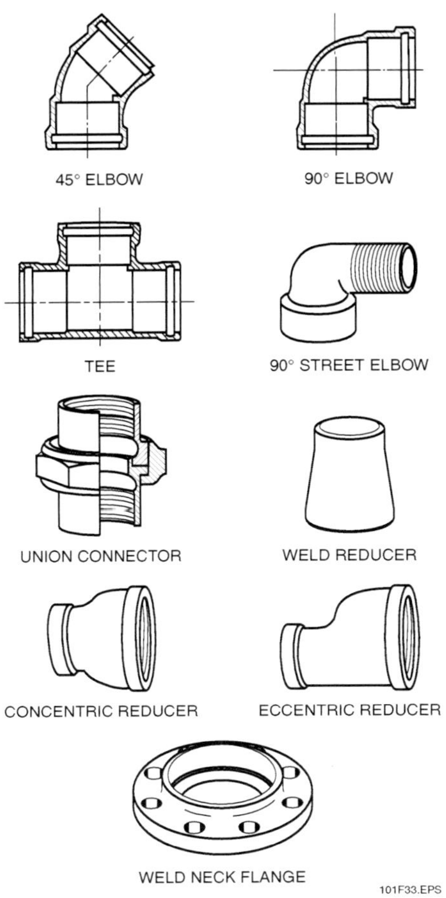

Figure 33 ◈ Types of fittings.

7.9.0 Manifolds

The pipeline uses manifolds to control the flow path of oil, gas, and other petroleum products. Manifolds, with valves and meters, branch off of the line to divert specific volumes of certain products to different locations or other pipelines. *Figure 34* shows a delivery manifold.

The complexity of manifold piping ranges from very simple to very complicated. A booster station manifold may merely allow bypassing of the station. A more extensive manifold at an originating station may connect the main line, field-gathering lines, tanks, and pumps. This manifold may perform any or all of the following operations:

- Pumping product through the main line with tanks inactive
- Receiving product from the field into any tank
- Delivering from the main line into any tank
- Transferring from one tank to another
- Pumping product from a tank into the main line
- Isolating tanks from main-line flow
- Injecting product from any tank into the main line

7.10.0 Traps

Traps in the pipeline are designed to launch or catch scrapers and other pigs that are inserted in the pipeline. Valves and/or manifolds are used to isolate the trap for scraper insertion or removal and then to realign the trap to the pipeline flow. *Figure 35* shows incoming and outgoing scraper/pig traps.

Automation of operation has been extended to the launching of scrapers in some instances. Some traps are arranged so that one or more scrapers may be loaded and then launched at any desired time by remote-controlled valve changes.

7.11.0 Measuring/Sampling Equipment

All volumes entering, leaving, and passing through the pipeline system must be measured and accounted for. Accurate measurement is required for custody transfer, task purposes, leak detection, and delivery schedules.

The primary method of measurement in the pipeline is metering. Meters, whose accuracy is documented with meter provers, measure and record the volume of product passing through. Factors affecting volume are the temperature, pressure, and density of the product being measured. Various instruments are used to measure these factors concurrently with the volume measurement. Product characteristics are also checked periodically with automatic and manual sampling.

7.11.1 Meters

Various meters are used along the pipeline to monitor the volume of fluid or gas. Most product measurements are made by turbine meters or positive-displacement meters, although coriolis-type mass flowmeters, vortex, and ultrasonic meters are also used. Meter readings taken before

Classroom

Discuss the types of operations a manifold may perform.

Explain the use of traps.

Figure 34 ◆ Delivery manifold.

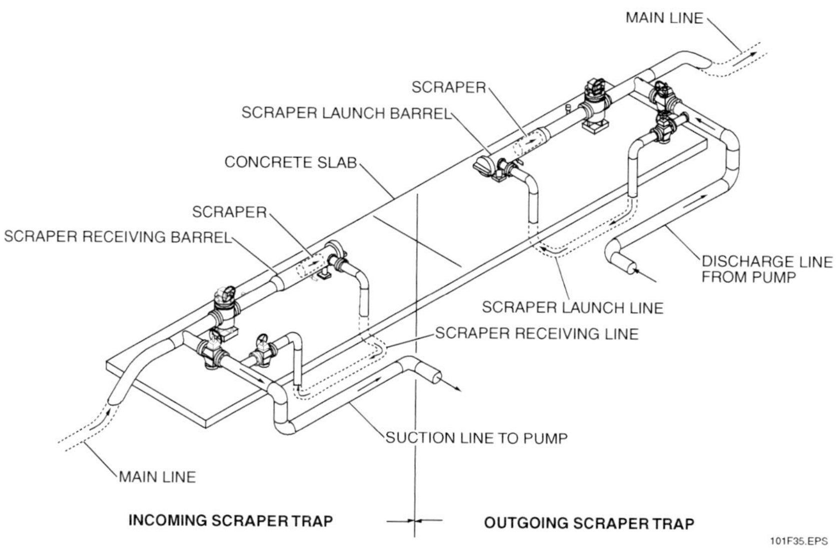

Figure 35 ◆ Incoming and outgoing scraper, or pig, traps.

Instructor's Notes:

and after deliveries are used to determine the volume delivered, or **throughput**. Meters are typically equipped with transmitting, remote-reading, and print-out accessories to facilitate the delivery of correct product volumes and for preparation of delivery tickets. Data is often transmitted to remote locations. *Figure 36* shows two types of meters.

Gas is typically measured by orifice, turbine, positive-displacement, ultrasonic, and rotary vane meters. These meters are generally automated to correct gas volumes for moisture content and temperature and to provide remote read-out and printing capability. Strainers are installed upstream of meters to remove any foreign material from the stream before it goes through the meter. The wire mesh in strainers is generally more coarse than filter materials are.

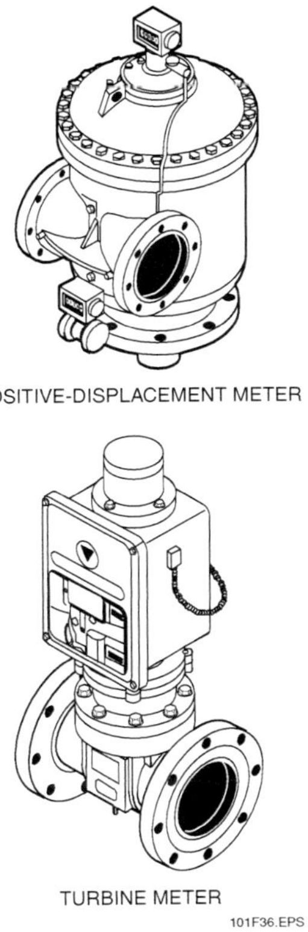

Figure 36 ◆ Two types of meters.

The orifice plate is the vital component of the orifice meter. It is the part that restricts the flow to develop the difference in pressure that is measured to calculate actual flow. The accuracy of orifice meters is dependent on the quality of the meter that is installed. Expert installation and maintenance are necessary to ensure the continued accuracy of the meter.

Orifice plates are secured between halves of flanges of machined carriers or in the body of a special orifice meter housing constructed so that the plate can be removed for inspection. Some meters allow the plate to be removed without stopping the flow. The orifice meter is made up of the carrier housing or flanges, the orifice plate, gaskets, O-rings, devices for removal, and pressure fittings. Pressure taps are drilled at a location exactly 1 inch on center from either side of the orifice plate. *Figure 37* shows some common orifice plate holders designed for easy removal.

Visual inspection of the housing unit involves close examination of the unit looking for any signs of rust, corrosion, cracks, or any buildup of foreign material around the outside housing joints, suggesting the possibility of a leak. Pressure fittings and sensor connections must also be inspected for leaks or damage of any kind. Plates vary from 2.5 inches to 30 inches in diameter.

7.11.2 Provers

A meter prover is used to calibrate meters for specific conditions of flow and product type. Meters are proved by passing a volume of product through the meter at a known rate and comparing this metered volume with the known volume prover. Two types of provers frequently used are the ball prover and the piston prover. A prover loop is a length of pipe or a calibrated U-shaped pipe with a known volume used to check or prove the accuracy of a meter.

8.0.0 ◆ ELECTRICAL POWER SYSTEMS

The electrical power system used by any pipeline company may be quite large or quite small. Power may be self-generated, or partially so, or may be purchased from another source. For the purposes of an introduction, a basic power system will be described from generation to delivery.

An electrical power system is an assembly of equipment and circuits for generating, transmitting, and distributing electrical energy. The elements of a power system form complex networks that require energy control centers to monitor and regulate their operation. The

Audiovisual

Show Transparency 17 (Figure 36). Explain the various types of meters and what they measure.

Classroom

Ask a trainee to explain what provers are used for. Have another trainee name the two types of provers that are frequently used.

Homework

Assign the reading of Sections 8.0.0–9.0.0 for the next session.

Materials

Ensure that you have everything required for demonstrations during this session.

Classroom

Discuss the principal elements of an electrical power system.

Explain the process of power generation and transmission. Answer any questions trainees may have.

reliability of the system is critical to pipeline operation. *Figure 38* shows an electrical power system. The principle elements of the system can be divided into the following:

- Power generation and transmission
- Power distribution
- Direct current (DC) power distribution
- Utilization equipment

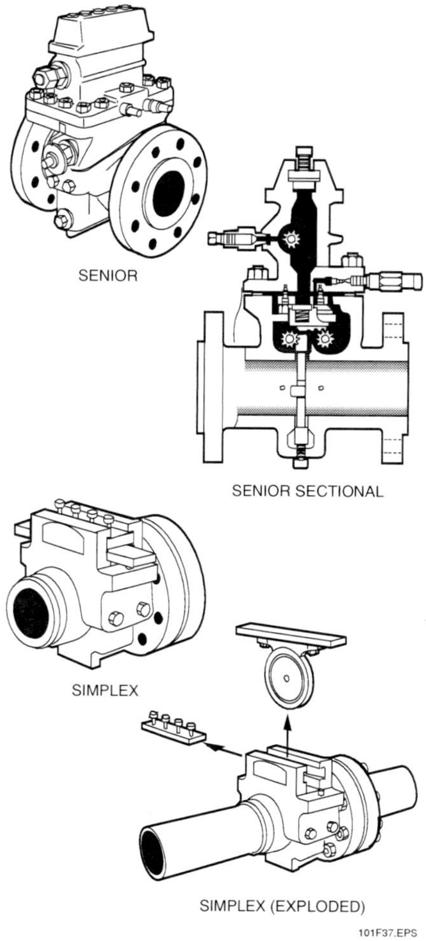

Figure 37 ◆ Common orifice plate holders.

8.1.0 Power Generation and Transmission

Electrical systems require generating stations to produce electric power, and transmission systems to carry it to areas of consumption. Along the way, the voltage of the power delivered must be adapted to the specific needs of consumers. The myriad of uses that consume this energy is the power system load.

8.1.1 Generation

The large quantities of electricity required to supply a system are produced in generating stations, commonly called power plants. These are actually conversion facilities where the chemical or nuclear energy of fuel, such as coal, oil, gas, and uranium, or the kinetic energy of falling water is converted to electricity.

The generators, which supply the electricity, can be driven by steam turbines, hydraulic turbines, gas turbines, or internal combustion engines. Coal, natural gas, and heavy fuel oil are commonly used fuel for steam turbines.

Three-phase 60 hertz (Hz) AC systems are used almost exclusively in the U.S. Consequently, power system generators are wound for three-phase output at a voltage usually limited by design features. The output of modern generating stations is usually stepped up by transformers to the voltage level of transmission circuits used to deliver power to distant load areas.

8.1.2 Transmission

The transmission system carries electric power efficiently and in large amounts from generating stations to consumption areas. Such transmission is also used to interconnect adjacent power systems for mutual assistance in case of emergency.

8.2.0 Power Distribution

Power is reduced and distributed through a series of substations to provide the voltage required in specific applications. A substation is an assembly of equipment through which electrical energy is passed for transmission, distribution, interconnection, transformation, conversion, or switching. Power distribution involves:

- Substations
- Disconnection means
- Transformers
- Cables/conduits
- Switchgear
- Motor control centers

Instructor's Notes:

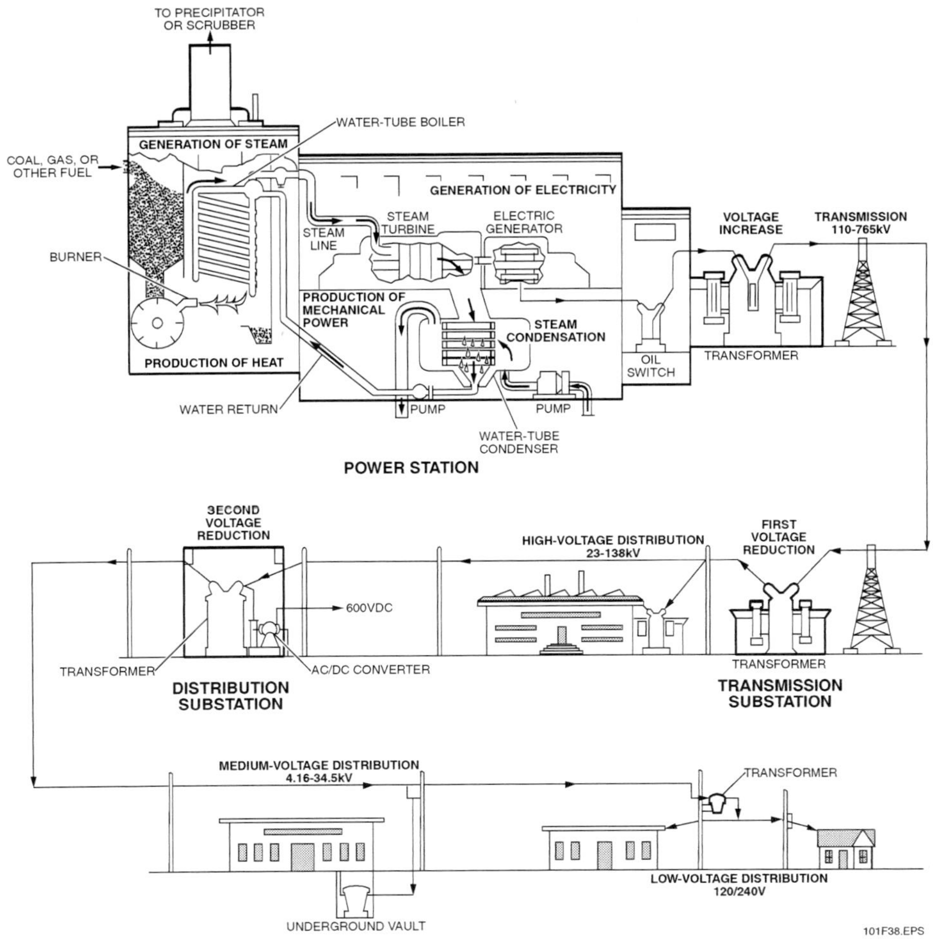

Figure 38 ◆ Electrical power system.

8.2.1 Substations

Substations are used for some or all of the following purposes:

- Connecting generators, transmission or distribution lines, and loads to each other
- Transforming power from one voltage level to another
- Interconnecting alternate sources of power
- Switching for alternate connections and isolating failed or overloaded lines and equipment
- Controlling system voltage and power flow
- Compensating for reactive power
- Suppressing overvoltage
- Detecting faults, monitoring, and recording information, power measurements, and remote communications

Classroom

Review the common purposes of substations. Provide examples, if relevant.

Classroom

Detail the components of a power distribution system. Ask trainees if they have any questions.

Distinguish between raceway and conduit.

Ask trainees to list the items an MCC cubicle may contain.

Minor distribution or transmission equipment installation is not referred to as a substation. Substations are referred to by the main duty they perform. Broadly speaking, they are classified as transmission substations when associated with high-voltage levels, and distribution substations when associated with low-voltage levels. However, they are also referred to in other ways, such as:

- Transformer substations, when the equipment includes transformers
- Switching substations, when the equipment is mainly for connections and interconnections
- Converter stations, when the main function is conversion from alternating current (AC) to DC and vice versa

8.2.2 Disconnection Means

Disconnect switches are commonly present so that sections or components can be isolated for maintenance. These have isolation and connection capability but may not be capable of interrupting load.

8.2.3 Transformers

Transformers use magnetic core and windings to transform power from one side of the transformer to the other at different voltages. Substation transformers range from small sizes of 1MVA to large sizes of 2,000MVA. Most are insulated and cooled by oil, and adequate precautions must be taken against fire hazard. These include adequate distance from other equipment, fire walls, fire extinguishing means, and pits and drains for containing leaked oil.

Current and potential transformers are for measuring currents and voltages. They are also for providing proportionately low-level currents and voltages for control and protection.

8.2.4 Cables and Conduits

Cables are two or more insulated wires and a grounding wire covered by an outer jacket, or sheath. Cable is usually classified by the type of covering it has, either nonmetallic or metallic. Cable may also be classified according to where it can be used. The *National Electrical Code* has definite guidelines on cables and their allowable uses. Because water can cause power loss or short circuits, cables are classified for either dry, damp, or wet locations. They are also classified for exposure to sunlight or rough use.

All cables are marked to show important properties and uses. These markings show the wire size, number of conductors, cable type, and voltage rating. The information is printed on nonmetallic sheathing and on tags attached to metal sheathing.

Raceway is a general term that refers to a wide variety of circularly and rectangularly enclosed channels used to house electrical wiring. Conduit is a raceway with a circular cross section similar to pipe that contains cables or wires. Conduit is used to provide protection for conductors and route them from one place to another. In addition, conduit makes it easier to replace or add wires to existing structures and, if it is metal, provides a permanent electrical path to ground.

8.2.5 Switchgear

Switchgear includes a wide variety of enclosed circuit breakers, disconnects, busways, and associated items that distribute power from a common point. An appropriate switching arrangement for connections of generators, transformers, lines, and other major equipment is basic to any substation design. Switchgears are use to provide this arrangement.

8.2.6 Motor Control Centers

The motor control center (MCC) contains the isolators for all the electrical equipment associated with a main line pump. Each MCC **cubicle** contains some or all of the following:

- Breaker
- Insulator
- Disconnect
- Isolator
- Reset button
- Hand/auto switch
- Run and stop light indicators
- Opened, closed, and stop light indicators

The isolator is the main isolator for the identified equipment and may be padlocked in the OFF position. The run and stop light indicators display the status of the associated equipment. Opened, closed, and stop light indicators are associated with motor-operated isolation valves.

Instructor's Notes:

8.3.0 DC Power Distribution

The DC power system provides the controls, monitor, and indicators of power required for control systems. Normal shutdown and automatic malfunction shutdown functions are also provided in the electrical system.

The DC power supply consists of batteries and chargers. A source of DC power is externally supplied by batteries and a battery charger. The charger and certain other equipment require an external power source. The battery charger will automatically charge the batteries. It does so by changing the batteries to full charge and then provides a float charge that will preserve the charge.

The charger will carry continuous and intermittent DC loads up to 80% of the maximum rated output. The charger has limiting characteristics to avoid exceeding its maximum output rating, and the output is further protected by fuses for short-circuit conditions.

9.0.0 ◈ CORROSION CONTROL

Corrosion is a natural process that, given the right conditions, can attack any metal or alloy. Most pipeline technicians are familiar with corrosion but do not have specific knowledge of the various types of corrosion. Everyone is surely familiar with rust, the most common form of corrosion which results in overall thinning of a metal. Most corrosion failures, however, are not caused by rust or general corrosion but by localized attack—pitting, crevice corrosion, intergranular corrosion, or stress corrosion cracking.

Each year unchecked corrosion can cost the pipeline industry millions of dollars in material damage and downtime. Corrosion can cause catastrophic failures with potentially severe consequences for the environment and serious injury or even death to personnel. It is necessary today to be able to identify potential problems and risks through a corrosion management program. An entire industry has built up around this need.

The largest organization in the world committed to the study of corrosion is the National Association of Corrosion Engineers (NACE). NACE was established in 1943 by eleven corrosion engineers in the pipeline industry. Today, NACE International has more than 50 years of experience in developing corrosion prevention and control standards. NACE is an excellent source of information for further study.

Federal guidelines have been implemented to support safety of both personnel and the environment. The complete text of the corrosion control requirements can be found in *49 CFR Part 192, Subpart I* and *49 CFR Part 195, Subpart H*. Companies are required to have written guidelines and procedures for most corrosion-related activities.

9.1.0 Types and Causes of Corrosion

Many factors contribute to metal corrosion; dissimilar metals that corrode at different rates and are sometimes incompatible, dissimilar soils, dissimilar aeration, bacteria, and stray current are factors that contribute to corrosion.

Pipeline corrosion is most prevalent when the failure of coatings, inhibitors, or cathodic protection occurs in a corrosive environment. These factors do not affect the pipeline equally at all locations, and corrosion does not grow at the same rate throughout a pipeline.

Internal corrosion of oil and gas pipelines is a growing concern to pipeline operators. The causes of internal corrosion of pipelines are chloride, carbon dioxide, hydrogen sulfide, oxygen, and microbiological activity. A recent severe case of internal corrosion caused several failures on a 20-inch oil line after only 2 years of service. The cause of the failure was high carbon dioxide concentrations (9%) in combination with high levels of brine.

Piping and pipeline systems are often great places to grow microbes that can produce corrosive conditions. Microbiologically influenced corrosion (MIC) often works in combination with other forms of corrosion, such as crevice corrosion and under-deposit corrosion, on in the pipe. It can also accelerate corrosion produced by impurities in the system environment, such as hydrogen sulfide and/or carbon dioxide.

For example, MIC can occur in the crevices produced by disbonded coatings on the outside of the pipe that are shielded from cathodic protection. Internally, sites for MIC are where water, debris, and other nutrients accumulate on the pipe wall. Both sites provide the local conditions that promote colonization of microbes, which can make existing corrosion problems worse by producing organic acids or acid-producing gases. There is a long list of microbes that can participate in the corrosion reaction; these include aerobic or anaerobic types (depending on whether they require oxygen or lack of oxygen to survive). They can also make use of nutrients in the environment (sulfates, nitrates, iron) to produce corrodants.

Classroom

Explain the components of a DC power system.

Laboratory

Divide trainees into small groups. Have trainees brainstorm a variety of possible consequences of corrosion. Ask each group to share their ideas with the class.

Classroom

Review *49 CFR Part 192, Subpart I* and *Part 195, Subpart H*. Locate and discuss the corrosion control requirements.

Explain the factors that may cause corrosion. Emphasize that there are other types of corrosion besides rust.

Classroom

Discuss the various methods available to detect MIC. Ask a trainee whether sessile or planktonic surveys are better.

Explain why injection and sampling systems are used. Answer any questions the trainees may have.

Audiovisual

Show Transparency 18 (Figure 39). Define corrosion coupons and explain how they are used.

Demonstration

Pass around several different coupons, both new and used if possible. Have trainees take turns examining each of them.

Classroom

Ask a trainee to explain what a smart pig can detect.

Stress why monitoring for corrosion is essential. Point out the importance of computers in the monitoring process.

9.2.0 Inspection and Testing

Pipeline integrity programs use tools, technologies, and strategies to ensure that pipeline systems are in a condition to perform safely and reliably. Increasingly sophisticated technology is available. Constant monitoring and inspection is essential to any pipeline integrity program.

9.2.1 Microbiologically Influenced Corrosion (MIC)

A variety of methods is available to detect MIC. Sampling and culturing can be performed for identification of the size of the infection and the active strains of organism present. This information is then used to identify preventative or remedial treatments. Chemical analysis and pH of corrosion products can also give important information relative to the nature of the problem. Evidence taken from the metal surface, called sessile surveys, has a better chance of properly assessing the situation versus sampling of the liquids, called planktonic survey.

9.2.2 Sampling

Injection and sampling systems are fundamental to corrosion control and process control programs. They are applicable to a large variety of processes in the petroleum, chemical, and water treatment industries. Injection systems are used to inject a wide range of chemicals into processes. Such chemicals include biocides, demulsifiers, corrosion inhibitors, oxygen scavengers, glycol and monoethylene glycol, dewaxers, methanol, odorizers, and product additives.

Sampling systems are used to take samples of the process product. Such samples are then analyzed in the laboratory for inhibitor concentration levels, the presence of metal ions, oxygen levels, scale forming compounds, and a wide range of other substances.

9.2.3 Corrosion Coupons

Coupons are specimens of materials that are exposed to tests or real environments to assess the effect of degradation on the material. They can be used inside a pipeline or outside if the pipe is underground. Corrosion coupons represent a widely used and relatively simple form of corrosion monitoring. Usually a coupon is exposed to a corrosive environment to represent an actual component/structure. After exposure for a specified amount of time, the coupons can be weighed and examined microscopically to determine and evaluate the type and average rate of corrosion damage. *Figure 39* shows typical coupons.

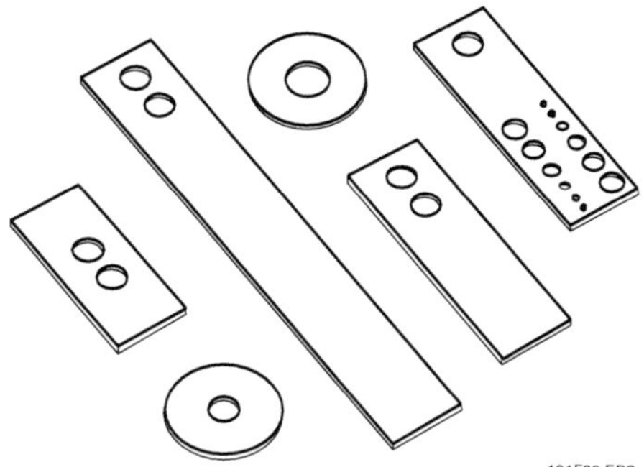

101F39.EPS

Figure 39 ◆ Typical coupons.

9.2.4 In-Line Inspection, or Smart, Pigs

Smart pigs are high-tech inspection tools covered with sensors that gather and record information in onboard computers as they pass through the pipeline. They check for various problems, such as thinning, pitting, and general corrosion. Some smart pigs can detect metal loss in areas around welds. *Figure 40* shows an ultrascan in-line inspection (ILI) pig.

9.3.0 Maintenance and Protection

Monitoring a pipeline for corrosion is essential in preventing failures. If an operator can identify active or growing corrosion defects, then the rate of corrosion can be individually predicted for each defect, allowing an opportunity for corrective measures before the failure occurs.

COURTESY OF PII GROUP LTD

101F40.EPS

Figure 40 ◆ Ultrascan in-line inspection pig.

Instructor's Notes:

As soon as production begins in a line, companies begin maintenance programs to reduce the risk of pipeline leaks. Sophisticated electronic control systems monitor changes in pressure, volume, and product flow. A variety of highly technical tools collect information about both the interior and exterior condition of pipelines. In-line inspection (ILI) tools, commonly called smart pigs, assess metal loss, strain, and cracking. This information becomes part of a computer database that is used by experts to identify and plan long-term maintenance programs.

To safeguard the outside surface of the pipe, protective coatings are applied when lines are built. In all cases, as a backup to the protective coating, cathodic protection is installed to help control external corrosion.

9.3.1 Inhibitors

A corrosion inhibitor is a chemical substance that, when added in small concentration to an environment, effectively decreases the corrosion rate. Inhibitors are chemicals that react with a metallic surface or with the environment the surface is exposed to, giving the surface a certain level of protection. Inhibitors often work by adsorbing themselves on the metallic surface, protecting the metallic surface by forming a film. Inhibitors are normally distributed in a solution and injected into a pipeline through a sampling/injection tool. Some are included in a protective coating formulation. Inhibitors slow corrosion processes by one of the following methods:

- Increasing the anodic or cathodic polarization behavior
- Reducing the movement or diffusion of ions to the metallic surface
- Increasing the electrical resistance of the metallic surface

A vapor corrosion inhibitor stops corrosion by releasing protective ions that are attracted to metal. These ions condense on the metal surface, forming a water-repellent layer that contains both positively and negatively charged ions. The layer is maintained and repaired by the constant condensation of new vapor.

Corrosion inhibitors used in deep-water treating of oil and gas wells and deep-water risers differ from the use of land-based inhibitors in several ways, including delivery of the chemical to the point of injection, selection of inhibitor, and monitoring for results.

Each situation must be carefully examined before the proper inhibitor is selected.

9.3.2 Cathodic Protection

Cathodic protection effectively protects underground or submerged metallic structures through the use of a negative potential applied by an external source to the pipeline or structure. Usually, once the pipeline has been made sufficiently negative, it will resist environmental (soil or moisture) corrosion. The method is typically applied to underground pipelines, storage tanks, the interior of water storage tanks, ocean pilings, and electrical transmission towers. Buried steel structures will revert back to their natural state as an iron oxide without proper protection.

Cathodic protection is the most widely applied electrochemical corrosion control technique that reduces the corrosion of a metal surface by making that surface the cathode of an electrochemical cell. Cathodic protection is accomplished by creating and controlling the action of large corrosion cells by connecting the negative terminals of DC power sources to the pipeline and the positive terminals to groundbed anodes (see *Figure 41*).

10.0.0 ◆ REGULATIONS, PROCEDURES, AND DOCUMENTATION

Each pipeline company has published operations and maintenance (O&M) and emergency procedures. These procedures are based on federal, state, and local regulations and on industry standards. Regulations have the effect of laws and are enforced by government agencies. Standards are generally accepted practices gained through experience. All pipeline employees are expected to become familiar with all procedures that are relevant to their job.

10.1.0 Regulations

The U.S. Department of Transportation (DOT) Office of Pipeline Safety also handles pipeline safety and the quality of pipeline design, construction, and operation. This office administers two sets of regulations, *Code of Federal Regulations (CFR) 49 Part 192* and *CFR 49 Part 195*, that address transportation of natural gas and other gases and liquids by pipeline. The pipeline industry is greatly affected by federal regulations. The Federal Energy Regulatory Commission (FERC) controls the natural gas and crude oil and products lines in addition to the regulations of other federal agencies. The Occupational Safety and Health Administration (OSHA) oversees every facet of worker safety and health, from the wearing of personal protective equipment to provision of hazard information to availability of emergency equipment.

Classroom

Have trainees write down the three methods by which inhibitors slow corrosion processes.

Audiovisual

Show Transparency 19 (Figure 41). Explain how cathodic protection works. Have trainees name typical applications of cathodic protection.

Homework

Assign the reading of Section 10.0.0 and have trainees complete the Review Questions for the next session.

Materials

Ensure that you have everything required for demonstrations and testing during this session.

Classroom

Clarify what the DOT Office of Pipeline Safety, FERC, and OSHA oversee and/or regulate.

Classroom

Have trainees list the AOCs that can occur in a pipeline system.

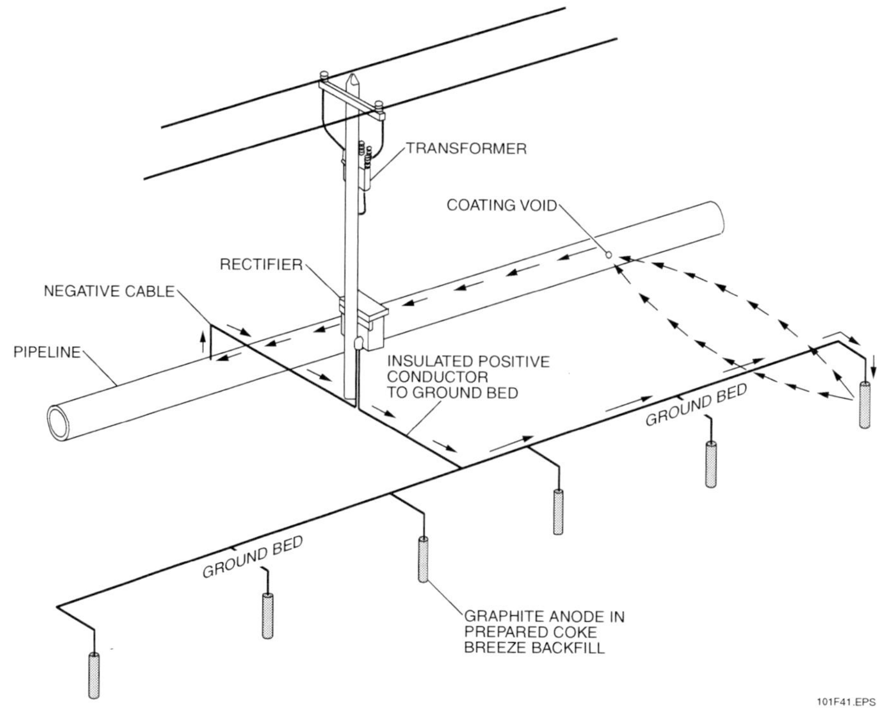

Figure 41 ◆ Cathodic protection arrangement.

State agencies can adopt federal regulations or enforce more stringent regulations than federal because each state has developed their own form of regulation regarding pipelines. Pipelines operate on an interstate or intrastate basis but not both. Interstate pipelines are located through more than one state and fall under federal jurisdiction and classification. Intrastate pipelines are located in a single state, and different states may classify the pipeline differently. Procedures should be relevant to the states involved.

10.2.0 Abnormal Operating Conditions

Abnormal operating conditions (AOCs) are specified by federal law in *49 CFR Part 195.503 and CFR 192.803*. The Operator Qualification Rule requires that an individual recognize and respond to any AOC that could happen while performing a task. An AOC refers to any condition identified by the individual that may indicate failure of a component, or deviation from normal operations that may indicate a condition exceeding design limits or that may result in a hazard to persons, property, or the environment. The following list identifies the AOCs that can occur when operating a pipeline system:

- Abnormal facility condition
- Activation of a safety device
- Communications failure
- Control system failure
- Power interruption
- Fire, explosion, or natural disaster
- Pipeline system damage
- Unexpected hazardous liquid or carbon dioxide encountered
- Unexplained flow rate deviation
- Unexplained pressure deviations
- Unexplained status change
- Other company specific circumstances

Instructor's Notes:

10.3.0 Company Policies and Procedures

Company policy and procedure manuals typically include:

- Operating procedures manual
- Maintenance procedures manual
- Safety procedures manual
- Emergency procedures manual

Each procedure will typically include procedure details, purpose, responsibilities, and other supporting information. There are many different requirements depending on the company and the location of the pipeline.

10.4.0 Qualifications

With the introduction of the new Operator Qualification Rule in *49 CFR 195.503*, pipeline operators are now required to have a written qualification program. Required elements of the qualification program include:

- Identification of covered tasks
- Qualification in the performance of covered tasks through operator evaluations
- Evaluation measures for individuals following a DOT-reportable accident or for other reasonable causes
- Communication processes
- Identification of intervals at which reevaluations will occur

10.5.0 Documentation

Documentation is an important part of the pipeline operation from origin to end point. The DOT requires that many operations and maintenance procedures be documented to record their correct and timely performance. Problems must be documented to track down the cause and to prevent recurrence.

Some documents are forms that must be completed by operations personnel and technicians. Others are forms, reports, and summaries that are printed out periodically through the control system. Some of the documentation required for operations includes:

- Accident/incident reports
- Operations logs
- Work orders
- Daily events/event logs
- Personnel qualifications
- Inspection and repairs

10.5.1 Accident/Incident Reports

Accident/incident reports are required, as specified in the Office Pipeline System (OPS) regulations, whenever there is a reportable incident along the pipeline. An incident is defined as an unexpected situation that causes, or could be expected to cause, an operational problem, harm to personnel, or damage to the environment. Reportable situations are defined in the regulations.

10.5.2 Operations Logs

Operations logs record the daily events of operation. These are usually summarized on reports printed from the control system; they include all activities communicated to and from the control room as associated with the pipeline.

10.5.3 Work Orders

Work orders and/or work permits are issued when specific maintenance or other tasks need to be performed. Each work order or permit has its own number and is recorded on a work order or permit log along with a description of the job and other details of the assignment.

These forms are required to ensure systematic maintenance of the pipeline, to assist with scheduling, and to record work done. Work orders and schedules for tasks done at specific time intervals can be triggered by the computer system. Not all pipeline companies use a work order system.

10.5.4 Daily Events and Event Logs

Daily events are recorded in the event log as specific tasks or activities that are performed during the day's operations. These might include repairs, inspections, AOCs, investigations, maintenance, or many other events.

10.5.5 Personnel Qualifications

The OPS regulation provides for formal documentation of the qualifications of personnel responsible for pipeline operations and maintenance in each company. The regulation was designed to reduce the risk of accidents on pipeline facilities caused by human error. Each company must have a written program specific to its own system for the qualification of individuals, including employees, contractors, and subcontractors who work on that system. These qualifications often cover about fifty different safety-related tasks, usually by position, such as pipeline controllers, and corrosion, maintenance,

Classroom

Note the typical components of company policy and procedure manuals.

Refer to *49 CFR 195.503*. Review the Operator Qualification Rule and what it means for pipeline operators.

Teaching Tip

See the Teaching Tip at the end of this module.

Classroom

Give examples of reportable situations, referring to the OPS regulations if necessary.

Ask a trainee why documentation of personnel qualifications is required.

Classroom

Have trainees identify the information that must be documented for repairs and for corrosion control.

and operations personnel. The rule requires that a number of specific elements be present in all operator plans, including:

- Personnel must demonstrate the ability to recognize and respond to abnormal operating conditions.
- Companies must keep records that demonstrate compliance to the OPS.

10.5.6 Inspection and Repairs

When repairs are made, the pipe specifications, defect description, stress levels, and action taken must be documented and maintained in case of an audit. Other information, in accordance with the requirements of regulatory authorities, must also be prepared. Some documents are forms that must be filled in by operations and maintenance personnel. Others are forms, reports, and summaries that are printed out periodically through the control system.

Federal guidelines require that all inspections, treatments, and repairs be documented. Some of the documentation required for corrosion control includes:

- Cathodic protection system inspections
- In-line inspection results
- Pipeline coating inspections
- Incident reports
- Maintenance reports

Record-keeping is a huge and important task. Tremendous amounts of inspection and monitoring data are generated over the life of a pipeline, including:

- Development of written pipeline integrity procedures and coordination with operation, maintenance, contingency, and emergency response procedures
- Identification and implementation of repairs and modifications
- Program evaluation

Many software programs are now available to make record-keeping easier and to help companies stay in compliance with regulations. Failure to maintain good documentation can result in significant penalties for your company.

Summary

The term *pipeline* has many meanings. In the simplest terms, it applies to the pipe and fittings used to construct facilities for transporting commodities. In a broader sense, it applies to all of the facilities in addition to the pipe and fittings that are needed to transport the commodities from the point of origin to the point of delivery. Pipeline facilities include the means of measuring, controlling, and accounting for quantities and qualities of product delivered from the custody of the original transporter to the shipper.

In order to efficiently operate and maintain the pipeline facilities, personnel must understand the basic principles of pipeline hydraulics when working with liquid pipelines, and pipeline pneumatics when working with gas pipelines. Personnel must also familiarize themselves with basic pipeline equipment and its proper use.

Electrical power systems supply power to the pipeline equipment and facilities. Electrical and instrumentation technicians maintain and repair this equipment to ensure that the pipeline remains in service. Another major task of pipeline personnel is to monitor corrosion and practice measures to prevent or minimize the presence of corrosion on the pipeline and equipment.

Each pipeline company has published operations and maintenance and emergency procedures. These procedures are based on federal, state, and local regulations and on industry standards. All pipeline personnel must familiarize themselves with these standards in order to operate the pipeline in a manner that will ensure the safety of the employees and the general public and protect the environment.

Instructor's Notes:

Review Questions

1. Oil and gas are formed from _____.
   a. decaying organic matter
   b. buried rocks
   c. salt domes
   d. groundwater

2. Different types of pumps or valves are used to cap the well, depending on the _____ of the fluid being withdrawn.
   a. temperature
   b. destination
   c. properties
   d. amount

3. Lines from the oil fields are called _____ lines.
   a. field
   b. main
   c. distribution
   d. gathering

4. The pipe must be _____ tested to verify the pipeline integrity.
   a. strength-
   b. X-ray-
   c. hydrostatically
   d. sonar-

5. Different petroleum products can be shipped consecutively in the same pipeline. This is called _____.
   a. bunching
   b. mixing
   c. batching
   d. blending

6. Large volumes of natural gas are typically stored in _____.
   a. underground storage fields
   b. surge tanks
   c. relief tanks
   d. open roof tanks

7. Oil containing a large amount of sulfur compounds is called a _____.
   a. sweet crude
   b. sour crude
   c. paraffin
   d. light crude

8. There is a set of _____ for each hazardous substance, listing its properties and hazards.
   a. chemical descriptions
   b. OSHA labels
   c. product papers
   d. MSDSs

9. The marine environment is particularly sensitive to _____.
   a. high pressure
   b. overproduction
   c. oil spills
   d. vapor accumulation

10. A(n) _____ depicts the functional layout of a piping system and its components as clearly and accurately as possible.
    a. P&ID
    b. alignment sheet
    c. strip map
    d. blueprint

11. _____ is the software used to monitor and control the flow of millions of barrels of oil and products through the pipeline.
    a. RTUS
    b. SCADA
    c. SEACA
    d. PLCS

12. The science of the behavior of fluids at rest or in motion is referred to as _____.
    a. pneumatics
    b. hydraulics
    c. compressibility
    d. viscosity

13. Boyle's law states that when temperature is kept constant and volume is decreased, pressure will _____.
    a. decrease
    b. increase
    c. stay the same
    d. not be affected

14. In a typical positive-displacement pump, a _____ slides back and forth in a cylinder.
    a. rotor
    b. vane
    c. piston
    d. blade

Classroom

Go over the Review Questions. Answer any questions trainees may have on materials covered to this point.

Examination

Administer the Module Examination. Be sure to record the results of the Exam on Craft Training Report Form 200 and submit the results to the Training Program Sponsor.

15. _____ are used to control the rate of flow in a line, to open or shut off a line, or to serve as automatic safety devices.
a. Motors
b. Valves
c. Pumps
d. Manifolds

16. Tanks must be calibrated for _____.
a. strength
b. sealing ability
c. capacity
d. wall thickness

17. _____ are used to join three pipe ends.
a. Flanges
b. Unions
c. Elbows
d. Tees

18. The pipeline uses _____ to control the flow path of oil and petroleum products.
a. manifolds
b. pumps
c. tanks
d. vents

19. Various _____ are used along the pipeline to monitor the volume of fluid.
a. gauges
b. tests
c. meters
d. provers

20. Communications failure is one example of an _____.
a. emergency
b. AOC
c. accident
d. interruption

Instructor's Notes:

GLOSSARY

Trade Terms Introduced in This Module

Anomaly: A defect in the pipeline or pipeline coating caused by corrosion, wear, or other external means.

Barrel: The common measure of petroleum products: one barrel equals 42 U.S. gallons or 0.15899 cubic meters or 9,702 cubic inches.

Cathodic protection: A means of preventing the destructive electrochemical process of corrosion of metal by using the metal as the cathode of a cell with a sacrificial anode.

Corrosion: The gradual destruction, dissolving, or wearing away of a substance, usually a metal or an alloy, caused by a chemical or electrochemical reaction with its environment. Corrosion can be a problem in different locations along the pipeline, both externally and internally.

Crude oil: Unrefined liquid petroleum. Crude oil ranges in color from yellow to black, and may have a paraffin, asphalt, or mixed base. Crude oils may be referred to as heavy or light according to API gravity, with the lighter oils having the higher gravities.

Cubicle: An enclosure for high-voltage equipment.

Department of Transportation (DOT): The federal agency that sets standards for, and regulates, transportation.

Hydrostatic testing: The most common final quality check on a complete pipeline. Water is pumped into the line and pressured up to a specified level. The pipe is checked for leaks or ruptures.

Main line: The primary pipe between distant points; it is also called the trunk line.

Manifold: An accessory system of piping to a main piping system that serves to divide the flow into parts, to combine several flows into one, or to reroute a flow to several possible locations.

Material safety data sheet (MSDS): A form containing information on a potentially hazardous material, including names, contents, test results, characteristics, potential hazards, and handling procedures.

Natural gas: A highly compressible, highly expandable mixture of hydrocarbons with a low specific gravity. It occurs naturally in a gaseous form. Besides hydrocarbon gases, it may contain nitrogen, helium, carbon dioxide, hydrogen sulfide, and water vapor. The gases are variable in form and may be found as gases or liquids under various conditions of pressure and temperature. Natural gas liquids include butane, propane, and natural gasoline.

Natural gas liquids (NGLs): The portions of reservoir gas that are liquefied at the surface in lease separators, field facilities, or gas plants.

Petroleum: Unrefined liquid hydrocarbons, including *natural gas liquids (NGLs).*

Pig: A scraping tool with blades or brushes that is forced through a pipeline to clean out accumulations of wax, scale, and debris from the pipe walls; it is also called a line scraper or go-devil.

Product: Any fluid, refined or not, that is transported through a pipeline.

Refined products: Manufactured petroleum products generated from crude oil.

Refinery: System of process units used to convert crude petroleum into fuels, lubricants, and other petroleum-derived products.

Seismic activity: The phenomena of earth movements, earthquakes, and earth vibrations, including those that are artificially induced.

Tank: A metal, plastic, or wooden container used to store a liquid. Three types include mud tanks for drilling, production tanks, and storage tanks.

Throughput: The volume of product that goes through a section of equipment during a specified amount of time.

Instructor's Notes:

MODULE 66101-02 — TEACHING TIPS

The following are suggested activities or instructional methods to help you teach the material in this AIG.

General

When you call on someone to answer a question, the rest of the class relaxes or even tunes out because they expect that the question and answer will take place only between you and the trainee you called on. Instead, use this technique to involve more trainees in answering questions and to keep them on their toes.

1. Ask trainees to define a term or explain a concept.
2. After one trainee has answered, ask a trainee seated nearby if the answer is right. Then ask whether a trainee in the back of the room agrees.
3. Ask trainees to explain why they think an answer is right or wrong.
4. Use the session to clear up incorrect ideas and encourage trainees to learn from their mistakes.

Section 2.2.0 *Refined Products*

This Quick Quiz will familiarize trainees with the types of refined products. You will need photocopies of the quiz provided on the following page. Trainees will need pencils. If you allow trainees to use the Trainee Guide, decrease the amount of time you give them to complete the quiz.

1. Make as many photocopies as you will need of the quiz.
2. Give trainees between 5 and 10 minutes to complete the quiz.
3. Go over the answers to the quiz.
4. Ask trainees if they have questions about the types of refined products.

Section 2.2.0 *Answers to Quick Quiz*

1. d. Consumer-ready natural gas
2. e. Fuel oil
3. c. Gasoline
4. b. Diesel
5. a. Jet fuel

Quick Quiz

Refined Products

For each description listed, identify the refined product the text best describes. Write the corresponding letter in the blank provided.

_____ **1.** Made up primarily of methane, this refined product's heat-value range is 1,020 to 1,045 Btus per cubic foot.

_____ **2.** This refined product comes in two grades: No. 1 and No. 2.

_____ **3.** This refined product is a volatile, flammable liquid hydrocarbon refined from crude oil.

_____ **4.** This refined product is a light hydrocarbon mixture, with a boiling range just above that of kerosene.

_____ **5.** This refined product is a highly refined kerosene petroleum distillate.

a. Jet fuel
b. Diesel fuel
c. Gasoline
d. Consumer-ready natural gas
e. Fuel oil

Section 2.3.0 *Special Products*

This exercise will familiarize trainees with special products. For this exercise, trainees will need their Trainee Guide, pencil, and paper. Allow 45 minutes for this exercise.

1. Divide trainees into small groups. Assign each group one or two special products (ensure that each special product is assigned).
2. Have each group write each of their special products at the top of a blank sheet of paper, one product per page.
3. Have each group fill in as many of the following categories as possible for each product:
   - Physical characteristics (color, odor, gas, or liquid)
   - Purpose (what it is used for)
   - Flammability
   - Safety concerns
   - Results of human exposure

   You may add more categories if you wish.
4. Post all the sheets on the board or wall. Have each group come to the front of the classroom to give a brief presentation on their products. Ask trainees if they have any information to add to each presentation. Answer any questions trainees may have.

Section 6.1.0 *The Ideal Gas Laws*

This exercise will help trainees understand the ideal gas laws. For this exercise, trainees will need their Trainee Guide, sheets of paper, posterboard (or other large sheet of paper), pencil, and colored markers. Allow 45 minutes for this exercise.

1. Divide trainees into small groups. Assign a gas law to each group (each law may be assigned to more than one group).
2. Ask each group to create a poster illustrating the assigned gas law. Emphasize the importance of a clear and simple design with a limited amount of text. Suggest using T, P, and V as symbols. Remind each group to include a legend explaining any symbols used. Suggest that trainees sketch their design on scrap paper first.
3. Invite each group to present their poster to the class. Ask the class to evaluate the poster: does it clearly and accurately explain the law? You may wish to hang the posters up for the remainder of the module.

Section 7.8.0 *Fittings*

This exercise will help trainees identify various types of fittings. For this exercise, you will need at least one sample of each type of fitting listed in *Figure 33* (a total of nine). Alternatively, you may use Transparency 16, *Figure 33,* with the callouts covered. Trainees may work alone or in groups. Trainees will need their Trainee Guide, pencil, and paper to take notes for discussion. Allow 45 minutes for this exercise.

1. Lay out an assortment of numbered fittings (at least one of each type) on a table. If possible, lay out several identical sets of fittings on several tables. Alternatively, you could show Transparency 16.
2. Have trainees identify the type of each fitting. Ask trainees to write one fact about each fitting.
3. Review the answers. If necessary, review any problem areas. Answer any questions trainees may have about fittings.

Section 10.5.0 *Documentation*

This exercise will familiarize trainees with various types of documentation and the information contained in each. If possible, obtain samples of each type of documentation listed in Section 10.5.0. For longer documentation types, use representative pages. Make one copy of each for every two trainees. Trainees will need their Trainee Guide, pencil, and paper to take notes for discussion. Allow between 30 and 45 minutes for this exercise.

1. Divide trainees into pairs. Distribute a copy of each type of documentation to each pair.
2. For each documentation sample, have trainees identify what type it is and take notes on the main information it requires.
3. Review trainees' findings. For each type of documentation, ask trainees why the documentation is important. What might happen if actions were not documented? Also ask why each piece of information on the forms is necessary or useful.
4. Answer any questions trainees may have about documentation.

MODULE 66101-02 — ANSWERS TO REVIEW QUESTIONS

Answer	Section
1. a	1.1.1
2. c	1.1.3
3. d	1.2.0
4. c	1.2.2
5. c	1.2.4
6. a	1.3.0
7. b	2.1.1
8. d	2.4.0
9. c	2.4.0
10. a	3.1.0
11. b	4.3.0
12. b	5.0.0
13. b	6.1.1
14. c	7.1.2
15. b	7.3.0
16. c	7.5.0
17. d	7.8.0
18. a	7.9.0
19. c	7.11.1
20. b	10.2.0

CONTREN™ LEARNING SERIES — USER UPDATES

The NCCER makes every effort to keep these textbooks up-to-date and free of technical errors. We appreciate your help in this process. If you have an idea for improving this textbook, or if you find an error, a typographical mistake, or an inaccuracy in NCCER's Contren™ textbooks, please write us, using this form or a photocopy. Be sure to include the exact module number, page number, a detailed description, and the correction, if applicable. Your input will be brought to the attention of the Technical Review Committee. Thank you for your assistance.

Instructors – If you found that additional materials were necessary in order to teach this module effectively, please let us know so that we may include them in the Equipment/Materials list in the Instructor's Guide.

Write: Curriculum Revision and Development Department
National Center for Construction Education and Research
P.O. Box 141104, Gainesville, FL 32614-1104

Fax: 352-334-0932

E-mail: curriculum@nccer.org

Craft ______________________ Module Name ______________________

Copyright Date ______________ Module Number ______________ Page Number(s) ______________

Description

____

____

____

(Optional) Correction

____

____

(Optional) Your Name and Address

____

____

Module 66102-02

Liquid Pipeline General Abnormal Operating Conditions

Liquid Pipeline General Abnormal Operating Conditions Annotated Instructor's Guide

Module 66102-02

MODULE OVERVIEW

This module reviews identification and response to abnormal operating conditions. Trainees will learn how to recognize and react to a variety of potentially dangerous conditions.

PREREQUISITES

Please refer to the Course Map in the Trainee Module. Prior to training with this module, it is recommended that the trainee shall have successfully completed the following:

Core Curriculum; Pipeline Core, Module 66101-02

OBJECTIVES

Upon completion of this module, the trainee will be able to:

1. Recognize and react to abnormal facility conditions.
2. Recognize and react to activation of a safety device.
3. Recognize and react to communications failures.
4. Recognize and react to power interruptions.
5. Respond appropriately to fire, explosions, and natural disasters.
6. Recognize and react to pipeline system damage.
7. Recognize and react to unexpected hazardous liquid or carbon dioxide (CO_2) leaks.
8. Recognize and react to unexplained pressure deviations.

PERFORMANCE TASKS

This is a knowledge-based module — there is no performance testing.

NCCER STANDARDIZED CRAFT TRAINING PROGRAM

The National Center for Construction Education and Research (NCCER) provides a standardized national program of accredited craft training. Key features of the program include instructor certification, competency-based training, and performance testing. The program provides trainees, instructors, and companies with a standard form of recognition through a National Craft Training Registry. The program is described in full in the *Guidelines for Accreditation*, published by the NCCER. For more information on standardized craft training, contact the NCCER by writing us at P.O. Box 141104, Gainesville, FL 32614-1104; calling 352-334-0911; or e-mailing info@nccer.org. More information may be found at our Web site, www.nccer.org.

HOW TO USE THIS ANNOTATED INSTRUCTOR'S GUIDE

Each page presents two sections of information. The larger section displays each page exactly as it appears in the Trainee Module. The narrow column ties suggested trainee and instructor actions to each page and provides icons to call your attention to material, safety, audiovisual, or testing requirements. The bottom of each page includes space for your notes.

Teaching Tip

If you see the Teaching Tip icon, that means there is a teaching tip associated with this section. Also refer to any suggested teaching tips at the end of the module.

SAFETY CONSIDERATIONS

Ensure that the trainees are equipped with appropriate personal protective equipment. Emphasize the importance of following all safety precautions and procedures when working with power tools.

PREPARATION

Before teaching this module, you should review the Module Outline, Objectives, and the Materials and Equipment List. Be sure to allow ample time to prepare your own training or lesson plan and gather all required equipment and materials.

MATERIALS AND EQUIPMENT LIST

- Transparencies
- Markers/chalk
- Blank acetate sheets
- Transparency pens
- Pencils and scratch paper
- Module Examinations*
- Overhead projector and screen
- Whiteboard/chalkboard
- Appropriate personal protective equipment
- Copies of your local code
- Photos of damaged pipeline supports or other pipeline components
- Copies of a sample Safety-Related Condition Report Form
- Copies of the *Code of Federal Regulations, Title 49, Part 195*

* Located in the Test Booklet packaged with this Annotated Instructor's Guide.

TEACHING TIME FOR THIS MODULE

An outline for use in developing your lesson plan is presented below. Note that each Roman numeral in the outline equates to one session of instruction. Each session has a suggested time period of 2½ hours. This includes 10 minutes at the beginning of each session for administrative tasks and one 10-minute break during the session. Approximately 5 hours are suggested to cover *Liquid Pipeline General Abnormal Operating Conditions.* You will need to adjust the time required for hands-on activity and testing based on your class size and resources.

Topic	Planned Time
Session I. Responding to Abnormal Operating Conditions, Part One	
A. Introduction	____________
B. Responding and Reacting to Abnormal Facility Conditions	____________
1. Laboratory	____________
C. Responding and Reacting to Activation of a Safety Device	____________
D. Recognizing and Reacting to Communications Failures	____________
E. Recognizing and Reacting to Power Interruptions	____________
1. Laboratory	____________
Session II. Responding to Abnormal Operating Conditions, Part Two	
A. Responding Appropriately to Fire/Explosion	____________
B. Recognizing and Reacting to Pipeline Damage	____________
1. Laboratory	____________
C. Recognizing and Reacting to Unexpected/Unintended Release or Encounter of Hazardous Liquid	____________
D. Recognizing and Reacting to Unexpected Carbon Dioxide Encounter	____________
E. Injuries As a Result of Unexpected/Unintended Release or Encounter	____________
1. Laboratory	____________
F. Recognizing and Reacting to Unexplained Pressure Deviations	____________
G. Summary	____________
1. Summarize module	____________
2. Answer questions	____________
H. Module Examination	____________
1. Trainees must score 70% or higher to receive recognition from the NCCER.	
2. Record the testing results on Craft Training Report Form 200 and submit the results to the Training Program Sponsor.	

Module 66102-02

Liquid Pipeline General Abnormal Operating Conditions

Instructor's Notes:

Instructor's Notes:

Introduction to the Pipeline Operator Qualification Rule, API, and NCCER

It could be said that petroleum products keep America running. Petroleum products like oil and natural gas are the principal sources of energy in the United States and are used in many manufacturing and chemical processes. Petroleum products are primarily transported by pipeline. There are more than a million miles of liquid and natural gas pipeline in this country, ranging in diameter from two inches to four feet. These pipelines bring crude oil and natural gas from production fields to processing plants and refineries—and ultimately to the consumer.

Pipeline Operator Qualification

Petroleum products are considered hazardous, so pipeline personnel must take extra precautions and follow specific procedures when working with them. To help ensure the safe handling and transportation of petroleum products, the United States Department of Transportation (DOT) mandates safe practices as federal regulations.

One such regulation, the Operator Qualification (OQ) rule, requires pipeline operators to develop a written qualification plan to evaluate their personnel and their contract personnel (see *49 CFR Part 192 Subpart N—Gas*, and *49 CFR Part 195 Subpart G—Liquid*). The evaluation hinges on the operator's ability to perform certain covered tasks and to be able to recognize and respond to abnormal operating conditions that may be encountered while performing these covered tasks.

Covered Tasks

A pipeline performance task is deemed *covered* if the activity passes the following four-part test:

1. Is it performed on a pipeline facility?
2. Is it an operations or maintenance task?
3. Is it performed as a requirement of *49 CFR Part 195 (Liquid)* or *49 CFR Part 192 (Gas)*?
4. Does it affect the operation or integrity of the pipeline facility?

If the answer to all four questions is *yes*, then the activity is considered *covered* and must figure into the evaluation process of pipeline personnel.

American Petroleum Institute

The OQ rule and its related covered tasks provided the catalyst to develop a unified pipeline training and assessment program—so that pipeline operators and owners have ready access to the assessment and training tools they need. This effort was the direct result of the vision and leadership provided by The American Petroleum Institute (API) and API members.

Efforts to unify the industry are nothing new to API. Indeed, the impetus for forming the Institute in 1919 was the need to standardize engineering specifications for drilling and production equipment. Today, the most pressing issues revolve around public perceptions and government policies toward the industry. Speaking with one voice on these issues has become as essential as having interchangeable parts in the field. API's members determine the petroleum industry's positions on public policy as well as the standards that govern its day-to-day operations worldwide.

Consortium on Operator Qualification

API assembled a focus group whose primary purpose was to identify covered tasks and provide compliance recommendations to pipeline operators.

The Consortium on Operator Qualification (COOQ) became the guiding force for the identification and development of covered tasks for the liquid pipeline industry. Their work product, API publication 1161, *Guidance Document for the Qualification of Liquid Pipeline Personnel*, formed the basis for the resulting Pipeliner Training and Assessment Program.

The covered task list created in API publication 1161 has now evolved to include gas covered tasks and is maintained by NCCER. The most current covered task list can be downloaded at www.nccer.org.

National Center for Construction Education and Research

With the guidance and support of API, its members, and the Consortium on Operator Qualification, the National Center for Construction Education and Research (NCCER) was able to design and implement the Pipeliner Training and Assessment Program.

NCCER is a not-for-profit 501(c)(3) education foundation established in 1995 by the world's largest and most progressive construction and maintenance contractors and national contractor associations. NCCER was founded to address the

severe workforce shortage facing the construction and maintenance industries and to develop a standardized training process and curricula. Today, NCCER is supported by hundreds of leading construction and maintenance companies, manufacturers, and national associations and developed more than 35 construction and maintenance titles.

NCCER also maintains a National Registry, which contains the records of trainees and instructors who have received NCCER training. All completion information for every trainee is sent to the NCCER and filed within the Registry. The Registry can then confirm training and skills for workers as they move from company to company, state to state, or even within their own company.

Standardized Curriculum (NCCER's Contren™ Learning Series)

NCCER curricula are developed using the expertise of industry subject matter experts (SMEs). The use of SMEs as an integral part of every new development or revision ensures that the curriculum is current, industry-driven, and indicative of industry practices. The development and/or revision process is facilitated from start to finish by NCCER project managers, who work closely with the SMEs and a select team of technical writer(s) on the development and presentation of the curriculum.

Pipeline: Five Training Paths

The pipeline curriculum focuses on essential skills and competencies within five skill areas:

- Corrosion Control
- Electrical and Instrumentation
- Maintenance
- Mechanical
- Operations (Field and Control Center)

SMEs from the pipeline industry have determined the skills and competencies needed based on API's covered tasks and the consensus of the group/industry. Specific company information is generally not incorporated into the curriculum unless the SMEs determine that the information is the industry standard.

Training Modules and Covered Tasks

The smallest unit of instruction in the curriculum is a *module*. Each curriculum is composed of modules (the equivalent of stand-alone chapters) that are logically sequenced into complete *levels* of instruction. Every module is task-oriented and requires the successful completion of a written (knowledge-based) examination and a performance (hands-on) test. Modules may be individually selected and/or combined for efficient task training. Modules in the pipeline curricula, to the extent possible, will align with API-developed covered tasks. While not all modules will relate to covered tasks, all covered tasks will relate to at least one module.

Benefits of Training to the Pipeline Marketplace

Applying NCCER's standardized training process allows pipeline owners and contractors in the United States to establish the benchmark for acceptable skill and competency levels within the pipeline industry. This process will assist training organizations in developing and/or improving standardized pipeline training programs that effectively match their resources and operating conditions while responding to regulatory demands and to the needs of the industry. The process will also recognize training organizations that meet the minimum standards of competence, warranting regulatory, public, and professional confidence. This in turn will foster national unity within the pipeline industry for the mutual benefit of individual technicians, owners, contractors, and ultimately pipeline-users and society.

Additional Resources

To review the Pipeliner Training and Assessment Program Implementation and Resource Guide and the latest version of pipeline covered tasks, visit NCCER's Web site at www.nccer.org, or contact:

National Center for Construction Education and Research
PO Box 141104
Gainesville, FL 32614-1104
(888) 622-3720
Web site: www.nccer.org

For more information on API, please contact:

American Petroleum Institute
1220 L Street, Northwest
Washington, D.C. 20005
(202) 682-8000
Web site: www.api.org

To order *API Pub 1161*, contact Global Engineering Documents by calling 1-800-854-7149 or visiting their Web site at www.global.his.com.

Additional sources of information on the pipeline operator qualification regulation include:

DOT – Office of Pipeline Safety
www.ops.dot.gov

Transportation Safety Institute
www.tsi.dot.gov

Instructor's Notes:

COURSE MAP

This course map shows each curriculum path in the Pipeliner Training program. The suggested training order begins on the left with the Pipeline Core modules and proceeds to the right. Which curriculum path you follow depends on your areas of specialty and interest. On a given curriculum path, skills levels increase as you advance to the right on the course map, although some are single-level curricula. The local Training Program Sponsor may adjust the training order.

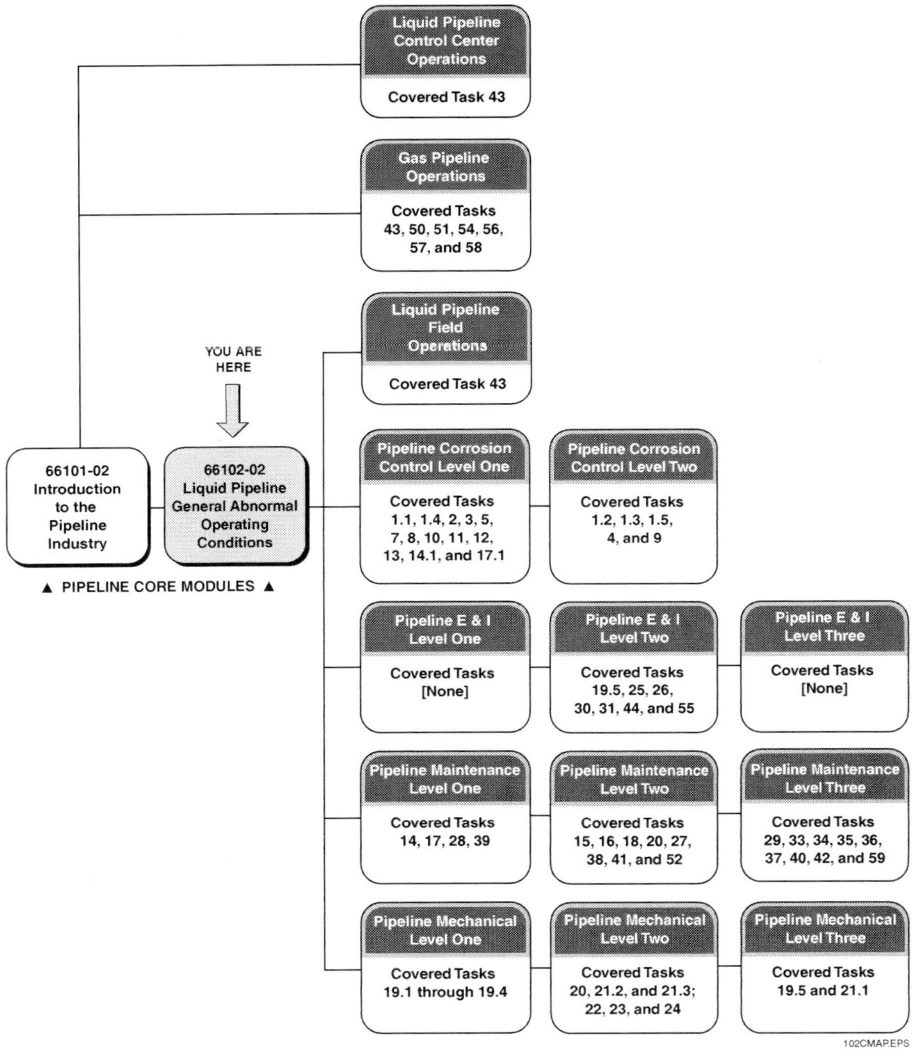

Homework

Assign reading of Module 66102-02.

Instructor's Notes:

MODULE 66102-02 CONTENTS

Figures

Table

Instructor's Notes:

MODULE 66102-02

Liquid Pipeline General Abnormal Operating Conditions

Objectives

When you have completed this module, you will be able to do the following:

1. Recognize and react to abnormal facility conditions.
2. Recognize and react to activation of a safety device.
3. Recognize and react to communications failures.
4. Recognize and react to power interruptions.
5. Respond appropriately to fire, explosions, and natural disasters.
6. Recognize and react to pipeline system damage.
7. Recognize and react to unexpected hazardous liquid or carbon dioxide (CO_2) leaks.
8. Recognize and react to unexplained pressure deviations.

Prerequisites

Before you begin this module, it is recommended that you successfully complete the following: Core Curriculum; Pipeline Core, Module 66101-02.

Required Trainee Materials

1. Pencil and paper
2. *Code of Federal Regulations, Title 49, Part 195*

1.0.0 ◆ INTRODUCTION

Abnormal operating conditions (AOCs) do not happen frequently on the pipeline. Companies have programs, procedures, training, safety equipment, warning devices, and fail-safe devices on practically every facet of operations. Computers and electronic controls are used to operate much of the pipeline and eliminate many problems and errors before they occur. However, the operator has the responsibility for making certain that when equipment does fail, the pipeline is kept in a safe condition. The operator must make use of training, skills, and good judgment until solutions are found to maintain safe operation.

Time is of the essence in such conditions. Personnel are expected to take decisive action to overcome the abnormal condition, maintaining the safety of pipeline operations. Personnel must follow company procedures. As regulation *49 CFR 195.402* states, each company must ensure its employees maintain a thorough knowledge of its operations, maintenance, and emergency manuals.

If it is necessary to deviate from company policy for the safety and protection of personnel, the public, the environment, or equipment, then the operator is expected to handle the situation. The operator goes through several logical procedures in making decisions and performing actions. Although not every procedure is included in every action to secure the safety and operation of the pipeline, the list of actions might be divided as follows:

- Recognize the AOC.
- Investigate the cause of the alert.
- Perform the actions planned.
- Make the appropriate notifications.

1.1.0 Definition

An AOC is defined by federal pipeline safety law *49 U.S.C. 60102* and pipeline safety regulations *49 CFR 192.605* and *49 CFR 195.402*. AOCs are also specified by federal law in *49 CFR 195.503*. This module will concentrate on AOCs related to *49 CFR 195 – Transportation of Hazardous Liquids by Pipeline.*

Materials

Ensure that you have everything required to teach the course. Check the Materials and Equipment List at the front of this module.

Classroom

Explain that terms shown in bold are defined in the Glossary at the back of this module.

Audiovisual

Show Transparency 1, Course Objectives.

Teaching Tip

See the general Teaching Tip at the end of this module.

Classroom

Explain the importance of proper reaction to abnormal operating conditions.

Define abnormal operating conditions. Refer to the appropriate sections of the CFR.

Classroom

Refer trainees to Table 1. Discuss the examples of and reactions to various abnormal operating conditions.

An AOC is any condition identified by the individual that may indicate failure of a component or deviation from normal operations. It may also be a condition exceeding design limits or resulting in hazards to persons, property, or the environment.

The Operator Qualification rule requires that an individual be able to recognize and respond to any AOC that could happen while performing a covered task. *Table 1* shows a layout of abnormal operating conditions.

Table 1 Layout of Abnormal Operating Conditions (1 of 2)

Abnormal operating condition (Recognize and react to...)	Example	Reaction
Abnormal facility condition	During the course of performing a covered task, the individual could encounter one of the following: • Exposed pipe • Missing or damaged pipeline supports • Incorrect ANSI pressure rating of a component or material type • Disbonded coatings • Unauthorized equipment on the right-of-way • Sunken floating tank roof	Determine the extent of damage, perform repairs if properly trained to do so, make the appropriate notifications as required by company policy, and complete any required documentation.
Activation of a safety device	During the course of a maintenance or operating activity, the following safety devices could activate: • Pressure relief valve relieves to tank/sump/atmosphere • Emergency or abnormal shutdown • High-pressure shutdown • Hazardous atmosphere alarm indicating the presence of an LEL(20%) or HEL(40%) explosive atmosphere • High sump level or shutdown alarm • High tank shutdown alarm These devices are typically designed to operate and reduce/eliminate a hazardous situation.	Contact the operations or control center and make the appropriate company notifications, minimize the situation if possible, and complete any required documentation.
Communications failure	During the course of operations, the operator could have a loss of the following communication devices: • Radio communications • Telephone • SCADA communication • Satellite	Immediately contact the control center, make the appropriate notifications as required by company policy, and complete any required documentation.
Power interruption	During the course of operating a pipeline component, the individual could experience a loss of commercial or local AC power due to the following: • Utility power failure due to equipment problems or storms • Station or terminal power problems (for example, motor overload, breaker trips)	Inform the power company of the outage if applicable, make the appropriate notifications as required by company policy, and complete any required documentation.

102T01A.EPS

Instructor's Notes:

Table 1 Layout of Abnormal Operating Conditions (2 of 2)

Abnormal operating condition (Recognize and react to...)	Example	Reaction
Fire, explosion, or natural disaster	During the course of a maintenance or operations activity, the individual could discover a fire or explosion. This fire or explosion could be caused by the individual during the performance of a task due to things such as: • Ignition of a hazardous vapor • Unintentional detonation Fire and explosion may also result from natural disasters, such as: • Lightning strike • Flooding • Landslide • Tornado	Notify local 911 emergency personnel; limit access to area as much as possible; make the appropriate notifications as required by company policy, and complete any required documentation.
Pipeline system damage	During the course of a maintenance or operations activity, the individual could encounter or cause damage to the pipeline such as: • Damaged or inoperable pipeline components (pipe, valves, pumps, tank gauges) from external force • Gouges/dents/pitting • Damaged pipe support • Wrinkles/buckle/cracks • Tank shell damage • Sunken roof, leaking pontoon • Cracked or significantly corroded flange or bolts • Damage due to flooding or land movement	Anticipate the proper response for unexpected hazardous liquid or carbon dioxide encounter and for fire/explosion if appropriate; determine the extent of the damage; make the appropriate notifications as required by company policy, and complete any required documentation.
Unexpected hazardous liquid or carbon dioxide encountered or unintended release	During the course of maintaining or operating the pipeline facility, the individual could encounter an unexpected presence of hazardous liquid or carbon dioxide, or could cause an unintended release during the performance of a task. This does not include the controlled transfer, sampling, venting, draining, or handling of hazardous liquid or carbon dioxide. The individual could be expected to respond to: • Hydrocarbon release • Vapors	Notify local 911 emergency personnel and/or appropriate persons/authorities according to company procedure, and complete any required documentation.
Unexplained pressure deviations	During the course of maintaining or operating the pipeline facility, the individual could observe the following pressure deviations: • Increased pressure • Decreased pressure • High-pressure limit • Low-pressure limit • Loss of line pressure	Follow troubleshooting procedures, perform shutdown of the facility if necessary, make the appropriate notifications as required by company policy, and complete any required documentation.

102T01B.EPS

Classroom

Ask a trainee to give several examples of abnormal facility conditions.

Audiovisual

Show Transparency 2 (Figure 1). Explain how RTDs work. Ask trainees to explain how the two other types of alarms work.

Classroom

Ask a trainee to explain the proper response to damaged pipeline supports.

Demonstration

Show photos of damaged pipeline supports or other pipeline components. Discuss the possible consequences of this damage.

2.0.0 ◆ RESPOND AND REACT TO ABNORMAL FACILITY CONDITIONS

Abnormal facility conditions relate to problems involving the physical structure of the pipeline. Examples of abnormal facility conditions include exposed pipe, missing or damaged pipeline supports, evidence of corrosion, missing signs/markers, equipment stalled over the line, hazardous atmosphere alarms, or resistance temperature device (RTD) alarms. Incorrect data caused by equipment malfunction, old or incorrect maps or survey records, as well as the incorrect interpretation of the data also create AOCs.

In almost all incidences of noted abnormal facility conditions, personnel should determine the extent of damage if applicable, perform repairs within a reasonable time if properly trained to do so, make the appropriate notifications as required by company policy, and complete the required documentation.

2.1.0 Respond to High Pump Case Temperature/Pressure Alarm

Main-line pumps are fitted with RTDs that give temperature readings for the pump case and product temperature. The devices are preset to signal an alarm in the control center when the temperature reaches a critical level and will shut the pump down automatically if some action cannot be taken to lower the temperature. When a pump case temperature reaches such a critical level and automatically shuts down, it is considered an AOC. *Figure 1* shows pump case temperature RTDs.

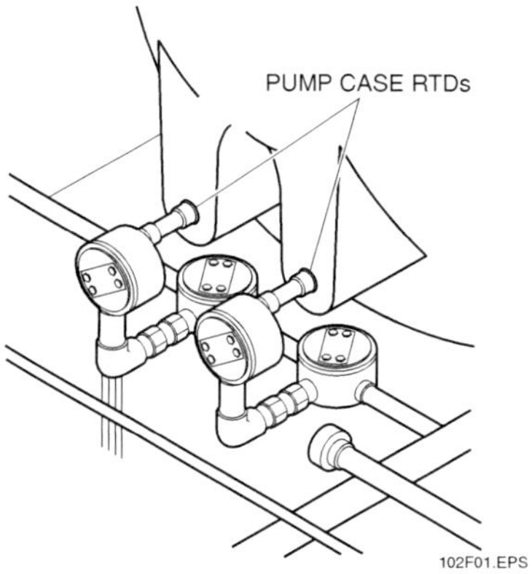

Figure 1 ◆ Pump case temperature RTDs.

Most main-line pumps are set to shut down on low pressure. Companies fit pumps with differential pressure switches and temperature switches to protect the equipment from damage caused by low pressure, no pressure, or no flow and to avoid exceeding **maximum operating pressure (MOP)**.

Pumps also have high discharge pressure alarms that will shut down the pump when the discharge pressure exceeds a preset limit. Many times, this indicates a downstream line obstruction that is not allowing the product to flow properly from the pump.

Another type of alarm that can be received from a pumping unit is a seal failure alarm. If responding to a seal failure alarm, operations and/or maintenance personnel should investigate the validity of the alarm, check the station sump level, and notify the appropriate personnel. This may not necessarily result in an AOC, but it would require control center personnel to document the alarm and notify the appropriate personnel to check and possibly replace the seal. The control center operator should never attempt to restart the pump before investigating the cause of the alarm and responding to the alarm per company policy.

2.2.0 Respond to Damaged Pipeline Supports

Damaged pipeline supports (see *Figure 2*) can be serious problems on the pipeline but may not require immediately shutting down the system. A report detailing the damage is normally submitted by maintenance or inspection personnel who are traversing the pipeline. Action is determined by the extent of the damage.

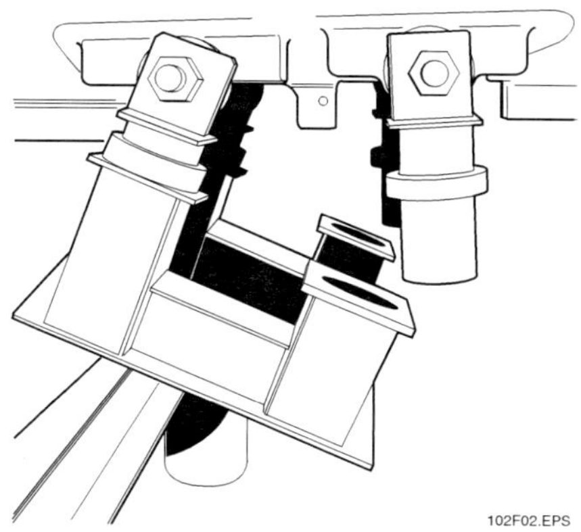

Figure 2 ◆ Damaged pipeline support.

Instructor's Notes:

If the support is damaged, the pipeline may have to be shut down. If the damage is not critical, it may be possible to continue operation under heightened awareness. Such a decision is a judgement call on the part of the operator in collaboration with qualified maintenance or inspection personnel. The operator must submit a report to appropriate personnel so that repairs can be scheduled.

2.2.1 Reaction

If the damage to the support is critical or indicative of a possible release, the individual making the discovery must contact the control center or the appropriate company personnel immediately if the pipe is in danger of excess movement. The operator and/or call taker should then follow company policy precisely to investigate the condition and dispatch qualified personnel to the scene. Until the arrival of qualified personnel, the individual at the scene should stay a safe distance away from the site, evacuate the area if necessary, and make the appropriate notifications as required by company policy.

2.3.0 Documentation

The documentation and records required following an AOC depend upon the size of the incident and whether a release occurred. Any material defect or physical damage that impairs the serviceability of a pipeline requires the operator to complete a safety-related condition report and the designated company official to submit the report to government agencies (*CFR 195.55*). *Figure 3* shows a typical form for a safety-related condition report.

> **NOTE**
> A safety-related condition report is only required if the condition is not corrected within a given amount of time, or occurs within a certain distance of residences, streams, or other public areas, or if the condition is indicative of general corrosion.

3.0.0 ◆ RESPOND AND REACT TO ACTIVATION OF A SAFETY DEVICE

Safety devices are installed to protect the safety of pipeline operations and equipment. Any time a safety device activates, an alarm is sent to the control center. Safety devices include pressure relief systems, pressure limiting devices, high pump case pressure alarms, overfill protection devices, and main-line pressure shut-down alarms.

Pumps also have high discharge pressure alarms that will shut down the pump when the discharge pressure exceeds a preset limit for the pipeline.

> **CAUTION**
> An individual's actions during the performance of a covered task could cause a safety device to activate. The activation represents a deviation from normal operations, resulting in an AOC.

3.1.0 Respond to the Activation of a Pressure Relief System

Activation of **relief devices** along the pipeline system is an abnormal occurrence. Relief valves can discharge into tanks or vessels that are designed and constructed in compliance with *API Standard 650* or *API 2510* for pressure vessels. Relief valves (see *Figure 4*) can be installed on pipeline or pressure **breakout tanks**.The purpose of a relief valve is to protect the pipeline or pressure tank from overpressure. These tanks can also be used to temporarily absorb **surge pressure** from the main-line and manifold pressure relief valves to lower pressure on the line. An abnormal operating condition exists when a pressure relief occurs.

3.1.1 Reaction

Each relief event should be recognized and documented by operations personnel. If relief becomes excessive, operations personnel must immediately begin to investigate and make adjustments to ensure that the pressure of the breakout tank or pipeline does not exceed 110% of the MOP.

If the alarm is signaled and the pressure remains high, control center operations personnel must take immediate action to reduce pressure. A pressure reduction is usually achieved by shutting down the upstream pumping station, closing a block valve, or shutting down the pipeline system.

3.1.2 Documentation

Any malfunction or operating error that causes the pressure of a pipeline to rise above 110% of its maximum operating pressure may require the operator to complete a safety-related condition report and the designated company official to submit the report to government agencies (*CFR 195.55*).

Classroom

Ask a trainee to explain when a safety-related condition report form is needed.

Have trainees name various types of safety devices.

Safety

Explain some of the actions that could cause a safety device to activate.

Classroom

Ask a trainee to explain what a relief valve is and what it is used for.

Discuss the appropriate reaction to overpressure. Ask trainees if they have any questions.

Refer trainees to 49 CFR 195. Discuss when reports of overpressure are required.

Safety-Related Condition Report

Date: ______________ Government Letter #: ________

APSC FILE #: ____________

Information Resources Manager
Office of Pipeline Safety
Research and Special Programs Administration
U.S. Department of Transportation
Room 2335
400 Seventh Street SW
Washington, D.C. 20590

Safety-Related Condition Report

49 CFR 195.55
Subject:

Operator:

Date of Report:

Person Submitting Report: Senior Vice President
Business Unit

Person Determining Condition:

Name:

Title:

Telephone:

Date Condition Discovered:

Date Condition Determined:

Location of Condition:

Description of Condition:

Corrective Action:

Sincerely,

Sr. Vice President, Business Unit

SAMPLE

102F03.EPS

Figure 3 ◈ Safety-related condition report form.

Instructor's Notes:

102F04.EPS

Figure 4 ◆ Relief valve.

When filing a safety-related condition report, management must include a description of the condition, including the circumstances leading to its discovery, any significant effects of the condition on safety, and the name of the commodity being transported or stored. Management must also describe the corrective action taken (including reduction of pressure or shutdown) and the planned follow-up or future corrective action, including the anticipated schedule for starting and concluding such action.

3.2.0 Respond to Unintended Valve Closure or Shutdown

The unintended closure of a main-line block valve, remote block valve, manual block valve, pump station suction or discharge valve, and certain terminal valves (manifold/tank) will result in an automatic shutdown of the pipeline. In these cases, the operator should inspect the equipment to determine why closure occurred and reset/open/close the valve when it is safe to do so. Such occurrences may not cause an AOC, but if the event causes the pipeline to exceed design limits in any way, its classification changes to abnormal.

3.3.0 Testing Safety Devices

Safety devices provided at each pump station must be tested under conditions approximating actual operations (*CFR 195.262*). If, for example, the control room calls to check the **lower explosive limit (LEL)** alarm at a pumping station, it must first shut down the station and then check the LEL alarm at the station.

If, while testing a safety device, an alarm indicates a malfunction not related to the device being tested, operations and/or maintenance personnel should contact control center operations, make appropriate company notifications, and complete the required documentation.

> **WARNING!**
> If the hazardous atmosphere alarm indicates the presence of a LEL of 20%, evacuate the area.

4.0.0 ◆ RECOGNIZE AND REACT TO COMMUNICATIONS FAILURES

Failure of communications usually refers to a loss of contact between the control center **supervisory control and data acquisition (SCADA)** system and remotely controlled operating components on the pipeline. Essentially, operations personnel lose the ability to remotely monitor and control the pipeline facility. Communication failures may include:

- Intermittent facility outages
- Loss of the SCADA system

Sometimes the communications with various components can be verified, but the affected control system will not work. In these cases, the problem involves other components of the control system that may be mechanical, electrical, or hydraulic.

The failure to communicate with a remote control valve operator is considered a communications AOC because the valve cannot be opened or closed by the operator. The valve could also have a mechanical failure that prevents it from being operated. Sometimes electrical components or connections burn out, and if this happens on some pump drivers, they would not be able to respond in order to start up or shut down. Control of equipment is lost in this situation and could happen in many others.

4.1.0 Reaction

If it is discovered that there is a loss of a communications device, personnel should immediately contact the control center to verify the condition. Operations and/or other appropriate personnel should then determine the exact cause and take decisive action to control the condition, while always following company policies and procedures. In some instances, operations personnel might be able to bypass the offending component until appropriate maintenance can be performed. In other cases, the situation may warrant a shutdown.

Laboratory

Refer to Figure 3. Distribute copies of a sample Safety-Related Condition Report form to pairs of trainees. Create a scenario and have them fill out the form. Discuss the results. See the teaching tip at the end of this module.

Classroom

Ask a trainee to explain the circumstances in which an unintended valve closure is considered to be an abnormal event.

Safety

Emphasize the importance of evacuation when the LEL reaches 20 percent.

Classroom

Have a trainee explain what a communications failure is. Ask for examples.

Discuss the appropriate reaction to the loss of a communications device.

Classroom

Ask a trainee when a pipeline must be shut down as a result of a communications device failure.

Discuss the effects of a power interruption on a pipeline.

Laboratory

Divide trainees into small groups. Present several scenarios of power interruptions. Have the groups determine the appropriate responses to each.

Homework

Assign the reading of Sections 6.0.0–11.0.0. If desired, have trainees complete the Review Questions for discussion during the next session.

Materials

Ensure that you have everything required for demonstration and testing during this session.

Classroom

Ask a trainee to explain how flame detectors work. Have others explain how fire alarms and fire detectors work.

4.2.0 Notifications

If the component can be operated manually, the problem can be alleviated until repairs can be made. Operations must monitor pressure, however, and if 110% of MOP is exceeded, the pipeline must be shut down without discussion. If the pressure remains within limits or the component is not one that will affect the operation directly, the condition must still be monitored closely if the pipeline is to stay in operation. Even after normal conditions have been restored, operations personnel must continue to monitor the condition to make certain that repairs have been successful.

5.0.0 ◆ RECOGNIZE AND REACT TO POWER INTERRUPTIONS

A loss of power impacts the abilities of the control center or facility operations to safely control and monitor the pipeline facility. Virtually every component of the pipeline is in some way dependent upon electricity. If power goes off, electronic indicators will be lost. The SCADA system may be off. Pumps will shut down. Remote control valves will not operate, making control of the system difficult.

The pipeline may have a problem if the power failure is of a local nature because only a part of the system will shut down without electricity. Company policy and procedures must be followed. Area-wide power failure may also cut telephone communications and isolate the affected station. At the receiving terminal, a shipment that is in process may not have storage and cannot be shipped further without pumps. A tank being filled that cannot be switched may reach a high level or possibly overflow.

When power failure occurs, some systems can be protected by direct current (DC) backup power supplies, but most stations, terminals, and equipment are operated on alternating current (AC) or are equipped with uninterruptible power systems (UPS), a battery backup that converts DC to AC. The UPS or back-up generators supply emergency AC power to allow minimum operating capabilities or at least the ability to safely shut down the facility until the power is restored.

5.1.0 Reaction

Reaction to power failure is determined by what part and how much of the pipeline is affected. If only a single component or station is involved, it may be possible to adjust line pressure and flow to keep the pipeline in operation.

Suppose power interruption occurs while operating a valve. In this scenario, the crew/individual should be prepared to operate the valve manually, either by hand or by another type of operator.

Suppose a pipeline system with multiple stations is flowing with all units running. Power is lost at a station that is needed to keep pressure below MOP. In this situation, the control center operator would have to shut down the pipeline system to avoid overpressurizing the system. The terminal operator must be sure to switch off any pumps that were running so that when power returns, they will not start up. Yard personnel must be stationed outside so valves can be operated manually. If a shutdown is not required, the shipment will continue to arrive, and the schedule must be followed and the tasks completed manually.

6.0.0 ◆ RESPOND APPROPRIATELY TO FIRE/EXPLOSION

Some pipelines and facilities are fitted with detection devices that send an alarm when certain conditions occur. Such equipment includes flame detectors, fire detectors, and seismic detectors. Reaction to these types of alarms may require evacuation of facilities.

Flame detectors (see *Figure 5*) are thermal detectors that have normally open contacts. They work by closing their contacts when exposed to rapidly rising temperature, as would be experienced in the presence of open flames. When the contacts close, an alarm is signaled on the console in the control center and on the fire control panel in the station.

Fire alarms are activated by ultraviolet (UV) detectors or by manual pull-switch. Fire detectors are UV detectors that are set off by the light given off by open flame. Their response is almost instantaneous, and they signal fires in the very beginning stages. UV detectors are connected to a firefighting system and immediately trigger a dump of a fire suppression material into the area.

> **NOTE**
> Verify that a hazardous alarm test has been performed before performing any work, such as welding or cutting, opening electrical boxes, or operating non-explosion-proof tools in hazardous classified areas, that may produce an ignition source.

Instructor's Notes:

102F05.EPS

Figure 5 ◆ Flame detector.

6.1.0 Reaction

Reaction to a flame detector or fire detector alarm must be immediate and decisive. Personnel must first verify that it is not a false alarm. A false alarm or an expected alarm due to maintenance activities requires no further action unless, in the judgment of operations, the system needs to be investigated. Alarms often appear on the control center control panel or SCADA screen.

Confirmation of the alarm gives the exact location of an active fire. Company policy dictates procedures for such emergency action and it must be followed. Procedures are site-specific and may allow for some flexibility in responding. However, maintenance and operations personnel should not need to consult a manual in case of an alarm. All personnel must be adequately trained and familiar with emergency procedures in advance. It is important to take necessary action, including a shutdown of the pipeline if required to control the emergency.

> **NOTE**
> According to *CFR 195.403*, operators shall establish and conduct a continuing training program to instruct emergency personnel in the proper use of firefighting procedures and equipment, in the identification of hazardous conditions likely to cause emergencies, in the steps necessary to control accidental releases, and in the characteristics of hazardous liquids or carbon dioxide transported.

The safety of personnel is always top priority. Protection of property is always second. Attention must be totally focused on the emergency. Response must be implemented relative to the type of fire in progress.

If the fire is a small or incipient (beginning) fire, personnel should use the available fire extinguishing equipment to control the situation. Personnel should respond in the way they have been trained. Such fires are usually the result of faulty electrical equipment or a smaller item such as a trash can fire. Responders should make certain to use the proper type of fire extinguisher for the type of fire.

If the fire is larger than the immediate workforce can contain, operations personnel should minimize the situation, notify appropriate personnel and agencies, and complete the required documentation. For example, during an electrical storm, if lightning strikes a full tank at the facility and the roof seal catches on fire, operations should follow company policy and contact the fire department immediately to contain the fire.

6.2.0 Respond Appropriately to Natural Disasters

Natural disasters include landslides, earthquakes, flooding, or other dangerous weather patterns that could be precursors to soil and pipeline movement. The impact of moving soil may result in forces sufficient enough to cause pipeline failure or product release.

Additional stresses are imposed on piping by soil movement or subsidence. Rains resulting in saturated soil conditions may also result in soil movement in some areas. Pipeline personnel should be alert to weather patterns and conditions that could cause soil movement and have personnel available for patrolling, preventive maintenance activities, and emergency response actions such as shutdown, isolation, and containment in case of a pipeline failure.

Patrols should be designed to effectively evaluate right-of-way (ROW) conditions in areas susceptible to soil movement that might jeopardize the operation of the pipeline. Personnel should evaluate areas of potential soil movement and, as appropriate, monitor soil subsidence and take preventive actions to ensure the safe operation of the pipeline.

Finally, pipeline personnel should evaluate the accessibility of pipeline facilities, such as valves, needed to isolate pipeline segments or sections of pipelines that might be jeopardized by soil subsidence.

Classroom

Discuss the appropriate reaction to a fire alarm. Ask a trainee to identify the top priority when responding to an alarm.

Ask trainees to identify some of the natural disasters that can affect pipelines. Discuss the preventive action to be taken for each.

Classroom

Discuss the various conditions that might require an emergency shutdown.

Have a trainee describe situations that require evacuation. Ask another trainee to describe the required steps for every company evacuation procedure.

Laboratory

Identify and discuss procedures for public evacuation and notification in the event of a major release.

Classroom

Ask trainees how to identify pipeline damage.

6.3.0 Perform Emergency Shutdown

Events occurring on the pipeline that present immediate hazard to persons or property require an emergency shutdown. An event might be caused by human error, mechanical failure, electrical failure, or an accident on the pipeline. It might also be brought about by threats from weather or civil disobedience. Regardless of the reason, the line must be shut down quickly and safely. Each company has a standard operating procedure for such an action. (In case of a pump shutdown from low pressure, there may be some time to discuss the move with appropriate management if the line has enough flexibility to withstand the stress.)

However, once the need is confirmed, decisive action must be taken immediately.

> **NOTE**
> In emergency conditions such as a possible release, the pipeline operator has full responsibility and authority to make shutdown decisions and take appropriate action. These actions are governed by the company release policy above all other considerations.

In the case of a fire, explosion, or release, discussion time is obviously not an option. The pipeline must be isolated by operations personnel by safely shutting down the originating station and continuing downstream in a safe manner until all pumps are offline. Operations personnel will then close all block valves to complete the shutdown. The responsible operations and maintenance personnel should be notified.

Threats from natural disaster or civil disobedience may come without warning. If an individual witnesses these kinds of threats to the pipeline, the supervisor or control center should be notified immediately. Such affected facilities should be isolated. Once facilities are isolated, pipeline shutdown may follow.

6.4.0 Evacuate the Facility

Safety of personnel and the public is the first priority in any event. Regardless of the emergency, if there is a threat of injury or death, all affected persons must be evacuated. The operator must evaluate every emergency to determine possible danger to the people involved. Events that could result in the need for an evacuation are fire, explosion, release of hydrocarbon or other hazardous vapors, seismic activity, and other natural phenomena.

The responsibility to maintain the safety of personnel and the public cannot be taken lightly. If the determination is made that a hazardous atmosphere or condition exists, the command to evacuate must be given. In the case of a release of product, the concentration of vapors downwind should be tested repeatedly to determine if there is a threat of explosion. If a hazardous atmosphere is traveling toward a populated area, the area must be warned and possibly evacuated.

6.4.1 Company Procedures

Companies have standard procedures in place to guide evacuations. By law, Emergency Response Plans must be in place and evaluated annually. Facility layouts with evacuation maps are required to be posted in every area or department. Safe areas are designated for all personnel. Once the command to evacuate is given, designated persons are required to call a roll of each employee name and check off who is present in the safe area. Every person must be accounted for.

6.4.2 General Public

In case of a release of product, the atmosphere downwind must be tested repeatedly. A vapor cloud traveling toward a populated area requires that the people be warned and possibly evacuated to a safe area. That area must be determined by the direction in which the hazardous vapors are traveling. Law enforcement and fire and rescue personnel must be solicited to aid in directing people away in a route that will give them the most protection.

In a major release, company records must be consulted so that potentially affected isolated people can be contacted in person. The path of the vapors or cloud should be plotted continuously so that if there are people, animals, or roads that the vapors will cross, the evacuation can be expanded. All roads leading to the area must be blocked until the danger has passed. All means should be employed to keep the product from reaching sewers, underground caverns, or waterways.

7.0.0 ◆ RECOGNIZE AND REACT TO PIPELINE DAMAGE

Pipeline damage occasionally occurs. Damaged and inoperable components, such as valves, pipe, or pumps, should be reported. One case may be a pipeline being hit or struck by excavation equipment. In this case, personnel may see gouges, scrapes, dents, or even buckling, but some signs are less obvious. A sheen on the water that covers submerged pipelines could be an indication of

Instructor's Notes:

valve damage. A frost ball or vapor cloud that is discovered along the right-of-way may indicate a pipeline release of **highly volatile liquid (HVL)**. Pipelines in some areas have been shot by hunters accidentally or by other people intentionally. Vandalism is also a potential cause of damage.

7.1.0 Reaction

The first reaction should be to determine the extent of the damage. Some of these incidents are serious enough to require an emergency shutdown. A shutdown would more likely be involved if the pipeline were struck or damaged in such a way as to involve a major release and/or explosion. If the pipeline were struck and there were no release, then the abnormal operating condition would indicate a pipeline anomaly.

The least desirable way to detect a release is by receiving notification from a third party, yet this is the most important call an operator can receive. Conditions of damaged pipe are frequently brought to the company's attention through a telephone contact from outside the company. Frequently, the call comes from an aerial patrol of the pipeline ROW or an individual landowner who contacts the One-Call notification system to report a release, damage, or other such problem.

Personnel should always anticipate the proper response to potential emergency situations, make the appropriate notifications per company policy, and complete the required documentation.

Examples of times that a pipeline would have to be shut down due to damage are if an excavator operating a bulldozer gouges the pipeline causing a release, or if a field operator loses control of a vehicle and dents a large section of the pipeline or runs into a valve. In these situations, control center operations should be notified immediately. In some cases when a known release is involved, the control center operator may be able to drain a portion of the pipeline into a delivery system to lessen the amount of product that could be released from the pipeline.

If qualified and trained to do so, personnel should perform an initial site assessment, determine the location of the release, isolate it, and contact 911 emergency personnel if necessary. Regardless, control center operations should dispatch qualified personnel to assess the conditions. If the investigation reveals an immediate hazard, the line must be shut down in a safe manner as rapidly as possible. On multiple pump station systems, the pump units must be shut down in the order of upstream units first and then downstream units.

7.2.0 Abnormal Movement or Loading of an In-Service Pipeline

Great care must be taken during the process of moving in-service pipelines. Additional stress could be placed on the pipeline and could lead to pipe failure. When in-service pipe is lowered, the ambient temperature is also a critical factor. As the weather changes, the pipe may expand or contract and expand.

8.0.0 ◆ RECOGNIZE AND REACT TO UNEXPECTED/UNINTENDED RELEASE OR ENCOUNTER OF HAZARDOUS LIQUID

An operator may suddenly observe product being released or sprayed from a manifold, valve, or ruptured piece of equipment. A pump could start accidentally during a repair process, resulting in an increase in pressure in the pipeline. An isolation device might fail. The pipe may be of insufficient wall thickness (corrosion). There could also be an occasion when product vapor is discovered to be exceedingly strong in an unexpected area or a pool of product is observed during ROW inspections.

Such discoveries may signal the presence of a release. A major release in the field, terminals, or tank farms demonstrates many of the same characteristics. Decisive action must be taken to control the event as quickly as possible. If it is not controlled quickly, explosion, fire, and injury is imminent, as well as pollution of the environment and loss of product.

8.1.0 Reaction

Reactions to product release are determined by the type of facility or location involved. If the hazard is discovered in a building or enclosed space, it must be handled somewhat differently than if it is discovered outside.

8.1.1 Enclosed Areas

If the hazard is in a building or enclosure, personnel must get to a safe location immediately. Manually operated doors and windows should be opened to allow vapors to dissipate. The source of the product release must be determined quickly.

The nearest cutoff valve should be closed. If a major line or equipment is involved and can be operated from the control center, it is better for personnel to move to a safe location and call the control center operator to have it closed remotely.

Classroom

Discuss proper reactions to specific types of pipeline damage. Ask trainees if they have any questions.

Ask trainees to name possible causes of unintended/unexpected release.

Teaching Tip

See the Teaching Tip at the end of this module.

Classroom

Refer trainees to *49 CFR 195.50*. Discuss circumstances in which accident reports are required.

No electrical equipment should be operated in the area. If power cannot safely be cut off at the switchgear, operations personnel must call the electrical supplier and have power cut off from outside the facility. Personnel should only be allowed to enter the area using appropriate personal protective equipment (PPE).

> **NOTE**
> To identify the hazards associated with the chemical being released, consult the material safety data sheets (MSDS).

8.1.2 Outdoor Areas

After first assuring the protection of the public, report the release to control center operations and/or appropriate personnel. The source of product being released must be discovered and cut off. Sources of ignition must be avoided and the power cut off. Personnel must remove themselves to a safe location. Vapor associated with the release must be allowed to dissipate before personnel attempt to enter the area. If an odor of gasoline has been detected even a few hundred feet from the source, control center operations must be contacted to shut down the line. If the leak is from equipment at a remote station, drivers should approach the equipment from the upwind side of the station.

> **WARNING!**
> If you are witnessing a release and there is a downed person in the area of the release, notify the dispatch control center to call 911 and clear the area.

> **NOTE**
> Process Safety Management (PSM) stipulates that employers must ensure that contract employees are trained in the work practices necessary to perform their job safely.

Personnel should only be allowed to enter the area using appropriate PPE. Repairs must be carried out with the permits and documentation as required. Product on the ground will require vacuum equipment and/or other means of cleanup.

8.1.3 Response

Responses to both enclosed and outdoor releases would include some or all of the following:

- Open all windows and doors.
- Find the source of the release.
- Use appropriate PPE.
- Avoid opening switchgear doors.
- Operate no electrical switches or equipment including lights.
- Exit any area where meters or smells indicate vapors are increasing.
- Have the power company cut power to shut down any pumps running if required before the problem is solved. (Go to an outside telephone to call the power company.)

8.2.0 Notifications

Appropriate personnel must be notified if conditions advance to dangerous levels. In case of a pipeline shutdown or other major action, appropriate personnel must be notified as soon as possible. If it is determined that product has escaped or if there has been an injury or major property damage, government agencies must receive appropriate reports.

An accident report is required for each failure in a pipeline system in which there is a release of hazardous liquid or carbon dioxide resulting in any of the following (see *CFR 195.50*):

- Any emergency that includes fire, explosion, or injury
- Releases of five gallons or more of HVLs
- All releases on water
- Releases that exceed five or more barrels of product
- Death of any person
- Bodily harm
- Estimated property damage exceeding $50,000

The accident report should be filed with the National Response Center (NRC) no later than 30 days after the discovery of the incident.

Instructor's Notes:

9.0.0 ◆ RECOGNIZE AND REACT TO UNEXPECTED CARBON DIOXIDE ENCOUNTER

Carbon dioxide is a colorless, odorless, gaseous compound used as a propellant in many domestic applications, such as fire extinguishing equipment, and is a by-product of combustion. Being heavier than air, it can collect in low-lying areas where it may displace oxygen; exposure can result in suffocation and death. A leak detector can detect a hazardous atmosphere, but without it, personnel can be overcome without realizing the problem. Carbon dioxide is often a by-product of incomplete burning and can be a deadly hazard if personnel are not aware of its possible existence. Some pipeline operations pump carbon dioxide commercially, and in such operations, the hazards are increased.

9.1.0 Reaction

In case of a fire, the operator must make certain that no one is allowed to enter the building or area after the fire is out unless the proper gas detection meter is carried along with appropriate PPE. The meter must be calibrated for detection of carbon dioxide immediately before entry. Of special concern until the atmosphere has been tested for carbon dioxide would be low-lying areas or lower rooms and compartments of buildings involved.

The operator must make certain that the first personnel to enter such questionable atmospheres are wearing self-contained breathing apparatus (SCBA). Cases have been noted in which co-workers have discovered associates lying in such areas unconscious. Rushing to their aid, the co-workers were also overcome by carbon dioxide.

In such emergencies, personnel who are trained and qualified in the use of SCBAs and are trained in rescue must enter first. Other workers must be trained to understand the danger and to stay a safe distance away until the deadly atmosphere has been cleared. In the case of a building or compartment, signs must be posted at the doors warning of the dangerous atmosphere. The operator must make certain that all such areas are plainly marked so that personnel will not enter.

10.0.0 ◆ INJURIES AS A RESULT OF UNEXPECTED/UNINTENDED RELEASE OR ENCOUNTER

It is the operator's duty to call for emergency rescue and medical assistance as soon as the incident is known. The operator must make certain that any injured or unconscious personnel are provided medical attention as quickly as possible.

Company procedure must be strictly followed in such incidents. In some cases, if there are physical injuries, the person must only be moved by trained and qualified personnel. In most companies, properly trained personnel are allowed to apply cardiopulmonary resuscitation (CPR) and first aid until an ambulance arrives.

10.1.0 Notifications

The supervisor must make certain that the incident and the injury are properly recorded and documented. In the case of personnel being injured, reports must be made to the Occupational Safety and Health Administration (OSHA) as required by the federal standard. The death of an employee requires notification of the NRC immediately and OSHA within a specified number of hours. A written accident report must be submitted to the NRC as soon as possible but no longer than 30 days from the time of the incident.

When the problem has been solved, the operator must complete the required documentation. Included in the list of documentation that must be completed are:

- Incident Investigation and Analysis Report
- Incident Report
- Event Notification forms
- Control center logs
- Station logs (if involved)
- Records of recommendations for corrective actions and training
- Documentation of such corrections and training after completion

Classroom

Discuss, from your own experience if possible, specific instances of abnormal operating conditions, such as fires or releases.

Emphasize the importance of following the proper procedure when dealing with fires.

Ask trainees what must be done in case of injuries resulting from an unexpected or unintended release of product from the pipeline.

Laboratory

Have trainees name the various pieces of documentation required for reporting injuries. Write their answers on the board and discuss.

Classroom

Ask a trainee to explain the consequences of an unexplained deviation in pressure. Ask another trainee to identify the appropriate response.

11.0.0 ◈ RECOGNIZE AND REACT TO UNEXPLAINED PRESSURE DEVIATIONS

Unexplained pressure deviations can include an absence of pressure, a sudden increase in pressure, or a sudden decrease in pressure. Deviations in pressure indicate a serious condition on the pipeline. Occurrences such as unintended valve closures cause AOCs that will endanger the pipeline. The condition must be recognized, and reaction must be quick and decisive. If not, an emergency condition could result very quickly.

11.1.0 Reaction

If a change in pressure occurs that is different from the expected pressure, personnel should immediately investigate the change. If the cause is not determined, the operating lines should be shut down, referring to the abnormal condition shutdown procedure. Personnel should then proceed with emergency response notification.

Summary

An AOC is a condition that may indicate a malfunction of a component or deviation from normal operations, which may exceed design limits or result in a hazard to persons, property, or the environment. As required by law, companies have standard operating procedures in place to give personnel direction as to how to alleviate these conditions. Time is of the essence in such conditions. Operations personnel should be able to recognize and react to these conditions, using knowledge of policies and procedures as well as their own skill and good judgment.

In the event of a fire, explosion, or other disaster, operations personnel must be prepared to aid in emergency operations. Whether the AOC is the result of a valve closed inadvertently or a major release of product, personnel must react quickly to ensure the protection of the public and to preserve the safety and efficiency of the pipeline.

Instructor's Notes:

Review Questions

1. An example of an abnormal facility condition would be _____.
   a. a high-pressure shutdown
   b. unauthorized equipment on the ROW
   c. transformer failure
   d. a lightning strike

2. The first reaction once pipeline damage has been discovered is to _____.
   a. shut down the pipeline facility
   b. contact the control center
   c. evacuate the surrounding area
   d. determine the extent of the damage

3. Main-line pumps have high discharge pressure alarms that will shut down the pump when the discharge pressure _____.
   a. exceeds 100% of the MOP
   b. exceeds a preset limit
   c. is too low
   d. is at null

4. If 110% of maximum operating pressure is being exceeded, operations personnel must _____.
   a. shut down the pipeline
   b. continue to monitor the situation closely
   c. file an accident report form
   d. contact OSHA

5. _____ failures are somewhat interlinked with communications failures because the SCADA system may be involved.
   a. Power
   b. Equipment
   c. Control system
   d. Pump

6. In the case of a release, pipeline operations personnel should first _____.
   a. shut down the originating station, continuing until all pumps are offline
   b. engage the emergency response team
   c. close all block valves to shut down the system
   d. debate the situation with their co-workers

7. An uninterruptible power system _____.
   a. is a device that converts alternating current (AC) to direct current (DC)
   b. provides DC backup in case of power failure
   c. breaks down electric power from a high-voltage source to appropriate levels for distribution
   d. provides protection against primary AC power failure

8. An accident report is required for a failure in a pipeline system in which there is a release of product resulting in _____.
   a. releases of carbon dioxide
   b. releases that exceed five barrels of product
   c. estimated property damage exceeding $10,000
   d. releases of three gallons of HVLs or more

9. Carbon dioxide can collect in low-lying areas and displace oxygen because _____.
   a. it is a by-product of combustion
   b. it is heavier than air
   c. it is 1.5 times less dense than air
   d. it is produced by respiration

10. If an odor of _____ has been detected even a few hundred feet from the source, control center operations must be contacted to shut down the line.
    a. carbon dioxide
    b. propane
    c. gasoline
    d. carbon monoxide

Classroom

Go over the Review Questions. Answer any questions trainees may have on materials covered to this point.

Examination

Administer the Module Examination. Be sure to record the results of the Exam on Craft Training Report Form 200 and submit the results to the Training Program Sponsor.

GLOSSARY

Trade Terms Introduced in this Module

Abnormal operating conditions (AOCs): Conditions within a pipeline system that exceed operating design limits as specified in *49 CFR 195*.

Breakout tank: A DOT term used to identify a tank that relieves surges in a jurisdictional hazardous pipeline system or receives and stores hazardous liquid transported by a pipeline for subsequent reinjection and continued transportation by the jurisdictional pipeline. The breakout tank is a jurisdictional tank.

Highly volatile liquid (HVL): A hazardous liquid that will form a vapor cloud when released to the atmosphere and has a vapor pressure exceeding 40 pounds per square inch absolute at 100°F.

Lower explosive limit (LEL): The minimum concentration of vapor in the air below which propagation of a flame does not occur in the presence of an ignition source.

Maximum operating pressure (MOP): The maximum internal gauge pressure at which a hazardous liquid or carbon dioxide pipeline system or segment of such may be operated. The highest level of pressure allowed for a particular pipeline as set by the Department of Transportation to regulate the range of safe operations for that pipeline. Broadly speaking, it is set at 80 percent of the original design pressure, considering all components, or the highest pressure the pipeline was operated at during the last five years.

Relief device: Device for diverting a stream of product into a tank to take pressure off the pipeline during an overpressure condition.

Supervisory Control and Data Acquisition (SCADA): Computer system that is used to monitor and control the operation of a pipeline.

Surge pressure: Pressure produced by a change in velocity of the moving stream that results from shutting down a pump station or pumping unit, closure of a valve, or any other change in velocity of the moving stream.

Instructor's Notes:

MODULE 66102-02 — TEACHING TIPS

The following are suggested activities or instructional methods to help you teach the material in this AIG.

General

When you call on someone to answer a question, the rest of the class relaxes or even tunes out because they expect that the question and answer will take place only between you and the trainee you called on. Instead, use this technique to involve more trainees in answering questions and to keep them on their toes.

1. Ask trainees to define a term or explain a concept.
2. After one trainee has answered, ask a trainee seated nearby if the answer is right. Then ask whether a trainee in the back of the room agrees.
3. Ask trainees to explain why they think an answer is right or wrong.
4. Use the session to clear up incorrect ideas and encourage trainees to learn from their mistakes.

Section 3.1.2 *Documentation*

This exercise will allow trainees an opportunity to practice reporting a safety-related condition. Trainees will need paper and pencil. Allow 20–30 minutes for this exercise.

1. Divide the class into pairs. Provide one member of each pair with a scenario involving a safety-related condition.
2. Have the trainees "report" to their partners on the safety-related condition.
3. The trainees who receive the reports should use the information to write a report, including all necessary information. They may refer to *Figure 3* as an example.
4. Have the pairs switch roles. Provide new scenarios to each pair.
5. Review the reports in class, and discuss the processes of collecting information and writing a complete and accurate report.

Section 8.1.3 *Response to Unexpected or Unintended Releases*

This exercise will familiarize trainees with proper response to unexpected and unintended releases. You will need an overhead projector, blank acetate sheets, and a transparency marker. Trainees will need pencils and paper to take notes for later discussion. Allow 30–45 minutes for this exercise.

1. Divide the class into two teams. Tell trainees that you will be stating several kinds of indoor and outdoor releases. The first team to identify the proper procedure for handling the type of release will be awarded one point. If they get it wrong, the other team will have an opportunity to answer.
2. No individual can answer more than twice.
3. Proceed until a predetermined score has been reached by one of the teams.
4. Discuss the results and review any unclear or inaccurate procedures that were presented during the exercise.

MODULE 66102-02 — ANSWERS TO REVIEW QUESTIONS

Answer	Section
1. b	1.1.0
2. d	2.0.0
3. b	3.0.0
4. a	4.2.0
5. a	5.0.0
6. a	6.3.0
7. d	5.0.0
8. b	8.2.0
9. b	9.0.0
10. c	8.1.2

CONTREN™ LEARNING SERIES — USER UPDATES

The NCCER makes every effort to keep these textbooks up-to-date and free of technical errors. We appreciate your help in this process. If you have an idea for improving this textbook, or if you find an error, a typographical mistake, or an inaccuracy in NCCER's Contren™ textbooks, please write us, using this form or a photocopy. Be sure to include the exact module number, page number, a detailed description, and the correction, if applicable. Your input will be brought to the attention of the Technical Review Committee. Thank you for your assistance.

Instructors – If you found that additional materials were necessary in order to teach this module effectively, please let us know so that we may include them in the Equipment/Materials list in the Instructor's Guide.

Write: Curriculum Revision and Development Department
National Center for Construction Education and Research
P.O. Box 141104, Gainesville, FL 32614-1104

Fax: 352-334-0932

E-mail: curriculum@nccer.org

Craft ______________________ Module Name ______________________

Copyright Date ____________ Module Number ____________ Page Number(s) ____________

Description

____

____

____

____

(Optional) Correction

____

____

____

(Optional) Your Name and Address

____

____

____

Module 61103-02

Locating Pipeline and Cable

(CT 14.1 and 17.1)

Locating Pipeline and Cable
Annotated Instructor's Guide

Module 61103-02

MODULE OVERVIEW

This module familiarizes trainees with procedures and requirements for locating underground pipeline and cable, including the One-Call notification system. Trainees will have an opportunity to practice using various methods to locate and trace pipeline and cable.

PREREQUISITES

Please refer to the Course Map in the Trainee Module. Prior to training with this module, it is recommended that the trainee shall have successfully completed the following:

Core Curriculum; Pipeline Core, Modules 66101-02 and 66102-02

OBJECTIVES

Upon completion of this module, the trainee will be able to:

1. Explain the One-Call notification system.
2. Identify and explain the types of pipe and cable found underground.
3. Identify and explain the types of equipment used to locate underground pipe and cable (CT 14.1 and 17.1).
4. Perform approved procedures to accurately locate underground pipe and cable (CT 14.1 and 17.1).

PERFORMANCE TASK

Under the supervision of the instructor, the trainee should be able to:

1. Locate line (CT 14.1 and 17.1).

NCCER STANDARDIZED CRAFT TRAINING PROGRAM

The National Center for Construction Education and Research (NCCER) provides a standardized national program of accredited craft training. Key features of the program include instructor certification, competency-based training, and performance testing. The program provides trainees, instructors, and companies with a standard form of recognition through a National Craft Training Registry. The program is described in full in the *Guidelines for Accreditation*, published by the NCCER. For more information on standardized craft training, contact the NCCER by writing us at P.O. Box 141104, Gainesville, FL 32614-1104; calling 352-334-0911; or e-mailing info@nccer.org. More information may be found at our Web site, www.nccer.org.

HOW TO USE THIS ANNOTATED INSTRUCTOR'S GUIDE

Each page presents two sections of information. The larger section displays each page exactly as it appears in the Trainee Module. The narrow column ties suggested trainee and instructor actions to each page and provides icons to call your attention to material, safety, audiovisual, or testing requirements. The bottom of each page includes space for your notes.

Teaching Tip

If you see the Teaching Tip icon, that means there is a teaching tip associated with this section. Also refer to any suggested teaching tips at the end of the module.

SAFETY CONSIDERATIONS

Ensure that the trainees are equipped with appropriate personal protective equipment. Emphasize the importance of following all safety precautions and procedures when working with power tools.

PREPARATION

Before teaching this module, you should review the Module Outline, Objectives, Performance Task, and the Materials and Equipment List. Be sure to allow ample time to prepare your own training or lesson plan and gather all required equipment and materials.

MATERIALS AND EQUIPMENT LIST

Whiteboard/chalkboard

Markers/chalk

Overhead projector and screen

Transparencies

Blank acetate sheets

Transparency pens

Pencils and scratch paper

Appropriate personal protective equipment

Copies of your company's policy and procedures manual

Copies of a pipeline strip map

Copies of a pipeline alignment sheet

Ferromagnetic locators

Stick metal detectors, split box locators, flag markers

Copies of *49 CFR 195.250* and *192.325*

Copies of the Performance Verifications for Covered Tasks 14.1 and 17.1**

Module Examinations*

Performance Profile Sheets*

* Located in the Test Booklet.

** Available for download at www.nccer.org.

TEACHING TIME FOR THIS MODULE

An outline for use in developing your lesson plan is presented below. Note that each Roman numeral in the outline equates to one session of instruction. Each session has a suggested time period of 2½ hours. This includes 10 minutes at the beginning of each session for administrative tasks and one 10-minute break during the session. Approximately 5 hours are suggested to cover *Locating Pipeline and Cable (CT 14.1 and 17.1)*. You will need to adjust the time required for hands-on activity and testing based on your class size and resources.

Topic	Planned Time
Session I. Notification Systems and Physically Locating Underground Pipes and Cables	
A. Introduction	____________
B. One-Call Notification Systems	____________
C. Types of Underground Pipe and Cable	____________
D. Locating Underground Pipes and Cables	____________
1. Reading Maps and Alignment Sheets	____________
2. Installing Temporary Markers	____________
3. Determining Pipeline and Cable Depth	____________
4. Physically Locating Pipes and Cables	____________
Session II. Electronically Locating Underground Pipes and Cables and Separation Between Underground Structures	
A. Electronically Locating Underground Pipes and Cables	____________
B. Other Methods	____________
C. Requirements for Separation	____________
D. Reasons for Maintaining Separation	____________
E. Summary	
1. Summarize module	____________
2. Answer questions	____________
F. Module Examination	____________
1. Trainees must score 70% or higher to receive recognition from the NCCER.	
2. Record the testing results on Craft Training Report Form 200 and submit the results to the Training Program Sponsor.	
G. Performance Testing	____________
1. Trainess must perform each task to the satisfaction of the instructor to receive recognition from the NCCER.	
2. Record the testing results on Craft Training Report Form 200 and submit the results to the Training Program Sponsor.	

Module 61103-02

Locating Pipeline and Cable

(CT 14.1 and 17.1)

Instructor's Notes:

Instructor's Notes:

Introduction to the Pipeline Operator Qualification Rule, API, and NCCER

It could be said that petroleum products keep America running. Petroleum products like oil and natural gas are the principal sources of energy in the United States and are used in many manufacturing and chemical processes. Petroleum products are primarily transported by pipeline. There are more than a million miles of liquid and natural gas pipeline in this country, ranging in diameter from two inches to four feet. These pipelines bring crude oil and natural gas from production fields to processing plants and refineries—and ultimately to the consumer.

Pipeline Operator Qualification

Petroleum products are considered hazardous, so pipeline personnel must take extra precautions and follow specific procedures when working with them. To help ensure the safe handling and transportation of petroleum products, the United States Department of Transportation (DOT) mandates safe practices as federal regulations.

One such regulation, the Operator Qualification (OQ) rule, requires pipeline operators to develop a written qualification plan to evaluate their personnel and their contract personnel (see *49 CFR Part 192 Subpart N—Gas,* and *49 CFR Part 195 Subpart G—Liquid*). The evaluation hinges on the operator's ability to perform certain covered tasks and to be able to recognize and respond to abnormal operating conditions that may be encountered while performing these covered tasks.

Covered Tasks

A pipeline performance task is deemed *covered* if the activity passes the following four-part test:

1. Is it performed on a pipeline facility?
2. Is it an operations or maintenance task?
3. Is it performed as a requirement of *49 CFR Part 195 (Liquid)* or *49 CFR Part 192 (Gas)*?
4. Does it affect the operation or integrity of the pipeline facility?

If the answer to all four questions is *yes*, then the activity is considered *covered* and must figure into the evaluation process of pipeline personnel.

American Petroleum Institute

The OQ rule and its related covered tasks provided the catalyst to develop a unified pipeline training and assessment program—so that pipeline operators and owners have ready access to the assessment and training tools they need. This effort was the direct result of the vision and leadership provided by The American Petroleum Institute (API) and API members.

Efforts to unify the industry are nothing new to API. Indeed, the impetus for forming the Institute in 1919 was the need to standardize engineering specifications for drilling and production equipment. Today, the most pressing issues revolve around public perceptions and government policies toward the industry. Speaking with one voice on these issues has become as essential as having interchangeable parts in the field. API's members determine the petroleum industry's positions on public policy as well as the standards that govern its day-to-day operations worldwide.

Consortium on Operator Qualification

API assembled a focus group whose primary purpose was to identify covered tasks and provide compliance recommendations to pipeline operators.

The Consortium on Operator Qualification (COOQ) became the guiding force for the identification and development of covered tasks for the liquid pipeline industry. Their work product, API publication 1161, *Guidance Document for the Qualification of Liquid Pipeline Personnel*, formed the basis for the resulting Pipeliner Training and Assessment Program.

The covered task list created in API publication 1161 has now evolved to include gas covered tasks and is maintained by NCCER. The most current covered task list can be downloaded at www.nccer.org.

National Center for Construction Education and Research

With the guidance and support of API, its members, and the Consortium on Operator Qualification, the National Center for Construction Education and Research (NCCER) was able to design and implement the Pipeliner Training and Assessment Program.

NCCER is a not-for-profit 501(c)(3) education foundation established in 1995 by the world's largest and most progressive construction and maintenance contractors and national contractor associations. NCCER was founded to address the

severe workforce shortage facing the construction and maintenance industries and to develop a standardized training process and curricula. Today, NCCER is supported by hundreds of leading construction and maintenance companies, manufacturers, and national associations and developed more than 35 construction and maintenance titles.

NCCER also maintains a National Registry, which contains the records of trainees and instructors who have received NCCER training. All completion information for every trainee is sent to the NCCER and filed within the Registry. The Registry can then confirm training and skills for workers as they move from company to company, state to state, or even within their own company.

Standardized Curriculum (NCCER's Contren™ Learning Series)

NCCER curricula are developed using the expertise of industry subject matter experts (SMEs). The use of SMEs as an integral part of every new development or revision ensures that the curriculum is current, industry-driven, and indicative of industry practices. The development and/or revision process is facilitated from start to finish by NCCER project managers, who work closely with the SMEs and a select team of technical writer(s) on the development and presentation of the curriculum.

Pipeline: Five Training Paths

The pipeline curriculum focuses on essential skills and competencies within five skill areas:

- Corrosion Control
- Electrical and Instrumentation
- Maintenance
- Mechanical
- Operations (Field and Control Center)

SMEs from the pipeline industry have determined the skills and competencies needed based on API's covered tasks and the consensus of the group/industry. Specific company information is generally not incorporated into the curriculum unless the SMEs determine that the information is the industry standard.

Training Modules and Covered Tasks

The smallest unit of instruction in the curriculum is a *module*. Each curriculum is composed of modules (the equivalent of stand-alone chapters) that are logically sequenced into complete *levels* of instruction. Every module is task-oriented and requires the successful completion of a written (knowledge-based) examination and a performance (hands-on) test. Modules may be individually selected and/or combined for efficient task training. Modules in the pipeline curricula, to the extent possible, will align with API-developed covered tasks. While not all modules will relate to covered tasks, all covered tasks will relate to at least one module.

Benefits of Training to the Pipeline Marketplace

Applying NCCER's standardized training process allows pipeline owners and contractors in the United States to establish the benchmark for acceptable skill and competency levels within the pipeline industry. This process will assist training organizations in developing and/or improving standardized pipeline training programs that effectively match their resources and operating conditions while responding to regulatory demands and to the needs of the industry. The process will also recognize training organizations that meet the minimum standards of competence, warranting regulatory, public, and professional confidence. This in turn will foster national unity within the pipeline industry for the mutual benefit of individual technicians, owners, contractors, and ultimately pipeline-users and society.

Additional Resources

To review the Pipeliner Training and Assessment Program Implementation and Resource Guide and the latest version of pipeline covered tasks, visit NCCER's Web site at www.nccer.org, or contact:

National Center for Construction Education and Research
PO Box 141104
Gainesville, FL 32614-1104
(888) 622-3720
Web site: www.nccer.org

For more information on API, please contact:

American Petroleum Institute
1220 L Street, Northwest
Washington, D.C. 20005
(202) 682-8000
Web site: www.api.org

To order *API Pub 1161*, contact Global Engineering Documents by calling 1-800-854-7149 or visiting their Web site at www.global.his.com.

Additional sources of information on the pipeline operator qualification regulation include:

DOT – Office of Pipeline Safety
www.ops.dot.gov

Transportation Safety Institute
www.tsi.dot.gov

Instructor's Notes:

COURSE MAP

This course map shows all of the modules in the first level of the Pipeline Corrosion Control curriculum. The suggested training order begins at the bottom and proceeds up. Skill levels increase as you advance on the course map. The local Training Program Sponsor may adjust the training order.

PIPELINE CORROSION CONTROL LEVEL ONE

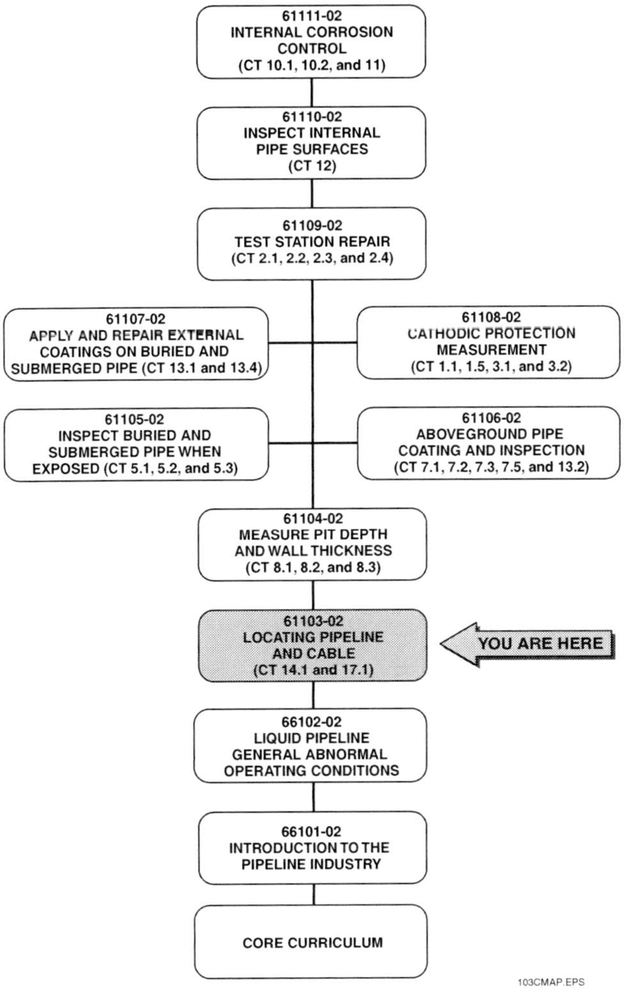

Homework

Assign reading of Module 61103-02.

Instructor's Notes:

MODULE 61103-02 CONTENTS

Figures

Tables

Instructor's Notes:

MODULE 61103-02

Locating Pipeline and Cable

Objectives

When you have completed this module, you will be able to do the following:

1. Explain the One-Call notification system.
2. Identify and explain the types of pipe and cable found underground.
3. Identify and explain the types of equipment used to locate underground pipe and cable.
4. Perform approved procedures to accurately locate underground pipe and cable.

Prerequisites

Before you begin this module, it is recommended that you successfully complete the following: Core Curriculum; Pipeline Core, Modules 66101-02 and 66102-02.

Required Trainee Materials

1. Pencil and paper
2. Appropriate personal protective equipment

1.0.0 ◆ INTRODUCTION

Encroachment on all utilities and hazardous or non-hazardous pipeline **rights-of-way (ROW)** and corridors is a serious and potentially dangerous matter. In order to protect pipeline rights-of-way, the federal government has developed standards to protect pipeline installations from harm. Federal law has mandated the development of One-Call notification systems as a way to protect people, property, and the environment from those who might damage it by digging, excavating, or blasting on or near all utilities.

Physical damage to pipelines and cables is an ongoing danger and pipeliners are vigilant in protecting their assets. Even the most careful excavation crews sometimes make mistakes. Excavation machinery can cause great damage. Pipeline companies require that excavations or encroachments proceed in very specific ways so that the pipeline, people, and the environment will be protected.

Companies and individuals must strictly observe applicable safety guidelines when performing work in the vicinity of underground pipelines and cables. A critical part of working safely around these underground installations is being able to accurately identify their location. Included in this process are:

- Complying with One-Call guidelines
- Identifying the types of pipes and cables found underground
- Following approved procedures for using underground pipe and cable locators
- Accurately locating underground pipes and cables

Trained and qualified inspection individuals should calibrate or verify the proper operation of pipe and cable location equipment prior to use, including checking the batteries.

2.0.0 ◆ ONE-CALL NOTIFICATION SYSTEMS

One-Call notification systems are federally mandated damage prevention programs for the use of contractors and others who perform any excavation activities. Excavation activities include digging, blasting, boring, trenching, tunneling, backfilling, the removal of aboveground structures

Materials

Ensure that you have everything required to teach the course. Check the Materials and Equipment List at the front of this module.

Classroom

Explain that terms shown in bold are defined in the Glossary at the back of this module.

Audiovisual

Show Transparency 1, Course Objectives.

Show Transparency 2, Performance Task.

Teaching Tip

See the general Teaching Tip at the end of this module.

Classroom

Ask a trainee to explain why the One-Call notification system was developed.

Have trainees name types of excavation activities. Write their answers on the board or an overhead.

Present several scenarios involving pipeline activities, including excavation activities. For each scenario, have trainees answer whether they must notify the One-Call system, and explain why or why not.

Classroom

Pair up trainees. Have trainee A pretend to call the One-Call system (trainee B) and provide the required information. Then have trainees switch roles.

Have a trainee explain the term *clearing the ticket*.

by explosive or mechanical means and other earthmoving operations. Any person or company planning to engage in any of these activities is required to call the One-Call notification system number and make notification of their intent to perform excavation activities. The One-Call notification system number can be found on pipeline **line markers**, as shown in *Figure 1*.

103F01.EPS

Figure 1 ◆ Typical pipeline marker.

2.1.0 Purpose of One-Call Systems

The basic purpose of One-Call systems is to protect buried utilities, including pipelines that carry hazardous materials, and thereby protect people and the environment by preventing accidental damage to the pipeline. Every company operating a pipeline system carrying hazardous materials must be covered by a One-Call notification system.

2.1.1 Reliability

A One-Call system must have a reliable way of receiving and recording contacts from excavators and citizens so that the system is dependable. When a call is received, the pipeline company is notified and qualified company personnel are required to clear the ticket or to be dispatched to the site to pinpoint and mark the line.

> **NOTE**
> Clear the ticket means confirming that there are no utilities present. This information is recorded on the One-Call ticket and returned to the issuing office where it is kept on file. If there is any doubt or confusion, call the excavator or meet with them on site to verify the location of the planned excavation.

Excavators are required to call the published One-Call telephone number covering the area of each intended excavation activity and provide the following information:

- Name of person notifying the system
- Name, address, and telephone number of the excavator
- Specific location
- Starting date
- Description of the intended excavation activity

Legal support for pipeline rights-of-way agreements grant companies the right to exercise authority over what activity is allowed on the right-of-way. Most companies that operate high-pressure underground pipelines require that a company representative be on site to monitor excavation and encroachments to ensure the safety of the pipeline. No excavation is allowed until a representative of the company is on site. It is the responsibility of the responding company representative to understand existing ROW agreements.

Each company that operates a buried pipeline transporting hazardous materials is responsible for ensuring that One-Call system guidelines are strictly observed by pipeline personnel. The company is also required to see that all appropriate records relating to One-Call contacts are completed, including proper documentation of all calls received, actions taken, and required reports filed.

Instructor's Notes:

3.0.0 ◆ TYPES OF UNDERGROUND PIPE AND CABLE

It has been said that buried utility cables and pipelines, and the products and services they carry, are like the veins and lifeblood of a community. Lack of these products and services can negatively impact a community, even to the point of paralysis. This can and does happen when buried lines and cables are not properly located and flagged before excavation work is started.

Companies and individuals engaged in work on or near these systems realize that precisely locating buried pipes and cables is a huge responsibility. There are no acceptable shortcuts. It is critical that you have accurate and complete information before initiating a job. Typical underground pipelines and cables one might encounter during the course of construction, landscaping, utility system repair, or major pipeline work include the following:

- Pipelines
  - Oil pipelines
  - Gas pipelines
  - Product pipelines
- Buried utility cables
  - Power cables
  - Telephone cables
  - Fiber-optic cables
  - Cable television (CATV) cables
- Other lines
  - Gas distribution lines
  - Metal, plastic, and concrete water pipes
  - Sewers and drains

4.0.0 ◆ LOCATING UNDERGROUND PIPES AND CABLES

When performing work in the vicinity of underground pipelines and cables, you must locate and mark the pipeline and right-of-way so that it is plainly discernable to persons planning excavation or other work. However, locating the pipeline is somewhat more complicated than just looking at a map and picking the spot. Locating underground pipelines and cable employs several basic procedures, including:

- Reading maps and **alignment sheets**
- Installing temporary markers
- Determining the depth of cover
- Using physical locating methods
- Using electronic equipment locating methods

4.1.0 Reading Maps and Alignment Sheets

Properly reading and interpreting maps and alignment sheets requires a basic understanding of maps and their related symbols. This section builds on the skills taught in the Core Curriculum module, *Introduction to Blueprints*. Trainees are advised to review this module prior to proceeding.

Individuals involved in locating underground pipelines and cables need to be proficient in reading maps and alignment sheets. Maps, strip maps, and alignment sheets are created to furnish pertinent, permanent information about the location of various pipeline and cable installations. If there is a question about the accuracy of the strip maps and alignment sheets being used, a surveyor's map or blueprint may be obtained to pinpoint the exact location of the pipeline or cable right-of-way. *Figure 2* shows a pipeline strip map and its related symbols.

> **NOTE**
> Unless the lines have been recently and accurately surveyed and have had the Global Positioning System coordinates established, the maps, alignment sheets, and as-built drawings may *not* be 100% accurate. The original drawings may have not have been exact. Since then, the lines may have been abandoned, moved, or new ones may have been built. If in doubt, always pinpoint and mark trade lines.

Alignment sheets show the location and size of each pipe installed, all crossings of public roads, railroads, rivers and streams, buried utilities, and pipelines operated by other companies. Locations should be accurate as shown on the alignment sheet. *Figure 3* shows a pipeline alignment sheet.

4.2.0 Installing Temporary Markers

The applicable standard, *49 CFR 195.410*, requires that each pipeline operator must accurately install and maintain line markers over buried pipelines. Upon receiving a One-Call notification, qualified individuals are dispatched to the area to investigate the proposed work site. Only trained and qualified individuals should be assigned inspection and repair tasks on or near pipelines.

Classroom

Have trainees name some types of underground pipelines and cables. Stress the importance of properly flagging such pipes and cables. Discuss the consequences of not doing so.

If trainees need a refresher on reading maps and alignment sheets, have them review the Introduction to Blueprints module.

Distribute copies of a pipeline strip map and a pipeline alignment sheet. Have trainees practice reading them.

Have trainees write out the steps involved in placing temporary markers after receiving a One-Call notification.

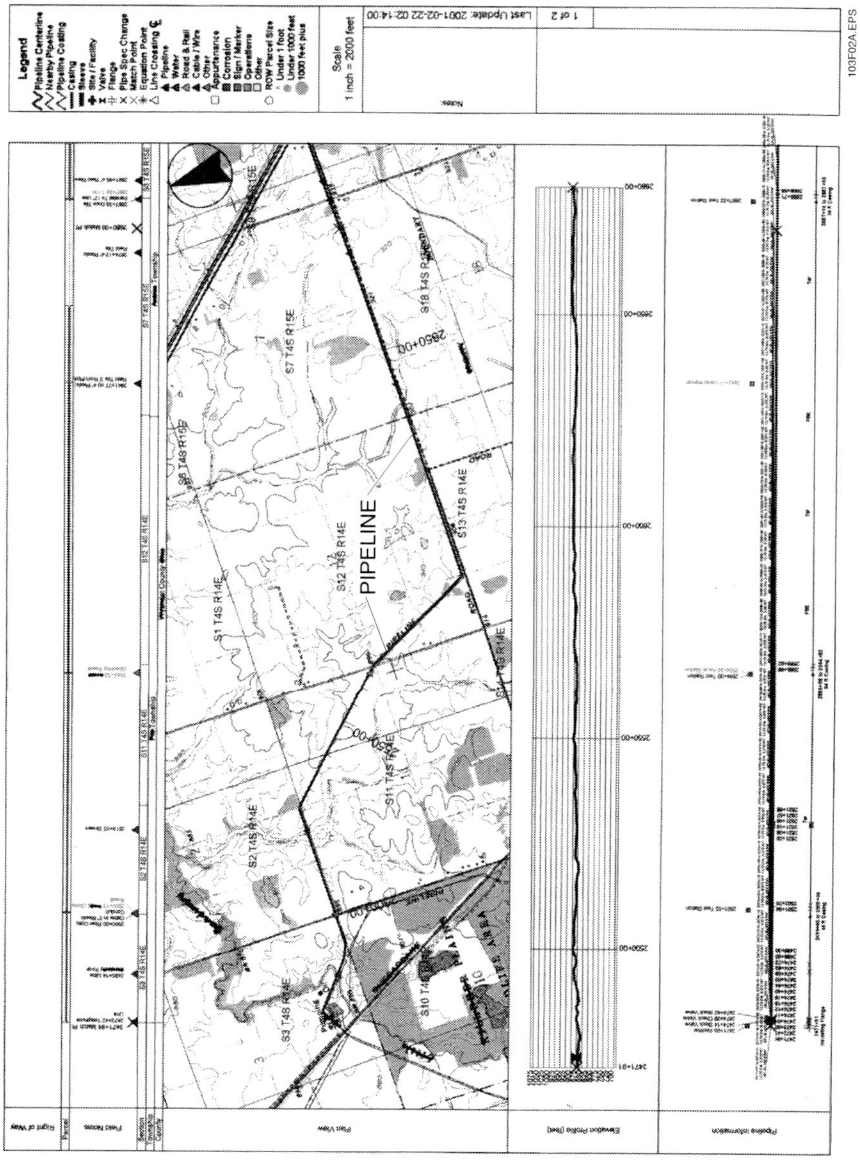

Figure 2 ◆ Pipeline strip map (1 of 2).

Instructor's Notes:

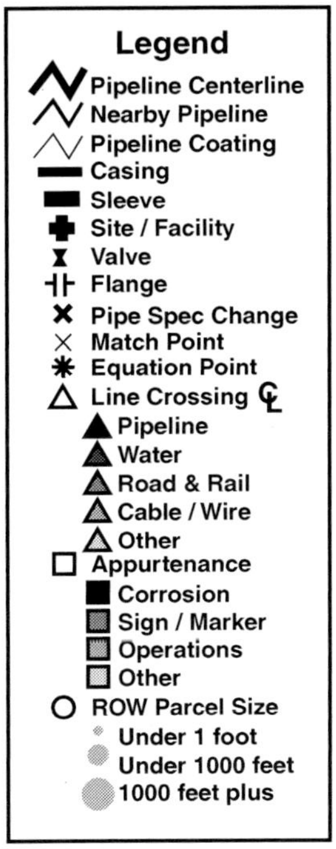

103F02B.EPS

Figure 2 ◈ Pipeline strip map (2 of 2 – legend detail).

The pipeline company's first responsibility is to physically locate the pipe. Temporary white markers or paint may be used to mark the proposed excavation or construction. Once the line is physically located, the work crew must replace any missing markers, repair damaged markers, and replace any incorrectly installed markers.

If the pipeline company determines that work should be allowed and a permit is issued for the work to proceed, small yellow flags, stakes, and/or paint are used to mark the pipeline right-of-way. The exact location can be plainly seen by anyone who will be performing excavation or blasting. Temporary pipeline markers should be evenly spaced along the location of the line or cable.

4.3.0 Determining Pipeline and Cable Depth

Depth of cover for pipes and cables is usually specified on engineering blueprints and elevations or as-built maps and typically varies according to the

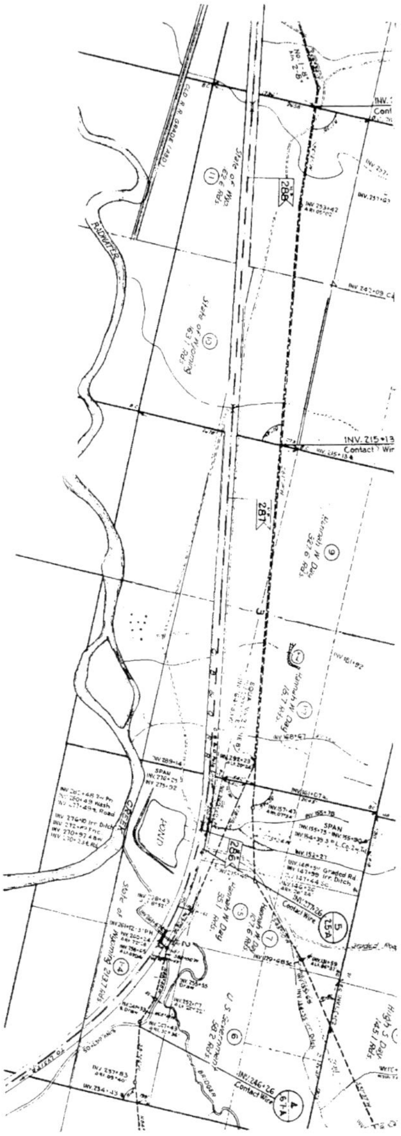

103F03.EPS

Figure 3 ◈ Pipeline alignment sheet.

Classroom

Ask a trainee to explain why depth of cover must be verified.

Audiovisual

Show Transparency 3 (Figure 4). Have a trainee explain the purpose of probe rods.

Safety

Stress the importance of probing carefully and slowly. Ask a trainee to explain how a worker can minimize the risk of electrocution.

Ask trainees to explain the hazards of pot-holing and how to offset them.

location and use of the area over the installed system. While some installations may follow set standards, logging, construction, and other uses may be approved on a case-by-case basis. Before excavation, the depth of cover must be verified by a qualified individual. The actual cover of a pipeline or cable may have changed after installation due to flooding or other influences. No excavation can be allowed until the pipeline has been definitely located.

Unless specifically exempted, all oil and gas new construction pipelines must be buried below the level of cultivation for protection. *Table 1* shows depth-of-cover requirements for new construction liquid pipelines in the locations shown. See *Table 2* to review depth-of-cover standards for gas pipelines.

Table 1 Depth-of-Cover Standards for Liquid Pipelines

Location	Cover (inches)	Notes
Industrial, commercial, and residential areas	36	
Inland water of at least 100 ft width high water to high water	48	Water crossings
Drainage ditches at public roads and railroads	36	Bar-Ditch
Deepwater port safety zone	48	
Gulf of Mexico and offshore water less than 12 ft at low tide	36	
All other areas	30	Railroads, highways, other
Underwater	up to 96	

103T01.EPS

Table 2 Depth-of-Cover Standards for Gas Pipelines

Location	Normal Soil (inches)	Consolidated Rock (inches)
Class 1 locations	30	18
Classes 2, 3, and 4 locations	36	24
Drainage ditches at public roads and railroads	36	24

103T02.EPS

NOTE

Many companies required more cover than these standards. Check your company's depth-of-cover requirements.

4.4.0 Physically Locating Pipes and Cables

Trained individuals use certain methods and tools for physically locating underground pipelines and cables. The physical location of pipes and cables can be accomplished by probing, **pot-holing** by hand, and by using vacuum units. Power equipment, such as a backhoe, is not normally used until the exact location of the system is established.

4.4.1 Using Probe Rods

Probe rods (see *Figure 4*) made of metal and fiberglass are used by some companies for locating underground pipelines and utility cables before undertaking any excavation work. Probe rods should have a smooth, rounded end. They should be used cautiously to minimize the possibility of damaging or breaking the pipeline coating and creating an opening through which corrosion could attack the pipeline. Probe rods should only be used by trained and qualified persons.

WARNING!

Proceed slowly and carefully when probing. An electrocution hazard from electrical lines may be present. To offset this risk, use only tools made of a nonconductive material (i.e. fiberglass) or tools equipped with insulated handles.

NOTE

Some companies do not allow probing at all. Check your company's policies.

4.4.2 Hand Digging and Pot-Holing

Pot-holing is a method of physically locating the pipeline by hand digging small holes. Pot-holing may be more hazardous than probing because tools with sharp edges could potentially cut into a powered electrical line.

In addition to this electrocution hazard, tools with sharp edges can easily damage the coating of the pipeline or other utility line. When pot-holing, individuals should proceed with caution until the pipeline is exposed.

Instructor's Notes:

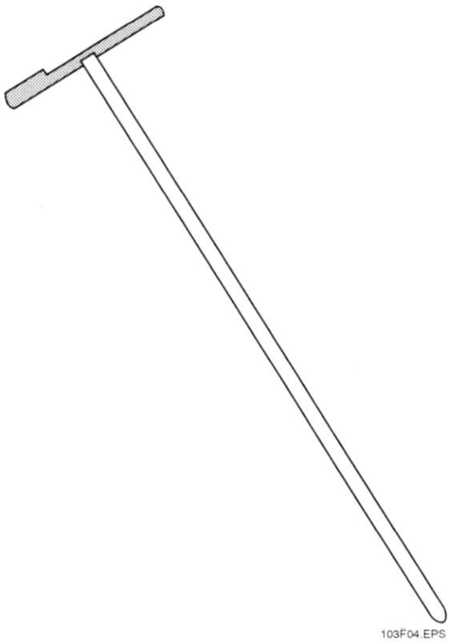

103F04.EPS

Figure 4 ◆ Typical probe rod.

> **WARNING!**
> When hand digging or pot-holing, use extreme caution and proceed slowly as electrical lines could be present. It is critical that you only use tools made of nonconductive materials or tools equipped with electrically insulated handles to offset the risk of electrocution.

4.4.3 Using Vacuum-Powered Tools

A **Water-Vac** unit (see *Figure 5*) is an effective tool available to hydraulically bore and air vacuum out a hole over a pipeline. It opens a clean, smooth hole with no damage to the pipeline or cable. Water-Vac units minimize disruption of other work on the site.

Air-Vac units operate on the same basic principle as Water-Vac equipment. Air-Vac vacuum trucks bore holes with compressed air rather than water and vacuum the debris from the hole over the pipeline.

Both types of vacuum units are designed to reduce the risk associated with using a pick, shovel, posthole digger, or other sharp metal tools and probes to expose the pipeline.

103F05.EPS

Figure 5 ◆ Water-Vac unit.

> **WARNING!**
> It is imperative that you wear appropriate personal protective equipment, especially eye protection, as an Air-Vac unit tends to propel more dirt and rock particles into the surrounding air than a Water-Vac unit.

4.4.4 Using a Backhoe

A backhoe is a large, powerful piece of construction equipment which can be safely used to quickly expose underground pipe and cables by moving large amounts of cover. The size and power of the backhoe can significantly reduce the time needed to perform the job.

> **CAUTION**
> Exercise care while digging near known pipeline and cable locations. Digging too deeply, or off center from the target area, could cause damage to pipes and cables before the backhoe operator is able to reverse the direction of the equipment. The initial excavation often destroys the temporary line markings such as flags, stakes, and paint. Therefore, to help the backhoe operator, the line should be periodically relocated in the hole as the hole deepens and the pipeline is exposed.

Classroom

Have trainees list the advantages and disadvantages of Water-Vacs and Air-Vacs.

Safety

Stress the importance of wearing personal protective equipment when operating Air-Vacs and Water-Vacs.

Ask trainees to identify the special considerations necessary when a backhoe is used to locate underground pipes and cables.

Classroom

Have trainees list the actions that should be taken before resorting to a backhoe.

Homework

Assign the review of Sections 4.5.0–5.2.1 and completion of the Review Questions for the next session.

Materials

Ensure that you have everything required for teaching and testing during this session.

Teaching Tip

See the Teaching Tip at the end of this module.

Before resorting to heavy equipment to expose a pipeline or utility service, make sure that other critical preliminary steps have been completed. These could include any or all of the following:

- Consult available strip maps and alignment sheets.
- Verify that locating equipment items are operational.
- Physically locate the line.
- Verify that the location and depth are accurate.
- Notify the state One-Call system.
- Ensure that appropriate supervisors are informed.
- Use alternate methods (hand digging or potholing) to perform the work, if necessary.
- Proceed with caution.

4.5.0 Electronically Locating Pipes and Cables

Electronic **line locators** are a class of lightweight, portable detectors used to locate a variety of underground utility lines. More advanced models can locate a specific type of utility, even in congested utility areas. They are accurate, but the physical pipe or cable location may need to be verified with a physical location tool before excavation is allowed to proceed. *Figure 6* shows how a typical electronic pipe and cable locator is used.

103F06.EPS

Figure 6 ◆ Using an electronic pipe and cable locator.

> **NOTE**
> Always check standard manufacturer's operating procedures.

Locating an underground utility line with an electronic locator unit typically requires that you follow these steps:

Step 1 Once in the field, at the site where the location of utility lines and cables needs to be pinpointed, review maps, alignment sheets, and as-built drawings.

Step 2 Look for indications of lines such as valves, valve box covers, manhole covers, pedestals, junction boxes, or utility poles with downward cables for visible clues of what might be underground.

Step 3 Use a ferromagnetic locator as an initial sweep to check for iron, metal pipes, anchors, and deadmen.

Step 4 Use the passive modes in the receiver, if available, to sweep the area for indications of signals on lines. These might include:
- 50/60 Hz-power
- Radio signals
- Cathodic protection (CP) signals
- Cable TV (CATV) signals

Step 5 Determine if induction, connection, induction clamp, or two-person sweep method is to be used.

Step 6 Set the transmitter controls.

> **NOTE**
> Always direct connect the transmitter when possible. The signal will be cleaner and will travel farther.

Step 7 Use the locator receiver to locate and mark the target pipes or cables.

Step 8 Verify the identification of the target.

Step 9 Determine the depth of the target.

Step 10 Notify concerned supervisors, planners, and work crews of these findings.

Instructor's Notes:

> **NOTE**
> These are generic steps for operating a typical set of equipment to locate a buried pipe or cable. They do not include enough information to use a specific locator. Detailed information must be obtained from the original equipment manufacturer's (OEM) operating and maintenance instruction manual or their trainers.

4.5.1 Stick Locators

These lightweight, portable detectors, also referred to as metal detectors, are quick and easy to use in the field to locate a variety of underground iron metal items, including the following:

- Pipe
- Deadman
- Anchor
- Tank
- Survey Pins

Stick locators (see *Figure 7*) typically consist of two encased sensors in a tube or shaft. These antenna sense the magnetic field from the iron metal and produce increased noise over the object or line. Some models also have a meter to help in noisy environments, a 50/60 Hz mode to indicate live electric current flow and/or an erase mode to eliminate detection of iron debris (nuts, bolts, and welding rods).

> **WARNING!**
> These detectors will locate both dead and live lines. Always assume the line is live until you confirm otherwise.

To effectively use a stick locator, walk across the area where the item you are seeking should be located, sweeping the locator back and forth above the ground in front of you. Once you get an indication of underground metal, move with the locator in a circle (approximately 6' out) around the spot to determine if the indication is a pipeline or some smaller metal item (deadman, anchor, barrel, or piece of iron trash).

> **CAUTION**
> Make sure you keep the locator away from steel-toe boots.

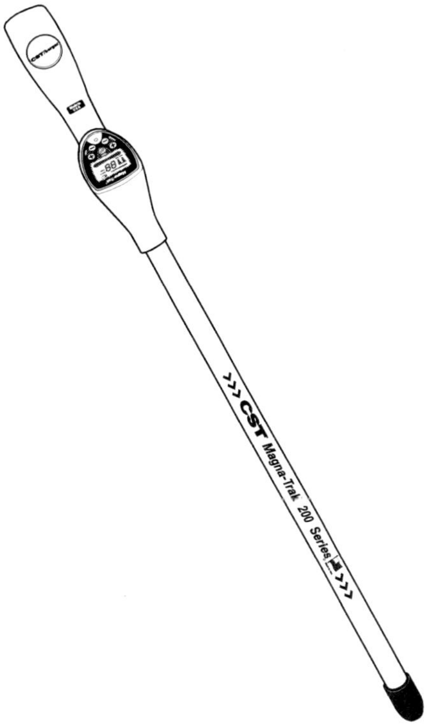

Figure 7 ◈ Typical stick locator.

Hold the locator upright while walking in a smaller circle to narrow the search and pinpoint the item. Use the erase mode, if available, and repeat the process to confirm the finding.

> **NOTE**
> These generic directions for operating a typical stick locator do not include enough information to use a specific equipment item. Detailed operating instructions can be obtained from the original equipment manufacturer's instruction manuals or from the factory trainer.

Audiovisual

Show Transparency 4 (Figure 7). Discuss proper procedures for using a stick locator.

Laboratory

Have trainees practice using stick locators.

Safety

Remind trainees to assume that all lines are live until otherwise confirmed. Ask trainees to explain why the locator should be kept away from steel-toe boots.

Audiovisual

Show Transparency 5 (Figure 8). Ask a trainee to explain how the transmitter operates. Have another trainee explain how the receiver works.

Safety

Explain how the operator can ensure that the signal is being applied to the correct pipeline. Ask a trainee to explain why the operator must verify this.

4.5.2 Electromagnetic (Battery-Operated) Locators

Electromagnetic locators have become the standard tools for locating and tracing buried pipe and utility cables. They range from simple units able to detect the presence of pipes and cables, to sophisticated equipment able to identify systems, depth, and system faults in the most complex situations. Locators contain two major components: a transmitter and a receiver. *Figure 8* shows an electronic locator.

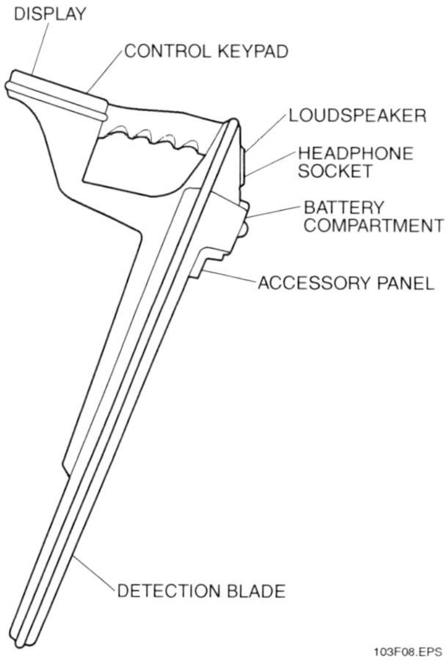

Figure 8 ◆ Electronic locator.

Transmitters apply an easily identifiable signal to the line with which you are working. The technique employed will vary with the site and situation. The user will gain a better understanding of which technique works best in a given situation through field experience. Techniques include direct connection, clamping, inductive coupling, and two-person sweep. *Figure 9* shows the two-person sweep.

When the signal is detected, the receiver provides a signal to the user, indicating the location of the pipe or cable being sought. Following this

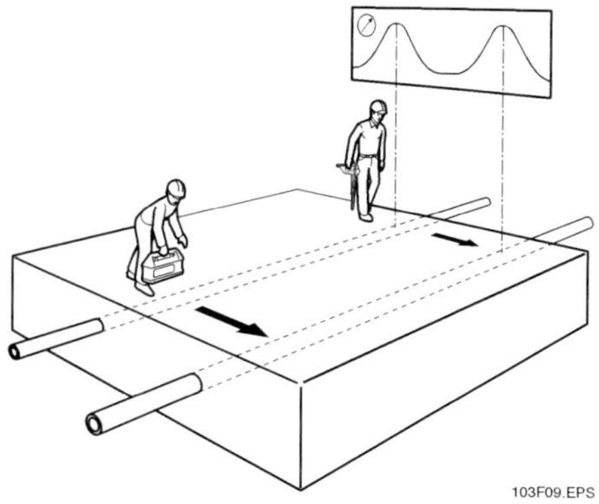

Figure 9 ◆ Two-person sweep.

signal enables the user to trace, pinpoint, and identify the target line, or locate other, unidentified lines in the target area. Receivers can also detect the presence of passive signals. Examples of passive signals include:

- 60 Hz AC electrical power
- Cathodic protection signals
- Radio frequency signals of 17–28 Hz
- Cable TV signals at 31 kHz

Transmitters typically operate over a wide frequency range from 4 Hz to 500 kHz. Transmitter signal power ranges from a very low 0.1 watt to 5 watts in a portable transmitter, to very strong signals of 150 watts output power.

On different electronic locators, transmitters are capable of transmitting signals in a variety of frequency ranges as shown in *Table 3*.

Table 3 Common Transmitter Frequency Settings

Connect Only[1]	Connect or Clamp Around	Induction[2]
4 Hz/8 Hz	8 kHz	33 kHz
98 Hz	9.8 kHz	65 kHz
440 Hz	22 kHz	83 kHz
512 Hz	29 kHz	200 kHz
577 Hz	33 kHz	400 kHz
815 Hz	65 kHz	500 kHz
1000 Hz		

[1] Best for wire or diagnostics; needs more power on pipes
[2] Best on tracer tape and tracer wire

103T03.EPS

Instructor's Notes:

The more powerful the signal, the greater the distance over which it can be detected and the deeper the lines can be and still be located. When pinpointing deeper lines, (+10') however, it can be more difficult to get accurate depth readings because of the fading signal and interference.

This equipment operates on the principle of induction. A coil antenna will produce an AC voltage signal proportional to the strength of the AC field or AC signal passing through it. Receiver antennas or aerials usually take this shape and are typically fitted with ferrite rods through the center of the coil to make them more effective at amplifying the small signals they pick up.

CAUTION

In the induction mode, the transmitter can put a signal on any buried line so the operator must ensure the signal is being applied to the company's pipeline.

The following are locator operating modes:

- *Null mode* – Select the Null mode and move along the suspected path of the target line. A quiet response, or no signal, will be indicated directly over the target line with an increased response on either side of the line. The numeric display makes it easier to interpret the response in noisy areas. Many units provide a visual indication by including right/left arrows that switch directly over the line.
- *Peak - Single Horizontal* – Provides a wider, maximum signal or noise over the line and tends to be more sensitive.
- *Peak - Dual Horizontal* – Provides a narrow, more precise pinpoint and gets rid of airborne interference if electronics/software are designed correctly. It provides a maximum or high noise/signal strength on top of the line.

Two problems can occur when locating with a null antenna:

- Varying amounts of airborne interference signals exist in the field along the line route. These signals range from VLF to electric line current (60 Hz) to radio frequencies. These interference signals can dramatically distort the locating signal that runs down the line from the transmitter. This distortion can cause the instrument to inaccurately locate up to 50" of the top center of the pipe or cable.

  For example, overhead power lines, especially high-power transmission lines, can cause the locator to hum, to have no left-to-right null arrow change and/or no clear null point, or no signal at the locale.
- If two lines run parallel and are relatively close together (approximately 8' or less depending on depth of cover), and/or bonded together, the transmitted signal can be on both lines. When this occurs, the signals from both lines can merge or blend together, causing major distortion. If a null vertical antenna is used to locate these lines, the receiver often inaccurately indicates that there is only one line present, not two, and the line will appear to be located between the actual location of the two lines.
- If the null vertical antenna receiver has left/right arrows, the arrow switchover can occur between the lines. In some cases, one arrow will pop up and the other never does. In neither case will you get an accurate result.

To accurately locate under these interference conditions, the receiver must be capable of operating in a dual horizontal peak antenna mode and have appropriate electronics and software. Changing to this mode of operation will reject the interference and accurately locate target lines. In this mode, lower transmission frequencies (below 10K Hz) and lower manual gain settings improve locator performance.

Depth readings taken with an electronic locator are most accurate when taken in the active transmitter mode.

Operating steps for this unit are listed below:

Step 1 Apply the transmitter signal.

Step 2 Turn unit on.

Step 3 Select frequency.

Step 4 Select peak and line modes.

Step 5 Adjust gain as necessary.

Step 6 Keeping locator unit in a vertical position, sweep the area.

Step 7 Mark point of the strongest signal.

Step 8 At the mark, rotate locator 90 degrees until minimum signal (null) is achieved.

Step 9 Rest the receiver on the ground. Depress the depth key. The display will show the target depth.

When measuring conductor depth by triangulation, begin by determining the centerline of the conductor. There are two ways to estimate depth. Once the centerline is known, position the

Classroom

Discuss the two problems that can occur when locating with a null antenna. Have a trainee explain how to accurately locate when there is interference.

Audiovisual

Show Transparency 6 (Figure 10). Ask a trainee to explain the process of triangulation.

Laboratory

Have trainees practice using an electromagnetic locator and both methods of triangulation to determine depth.

Teaching Tip

See the Teaching Tip at the end of this module.

receiver (split box in this example) directly over and parallel to the line. Holding the receiver as close to the ground as possible, tilt the receiver at a 45-degree angle.

> **NOTE**
> Perform this step carefully as any deviation from the 45-degree angle will affect the depth analysis.

With the receiver at 45 degrees and facing the centerline of the target conductor, back away and mark the point where the minimum signal is present. Complete the same process on the opposite side of the target line. Measure the distance from the receiver to the centerline. Subtract the height of the receiver above the ground (approx. 5"). This value will be the distance of the target line below ground level. *Figure 10* shows measuring depth by triangulation.

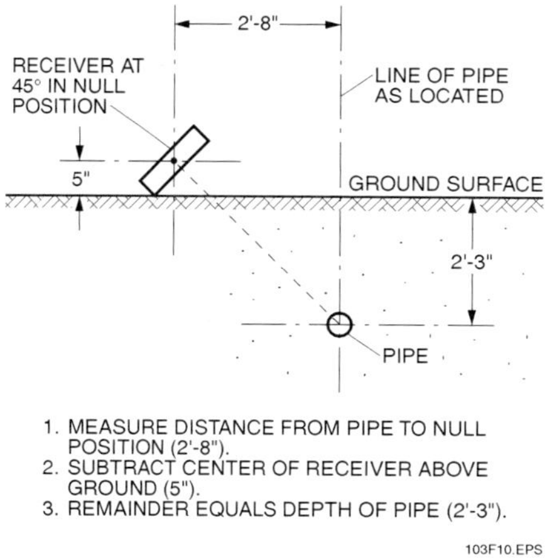

Figure 10 ◆ Measuring depth by triangulation.

A second method of triangulation, with no math or 45-degree angles, is the 100.0-70.0-100.0 method using a dual horizontal peak antenna. Put the receiver on the top centerline and adjust the sensitivity to exactly 100.0 (or to 99.8 if it is a digital display that does not go to 100). Then move perpendicular to the right of the line (see *Figure 11*) until the sensitivity is exactly 70.0. Mark the spot. Move the receiver back to the centerline and make sure it is still reading 100.0 (99.8). Move left

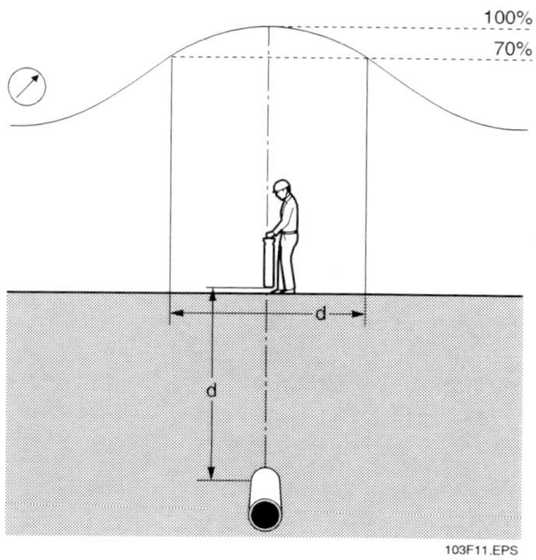

Figure 11 ◆ 100-70-100 measuring method.

of the line in the same manner until the reading is 70.0. Mark the spot. The distance between the two spots is the depth to the center of the pipe/signal.

> **NOTE**
> All locators sense depth to the center of the signal, roughly the center of the pipe or cable. Subtract half of the pipe diameter from the depth measured between the spots to get actual cover depth.

4.5.3 Transmitter Signal Effectiveness

The effectiveness of locator-transmitted signals and, therefore, the quality and distance of the signal, varies according to a number of signal characteristics. These variables include the following:

- *Conductor metal* – Signals from a locator transmitter unit are impacted by a number of different factors. Transmitted signals will travel a greater distance if the metal is a good conductor. A list of conducting materials that must often be located or traced include gold, silver, copper, aluminum, steel, cast iron, and ductile iron. Gold is the best conductor and ductile iron is the least effective conductor.
- *Line size* – The physical size of the line to which the signal is applied also determines the distance the signal travels along that line. The

Instructor's Notes:

smaller the line, the further the signal runs. Smaller lines offer less surface area to ground and less resistance to current flow than do larger lines.

- *Line continuity* – The better the continuity of a line, the further a signal is able to travel along the line. More efficient lines are those that contain fewer splices, fewer welded connections, and a limited number of mechanical or physical couplings tying smaller sections of line together.
- *Coating integrity and quality* – The quality and integrity of line coating affects how well the line carries a signal. The better the coating/sheath, the further the signal runs.
- *Signal frequency* – Lower frequency signals travel better than higher frequencies, assuming the signal will travel down the carrier line. Higher frequency signals tend to travel shorter distances and often experience a greater loss of signal strength to other lines, fences, and rebar in nearby structures.
- *Soil type* – Different soil types have a different effect on signal travel. The more organic matter, the further the signal runs, up to a point (signal distance is shorter in sand). Beyond that point, too much organic matter decreases the effective travel distance.
- *Soil moisture* – Soil moisture also impacts transmitter signal travel. The more water in the soil, the further the signal travels. As with soil types, this is true only up to a point. Very dry soil or very wet, swampy conditions increase signal loss, reduce effectiveness, and shorten the distance a signal will travel.
- *Transmitter output* – As the transmitter output signal strength increases, the signal travels further and can be detected at a greater distance from the input location. Signal travel is even more effective with proper **impedance matching**.
- *Signal interference* – As interference from other signals and other lines decreases, the further the transmitter signal runs and can be detected. Dual antennas increase the unit's capability to pick up weak signals.
- *Input connection* – The type of signal input connection has a strong relationship to signal travel. Signals run further if they are input through a direct connection, which results in 100% of the transmitter signal strength being input onto the line. In comparison, clamping results in only 60-90% of the signal from the transmitter entering the line. Induction, the least effective, only transfers between 10-50% of the signal from the transmitter to the line.

CAUTION

Be sure to secure the transmitters against theft.

4.5.4 Finding Foreign Lines in a Right-of-Way (ROW) – Parallel or Crossing

There are three ways to detect foreign lines in the ROW, whether they are parallel or crossing. Some companies require the use of two, or all three, methods.

The first method involves the use of one of the stick metal detectors. Sweep the area wider than you will be excavating. Follow these general directions plus the manufacturer's instructions and training. Remember that these locators detect metal pipes and objects, not "hot" electric cables, telephone lines, or fiber-optic cables.

CAUTION

Make sure you keep the locator away from steel toe boots.

Second, when using a split box locator to find lost or unknown metal pipes or cables in the inductive mode, the locator transmitter emits a signal that induces an electromagnetic field around the object you are seeking, which can be detected by the receiver. The locator is carried on a long handle provided for that purpose when initially locating the pipe or cable.

As you cross over the target, the locator receiver provides the user with both sound and meter indications that you are standing over a conductor. Beyond the target, both the sound and the meter reading indications decrease. To establish the path of the target, move 5-10 yards up and down the line and take additional readings. Drawing a line along these three points will show you the path of the conductor.

The third method (refer back to *Figure 9*) is the two-person sweep. Start 25'-30' apart with both transmitter and a receiver tuned to higher induction frequency (over 60 kHz). Walk along parallel with each other. When the individual with the receiver gets an indication of a signal, the individual with the transmitter should place it on the ground. The receiver is then used to pinpoint the target line. Mark the spot. The transmitter is then moved back and forth to locate the strongest signal point and that spot is marked.

Safety

Discuss various ways to keep transmitters from being stolen.

Remind trainees to keep the locator away from steel-toe boots.

Laboratory

Divide class into groups of three. Have each group member describe to the other members one of the three methods of detecting foreign lines in the right-of-way.

Teaching Tip

See the Teaching Tip section at the end of this module

Classroom

Ask a trainee to explain how to use an induction set. Have other trainees name advantages and disadvantages of this process.

Have a trainee explain how to use the direct connection method. Ask other trainees to explain how to improve earth grounding.

At this point both the transmitter and receiver are over the target pipe or cable. Continue this process until the entire area is swept in this direction. The area is then swept in a direction that is perpendicular to the first sweep. To ensure that no lines are overlooked, two additional sweeps oriented diagonally to the first two must be completed.

4.5.5 Using an Induction Set

Once you know the location of a starting point for the target pipe or cable, you can perform line tracing by placing the transmitter unit directly over and in line with the target. The transmitter and receiver should be at least 30' apart to prevent **air coupling** of the signal between the transmitter and receiver.

The stronger the transmitter signal, the further from its location you will be able to detect it with the receiver. Begin tracing the precise path of the target by walking along the supposed path and moving the receiver back and forth across the area. Peak signal strength will be present directly over the target, allowing you to accurately mark its location.

Continue marking the line in this manner until the signal is no longer clear. Move the transmitter down the line (referred to as **leapfrogging**) and resume the process.

Advantages gained from using this process include:

- The process is quick and easy.
- The process is often the only alternative.

Remember to use passive modes in the receiver to help get a starting point. Signals that can be located in the passive mode include:

- 60 Hz AC electrical power
- Cathodic protection signals
- VLF radio broadcast signals
- Cable TV signals at 31 kHz

Disadvantages encountered while using this process may stem from:

- Very little of the transmitted signal is detectable (10-50% only).
- Signal coupling on all lines prevents identification of a specific target line.

4.5.6 Using Inductive Couplers/Clamps

Clamping is a safe, convenient method of locating and tracing. A clamp or coupler is either standard equipment or an option with most locators.

An advantage to this approach is that clamping safely applies an easy-to-read signal to a live cable without interrupting the supply. A distinct disadvantage is that clamp-applied signals do not travel far when compared to direct-connected signals. This approach can be used on:

- Direct buried cables (energized or de-energized)
- Non-metallic conduit with cables inside

Surround the cable or non-metallic conduit with the coupler. Apply the signal as close to the dirt as possible. Remember that the signal travels away from the closest good ground, so you may have to go to the other end and apply the signal to get it to run.

> **NOTE**
> If cables are in metal conduit, connect directly to the metal and locate it just like a pipe. The use of a button magnet with a bolt or vise grips can be helpful to connect the transmitter lead to the metal conduit/pipe.

4.5.7 Using Direct Connection

Direct connection provides you with the strongest and cleanest signals for the receiver to detect. Clip directly to test leads, valves, fittings, or couplers.

> **NOTE**
> Be aware that if an electrical insulator kit has been installed, it will stop the signal and prevent you from continuing down the line.

Advantages in using this method include:

- *Frequency choice varies* – Lower frequencies are better as they are more easily detected. Check for the best current measurement frequency or signal strength for the location and conditions.
- *Signal power can be set at the optimum level* – Using the correct power level is important. Excess power might create a stronger signal that can be detected at a greater distance, but it also can spread onto other lines, creating possible confusion in detecting and interpreting the signal.

The ground cable and rod should be run out 10'-15' perpendicular to the side of the line to be located without crossing any other lines. The transmitter signal will couple onto any lines you cross. Use the passive signals to sweep the area for

Instructor's Notes:

foreign lines. Look for existing temporary line markers, permanent line markers, paint, flags, stakes or signs that might indicate the location of the target line. In the absence of any of these indicators, use the two-person sweep to locate the target line.

Often, the limiting factors for signal connection quality or signal distance include improper connections and/or improper grounding. To correct improper connections, make sure the connection lead is attached to shiny metal, not to a painted surface, rusty surface, or mud-covered area. On larger lines, a C-clamp, vise grips or a magnet fitted with a bolt can be used to connect the transmitter lead to the pipe.

To improve earth grounding, try one or more of the following:

- Use a longer ground rod (never use rebar as it is a poor conductor).
- Run water down the ground stake hole.
- Drive the ground stake in at an angle.
- Use multiple ground stakes jumpered, or bonded, together.
- Use an isolated anode as a ground.
- Use a totally isolated pipe as a ground.

4.5.8 Other Locating Techniques

To find a side leg, tee, or intersection, use the transmitted signal on the line or tune the receiver to one of the passive modes (60 Hz power or radio, CP signal or CATV). Put the receiver on dual horizontal peak antennas. Then move 4'-6' off the line and turn the receiver antenna blades parallel to the route of the line. Walk down the line and when you get to a tee or intersection, the receiver will indicate signal and noise to pinpoint the line.

There are several methods or features for accurately pinpointing lines in highly congested areas:

- *Double-ended connection* – The use of double-ended connections is invaluable in locating bonded, plumbed, and common lines in plants, stations, and yards (see *Figure 12*). The transmitter is connected as usual to the pipe or cable. Instead of attaching the ground lead on a ground stake 10'-15' perpendicular to the line route, use an extra-long extension ground wire spool (30'-1000'). Run the extension ground parallel to the line and 10'-15' off the suspected route of the line. Attach it to the end of the same line. The transmitter is now connected or hardwired to both ends of the line.

Classroom

Have trainees write out the steps for finding a side leg, tee, or intersection.

Audiovisual

Show Transparency 7 (Figure 12). Ask a trainee to explain the use of double-ended connections in pinpointing lines in congested areas.

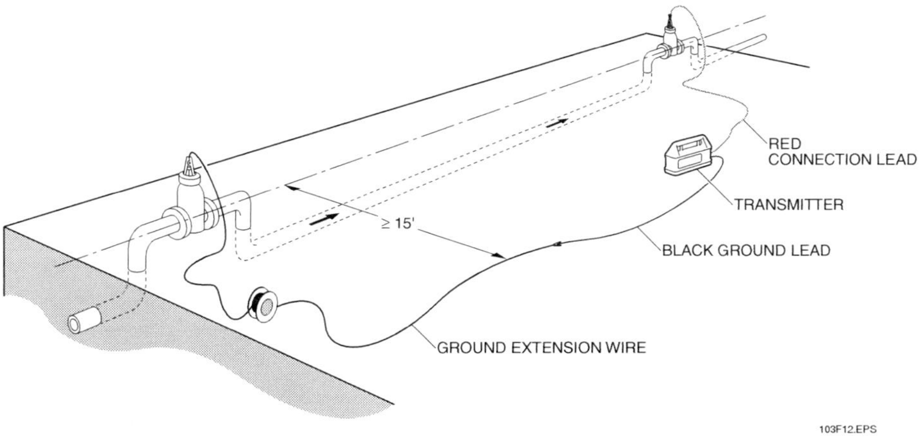

Figure 12 ◆ Double-ended connection.

Teaching Tip

See the Teaching Tip at the end of this module.

Safety

Ask a trainee to explain why dowsing must be physically verified.

Classroom

Write the various methods of locating and tracing on the board or an overhead. Ask trainees to list pros and cons for each method. Create a master list and discuss.

Performance Testing

Have each trainee perform Performance Task 1 to your satisfaction. Fill out a Performance Profile Sheet for each trainee.

- *Current measurement* – If the receiver has a current measurement of the transmitter signal, it can be used to pinpoint a specific line. Activate the current measurement function on the receiver and the line with the highest current measurement will usually be the line the transmitter is connected to or clamped around.
- *Current direction* – Some locators have a help mode called current direction. The theory is that the current will run in one direction on the line to which the transmitter is connected. Signal current will run in the opposite direction on all other lines, even if bonded or connected. Regardless of the number of lines in the ROW, the current on the target line will run in the opposite direction from all other lines.
- *Two transmitters on two different frequencies* – Finally, two transmitters can be used at the same time to pinpoint difficult close lines. Connect one transmitter, at a low frequency, to one of the lines. Using a different frequency, connect a second transmitter to the opposite end of the line. Use the receiver to toggle back and forth between the two frequencies to pinpoint each of the two lines. Accurate depth is usually impossible to determine in these congested situations. Use current measurement if a frequency is on both lines. The higher current will be measured on the line with the transmitter tuned to that frequency. For example, with 8k Hz on both lines, the higher current measurement should be detected on the line connected to the transmitter sending the 8k Hz signal.

4.5.9 Underwater Locating

When locating in shallow water, the water is invisible to the signal and has no serious impact on signal strength. However, the signal will dissipate over distance. Three methods are available to employ a submersible antenna for underwater locating:

- Wading with antenna
- Antenna taped to a plastic pole and suspended from a boat
- Antenna carried by a diver

Figure 13 shows a method of underwater locating.

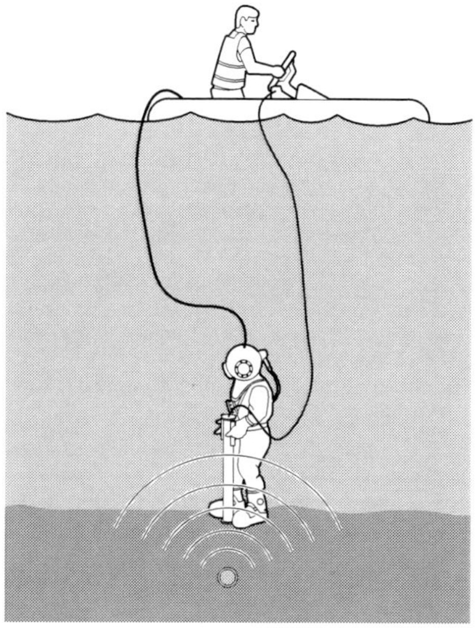

103F13.EPS

Figure 13 ◆ Method of underwater locating.

4.6.0 Other Methods

The oldest locating technique is dowsing. This method has often been used in the field to locate non-metallic lines without tracer wire or tracer tape. It works much better for some individuals than others.

Wires, welding rods, or other variants are used as locators in the performance of this obscure art. This alternate method of locating pipes and cables will not hold up in court or satisfy Federal or State One-Call requirements. While it is difficult to place a value on this method, it is cheap and easy to implement, but does not work equally well for everyone. Many companies will not allow dowsing.

> **WARNING!**
> Dowsing must always be physically verified using another method such as probing, pot holing, or vacuum excavating.

Instructor's Notes:

5.0.0 ◆ SEPARATION BETWEEN UNDERGROUND STRUCTURES

Crossing a pipeline with another pipeline or other construction is necessary in some locations. To protect pipelines carrying hazardous products, separation standards have been established by federal, state, and local governments. Companies have set up their own policies and standards in order to maintain safe operating conditions and to allow adequate room for maintenance and repairs.

5.1.0 Requirements for Separation

Requirements for separation between pipelines and other installations are divided into DOT minimum requirements and pipeline company requirements. DOT minimum requirements are mandated by law. Company requirements are set by company policy.

5.1.1 DOT Minimum Requirements

In a general description, *49 CFR 195.250* states that any pipeline installed underground must have a minimum of 12" clearance between the outside of the pipe and the extremity of any other underground structure. In situations where 12" of clearance is not practical, the distance may be reduced if the provisions for corrosion control are adequate. Clearance for drainage tile may be less than 12" but not less than 2".

Each main must be installed with enough clearance to allow proper maintenance and to protect it from damage that could result from being too close to another structure according to *49 CFR 192.325*. In the case of plastic pipeline, it must be installed with sufficient clearance and insulation to protect it from any heat source that might interfere with the serviceability of the pipeline.

5.1.2 Company Requirements

Pipeline companies holding rights-of-way agreements with landowners exercise authority over what is done in relation to pipeline safety within the boundaries of the right-of-way. When other companies wish to cross the original pipeline, they must secure the permission of the company holding the right-of-way agreement. In order to obtain permission to cross the pipeline right-of-way, the encroaching party must agree to the requirements and policies of the company holding the original agreement.

Although the legal requirement for separation is only 12", some companies require separation to be at least 24". Crossings by agricultural field drain tiles may be granted exceptions by company officials. It is often required that the new installation crosses underneath the original pipeline. Some companies also require the angle of crossing the pipeline to be as close to 90 degrees as possible to make it easier to locate the line and decrease interference problems.

5.1.3 Electrical Crossings

Low-voltage electrical cables are required to maintain a distance of 24" of separation and to be encased in rigid, non-metallic conduit. They are also required to have a color-coded tape buried above their installation.

High-voltage electrical cable separation requirements vary with the amount of voltage the cable carries. Distances range from a 24" separation for 600 volts or less to 42" or more for cable carrying over 40,000 volts. The company may require the encasement of electrical cables in red concrete.

Some companies do not allow electrical poles or guy anchors to be placed on the right-of-way. If poles and anchors are allowed, companies may require a 20' separation, or more, from the physical pipeline. Poles are only allowed if it is certain they will not interfere with maintenance or equipment. Companies also have requirements to make certain that overhead electrical lines are high enough not to interfere with pipeline maintenance equipment.

5.2.0 Reasons for Maintaining Separation

Pipeline companies require new installations to be separated from their pipeline by specified distances. These distances are the minimum allowed for the protection of the pipeline due to reduced cathodic protection, galvanic interference, and physical damage during excavation. The required distance is also an allowance for enough room to perform maintenance.

5.2.1 Reduction of Physical Damage During Excavation

The process begins with physically locating the pipe. Some companies require their representatives to be on site to oversee excavation being done on or near their lines. If the company representative must leave, work must cease until the individual returns. Company-specific excavation practices must be strictly adhered to.

Classroom

Distribute copies of *49 CFR 195.250* and *192.325*. Review the DOT minimum requirements.

Discuss the additional separation requirements that companies may impose. Ask trainees to explain the reasons for them.

Have trainees write out the requirements for electrical crossings. Ask trainees if they have any questions.

Ask a trainee to explain why new installations must be separated from existing pipelines.

Ask trainees to suggest reasons why companies require their representatives to oversee excavation being done on or near their lines.

Classroom

Go over the Review Questions. Answer any questions trainees may have on materials covered to this point.

Examination

Administer the Module Examination. Be sure to record the results of the Exam on Craft Training Report Form 200 and submit the results to the Training Program Sponsor.

Performance Testing

Ensure that the Performance Task has been completed and Performance Profile Sheets for each trainee are filled out. Be sure to record the results of the testing on Craft Training Report Form 200 and submit the results to the Training Program Sponsor.

Summary

Buried pipeline and cable must be protected. Federal and state governments provide legal support for company procedures so that people, property, and the environment are protected. Appropriate training, thorough inspection of work sites where underground pipes and cables are installed, investigation of problems, implementation of preventive measures, and timely response to abnormal operating conditions are all critical steps toward effective protection of lives and the environment.

Review Questions

1. The basic purpose of a One-Call system is to _____.
   a. protect people, property, and the environment
   b. protect company-owned pipeline systems
   c. ensure safety guidelines are followed
   d. help excavators locate all lines in an area

2. Excavating activities *do not* include _____.
   a. digging
   b. locating and tracing cables
   c. removal of aboveground structures by blasting
   d. backfilling

3. Typical underground pipes and cables include _____.
   a. only metallic pipelines and conduit
   b. pipelines over 36" in diameter and all cables
   c. all product lines except sewers and drains
   d. pipelines, utility cables, and other lines

4. When a One-Call notification is received, pipeline operators must dispatch someone to investigate the planned work site who is _____.
   a. bonded by the company
   b. a full-time employee of the company
   c. trained and qualified
   d. on-duty and available

5. To identify the proposed excavation site, temporary markers must be _____.
   a. white markers or white paint
   b. yellow flags
   c. visible from the edge of the proposed excavation site
   d. spaced no more than 24" apart

6. Pot-holing is more hazardous than probing because _____.
   a. of the potential for back injury
   b. of the potential for eye injury from flying debris
   c. sharp tools could cut into an electrical cable
   d. the tools are heavy

7. Lightweight, portable electronic locators are typically used to _____.
   a. locate nonmetallic objects
   b. locate underground utility lines
   c. trace energized utility lines
   d. determine depth of cover over fiber-optic cables

8. The two major components of a battery-powered electromagnetic locator are the _____.
   a. detector blade and receiver
   b. transmitter and receiver
   c. transmitter and control keypad
   d. receiver and digital display unit

9. Electromagnetic transmitters usually operate over a frequency range of _____.
   a. 17 Hz to 28 Hz
   b. 110 Hz to 220 Hz
   c. 9.8 kHz to 65 kHz
   d. 4 Hz to 500 kHz

10. If two lines run parallel under a moderate depth of cover and are relatively close together, _____.
    a. the transmitted signal will only couple to the target line
    b. signals from the target line will not blend with signals from the other line
    c. a null antenna will always accurately locate the target line
    d. the target line may appear to be located between the actual locations of the two lines

Instructor's Notes:

GLOSSARY

Glossary

This glossary will help you learn some terms specific to locating pipeline and cable. Trade Terms introduced in this module appear in italic text.

Air coupling: When using an induction set to locate target pipe or cable, air coupling of the signal between the transmitter and the receiver can occur if the two units are too close together. At distances of less than 30' apart, the airborne signal tends to be stronger than the signal coupled to the target pipeline and results in a false reading at the receiver.

Alignment sheets: Maps and drawings that identify the pipeline and its route in relation to direction, physical features, and political features.

Encroachment: Encroachment is the act of intruding upon the property of others. The federally mandated One-Call system was developed to ensure that any individual planning excavation in the area of pipeline rights-of-way follows government developed standards to protect pipeline installations from harm.

Impedance matching: Condition in which the external impedance of the connected load is equalized with the internal impedance of the signal source, resulting in a maximum transfer of energy. On some pipe and cable locator transmitters, built-in electronics can adjust the transmitted signal to optimize the available power output by automatically matching it to the impedance created by the line and soil conditions. Other units are manually adjusted.

Leapfrogging: Leapfrogging is the process of tracing the location of a target line by periodically moving the transmitter when the transmitted signal strength becomes too weak to provide accurate readings. Moving the transmitter down the line to a known location (leapfrogging) and continuing the location process, allows you to accurately trace and mark pipelines over longer distances than possible with a stationary transmitter.

Line locator: An electronic device used to define the lateral limits of the pipeline.

Line marker: A highly visible device used to communicate the location of the pipeline.

Pot holing: Pot-holing is a method of physically locating a pipeline by hand digging small holes until you contact the surface of the pipe.

Probe rods: Probe rods are hand tools made of metal or fiberglass that are used by some companies for locating underground pipelines and utility cables before undertaking any excavation work. Probe rods should have a smooth, rounded end to help minimize damage to pipeline coatings.

(Right-of-Way) ROW: Area specifically set aside to contain the pipeline and provide access for maintenance and inspection.

Test station: Specific area or location where structure-to-soil potential is measured.

Water-Vac (also hydro-vac): A Water-Vac unit is a tool developed to hydraulically bore and vacuum out a hole over a pipeline. It opens a clean, smooth hole with no damage to the pipeline or cable.

ACKNOWLEDGMENTS

Figure Credits

CST/Berger	103F05
Radiodetection www.radiodetection.com	103F06

Instructor's Notes:

MODULE 61103-02—TEACHING TIPS

The following are suggested activities or instructional methods to help you teach the material in this AIG.

General

When you call on someone to answer a question, the rest of the class relaxes or even tunes out because they expect that the question and answer will take place only between you and the trainee you called on. Instead, use this technique to involve more trainees in answering questions and to keep them on their toes.

1. Ask trainees to define a term or explain a concept.
2. After one trainee has answered, ask a trainee seated nearby if the answer is right. Then ask whether a trainee in the back of the room agrees.
3. Ask trainees to explain why they think an answer is right or wrong.
4. Use the session to clear up incorrect ideas and encourage trainees to learn from their mistakes.

Section 4.5.0 *Electronically Locating Pipes and Cables*

This exercise gives trainees the opportunity to practice locating underground utility lines. For this exercise, trainees will need a ferromagnetic locator, pencil, paper, appropriate personal protective equipment, and their Trainee Manuals. You will need to coordinate access to a field site with a buried utility line. Allow 45 to 60 minutes for this exercise, not including travel time to and from the site.

1. Review the steps for electronically locating an underground utility line. Familiarize trainees with the operation of the specific locator they will be using. Emphasize safety considerations.
2. At the field site, pair up trainees and assign each pair to an area. Have each pair electronically locate the utility line and write a brief report on their findings.
3. Ask trainees if they have any questions.

Section 4.5.3 *Transmitter Signal Effectiveness*

This exercise will familiarize trainees with signal characteristics. For this task, you will need an overhead projector, acetate sheets, and transparency markers. Trainees will need their Trainee Guides, pencils, and paper to take notes for discussion. You will need to refer to Section 4.5.3 of your Instructor Manual for a list of variables. Allow 30 to 45 minutes for this exercise.

1. Divide the class in half. On an acetate sheet, write "Team One" on one half and "Team Two" on the other.
2. Explain that you will name, one at a time, the variables that can affect locator-transmitted signals. Start with Team One. If a member of Team One can correctly explain the effect the variable has on a signal, award a point to Team One. Then it is Team Two's turn.
3. To ensure complete participation, each team member can answer no more than two times (adjust according to the number of participants and items to be identified). Each right answer is marked as a check on the acetate sheet. A wrong answer means the other team has an opportunity to answer and score a point.
4. Keep score and encourage class participation. When the game is finished, ask trainees whether they have any questions about transmitter signal effectiveness.

Section 4.5.4 *Finding Foreign Lines in a Right-of-Way—Parallel or Crossing*

This exercise will give trainees an opportunity to practice the three methods of detecting foreign lines. For this exercise, trainees will need a stick metal detector, a split box locator, flag markers, pencil, paper, and their Trainee Guides. You will need to provide access to a field site with parallel and crossing foreign lines in a right-of-way. Allow 60 to 75 minutes for this exercise, not including travel time to and from the site.

1. Review the three methods and remind trainees of safety considerations.
2. Pair up trainees. Have each pair take turns practicing the first and second methods. Then have them work together to do the third method, the two-person sweep.
3. Have trainees document their findings.
4. Check their findings. Answer any questions trainees may have.

Sections 4.5.5–4.5.8 *Induction Set, Direct Connection, and Other Locating Techniques*

This exercise will give trainees an opportunity to practice various methods of locating and tracing lines. For this exercise, trainees will pencils, paper, and their Trainee Guides. You will need to provide access to a field site and necessary equipment. Allow 60 to 75 minutes for this exercise, not including travel time to and from the site.

1. Review the process for using an induction set and direct connection as well as the other locating techniques explained in Section 4.5.8.
2. Have trainees practice using these techniques. Have them document their findings.
3. Check their findings. Answer any questions trainees may have.

MODULE 61103-02—ANSWERS TO REVIEW QUESTIONS

Answer	Section
1. a	2.1.0
2. b	2.0.0
3. d	3.0.0
4. c	4.2.0
5. a	4.2.0
6. c	4.4.2
7. b	4.5.0
8. b	4.5.2
9. d	4.5.2
10. d	4.5.2

CONTREN™ LEARNING SERIES — USER UPDATES

The NCCER makes every effort to keep these textbooks up-to-date and free of technical errors. We appreciate your help in this process. If you have an idea for improving this textbook, or if you find an error, a typographical mistake, or an inaccuracy in NCCER's Contren™ textbooks, please write us, using this form or a photocopy. Be sure to include the exact module number, page number, a detailed description, and the correction, if applicable. Your input will be brought to the attention of the Technical Review Committee. Thank you for your assistance.

Instructors – If you found that additional materials were necessary in order to teach this module effectively, please let us know so that we may include them in the Equipment/Materials list in the Instructor's Guide.

Write: Curriculum Revision and Development Department
National Center for Construction Education and Research
P.O. Box 141104, Gainesville, FL 32614-1104

Fax: 352-334-0932

E-mail: curriculum@nccer.org

Craft ______________________ Module Name ______________________

Copyright Date __________ Module Number __________ Page Number(s) __________

Description

____

____

____

(Optional) Correction

____

____

(Optional) Your Name and Address

____

____

Module 61104-02

Measure Pit Depth and Wall Thickness

(CT 8.1, 8.2, and 8.3)

Measure Pit Depth and Wall Thickness Annotated Instructor's Guide

Module 61104-02

MODULE OVERVIEW

This module introduces trainees to the tools used to determine the severity of corrosion damage. Trainees will learn to measure the pits and the remaining wall thickness to determine the urgency of the need for repair.

PREREQUISITES

Please refer to the Course Map in the Trainee Module. Prior to training with this module, it is recommended that the trainee shall have successfully completed the following:

Core Curriculum; Pipeline Core, Modules 66101-02 and 66102-02; Pipeline Corrosion Control Level One, Module 61103-02

OBJECTIVES

Upon completion of this module, the trainee will be able to:

1. Using a pit gauge, measure pitting or other metal loss features to pipeline structures and identify conditions requiring corrective action (CT 8.1).
2. Use a D-Scan Ultrasonic Thickness (UT) Meter to accurately collect and record nominal wall thickness readings on a pipeline or other coated structure (CT 8.2).
3. Obtain accurate measurements of localized corrosion to be used for calculating remaining pipe strength per B31G or RSTRENG (CT 8.3).

PERFORMANCE TASKS

Under the instructor's supervision, the trainee should be able to:

1. Accurately measure pit depth using a pit gauge (CT 8.1).
2. Properly measure a corroded area of pipe (CT 8.3).
3. Using a handheld ultrasonic meter, accurately measure wall thickness (CT 8.2).

NCCER STANDARDIZED CRAFT TRAINING PROGRAM

The National Center for Construction Education and Research (NCCER) provides a standardized national program of accredited craft training. Key features of the program include instructor certification, competency-based training, and performance testing. The program provides trainees, instructors, and companies with a standard form of recognition through a National Craft Training Registry. The program is described in full in the *Guidelines for Accreditation*, published by the NCCER. For more information on standardized craft training, contact the NCCER by writing us at P.O. Box 141104, Gainesville, FL 32614-1104; calling 352-334-0911; or e-mailing info@nccer.org. More information may be found at our Web site, www.nccer.org.

HOW TO USE THIS ANNOTATED INSTRUCTOR'S GUIDE

Each page presents two sections of information. The larger section displays each page exactly as it appears in the Trainee Module. The narrow column ties suggested trainee and instructor actions to each page and provides icons to call your attention to material, safety, audiovisual, or testing requirements. The bottom of each page includes space for your notes.

Teaching Tip

If you see the Teaching Tip icon, that means there is a teaching tip associated with this section. Also refer to any suggested teaching tips at the end of the module.

SAFETY CONSIDERATIONS

Ensure that the trainees are equipped with appropriate personal protective equipment. Emphasize the importance of following all safety precautions and procedures when working with power tools.

PREPARATION

Before teaching this module, you should review the Module Outline, Objectives, Performance Tasks, and the Materials and Equipment List. Be sure to allow ample time to prepare your own training or lesson plan and gather all required equipment and materials.

MATERIALS AND EQUIPMENT LIST

Transparencies
Markers/chalk
Blank acetate sheets
Transparency pens
Pencils and scratch paper
Module Examinations*
Performance Profile Sheets*
Overhead projector and screen
Whiteboard/chalkboard
Appropriate personal protective equipment
Copies of your company's policies and procedures manual
Copies of *49 CFR Part 195 (Liquid)* and *49 CFR 192 (Gas)*
Copies of Perfomance Verifications for Covered Tasks 8.1, 8.2, 8.3**
Mechanical and dial pit gauges, some with worn points
Sections of pitted pipe, corroded, and coated or uncoated pipe
Dial pit gauges with manufacturer's instruction
Mechanical pit gauges with manufacturer's instruction
Ultrasonic meters with operating manuals
Velocity of sound charts
Couplants
Graph paper
Carbon paper
Lumber pencils
Pipeline inspection report forms
Pipe cleaning materials

* Located in the Test Booklet.
** Available for download at www.nccer.org.

TEACHING TIME FOR THIS MODULE

An outline for use in developing your lesson plan is presented below. Note that each Roman numeral in the outline equates to one session of instruction. Each session has a suggested time period of 2½ hours. This includes 10 minutes at the beginning of each session for administrative tasks and one 10-minute break during the session. Approximately 5 hours are suggested to cover *Measure Pit Depth and Wall Thickness (CT 8.1, 8.2, and 8.3)*. You will need to adjust the time required for hands-on activity and testing based on your class size and resources.

Topic	**Planned Time**
Session I. Pit Gauges	
A. Types of Pit Gauges	______
1. Mechanical Pit Gauges	______
2. Laboratory	______
3. Dial Pit Gauges	______
B. Measuring Pit Depth	______
1. Using Mechanical Pit Gauges	______
2. Using Dial Pit Gauges	______
3. Performance Testing	______
C. Measuring Pit Length	______
D. Gathering Remaining Wall Strength Data	______
E. Evaluating Localized Corrosion Defects	______
F. Interaction Rule	______
1. Example Corrosion Array #1	______
2. Example Corrosion Array #2	______
3. Example Corrosion Array #3	______
4. Interaction Between Multiple Defect Arrays	______
5. Performance Testing	______
Session II. Ultrasonic Thickness (UT) Measurements	
A. Basic Ultrasonic Meter Operation	______
1. Digital Ultrasonic Thickness Gauge Function	______
2. Through-Coating Measurement	______
3. Using Couplants	______
4. Data Analysis	______
B. Meter Calibration	______
1. Calibration of Sound Velocity	______
2. Effects of Temperature on Calibration	______
3. Transducer Selection	______
4. Calibration to a Known Thickness, 1-Point	______
5. Calibration to a Known Thickness, 2-Point	______
6. Calibration to a Known Velocity	______
7. Performance Testing	______
C. Document Extent and Location of Damage	______
D. Summary	______
1. Summarize module	______
2. Answer questions	______

E. Module Examination
   1. Trainees must score 70% or higher to receive recognition from the NCCER.
   2. Record the testing results on Craft Training Report Form 200 and submit the results to the Training Program Sponsor.

F. Performance Testing
   1. Trainees must perform each task to the satisfaction of the instructor to receive recognition from the NCCER.
   2. Record the testing results on Craft Training Report Form 200 and submit the results to the Training Program Sponsor.

Module 61104-02

Measure Pit Depth and Wall Thickness

(CT 8.1, 8.2, and 8.3)

Instructor's Notes:

Instructor's Notes:

Introduction to the Pipeline Operator Qualification Rule, API, and NCCER

It could be said that petroleum products keep America running. Petroleum products like oil and natural gas are the principal sources of energy in the United States and are used in many manufacturing and chemical processes. Petroleum products are primarily transported by pipeline. There are more than a million miles of liquid and natural gas pipeline in this country, ranging in diameter from two inches to four feet. These pipelines bring crude oil and natural gas from production fields to processing plants and refineries—and ultimately to the consumer.

Pipeline Operator Qualification

Petroleum products are considered hazardous, so pipeline personnel must take extra precautions and follow specific procedures when working with them. To help ensure the safe handling and transportation of petroleum products, the United States Department of Transportation (DOT) mandates safe practices as federal regulations.

One such regulation, the Operator Qualification (OQ) rule, requires pipeline operators to develop a written qualification plan to evaluate their personnel and their contract personnel (see *49 CFR Part 192 Subpart N—Gas,* and *49 CFR Part 195 Subpart G—Liquid*). The evaluation hinges on the operator's ability to perform certain covered tasks and to be able to recognize and respond to abnormal operating conditions that may be encountered while performing these covered tasks.

Covered Tasks

A pipeline performance task is deemed *covered* if the activity passes the following four-part test:

1. Is it performed on a pipeline facility?
2. Is it an operations or maintenance task?
3. Is it performed as a requirement of *49 CFR Part 195 (Liquid)* or *49 CFR Part 192 (Gas)*?
4. Does it affect the operation or integrity of the pipeline facility?

If the answer to all four questions is *yes,* then the activity is considered *covered* and must figure into the evaluation process of pipeline personnel.

American Petroleum Institute

The OQ rule and its related covered tasks provided the catalyst to develop a unified pipeline training and assessment program—so that pipeline operators and owners have ready access to the assessment and training tools they need. This effort was the direct result of the vision and leadership provided by The American Petroleum Institute (API) and API members.

Efforts to unify the industry are nothing new to API. Indeed, the impetus for forming the Institute in 1919 was the need to standardize engineering specifications for drilling and production equipment. Today, the most pressing issues revolve around public perceptions and government policies toward the industry. Speaking with one voice on these issues has become as essential as having interchangeable parts in the field. API's members determine the petroleum industry's positions on public policy as well as the standards that govern its day-to-day operations worldwide.

Consortium on Operator Qualification

API assembled a focus group whose primary purpose was to identify covered tasks and provide compliance recommendations to pipeline operators.

The Consortium on Operator Qualification (COOQ) became the guiding force for the identification and development of covered tasks for the liquid pipeline industry. Their work product, API publication 1161, *Guidance Document for the Qualification of Liquid Pipeline Personnel,* formed the basis for the resulting Pipeliner Training and Assessment Program.

The covered task list created in API publication 1161 has now evolved to include gas covered tasks and is maintained by NCCER. The most current covered task list can be downloaded at www.nccer.org.

National Center for Construction Education and Research

With the guidance and support of API, its members, and the Consortium on Operator Qualification, the National Center for Construction Education and Research (NCCER) was able to design and implement the Pipeliner Training and Assessment Program.

NCCER is a not-for-profit 501(c)(3) education foundation established in 1995 by the world's largest and most progressive construction and maintenance contractors and national contractor associations. NCCER was founded to address the

severe workforce shortage facing the construction and maintenance industries and to develop a standardized training process and curricula. Today, NCCER is supported by hundreds of leading construction and maintenance companies, manufacturers, and national associations and developed more than 35 construction and maintenance titles.

NCCER also maintains a National Registry, which contains the records of trainees and instructors who have received NCCER training. All completion information for every trainee is sent to the NCCER and filed within the Registry. The Registry can then confirm training and skills for workers as they move from company to company, state to state, or even within their own company.

Standardized Curriculum (NCCER's Contren™ Learning Series)

NCCER curricula are developed using the expertise of industry subject matter experts (SMEs). The use of SMEs as an integral part of every new development or revision ensures that the curriculum is current, industry-driven, and indicative of industry practices. The development and/or revision process is facilitated from start to finish by NCCER project managers, who work closely with the SMEs and a select team of technical writer(s) on the development and presentation of the curriculum.

Pipeline: Five Training Paths

The pipeline curriculum focuses on essential skills and competencies within five skill areas:

- Corrosion Control
- Electrical and Instrumentation
- Maintenance
- Mechanical
- Operations (Field and Control Center)

SMEs from the pipeline industry have determined the skills and competencies needed based on API's covered tasks and the consensus of the group/industry. Specific company information is generally not incorporated into the curriculum unless the SMEs determine that the information is the industry standard.

Training Modules and Covered Tasks

The smallest unit of instruction in the curriculum is a *module*. Each curriculum is composed of modules (the equivalent of stand-alone chapters) that are logically sequenced into complete *levels* of instruction. Every module is task-oriented and requires the successful completion of a written (knowledge-based) examination and a performance (hands-on) test. Modules may be individually selected and/or combined for efficient task training. Modules in the pipeline curricula, to the extent possible, will align with API-developed covered tasks. While not all modules will relate to covered tasks, all covered tasks will relate to at least one module.

Benefits of Training to the Pipeline Marketplace

Applying NCCER's standardized training process allows pipeline owners and contractors in the United States to establish the benchmark for acceptable skill and competency levels within the pipeline industry. This process will assist training organizations in developing and/or improving standardized pipeline training programs that effectively match their resources and operating conditions while responding to regulatory demands and to the needs of the industry. The process will also recognize training organizations that meet the minimum standards of competence, warranting regulatory, public, and professional confidence. This in turn will foster national unity within the pipeline industry for the mutual benefit of individual technicians, owners, contractors, and ultimately pipeline-users and society.

Additional Resources

To review the Pipeliner Training and Assessment Program Implementation and Resource Guide and the latest version of pipeline covered tasks, visit NCCER's Web site at www.nccer.org, or contact:

National Center for Construction Education and Research
PO Box 141104
Gainesville, FL 32614-1104
(888) 622-3720
Web site: www.nccer.org

For more information on API, please contact:

American Petroleum Institute
1220 L Street, Northwest
Washington, D.C. 20005
(202) 682-8000
Web site: www.api.org

To order *API Pub 1161*, contact Global Engineering Documents by calling 1-800-854-7149 or visiting their Web site at www.global.his.com.

Additional sources of information on the pipeline operator qualification regulation include:

DOT – Office of Pipeline Safety
www.ops.dot.gov

Transportation Safety Institute
www.tsi.dot.gov

Instructor's Notes:

COURSE MAP

This course map shows all of the modules in the first level of the Pipeline Corrosion Control curriculum. The suggested training order begins at the bottom and proceeds up. Skill levels increase as you advance on the course map. The local Training Program Sponsor may adjust the training order.

PIPELINE CORROSION CONTROL LEVEL ONE

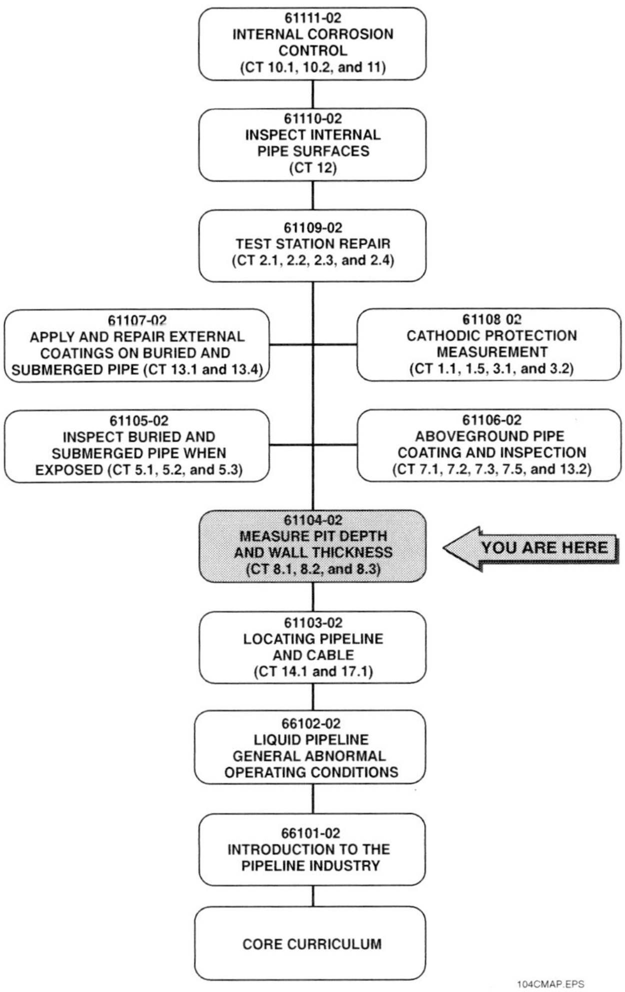

Homework

Assign reading of Module 61104-02.

Instructor's Notes:

MODULE 61104-02 CONTENTS

Figures

Tables

Instructor's Notes:

MODULE 61104-02

Measure Pit Depth and Wall Thickness (CT 8.1, 8.2, and 8.3)

Objectives

When you have completed this module, you will be able to do the following:

1. Using a pit gauge, measure pitting or other metal loss features to pipeline structures and identify conditions requiring corrective action *(CT 8.1)*.
2. Use a D-Scan Ultrasonic Thickness (UT) Meter to accurately collect and record nominal wall thickness readings on a pipeline or other coated structure *(CT 8.2)*.
3. Obtain accurate measurements of localized corrosion to be used for calculating remaining pipe strength per B31G or RSTRENG *(CT 8.3)*.

Prerequisites

Before you begin this module, it is recommended that you successfully complete the following: Core Curriculum; Pipeline Core, Modules 66101-02 and 66102-02; Pipeline Corrosion Control Level One, Module 61103-02.

Required Trainee Materials

1. Pencil and paper
2. Appropriate personal protective equipment

1.0.0 ◆ INTRODUCTION

Pipeline technicians must be aware of **localized corrosion** in the form of **pitting**. A single pit in an otherwise healthy system can cause the entire system to fail. Pitting is often difficult to detect because it can be covered by coatings or only a small hole may show on the surface of the piping or equipment. Even when general corrosion rates are low, microbes can accelerate localized pitting and cause pitting to go through a steel pipeline wall fairly quickly. Pits can undercut the surface, eating into the metal so that only a fragile layer remains on the surface; they can cut deep fissures into the piping with only tiny cracks visible at the surface. Identifying these areas requires skill and vigilance. Once identified, both the pits and the remaining wall thickness must be measured to determine the urgency of the need for repair.

Among the commonly used tools are pit gauges for testing the actual pit depth and ultrasonic gauges for determining wall thickness. This module explains types of pits and the tools used to determine the severity of corrosion damage.

2.0.0 ◆ PIT GAUGES

One of the most common tools used to check the extent of corrosion to a pipe is the pit gauge. Most pit gauges are easy to use and to carry. These gauges, if properly maintained, can provide accurate measurements. However, the accuracy of any pit measurement is only as reliable as the instrument that is used to take the measurement. Before a pit gauge is used for measurement, it must be inspected for any damage or worn points. During this inspection, always closely inspect the point of the instrument for wear. If a pit gauge does have a worn point, the pit gauge should be replaced. Only use an instrument that is in excellent condition; otherwise, the measurements taken during your inspection may not be accurate. Before using the pit gauge, clean the pit in accordance with your company's approved policy.

Materials

Ensure that you have everything required to teach the course. Check the Materials and Equipment List at the front of this module.

Classroom

Explain that terms shown in bold are defined in the Glossary at the back of this module.

Audiovisual

Show Transparency 1, Course Objectives.

Show Transparency 2, Performance Tasks.

Teaching Tip

See the general Teaching Tip at the end of this module.

Classroom

Ask trainees to describe the dangers of pitting. Have a trainee explain why pitting is often difficult to detect.

Laboratory

Provide a variety of pit gauges, some with worn points, and have trainees inspect the points for wear. Ask trainees to determine which gauges should be replaced.

Audiovisual

Show Transparency 3 (Figure 1). Ask a trainee to explain how to use the depth measurement scale. Have a trainee point out the formulas on the pit gauge.

Classroom

Ask a trainee to identify whether the dial pit gauge or mechanical pit gauge is more accurate. Ask a trainee to explain the drawbacks of dial pit gauges.

2.1.0 Types of Pit Gauges

Throughout the petroleum industry, several types of mechanical pit gauges are used to measure corrosion damage. Two commonly used types are discussed in the following sections.

2.1.1 Mechanical Pit Gauges

When corrosion or damage is discovered, the corrosion individual or maintenance team member must determine the remaining wall thickness of the pipe in order to set the priority for repair or replacement of the corroded section of pipe. The common mechanical pit gauge *(Figure 1)* is a small, sturdy tool. It is easily used to make a quick determination of the extent of corrosion damage present. These gauges are usually made of polished stainless steel and can easily be carried in the tool kit. Useful formulas, tables, and data giving outside diameter (OD), wall thickness, and weight per foot for various sizes of pipe are usually included with mechanical pipe pit gauges. Care must be taken to use these gauges properly and to recognize and understand their limitations.

Inaccurate readings can result unless the gauge is aligned properly and placed on an even surface. Where severely pitted or partially coated pipe and associated structures are involved, these handheld pit gauges provide the user with only an approximation of pit depth. The depth measurement scale on the mechanical pit gauge is marked in sixteenths of an inch and in **mils** (thousandths of an inch).

2.1.2 Dial Pit Gauges

When pit depth measurements are critical and accuracy is a priority, dial pit gauges, designed for a more accurate measurement of corrosion pit depth, may be used. Most dial pit gauges are capable of measuring depth to extremely close tolerances. On the down side, these indicators are more fragile than mechanical gauges and should be handled with care to ensure their repeated accuracy. *Figure 2* shows a dial pit gauge positioned on a pipe.

2.2.0 Measuring Pit Depth

Use company-specified methods and procedures to clean the pit to ensure an accurate measurement with the pit gauge. To obtain an accurate reading, clean the pit gauge thoroughly before using it.

2.2.1 Using Mechanical Pit Gauges

Align the gauge lengthwise with the pipe to be inspected. To provide an accurate reading of the pits, the gauge must rest squarely on the pipe surface. Position the arm on the gauge so that the pointer is at the deepest part of the pit. Read and record the corresponding number from the scale of the gauge. Repeat this process to verify the deepest pit measurement. This also helps you obtain the average pit depth. Document your findings and make notifications according to your company's policies. *Figure 3* shows the proper technique for measuring the pit depth.

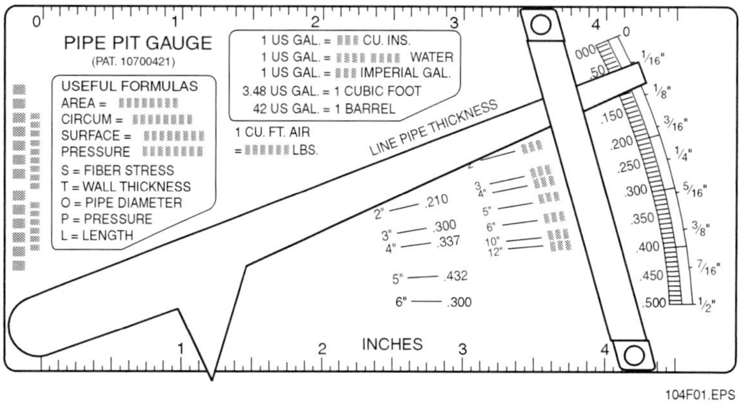

Figure 1 ◆ Mechanical pipe pit gauge.

Instructor's Notes:

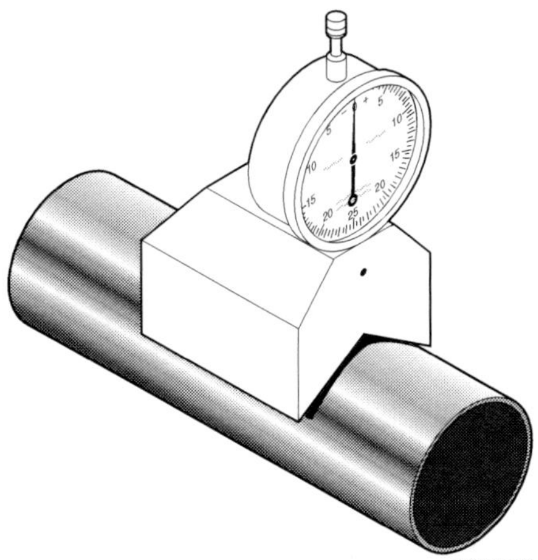
104F02.EPS

Figure 2 ◆ Dial pit gauge positioned on pipe.

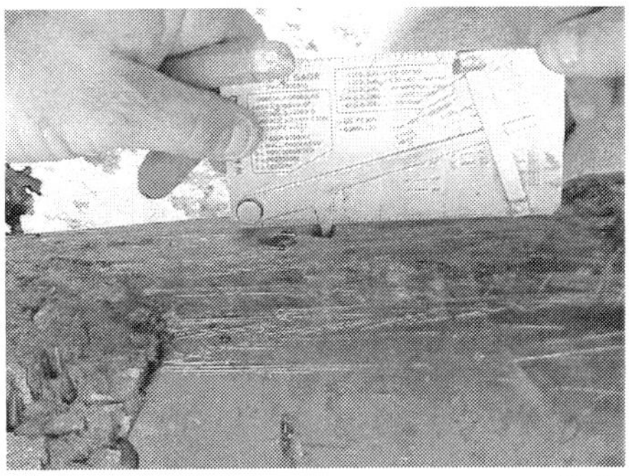
104F03.EPS

Figure 3 ◆ Proper technique for measuring pit depth.

2.2.2 Using Dial Pit Gauges

First, **calibrate** the dial pit gauge according to the manufacturer's instructions. Once the gauge is calibrated, place it squarely on the pipe surface. Push the pointer into the pit to be measured such that the control button is on the side of the gauge. This should register a reading on the gauge. Record this measurement. Continue this process until you have recorded an adequate sample of readings to establish a pit **profile**. When measuring multiple defects, always follow company guidelines. When you are finished, record your findings and make notifications according to your company's policies.

> **CAUTION**
> If pit depth is equal to 80% or more of the pipe company specification, the affected area of the pipe must be replaced immediately.

2.3.0 Measuring Pit Length

With the pointer of the pit gauge in the pit, observe the length of the pit on the bottom scale of the mechanical pit gauge (refer back to *Figure 1*). After this measurement is taken, make sure that the measurement is recorded on the appropriate company forms.

2.4.0 Gathering Remaining Wall Strength Data

To evaluate the structural integrity of pipelines with corrosion-based metal loss, pipeline operators use corrosion assessment methods called B31G and RSTRENG (Remaining Strength). The basis for these two calculation approaches has been in existence for more than 25 years. Both methodologies have been validated through burst testing of piping with corrosion and with in-service piping failures.

While acceptable, the B31G method is recognized to be very conservative in nature. The RSTRENG criterion was developed to offset this conservative characteristic. Pipeline operators are required by *49 CFR 192* and *195* to evaluate the remaining strength in any corroded pipeline using either ASME/ANSI B31G or RSTRENG. The B31G method calculates predicted failure points based on the maximum length and maximum depth of the corrosion defect on the pipe.

The RSTRENG approach combines this method with a more detailed profile of the defect by evaluating a greater number of depth measurements (minimum of three with one being the maximum depth in the defect) taken along the horizontal length of the defect. RSTRENG works equally well with single defects or with several interacting corrosion arrays. *Figure 4* shows a typical **river bottom profile** of the corrosion defect as determined through the RSTRENG method.

2.5.0 Evaluating Localized Corrosion Defects

If it is determined that the corrosion loss is localized, profile measurements of the corroded area must be obtained. A method to determine the appropriate number of measurements to profile is by creating a measurements grid of the corrosion region (see *Figure 5*).

Demonstration

Show trainees how to calibrate the dial pit gauge and place it on the pipe surface.

Safety

Stress the importance of replacing an area of pipe when the pit depth is equal to 80% or more of the pipe company specification.

Teaching Tip

See the Teaching Tip at the end of this module.

Performance Testing

Have each trainee perform Performance Task 1 to your satisfaction. Fill out a Performance Profile Sheet for each trainee.

Classroom

Ask a trainee to explain the B31G corrosion assessment method. Have a trainee explain the RSTRENG approach. Have trainees identify the pros and cons of each approach.

Audiovisual

Show Transparency 4 (Figure 4). Ask a trainee to point out the thinnest part of the pipe. Have a trainee point out the thickest part of the pipe.

Show Transparency 5 (Figure 5). Ask a trainee to explain how to interpret the measurement grid. Ask trainees if they have any questions.

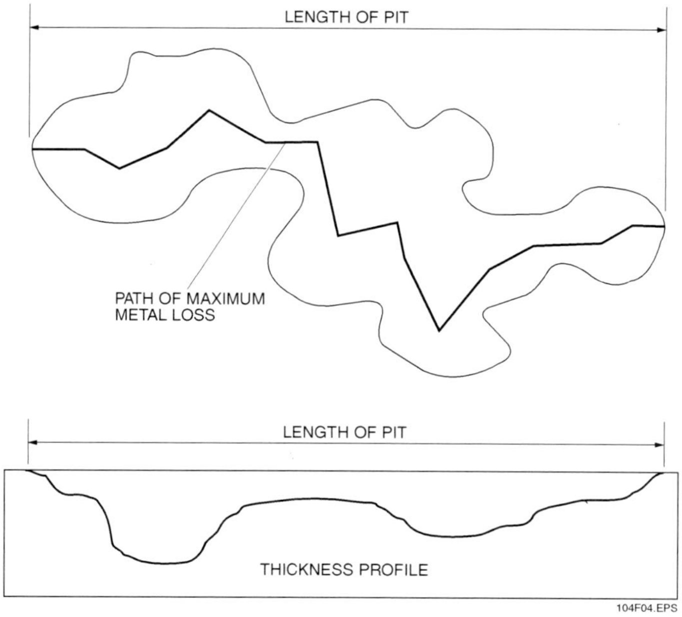

Figure 4 ◈ Typical river bottom profile.

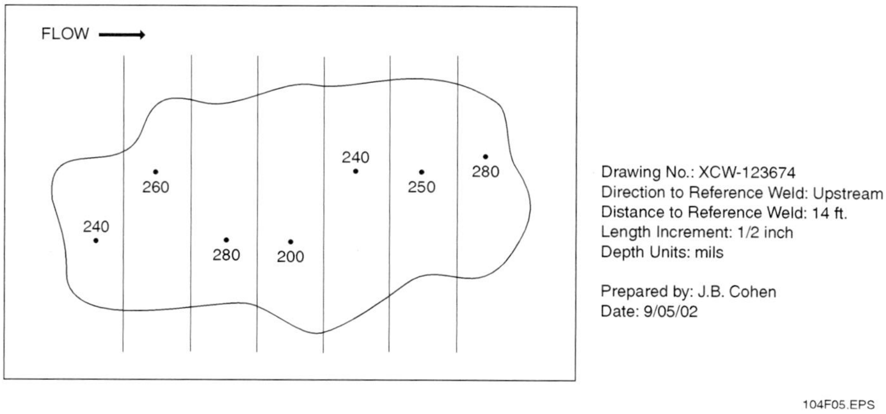

Figure 5 ◈ Corrosion measurement grid.

Instructor's Notes:

The following steps describe how to create a measurement grid of the external corroded area.

Step 1: Clean the area completely to aid in obtaining accurate information.

Step 2: Use graph paper (chart paper with grid lines) to make a transfer of the corroded area by rubbing the graph paper with carbon paper and a lumber pencil.

Step 3: Using the graph paper lines as a guide, measure and record the deepest pits in each segment. It is important to include flow direction, orientation, distance to the reference weld, date, and other company-required data on the measurement grid.

Step 4: Determine defect length using the interaction rule.

2.5.0 Interaction Rule

The interaction rule is the most widely used method of determining defect length, but there are other options. The version of the interaction rule illustrated in *Figure 6* is a guide to help determine if two or more metal loss defects are close enough together to interact as a single defect. The rule states that two or more defects may interact as one if the distance between them is less than the **longitudinal** length of the smallest defect.

When measuring multiple defects, remember that there is no one standard rule, so always follow your company's guidelines.

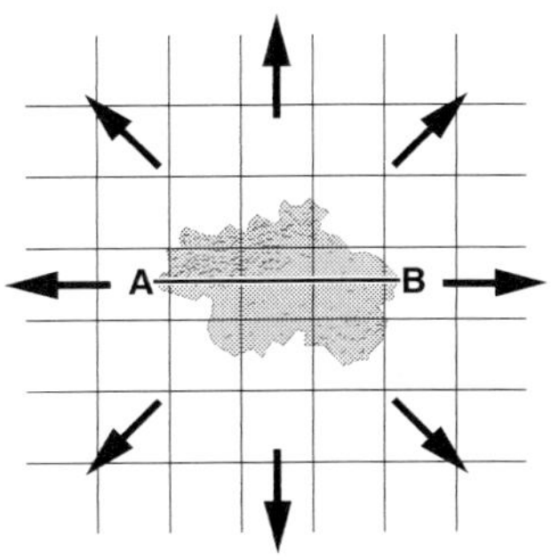

Figure 6 ◈ Interaction rule.

2.6.1 Example Corrosion Array #1

The longitudinal length of the shortest defect (B) in corrosion array #1 *(Figure 7)* is ¾". The distance between the two defects (G) is shown as ½". The two defects interact because G is less than the longitudinal length of the shortest defect. The total length of array #1 is 2¼".

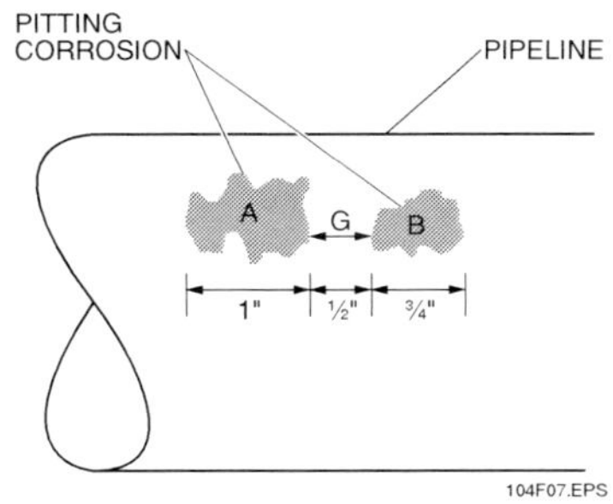

Figure 7 ◈ Corrosion array #1.

2.6.2 Example Corrosion Array #2

The longitudinal length of the shortest defect (F) in corrosion array #2 *(Figure 8)* is ½". The distance between the two defects (G) is shown as ¼". The two defects interact because G is less than the longitudinal length of the shortest defect. The total longitudinal length of array #2 is 1⅜".

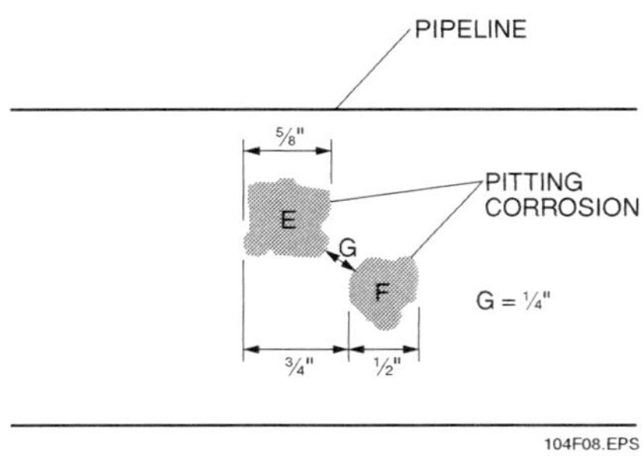

Figure 8 ◈ Corrosion array #2.

2.6.3 Example Corrosion Array #3

The longitudinal length of the shortest defect (D) in corrosion array #3 *(Figure 9)* is ¼". The distance between the two defects (G) is shown as ¼". The two defects interact because G is less than the longitudinal length of the shortest defect. The total longitudinal length of array #3 is ½".

Teaching Tip

See the Teaching Tip for Section 2.5.0 at the end of this module.

Audiovisual

Show Transparency 6 (Figure 6). Have a trainee explain when and how two defects may interact as one. Ask a trainee to explain the significance of Line AB.

Show Transparency 7 (Figure 7). Have a trainee use a transparency marker to label the areas of corrosion as A and B. Ask a trainee to label the distance between the two defects as G and measure the total length of the array. Ask a trainee whether the two defects interact, and explain why or why not.

Show Transparency 8 (Figure 8). Have a trainee point out the length of the shortest defect. Ask a trainee whether the two defects interact, and explain why or why not.

Show Transparency 9 (Figure 9). Have trainees, using pencil and paper, do the calculations to determine whether or not the two defects interact. Have a trainee explain to the class how the answer was obtained.

Audiovisual

Show Transparency 10 (Figure 10). Have a trainee explain whether or not the multiple defect arrays interact, and why.

Performance Testing

Have each trainee perform Performance Profile Task 2 to your satisfaction. Fill out a Performance Profile Sheet for each trainee.

Homework

Have trainees review Sections 3.0.0–4.0.0 and answer the Review Questions for the next session.

Materials

Ensure that you have everything required for teaching and testing during this session.

Classroom

Refer to Figure 11. Have a trainee explain how the digital ultrasonic gauge measures internal corrosion. Ask a trainee to explain how to use charts that show velocity of sound.

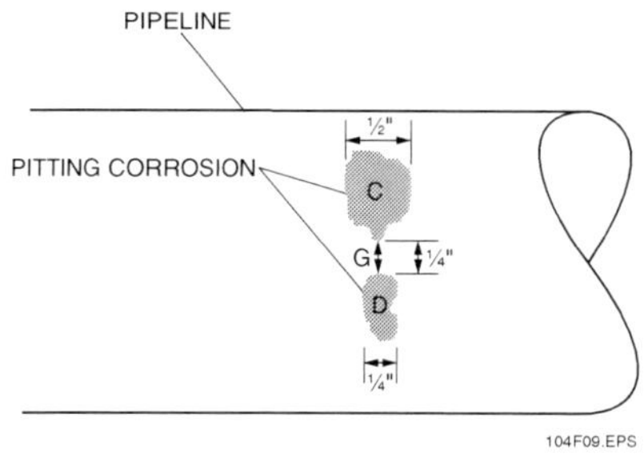

Figure 9 ◆ Corrosion array #3.

> **NOTE**
> The most conservative approach is to assume that the two corrosion defects, C and D, interact. In the example, the longitudinal length of D is equal to G. Although G is not less than the length of the shortest defect, as spelled out in the rule, it is usually assumed that the defects interact.

2.6.4 Interaction Between Multiple Defect Arrays

The interaction rule addresses interaction between defects, not defect arrays. *Figure 10* shows multiple corrosion defect arrays on a pipeline. These arrays, as shown, do not interact. The distances between the separate arrays are greater than the longitudinal length of the shortest defects in each array.

3.0.0 ◆ ULTRASONIC THICKNESS (UT) MEASUREMENTS

Testing on-line equipment is a common occurrence in most pipeline operations. A common testing method is digital ultrasonic thickness gauging for the measurement of remaining pipe wall thickness. Ultrasonic thickness gauging is widely used to determine remaining wall thickness, one factor in overall equipment integrity. This is especially useful where internal access to an equipment item is impractical.

Using a digital ultrasonic thickness gauge and available **transducers** (probes), the pipeline maintenance individual can quickly and easily measure and record the wall thickness on systems and equipment subject to corrosion damage. To avoid unnecessary downtime, and loss of production, these inspections can be made on in-service lines.

The use and calibration of general ultrasonic meters are discussed in the following sections.

3.1.0 Basic Ultrasonic Meter Operation

Ultrasonic thickness gauges, their basic operation, and management of gathered data is discussed in the following sections:

- Digital ultrasonic thickness gauge function
- Thru-coating measurement
- Data analysis

3.1.1 Digital Ultrasonic Thickness Gauge Function

The basic operation of the ultrasonic thickness gauge begins with the generation of an electrical energy pulse. The transducer probe converts this pulse into a high frequency sound wave. The sound wave emitted by the transducer probe as it rests on the surface of the pipe being measured, travels through the pipe wall and is reflected from the interior surface (**backwall**) of the pipe. *Figure 11* shows an ultrasonic thickness gauge in use.

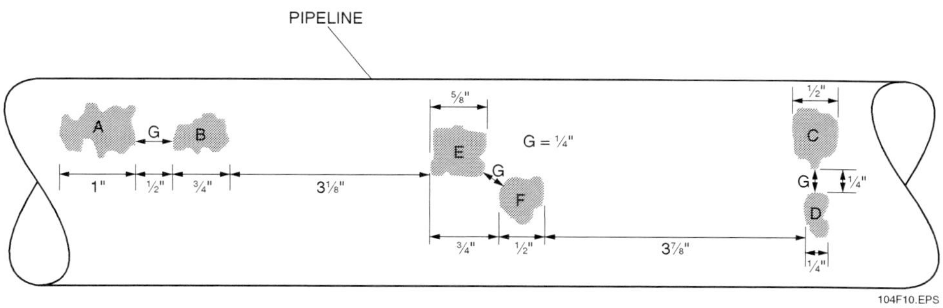

Figure 10 ◆ Multiple corrosion defect arrays.

Instructor's Notes:

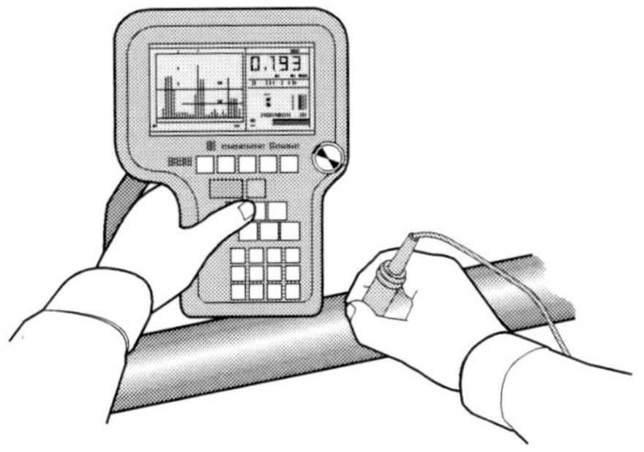

COURTESY OF ART LEACH, KRAUTKRAMER 104F11.EPS

Figure 11 ◆ Typical ultrasonic wall thickness gauge with transducer probe.

The reflected wave is picked up by the transducer probe upon its return to the surface side of the pipe. The wave is converted back into electrical energy and passed on to the gauge. Charts are available that show the **velocity of sound** through a given material. Using that value for the pipe material of construction, the time required for the emitted sound wave to travel to the interior wall of the pipe and return to the probe can be converted into a thickness value. Using the correct velocity of sound for the pipe material being measured is critical for accurate results.

Ultrasonic gauges are reliable tools for measuring equipment subject to internal corrosion. Where corrosion damage exists, the damaged area has a different molecular structure and is not a true part of the pipe or equipment material. The transmitted signal reflects off the line between the good damaged material and can be used to determine the thickness of the good material.

3.1.2 Through-Coating Measurement

Pipeline systems and equipment can also fall victim to environmental corrosion damage. Typically, aboveground equipment is protected by a thick layer of paint or one of many coating materials. This poses a problem for ultrasonic measurement because the operator does not know the exact thickness of the coating.

Another problem is caused by the extremely slow velocity of sound values for most coating materials. Because of this difference in velocity of sound values between the coating and the underlying metal, the ultrasonic thickness gauge measures the typical coating as being two to three times thicker than the actual thickness.This error could give the maintenance individual false data to work with.

In the past, accurate thickness readings could only be obtained by removing the coating and measuring only the pipe material itself. The paint-stripped area was then cleaned and re-coated, a costly process in terms of material and labor hours.

An improved method for measuring through-coating thickness is available with the use of dual-element transducer probes. These probes are designed to emit a more focused signal beam that is capable of more accurate readings in areas of varying thicknesses than single beam probes. *Figure 12* shows how the transducer probe is placed on the pipe to generate a readout of wall thickness.

Audiovisual

Show Transparency 11 (Figure 12). Ask a trainee to explain how the transducer probe aids in accurate measurement.

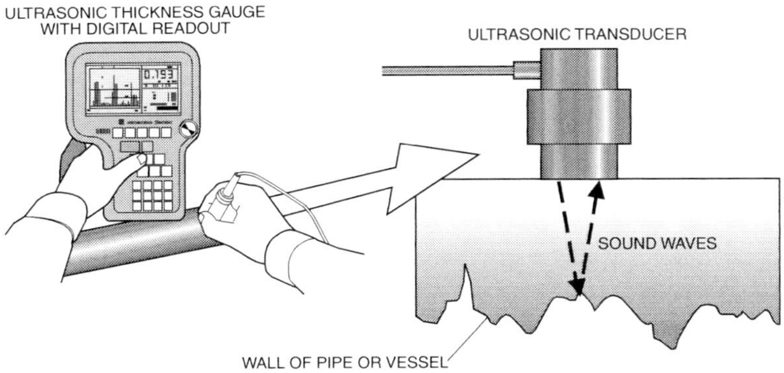

COURTESY OF ART LEACH, KRAUTKRAMER 104F12.EPS

Figure 12 ◆ Placement of a dual element transducer.

Audiovisual

Show Transparency 12 (Figure 13). Have a trainee explain why coatings make accurate pipe measurements difficult. Ask a trainee to interpret the readout of the transducer.

Classroom

Ask a trainee to explain the purpose of couplants. Have a trainee explain how couplants work.

Ultrasonic thickness gauges can be equipped with a multiple echo readout capability. This type of thickness gauge is used to identify and measure metal thickness apart from coating thickness and display the results in a single readout. *Figure 13* shows in detail how dual-element transducer signals are converted into digital values for both the coating and the pipe material.

Operators may need to adjust the gauge for starting point and gate width for accurate measurement of pipe or equipment material thickness.

3.1.3 Using Couplants

Operators must be familiar with the use of ultrasonic **couplants**. Properly used, they are placed between the pipe and the ultrasonic transducer. Testing skills must be developed so that couplants are used and applied in a consistent manner to minimize variations in couplant layer thickness and errors in test results.

Couplants are designed to provide improved probe coupling and acoustic transmission properties to ultrasonic testing by focusing the high frequency **sound waves** onto the target pipeline. The resulting penetration of the pipe wall with higher amplitude signals provides stronger echos back to the transducer and yields more accurate test results.

3.1.4 Data Analysis

Depending on the size of the equipment being tested, the number of thickness readings could be hundreds or thousands. The use of software programs to analyze large amounts of thickness data,

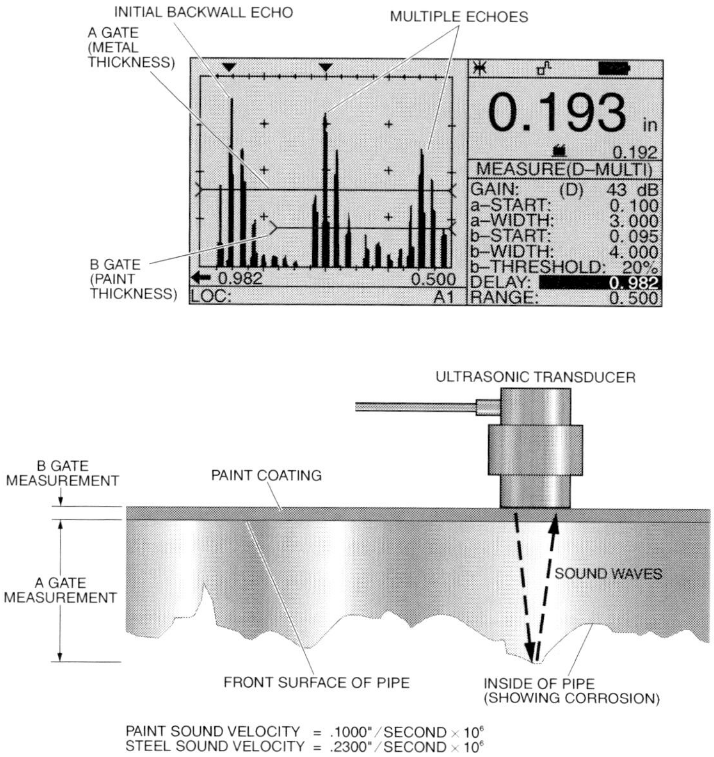

COURTESY OF ART LEACH, KRAUTKRAMER

104F13.EPS

Figure 13 ◆ Ultrasonic transducer placement and sample reading.

Instructor's Notes:

schedule future inspections, and predict the service life of the equipment is vital to any mechanical integrity program.

The data management software program can interpret massive amounts of data to make decision-making easy. By viewing the data in color levels related to thickness, it is easy to see that part of a component is below the minimum thickness value.

High-level software programs are typically used in large refineries, chemical processing plants, or power generation plants. Database history and complex analysis calculations within the program can predict the remaining service life of an in-service equipment item or component and calculate when the item needs re-inspection.

Documentation of testing results, record retention and comparison to historical data are other added benefits of these programs. Operators can store historic thickness data, make notes of the inspection, and identify their work. Storage of this history and analysis of equipment performance is an essential part of some code performance.

The high-level software programs that aid in data analysis provide the user with many benefits, including the following:

- Predict service life of an equipment or component item.
- Predict and establish inspection intervals.
- Review plant wide inspection intervals and schedule work accordingly.
- Track all equipment that is overdue for inspection.
- Alert operations to equipment that is potentially unsafe.
- Document all inspections on equipment.
- Document recommended repairs on equipment.
- Document the inspector that performed the inspection.
- Report minimum required thickness and maximum allowable operating pressure of the equipment.

3.2.0 Meter Calibration

Before any measurement is taken, the ultrasonic thickness gauge must be calibrated in order to provide accurate data. When calibrating the ultrasonic thickness gauge, always follow the proper procedures as recommended by the manufacturer. These procedures will vary depending on the manufacturer of the test instrument. The general calibration requirements are discussed in the following sections.

3.2.1 Calibration of Sound Velocity

Operators should be aware that sound velocity may not be constant in the material being tested. Heat treating, for example, can cause significant changes in sound velocity. This must be considered when evaluating the accuracy of the thickness provided by this instrument. Instruments should always be calibrated before testing, and the calibration should be checked after testing, to minimize testing errors.

3.2.2 Effects of Temperature on Calibration

Temperature variations will change the sound velocity of materials and transducer delay lines. This also changes calibration requirements. All calibrations should be performed on-site, and with test blocks at the same temperature as the test piece, to minimize errors due to temperature variations.

When calibrating a new transducer, or when ultrasonically measuring a test piece for the first time, the corrosion tester should calibrate the instrument/transducer combination on reference blocks that represent the complete range of possible thickness that may be encountered during the test activity. This is particularly important where the thickness history of the item to be tested is unknown and any measurements taken will serve as a baseline.

3.2.3 Transducer Selection

The transducer used in testing must be in good condition without noticeable wear of the front surface. Badly worn transducers will have a reduced effective measuring range. The specified range of the transducer must include the complete range of thickness to be tested. The temperature of the material to be tested must be within the transducer's temperature range. Transducers must be used such that they are firmly in contact with, and perpendicular to, the surface.

3.2.4 Calibration to a Known Thickness, 1-Point

Follow this procedure to calibrate the ultrasonic thickness gauge to a known thickness. These are the general steps required for this calibration. Always refer to your test instrument's operating manual for the procedure required for your specific instrument.

Classroom

Have a trainee explain how software can help analyze corrosion calculations.

Ask a trainee to explain why sound velocity may not be constant in the material being tested.

Classroom

Have a trainee describe a transducer in good condition.

Teaching Tip

See the Teaching Tip for Sections 3.1.0–3.2.7 at the end of this module.

Performance Testing

Have each trainee perform Performance Profile Task 3 to your satisfaction. Fill out a Performance Profile Sheet for each trainee.

Classroom

Ask a trainee to explain how an alignment sheet is used when determining the thickness of pipe. Discuss the information required on a pipeline inspection report.

Teaching Tip

See the Teaching Tip for Section 4.0.0 at the end of this module.

A calibration standard of the same material and velocity of sound as the material to be measured is required. For best results, the thickness of the calibration standard should be equal to, or slightly greater than the thickest area to be measured.

3.2.5 Calibration to a Known Thickness, 2-Point

Follow this procedure to calibrate the ultrasonic thickness gauge to a known thickness. Always refer to your test instrument's operating manual for the procedure required for your instrument.

This procedure requires two calibration standards (or one standard with two different thicknesses) of the same material, velocity, and curvature as the material under test.

For best results, the thicker standard should be equal to, or slightly greater than, the maximum thickness to be measured. The thickness of the thinner standard should be as close as possible to the lower end of the expected measurement range. Calibrate the gauge to both standards.

3.2.6 Calibration to a Known Velocity

To calibrate an ultrasonic thickness gauge to a known velocity, refer to the test instrument's operating manual to review the procedure required for the specific instrument you are using.

If you know the velocity of sound of the material to be measured, the instrument can be calibrated to the known velocity without using a calibration standard. However, the calibration should be always be verified by measuring a known thickness (or calibration standard) of the same material and velocity as the material to be measured. *Table 1* shows a chart of calculated sound velocities for different materials.

Table 1 Typical Sound Velocities in Various Metals

Material	in/µs	m/s
Aluminum	0.2500	6300
Brass	0.1700	4300
Cadmium	0.1100	2800
Cast Iron	0.1800	4600
Copper	0.1800	4600
Epoxy Resin	0.1100	2800
Glass (Crown)	0.2200	5600
Glass (Window)	0.2700	6800
Gold	0.1300	3300
Inconel	0.2200	5600
Lead	0.0850	2200
Magnesium	0.2300	5800
Manganese	0.1800	4600
Molybdenum	0.2500	6300
Monel	0.2100	5300
Neoprene	0.0630	1600
Nickel	0.2200	5600
Nylon	0.1000	2500
Phenolic	0.0560	1400
Platinum	0.1600	4100
Plexiglas	0.1100	2800
Polyethylene	0.0700	1800
Polystyrene	0.0930	2400
Polyurethane	0.0700	1800
Porcelain	0.2200	5600
Rubber (Butyl)	0.0730	1900
Rubber (Vulc.)	0.0900	2300
Silver	0.1400	3600
Steel	0.2300	5800
Tin	0.1300	3300
Titanium	0.2400	6100
Tungsten	0.2100	5300
Zinc	0.1600	4100
Zircaloy 2	0.1900	4700

104T01.EPS

3.2.7 Data Loggers/Recorders

Some models of digital ultrasonic thickness gauges have internal memory that allows the storage of thickness readings. Consult your manufacturer's literature for proper usage procedures related to this specialized function.

4.0.0 ◆ DOCUMENT EXTENT AND LOCATION OF DAMAGE

When determining the thickness of pipe, obtain the alignment sheet for the pipeline you are inspecting. Determine your exact location on the pipeline alignment sheet. Obtain the required information from the alignment sheet.

Obtain the required company pipeline inspection report form. The form for each company will differ but most will have the same information to be completed. As a minimum, the form must contain the exact location of the inspection and the results that you find during your inspection, along with your contact information. Always follow your company procedures when completing an inspection report. This will ensure that you provide your company with the information needed to evaluate the extent of the corrosion.

In addition, reports can be used to compare to your previous readings to see the rate of any corrosion that may be visible. *Figure 14* shows a typical pipeline inspection form.

Instructor's Notes:

PIPELINE INSPECTION REPORT

Frequency: At Time of Inspection

PART A — LOCATION

Reason For Inspection ______

1. Location: Division ______ County ______ State ______ Date of Inspection ______

A) Mainline
Alignment Sheet ______ Pipeline Size/Segment ______
Beg. M.P. ______ End M.P. ______
Beg. Sta. No. ______ End Sta. No. ______

Note: Station numbers shall be measured from a known landmark (Road Crossing, etc.)

B) Company Facility: Pump Station, Tank Farm, etc.
Location Name ______
Physical Location of Pipe Inspected ______
Milepost ______

PART B — MATERIAL/CONDITIONS

2. Piping:
Pipe Size ______ In. O.D. Nom. W.T. ______ In. Actual W.T. ______ Grade ______

3. External Pipe Coating:

A) Coating: ☐ Bare ☐ LAC 450 ☐ Taped ☐ Insulated ☐ Other ______

B) Application: ☐ Mill Coated ☐ Yard Coated ☐ Field Coated ☐ Other

C) Coating Thickness: Factory Applied ____ mile; Field Applied ____ mile

D) Condition of Coating: ☐ Good ☐ Fair ☐ Poor

E) Bond: ☐ Good ☐ Fair ☐ Poor

F) Cause of Coating Failure: ☐ Damage ☐ Disbondment ☐ Defective Application ☐ Decomposition ☐ Other ______

4. Environment
☐ Clay ☐ Loam ☐ Sand ☐ Gumbo ☐ Caliche ☐ Rock ☐ Other ______
☐ Wet ☐ Dry ☐ Moist Depth of Cover (Inches) ______
Soil Resistivity Measurement ______ (ohm-cm) Measurement Method ______

5. Coating Repaired ☐ No ☐ Yes
Repair System ______

6. Pipe-to-Soil Potential at Inspected Location (At Grade) ______ (volts) (At Pipe/Soil Interface) ______ (volts)

7. Were any Non-Destructive Tests Completed? ☐ No ☐ Yes Type ______

8. Evidence of: ☐ Corrosion ☐ Defect ☐ Damage ☐ Stress Corrosion Cracking (S.C.C.) ☐ Calcareous Deposits

9. Test Station Added ☐ No ☐ Yes I.D. No. ______

NOTES:
1) If S.C.C. or Corrosion is Noted, Complete Below and Back of Report (External Only, See Note 4)
2) If Defect or Damage is Noted, Complete Back of Report
3) If Repairs are Required, AAPL Form 502 "REPAIR REPORT FOR LEAK DAMAGE, CORROSION OR TEST FAILURE" Must be Attached.
4) AAPL Form 508 "INTERNAL CORROSION INSPECTION REPORT" Should be Completed When Internal Corrosion Is Detected.

SAMPLE

PART C — EXTERNAL CORROSION

10. General Corrosion Information:

A. DESCRIPTION
☐ Less than 10% of Nom. Wall Thickness, Complete Front of Report
☐ Greater than 10% of Nom. W.T., Complete Front and Back of Report
☐ Pitting ☐ General ☐ Localized
Max Depth ______ (In.) Longitudinal Length ______ (In.)

B. STATUS: ☐ Active ☐ Inactive ☐ Unknown

C. CAUSE: ☐ Galvanic ☐ Stray Current ☐ Unknown ☐ Other ______

11. Cathodic Protection: ☐ Impressed ☐ Galvanic

12. Pipe-to-Soil Potential:
a) Last Pipe-to-Soil Measurement at Nearest Point Each Side of Corroded Area
Upstream ______ (volts) Downstream ______ (volts)
b) Distance from Corroded Area to Measurement Points:
Upstream ______ (feet) Downstream ______ (feet)
c) Date of Measurements: Upstream ______ Downstream ______

13. pH of Soil Near Corroded Area ______ pH of Moisture Under Coating ______

14. Pipe Surface Temperature if Fluid is Heated ______ (deg F)

Remarks ______

Distribution: (Original) Corr. Ctrl./Eng (Copy) Division Office (Copy) Other ______

______ Technician or Inspector ___/___/___ Date ______ Supervisor ___/___/___ Date

104F14A.EPS

Figure 14 ◆ Typical pipeline inspection form. (1 of 2)

PIPELINE INSPECTION REPORT

Frequency: At Time of Inspection

PART D — DEFECT, DAMAGE OR CORROSION INVESTIGATION

Notes:
1) Tabulate all Dents, Gouges, Slivers, Corrosion, S.C.C., etc.
2) List Reference Point - Valve, Tee, Center Road etc., as shown on Alignment Sheet with appropriate Station Number, if available.
3) Indicate Location of Longitudinal Seam. If Pipe is Vertical Substitute Compass direction for o'clock Position. For Spiral Weld Pipe, Note as Such.
4) Defects Such as Dents, Slivers, Gouges, etc., shall be tabulated individually. Corrosion Cracks and/or Corrosion Pits Closely Grouped May be Tabulated as One Defect.
5) Defect Width and Length Will be Measured Along the Girth and Longitudinal Axis Respectively.

Reference Point: ☐ Upstream End (Station No.) ________ Description ________

☐ Downstream End (Station No) ________ Description ________

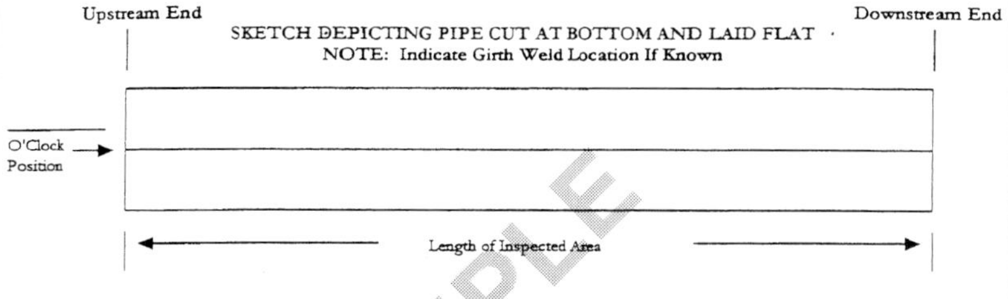

PART E — DEFECT, DAMAGE OR CORROSION MEASUREMENTS AND LOCATION

No.	Type	Distance from Girth Weld	Distance from Long Seam	Width (Circumferencial)	Length (Longitudinal)	Maximum Depth	Number of Anomalies
1.							
2.							
3.							
4.							
5.							
6.							
7.							
8.							
9.							
10.							
11.							
12.							
13.							
14.							

If Repair is Required, send to Division Manager for Approval.

Repair Required ☐ No ☐ Yes Type of Repair ________

Approved by ________ Title ________

Remarks: ________

104F14B.EPS

Figure 14 ◈ Typical pipeline inspection form. (2 of 2)

Instructor's Notes:

Summary

One of the most common tests on a pipeline is the inspection of the pipe for pits. A pit gauge is used to check the pipeline by measuring remaining wall thickness. Before any pit gauge is used for an inspection, the instrument must be fully inspected to ensure its serviceability. Do not use an instrument that is in poor condition.

The remaining strength (RSTRENG) of corroded pipe is one of the approved methods used to predict a minimum failure pressure. RSTRENG provides a more accurate prediction than the ASME/ANSI B31G requirements.

Ultrasonic thickness gauging is a commonly used method for measurement of remaining wall thickness in pipe and other equipment. This method of inspection has become widely used as a method of assuring mechanical integrity of equipment items that are prone to internal erosion and/or corrosion.

After the surface is prepared and an instrument is properly calibrated for use on the metal to be inspected, the operator uses a couplant as a medium to enable the ultrasonic waves to pass from the probe into the test material. The probe is placed firmly on the test location of the equipment and perpendicular to the surface, and a digital thickness value is displayed on the instrument. The inspector can store the reading into the internal memory of the instrument or print the data out depending on the type of instrument that is used. This data is used to determine the amount of corrosion that has occurred and the remaining wall thickness of the equipment. An informed decision can then be made as to what action is required now or at a future date.

One important aspect of the inspection is an inspection report completed in a timely manner. This is required by your company and usually there is a national, state, or local requirement as well. When completing the inspection report, always follow your company's procedures so that your company will have a useful document.

Review Questions

1. A common testing method for pipeline corrosion uses _____.
   a. frequency meters and calculators
   b. ID micrometers and rules
   c. pit gauges and ultrasonic thickness gauges
   d. OD micrometers and software

2. One of the most common tools used to check the extent of corrosion defects is a(n) _____.
   a. RSTRENG calculator
   b. optical gauge
   c. micrometer
   d. pit gauge

3. Dial pit gauges are designed for _____.
   a. use where common pit gauges will not fit
   b. more accurate measurement of pit depth
   c. use in areas where access is limited
   d. training use only

4. When using a dial pit gauge, the first step is to _____.
   a. take enough readings to provide a valid sample
   b. measure the depth of the largest pit
   c. clean the corroded area
   d. calibrate the gauge

5. The B31G corrosion assessment method _____.
   a. is the preferred method for evaluating corrosion
   b. is no longer approved for use by API
   c. is conservative in nature
   d. was developed to correct faults with the RSTRENG method

6. To calculate RSTRENG, the minimum acceptable number of depth measurements is _____.
   a. two
   b. three
   c. five
   d. six

7. The interaction rule is a guide for determining _____.
   a. whether company policies apply to the corrosion task
   b. whether corrosion defects require immediate remediation
   c. the estimated cost of corrosion repairs
   d. whether two or more metal loss defects are likely to interact

Classroom

Go over the Review Questions. Answer any questions trainees may have.

Examination

Administer the Module Examination. Be sure to record the results of the Exam on Craft Training Report Form 200 and submit the results to the Training Program Sponsor.

Performance Testing

Ensure that all Performance Tests have been completed and Performance Profile Sheets for each trainee are filled out. Be sure to record the results of the testing on Craft Training Report Form 200 and submit the results to the Training Program Sponsor.

8. For corrosion defects to interact, the distance between the defects must be less than _____.
   a. ½"
   b. the longitudinal length of the smallest defect
   c. the depth of the deepest pit
   d. the lateral measurement of the largest defect

9. Pipe wall thickness is monitored _____.
   a. because of potential excavation damage
   b. to prevent corrosion
   c. to ensure product changes go smoothly
   d. to assure mechanical integrity

10. Ultrasonic thickness gauges measure the material remaining in a pipe wall using _____.
    a. high-frequency sound waves
    b. a high-frequency filter circuit
    c. low-frequency filter circuit
    d. low-frequency sound waves

11. Couplants are designed to provide _____.
    a. reduced acoustic transmission
    b. air space between the probe and the test surface
    c. reduced echo strength
    d. improved probe coupling

12. Calibration of an ultrasonic thickness gauge is best if done with _____.
    a. published sound tables
    b. two or more test readings
    c. a new transducer
    d. a reference block of the same material as the item to be tested

13. Calibration to a known two-point thickness _____.
    a. requires two calibration standards or one standard with multiple thicknesses
    b. requires the corrosion tester to read each test point twice
    c. is only used when the item history is unknown
    d. uses the average of two tests for the result

14. Calibration of an ultrasonic thickness gauge to a known velocity _____.
    a. is required only when transducers are changed
    b. should be verified by measuring a standard of a known thickness
    c. requires two calibration standards
    d. is required every 12 months by API

15. As a minimum, a pipeline inspection form should provide for recording _____.
    a. pipeline map reference numbers
    b. corrosion repair estimates
    c. record of test readings
    d. exact location of the inspection and detailed findings

Instructor's Notes:

GLOSSARY

Glossary

This glossary will help you learn some terms specific to measuring pit depth and wall thickness. Trade Terms introduced in this module appear in italic text.

Backwall: Sidewall of the material being tested that is opposite from the spot where the ultrasonic probe emits the thickness measurement signal. Sound waves from the probe reflect off this spot and back to the probe where it is measured to determine material thickness.

Calibrate: A comparison of a standard to a reading to set-up equipment properly.

Circumferential: Area going around the pipe.

Couplant: Used to facilitate ultrasonic testing, such as testing wall thickness in a pipeline. A couplant is a liquid, grease, paste, or pliable solid placed between the transducer of the ultrasonic tester and the surface of the test piece to permit transmission of ultrasonic energy between them. Couplants, often referred to as coupling medium or coupling film, retard sound transmission by air between the transducer and the work surface.

Crosstalk barrier: Limit of undesired coupling of noise.

Dent: A depression that produces a reversal in the curvature of the pipe wall without reducing the wall thickness.

Doubling: An effect known to occur when the instrument is used on materials below the minimum thickness range of the probe. Results in an indication of a value twice the actual thickness.

Gouge: Surface damage that reduces the pipe wall thickness by displacement or removal of metal.

Lamination: A material made of several bonded layers.

Localized Corrosion: Corrosion in adjacent or isolated areas.

Longitudinal: Pertaining to line of the greatest lengthwise dimension of an object.

Mils: Unit of measure equal to 0.001 of an inch.

**Pitting:* Localized corrosion of a metal surface that is confined to a small area and takes the form of cavities called pits.

**Profile:* Anchor pattern on a surface produced by abrasive blasting or acid treatment.

Pulse-echo: A non-destructive test in which pulses of energy are sent into the material and the time for return echo from one or more surfaces is measured.

River bottom profile: The collection of more than one set pit depth depicting "peaks and valleys,"or the actual profile of the defect.

Sound waves: Particle displacement that propagates through a medium. Sound waves above the audible range are called ultrasonic.

Transducer: A device that converts an input signal into an output signal of a different form. Examples include microphones, loudspeakers, and ultrasonic test probes. An ultrasonic transducer converts alternating current energy above 20 kilohertz into mechanical vibrations of the same frequency and uses the vibrations to measure the thickness of pipeline walls.

Velocity of sound: Used to describe the rate at which sound travels through different materials. Referring to a list of common materials and their respective velocity of sound rating will provides the value for calibrating an ultrasonic thickness gauge before measuring the wall thickness of a pipeline of a known material.

REFERNCES AND ACKNOWLEDGMENTS

Additional Resources

This module is intended to present thorough resources for task training. The following reference works are suggested for further study. These are optional materials for continued education rather than for task training.

Control of Pipeline Corrosion, 1967. Peabody, Houston, TX; National Association of Corrosion Engineers (NACE).

Developing Simple Solutions for the Complexities of Wall Thickness Measurement Applications, 2000. Art Leach, Software Product Manager. Krautkramer.

Mechanical Integrity Using Remaining Wall Thickness and Data Analysis, 2000. Art Leach, Software Product Manager. Krautkramer.

Figure Credits

Art Leach, Krautkramer	104F10, 104F11, 104F12
Allen Carlile	104F03

Instructor's Notes:

MODULE 61104-02—TEACHING TIPS

The following are suggested activities or instructional methods to help you teach the material in this AIG.

General

When you call on someone to answer a question, the rest of the class relaxes or even tunes out because they expect that the question and answer will take place only between you and the trainee you called on. Instead, use this technique to involve more trainees in answering questions and to keep them on their toes.

1. Ask trainees to define a term or explain a concept.
2. After one trainee has answered, ask a trainee seated nearby if the answer is right. Then ask whether a trainee in the back of the room agrees.
3. Ask trainees to explain why they think an answer is right or wrong.
4. Use the session to clear up incorrect ideas and encourage trainees to learn from their mistakes.

Sections 2.2.1–2.2.2

Using Mechanical and Dial Pit Gauges

This exercise will give trainees practice using mechanical and dial pit gauges. Trainees will need pencils and paper to take notes and their Trainee Guides. You will need to provide sections of pitted pipe, dial pit gauges, mechanical pit gauges, and manufacturer's calibration instructions. Allow 30 to 45 minutes for this exercise.

1. Set up several workstations with a section of pitted pipe, a dial pit gauge, and a mechanical pit gauge at each. Divide the class into small groups.
2. Have trainees take turns calibrating the dial pit gauge according to manufacturer's instructions.
3. Ask trainees to use each type of pit gauge on samples of pitted pipe to find the remaining wall thickness. Have trainees refer to *Figure 3* for guidance.
4. Have the trainees at each workstation compare their measurements. Ask them to note which measurement seems more accurate.
5. Answer any questions trainees may have.

Section 2.5.0

Evaluating Localized Corrosion Defects

This exercise will give trainees an opportunity to make a measurement grid of the external corroded area of a pipe. Trainees will need pencils, paper, and their Trainee Guides. You will need to provide sections of corroded pipe, material for cleaning the pipe, graph paper, carbon paper, and lumber pencils. Allow 30 minutes for this exercise.

1. Have trainees follow the steps in Section 2.5.0 of the Trainee Guide.
2. After they have made the transfer of the corroded area (Step 2), have trainees check each other's work.
3. After trainees measure and record the deepest pits (Step 3), ask trainees why it is important to note flow direction, orientation, distance to the reference weld, and other company-required data.
4. Review trainees' work. Offer suggestions and corrections as needed. Answer any questions the trainees may have.

Sections 3.1.0–3.2.7 *Meter Calibration*

This exercise will give trainees an opportunity to practice calibrating and operating ultrasonic thickness meters. Trainees will need appropriate personal protective equipment, pencils and paper to take notes, and their Trainee Guides. You will need to provide sections of coated or uncoated pipe; appropriate ultrasonic meters, operating manuals, and velocity of sound charts; and couplants. Alternatively, have trainees simulate the procedures. Allow 45 to 60 minutes for this exercise.

1. Have trainees calibrate the ultrasonic thickness gauge according to the operating manual.
2. Supervise trainees as they use the ultrasonic gauge to measure the thickness of the pipe wall.
3. Check the accuracy of the trainees' calculation of wall thickness. Ask trainees if they have any questions.

Section 4.0.0 *Document Extent and Location of Damage*

This exercise will give trainees practice in reporting and documenting corrosion. Trainees will need appropriate personal protective equipment, pencils, paper, and their Trainee Guides. You will need to provide sections of corroded pipe, pit gauges, ultrasonic meters, couplants, and pipeline inspection report forms. Allow 45 to 60 minutes for this exercise.

1. Have trainees read through the pipeline inspection report form.
2. Have trainees inspect the section of corroded pipe and measure pit depth and wall thickness in at least one spot. Remind them to note their findings and measurements.
3. Ask trainees to fill out as much of the pipeline report form as possible.
4. Review trainees' completed forms. Answer any questions the trainees have.

MODULE 61104-02 — ANSWERS TO REVIEW QUESTIONS

Answer	Section
1. c	1.0.0
2. d	2.0.0
3. b	2.1.2
4. d	2.3.0
5. c	2.5.0
6. b	2.5.0
7. d	2.6.0
8. b	2.6.0
9. d	3.0.0
10. a	3.1.0
11. d	3.1.2
12. d	3.2.1
13. a	3.2.6
14. b	3.2.7
15. d	4.0.0

CONTREN™ LEARNING SERIES — USER UPDATES

The NCCER makes every effort to keep these textbooks up-to-date and free of technical errors. We appreciate your help in this process. If you have an idea for improving this textbook, or if you find an error, a typographical mistake, or an inaccuracy in NCCER's Contren™ textbooks, please write us, using this form or a photocopy. Be sure to include the exact module number, page number, a detailed description, and the correction, if applicable. Your input will be brought to the attention of the Technical Review Committee. Thank you for your assistance.

Instructors – If you found that additional materials were necessary in order to teach this module effectively, please let us know so that we may include them in the Equipment/Materials list in the Instructor's Guide.

Write: Curriculum Revision and Development Department
National Center for Construction Education and Research
P.O. Box 141104, Gainesville, FL 32614-1104

Fax: 352-334-0932

E-mail: curriculum@nccer.org

Craft ____________________ Module Name ____________________

Copyright Date __________ Module Number __________ Page Number(s) __________

Description

____

____

____

____

(Optional) Correction

____

____

____

(Optional) Your Name and Address

____

____

____

Module 61105-02

Inspect Buried and Submerged Pipe When Exposed

(CT 5.1, 5.2, and 5.3)

Inspect Buried and Submerged Pipe When Exposed
Annotated Instructor's Guide

Module 61105-02

MODULE OVERVIEW

This module familiarizes trainees with types of coatings and coating damage. Trainees will learn how to inspect buried and submerged pipe for physical damage, corrosion, and coating damage.

PREREQUISITES

Please refer to the Course Map in the Trainee Module. Prior to training with this module, it is recommended that the trainee shall have successfully completed the following:

Core Curriculum; Pipeline Core, Modules 66101-02 and 66102-02; Pipeline Corrosion Control Level One, Modules 61103-02 through 61104-02

OBJECTIVES

Upon completion of this module, the trainee will be able to:

1. Perform inspection of the pipe and coating for evidence of damage and/or abnormalities (CT 5.1).
2. Examine exposed pipe to identify any areas of corrosion (CT 5.2).
3. Perform inspection of the pipe coating, noting its condition whenever the pipe is exposed (CT 5.3).
4. Document the location of corrosion and damage to a buried pipeline.

PERFORMANCE TASKS

Under the supervision of the instructor, the trainee should be able to:

1. Inspect for physical damage on buried or submerged pipe (CT 5.1).
2. Inspect for external corrosion on buried or submerged pipe (CT 5.2).
3. Inspect the condition of external coating on buried or submerged pipe (CT 5.3).

NCCER STANDARDIZED CRAFT TRAINING PROGRAM

The National Center for Construction Education and Research (NCCER) provides a standardized national program of accredited craft training. Key features of the program include instructor certification, competency-based training, and performance testing. The program provides trainees, instructors, and companies with a standard form of recognition through a National Craft Training Registry. The program is described in full in the *Guidelines for Accreditation*, published by the NCCER. For more information on standardized craft training, contact the NCCER by writing us at P.O. Box 141104, Gainesville, FL 32614-1104; calling 352-334-0911; or e-mailing info@nccer.org. More information may be found at our Web site, www.nccer.org.

HOW TO USE THIS ANNOTATED INSTRUCTOR'S GUIDE

Each page presents two sections of information. The larger section displays each page exactly as it appears in the Trainee Module. The narrow column ties suggested trainee and instructor actions to each page and provides icons to call your attention to material, safety, audiovisual, or testing requirements. The bottom of each page includes space for your notes.

If you see the Teaching Tip icon, that means there is a teaching tip associated with this section. Also refer to any suggested teaching tips at the end of the module.

SAFETY CONSIDERATIONS

Ensure that the trainees are equipped with appropriate personal protective equipment. Emphasize the importance of following all safety precautions and procedures when working with power tools.

PREPARATION

Before teaching this module, you should review the Module Outline, Objectives, Performance Tasks, and the Materials and Equipment List. Be sure to allow ample time to prepare your own training or lesson plan and gather all required equipment and materials.

MATERIALS AND EQUIPMENT LIST

Overhead projector and screen
Transparencies
Blank acetate sheets
Transparency pens
Whiteboard/chalkboard
Markers/chalk
Appropriate personal protective equipment
Pencils and scratch paper
Copies of your local code
Section of pipe with alligatored coal tar coating
Access to a buried pipeline and necessary exterior inspection equipment
Access to an opened, buried pipeline and necessary interior inspection equipment
Mechanical pit gauges, dial pit gauges, and micrometer pit gauges
Sections of pitted pipe
Equipment needed to inspect a buried pipe for physical damage and external corrosion
Equipment needed to inspect the condition of external coatings on buried or submerged pipe
Module Examinations*
Performance Profile Sheets*
Copies of Quick Quiz**
Copies of the Performance Verifications associated with this module (5.1, 5.2, and 5.3)†

* Located in the Test Booklet.
** Located at the end of this module.
† Available for download at www.nccer.org.

TEACHING TIME FOR THIS MODULE

An outline for use in developing your lesson plan is presented below. Note that each Roman numeral in the outline equates to one session of instruction. Each session has a suggested time period of 2½ hours. This includes 10 minutes at the beginning of each session for administrative tasks and one 10-minute break during the session. Approximately 5 hours are suggested to cover *Inspect Buried and Submerged Pipe When Exposed (CT 5.1, 5.2, and 5.3)*. You will need to adjust the time required for hands-on activity and testing based on your class size and resources.

Topic	Planned Time
Session I. Coating Types and Coating Damage	
A. Introduction	
B. Identify and Explain Coating Types	____________
1. Coal Tar, Asphalt, and Tape	____________
2. FBE, Extruded Polyethylene Jacket, and Concrete	____________
3. Wax, Fiberglass, Mastic, and Copolymer	____________
C. Types and Causes of Coating Damage	____________
1. Sagging, Cracking, and Wrinkling	____________
2. Blistering, Peeling, and Flaking	____________
3. Intercoat Delamination and Undercutting	____________
Session II. Visual Inspection, Documentation, and Procedures	
A. Visually Inspect Buried Pipeline for Corrosion and Mechanical Damage	____________
1. Laboratories	____________
2. Types of Pit Gauges	____________
3. Calibrate Gauge and Check for Worn Point	____________
4. Use Pit Gauge to Measure Pit Depth	____________
5. Laboratory	____________
6. Use Pit Gauge to Measure Pit Length	____________
7. Laboratory	____________
8. Take Multiple Readings to Gather RSTRENG Data	____________
B. Document the Location of the Corrosion or Damage	____________
C. Procedures	____________
1. Inspecting for Physical Damage on Buried or Submerged Pipe	____________
2. Laboratory	____________
3. Performance Testing	____________
4. Inspecting for External Corrosion on Buried or Submerged Pipe	____________
5. Laboratory	____________
6. Performance Testing	____________
7. Inspecting the Condition of External Coatings on Buried or Submerged Pipe	____________
8. Laboratory	____________
D. Review Module	____________
E. Module Examination	____________
1. Trainees must score 70% or higher to receive recognition from the NCCER.	
2. Record the testing results on Craft Training Report Form 200 and submit the results to the Training Program Sponsor.	
F. Performance Testing	____________
1. Trainees must perform each task to the satisfaction of the instructor to receive recognition from the NCCER.	
2. Record the testing results on Craft Training Report Form 200 and submit the results to the Training Program Sponsor.	

Module 61105-02

Inspect Buried and Submerged Pipe When Exposed

(CT 5.1, 5.2, and 5.3)

Instructor's Notes:

Instructor's Notes:

Introduction to the Pipeline Operator Qualification Rule, API, and NCCER

It could be said that petroleum products keep America running. Petroleum products like oil and natural gas are the principal sources of energy in the United States and are used in many manufacturing and chemical processes. Petroleum products are primarily transported by pipeline. There are more than a million miles of liquid and natural gas pipeline in this country, ranging in diameter from two inches to four feet. These pipelines bring crude oil and natural gas from production fields to processing plants and refineries—and ultimately to the consumer.

Pipeline Operator Qualification

Petroleum products are considered hazardous, so pipeline personnel must take extra precautions and follow specific procedures when working with them. To help ensure the safe handling and transportation of petroleum products, the United States Department of Transportation (DOT) mandates safe practices as federal regulations.

One such regulation, the Operator Qualification (OQ) rule, requires pipeline operators to develop a written qualification plan to evaluate their personnel and their contract personnel (see *49 CFR Part 192 Subpart N—Gas,* and *49 CFR Part 195 Subpart G—Liquid*). The evaluation hinges on the operator's ability to perform certain covered tasks and to be able to recognize and respond to abnormal operating conditions that may be encountered while performing these covered tasks.

Covered Tasks

A pipeline performance task is deemed *covered* if the activity passes the following four-part test:

1. Is it performed on a pipeline facility?
2. Is it an operations or maintenance task?
3. Is it performed as a requirement of *49 CFR Part 195 (Liquid)* or *49 CFR Part 192 (Gas)*?
4. Does it affect the operation or integrity of the pipeline facility?

If the answer to all four questions is *yes,* then the activity is considered *covered* and must figure into the evaluation process of pipeline personnel.

American Petroleum Institute

The OQ rule and its related covered tasks provided the catalyst to develop a unified pipeline training and assessment program—so that pipeline operators and owners have ready access to the assessment and training tools they need. This effort was the direct result of the vision and leadership provided by The American Petroleum Institute (API) and API members.

Efforts to unify the industry are nothing new to API. Indeed, the impetus for forming the Institute in 1919 was the need to standardize engineering specifications for drilling and production equipment. Today, the most pressing issues revolve around public perceptions and government policies toward the industry. Speaking with one voice on these issues has become as essential as having interchangeable parts in the field. API's members determine the petroleum industry's positions on public policy as well as the standards that govern its day-to-day operations worldwide.

Consortium on Operator Qualification

API assembled a focus group whose primary purpose was to identify covered tasks and provide compliance recommendations to pipeline operators.

The Consortium on Operator Qualification (COOQ) became the guiding force for the identification and development of covered tasks for the liquid pipeline industry. Their work product, API publication 1161, *Guidance Document for the Qualification of Liquid Pipeline Personnel,* formed the basis for the resulting Pipeliner Training and Assessment Program.

The covered task list created in API publication 1161 has now evolved to include gas covered tasks and is maintained by NCCER. The most current covered task list can be downloaded at www.nccer.org.

National Center for Construction Education and Research

With the guidance and support of API, its members, and the Consortium on Operator Qualification, the National Center for Construction Education and Research (NCCER) was able to design and implement the Pipeliner Training and Assessment Program.

NCCER is a not-for-profit 501(c)(3) education foundation established in 1995 by the world's largest and most progressive construction and maintenance contractors and national contractor associations. NCCER was founded to address the

severe workforce shortage facing the construction and maintenance industries and to develop a standardized training process and curricula. Today, NCCER is supported by hundreds of leading construction and maintenance companies, manufacturers, and national associations and developed more than 35 construction and maintenance titles.

NCCER also maintains a National Registry, which contains the records of trainees and instructors who have received NCCER training. All completion information for every trainee is sent to the NCCER and filed within the Registry. The Registry can then confirm training and skills for workers as they move from company to company, state to state, or even within their own company.

Standardized Curriculum (NCCER's Contren™ Learning Series)

NCCER curricula are developed using the expertise of industry subject matter experts (SMEs). The use of SMEs as an integral part of every new development or revision ensures that the curriculum is current, industry-driven, and indicative of industry practices. The development and/or revision process is facilitated from start to finish by NCCER project managers, who work closely with the SMEs and a select team of technical writer(s) on the development and presentation of the curriculum.

Pipeline: Five Training Paths

The pipeline curriculum focuses on essential skills and competencies within five skill areas:

- Corrosion Control
- Electrical and Instrumentation
- Maintenance
- Mechanical
- Operations (Field and Control Center)

SMEs from the pipeline industry have determined the skills and competencies needed based on API's covered tasks and the consensus of the group/industry. Specific company information is generally not incorporated into the curriculum unless the SMEs determine that the information is the industry standard.

Training Modules and Covered Tasks

The smallest unit of instruction in the curriculum is a *module*. Each curriculum is composed of modules (the equivalent of stand-alone chapters) that are logically sequenced into complete *levels* of instruction. Every module is task-oriented and requires the successful completion of a written (knowledge-based) examination and a performance (hands-on) test. Modules may be individually selected and/or combined for efficient task training. Modules in the pipeline curricula, to the extent possible, will align with API-developed covered tasks. While not all modules will relate to covered tasks, all covered tasks will relate to at least one module.

Benefits of Training to the Pipeline Marketplace

Applying NCCER's standardized training process allows pipeline owners and contractors in the United States to establish the benchmark for acceptable skill and competency levels within the pipeline industry. This process will assist training organizations in developing and/or improving standardized pipeline training programs that effectively match their resources and operating conditions while responding to regulatory demands and to the needs of the industry. The process will also recognize training organizations that meet the minimum standards of competence, warranting regulatory, public, and professional confidence. This in turn will foster national unity within the pipeline industry for the mutual benefit of individual technicians, owners, contractors, and ultimately pipeline-users and society.

Additional Resources

To review the Pipeliner Training and Assessment Program Implementation and Resource Guide and the latest version of pipeline covered tasks, visit NCCER's Web site at www.nccer.org, or contact:

National Center for Construction Education and Research
PO Box 141104
Gainesville, FL 32614-1104
(888) 622-3720
Web site: www.nccer.org

For more information on API, please contact:

American Petroleum Institute
1220 L Street, Northwest
Washington, D.C. 20005
(202) 682-8000
Web site: www.api.org

To order *API Pub 1161*, contact Global Engineering Documents by calling 1-800-854-7149 or visiting their Web site at www.global.his.com.

Additional sources of information on the pipeline operator qualification regulation include:

DOT – Office of Pipeline Safety
www.ops.dot.gov

Transportation Safety Institute
www.tsi.dot.gov

Instructor's Notes:

COURSE MAP

This course map shows all of the modules in the first level of the Pipeline Corrosion Control curriculum. The suggested training order begins at the bottom and proceeds up. Skill levels increase as you advance on the course map. The local Training Program Sponsor may adjust the training order.

PIPELINE CORROSION CONTROL LEVEL ONE

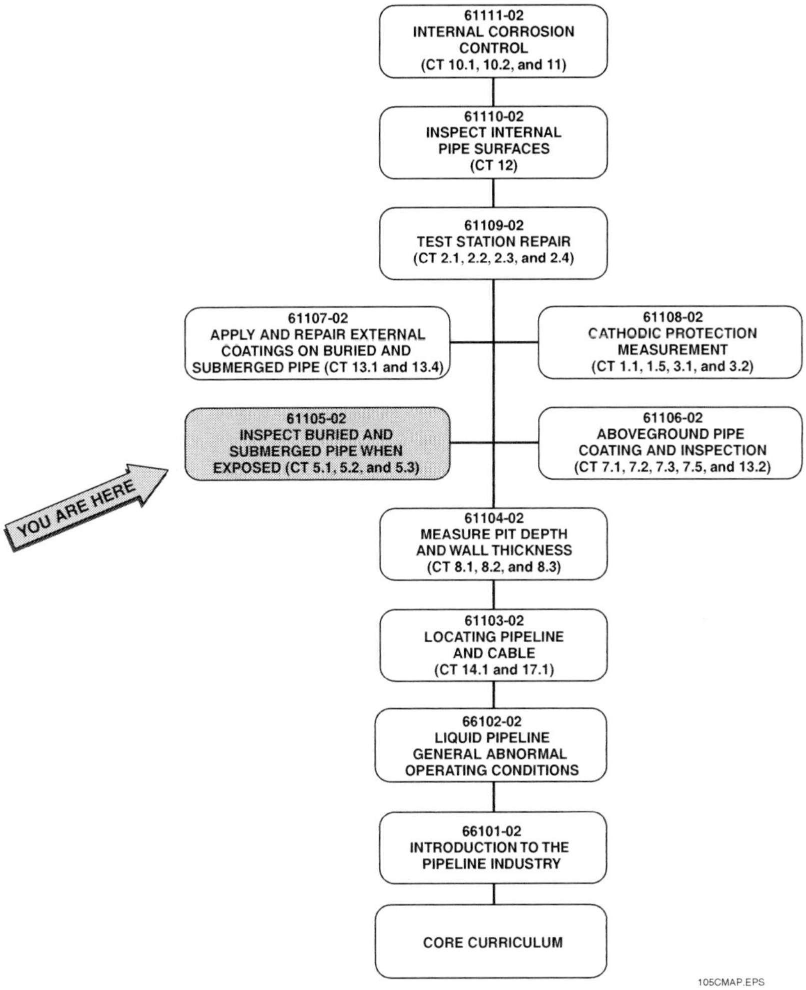

105CMAP.EPS

Homework

Assign reading of Module 61105-02.

Instructor's Notes:

MODULE 61105-02 CONTENTS

Figures

Instructor's Notes:

MODULE 61105-02

Inspect Buried and Submerged Pipe When Exposed (CT 5.1, 5.2, and 5.3)

Objectives

When you have completed this module, you will be able to do the following:

1. Perform inspection of the pipe and coating for evidence of damage and/or abnormalities (CT 5.1).
2. Examine exposed pipe to identify any areas of corrosion (CT 5.2).
3. Perform inspection of the pipe coating, noting its condition whenever the pipe is exposed (CT 5.3).
4. Document the location of corrosion and damage to a buried pipeline.

Prerequisites

Before you begin this module, it is recommended that you successfully complete the following: Core Curriculum; Pipeline Core, Modules 66101-02 and 66102-02; Pipeline Corrosion Control Level One, Modules 61103-02 and 61104-02.

Required Trainee Materials

1. Pencil and paper
2. Appropriate personal protective equipment

1.0.0 ◆ INTRODUCTION

Buried and submerged pipe is not easily inspected for **coating** and mechanical damage because there is a lot of effort involved in exposing the pipe for inspection. This module discusses the inspection of buried and submerged pipe.

2.0.0 ◆ IDENTIFY AND EXPLAIN COATING TYPES

Some of the many coating materials available for buried and submerged pipe are **phenolic**, **epoxies**, and **polyesters**. Thermosetting coatings are widely used in measurement equipment such as positive-displacement meter provers. Automatic dump tanks, which measure oil by counting the times a tank with a carefully calibrated volume is filled and emptied, require linings that are resistant not only to corrosion but to paraffin deposition as well.

Some thermosetting, catalytic-setting, and polymerization-setting coatings are so tough and moisture resistant that films 10 to 40 mils thick can provide good protection to buried pipelines. Thermoplastic pipe coatings can be applied hot in the liquid state to form seamless sheaths over the pipe. For buried pipelines and other buried structures pre-applied adhesives are also widely used. Coatings suitable for use underground or underwater might not be suitable for exposure to sunlight and weather.

> **WARNING!**
> Asbestos may be on pipe in the ground. Always follow proper procedures to safely work with asbestos. Never use power tools on asbestos coatings and never cause the asbestos to become airborne.

INSPECT BURIED AND SUBMERGED PIPE WHEN EXPOSED — TRAINEE MODULE 61105-02 5.1

Materials

Ensure that you have everything required to teach the course. Check the Materials and Equipment List at the front of this module.

Classroom

Explain that terms shown in bold are defined in the Glossary at the back of this module.

Audiovisual

Show Transparency 1, Course Objectives.

Show Transparency 2, Performance Tasks.

Teaching Tip

See the general Teaching Tip at the end of this module.

Classroom

Discuss various types of coating materials and common uses of each.

Safety

Ask a trainee to explain the dangers of asbestos. Have another trainee explain why power tools should not be used on asbestos coatings.

Audiovisual

Show Transparency 3 (Figure 1). Ask a trainee to explain the causes of alligatoring.

Demonstration

Bring in a section of pipe with coal tar coating that has alligatored. Ask a trainee to explain how to prevent alligatoring.

2.1.0 Coal Tar

There are a number of **coal tar coatings** made by dissolving processed coal tar, or a blend of these pitches, in suitable solvents. They dry entirely by evaporation of the solvents. Generally, they are all quite similar. The outstanding property of coal tar coatings is their extremely low permeability to moisture and their high **dielectric resistance**, both of which contribute to their corrosion resistance.

Coal tar coatings are made in different consistencies; those without any inert fillers, and those that contain inert materials in order to build film thickness. Coal tar coatings, in general, are not affected by mineral oil, but can be dissolved by a vegetable or animal oil, grease, and detergents. They have good resistance to weak acids, alkalis, salts, seawater, and other aggressive elements. They are corrosion-resistant because they are impermeable. They provide protection by exclusion of moisture and air from the underlying surface. They can be applied to concrete as well as to steel.

One of the problems with coal tar coatings for aboveground pipe is their tendency to **alligator** when exposed to direct sunlight. *Figure 1* shows alligatoring.

Alligatoring is brought about by the hardening of the upper layer of the coal tar film because of exposure to the sun's ultraviolet rays. The upper layer of the film contracts and slips over the softer underlayer, causing marks that resemble alligator hide. Coal tar cutback coatings should be protected from the sun in order to retain their corrosion-resistant properties.

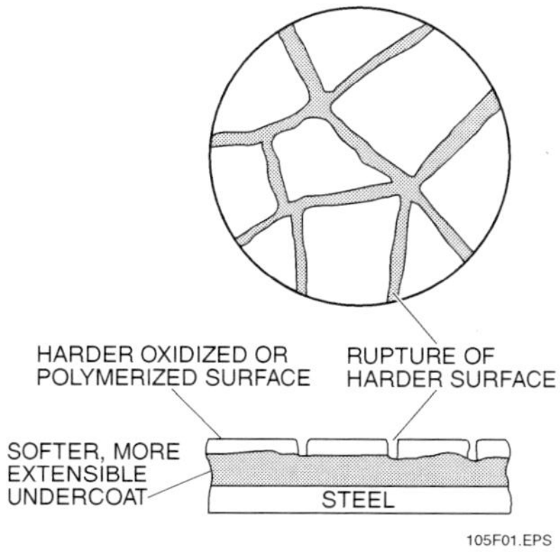

Figure 1 ◆ Alligatoring.

2.2.0 Asphalt

Asphalts are also extensively used as corrosion-resistant coatings. They not only are made with distilled asphalt from petroleum, but also are made from natural asphalt, or **Gilsonite®**. It is a very hard, asphaltic material and is mined in the same way that coal is mined. It has excellent chemical resistance and good weather resistance. Most of the better grades of asphaltic coatings will contain a percentage of Gilsonite® in order to upgrade the petroleum-derived asphalt. In general, the higher the percentage of Gilsonite®, the more weather resistant the coating will be.

Asphalt coatings are more weather-resistant than the corresponding coal tar coating since they do not tend to alligator. On the other hand, they are inferior in water resistance to the coal tar products.

Asphaltic coatings also are combined with higher polymer petroleum resins which increase the strength and provide more flexibility to the asphalt Gilsonite® coatings. Asphaltic coatings are also made with inert fillers, which add to the thickness of the coating and, in many cases, to its impervious characteristics. Some of these asphaltic-type coatings have excellent resistance to industrial fumes, water, moisture, condensation, and exterior weather exposure. They tend to lose gloss and to chalk when exposed to the weather; this does not detract from their other good characteristics or reduce their overall weather resistance.

2.3.0 Tape

Tape materials are being used frequently as a full **coating system**. Tapes normally used are plastic films of polyvinyl chloride (PVC) or polyethylene with a self-adhesive backing applied to primed pipe surfaces. Also common are plastic films with butyl rubber backings and plastic films with various bituminous backings or combinations of bituminous material and chemical resins.

Protective results are obtained with applications that are typically 35 mils or greater with proper overlap and tension.

Tape can be used as easy field repair for other coating systems and for the coatings of joints and fittings when factory-coated pipe is used. This type of coating system can also be used where existing pipe is being excavated and exposed. When applying a tape coating, the coating should be applied according to the manufacturer's specifications. Also refer to the manufacturer's specifications for specific recommendations and information concerning compatibility with hydrocarbon fuels. *Figure 2* shows a hot tape coating application.

Instructor's Notes:

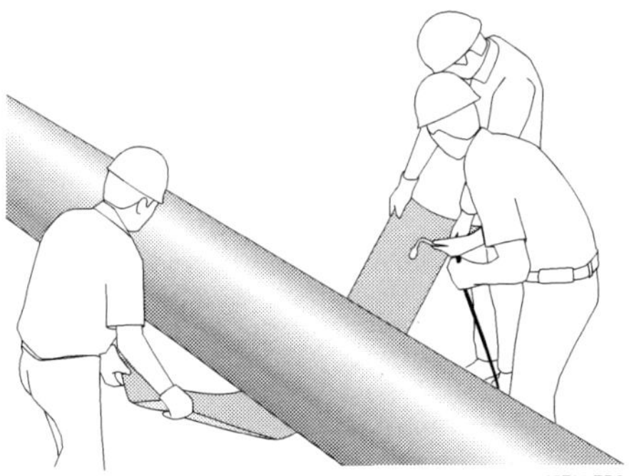

105F02.EPS

Figure 2 ◆ Hot tape coating application.

2.4.0 Fusion Bonded Epoxy (FBE)

Fusion bonded epoxy (FBE) coatings will generally perform well in steel underground pipe protection systems. FBE coatings may be applied in the shop or in the field by applying epoxy powder to a hot metal surface. The procedure fills in all small crevices on the pipe surface with a very hard, dense, abrasion-resistant coating with all the properties required of a good coating system. Field repairs and joints should be coated with liquid product repair kits or coated as specified by the manufacturer.

2.5.0 Extruded Polyethylene Jacket

When new piping is being ordered, pipe that is extrusion-coated with a polyethylene jacket can also be used. This is a factory-applied, extruded poly coating used to prevent corrosion. The coating is usually applied in a thicker film than conventional pipe coatings. The pipe joints and welds must be hand-wrapped with a compatible tape or **shrink sleeve** over cleaned and primed surfaces.

2.6.0 Concrete

Concrete used as a pipeline coating is a compounded concrete (or cement mortar) that will protect steel against corrosion very effectively. When the concrete or mortar is cast directly on the bare steel, the steel assumes a cathodic potential. With good concrete, the steel polarizes to resist either current collection or current discharge.

Concrete has proved to be an excellent protective material on distribution piping, in limited applications on transmission lines, and for shielded areas where cathodic protection cannot be used effectively. It is relatively expensive, but results in a strong and long-lived coating for specialty applications.

When used as the sole means of protection, there can be a corrosion hazard if a section of the concrete encasement breaks away from the pipe surface, exposing the pipe steel to contact with soil and water. Steel encased in suitable concrete can be protected cathodically and without difficulty because it polarizes readily.

2.7.0 Wax

Wax is a combination of paraffin, plastics, and inhibitors, and is commonly applied hot or cold in the field. Wax is normally used with a **primer** designed for wax coating. Hot wax has been mill-applied in the past but today it is applied in the field. There is an outer wrap that usually consists of a snug-fitting uniform layer of fiber-reinforced paper, rag felt, or a plastic film over the coating.

The operating temperature limitations are somewhat lower for this system than for other coatings.

2.8.0 Fiberglass

Fiberglass materials are a combination of a soft matrix material with strong fibrous material. Fiberglass materials combine the strength of the fibers with the ductility of the matrix materials. They can be very strong and rigid and can have a very high strength-to-weight ratio. They can also be manufactured so that their mechanical properties are directional and made to be strongest in the direction of maximum stress.

Fiberglass material's resistance to deterioration is limited by the resistance of their components. Any environment that attacks the matrix materials will significantly affect the entire fiberglass material. Fiberglass coating on buried pipe resists moisture absorption and is resistant to bacterial organisms.

2.9.0 Mastic

Mastic is a term commonly used to refer to materials formulated with selected sands and other inert materials bound with an insulating compound (commonly asphalt). These materials are applied hot or cold and are normally thicker than other coatings in common use. Protective wrappers or dielectric wrappers (for increased electrical resistance) are not needed under normal pipelining

Classroom

Discuss typical uses of each type of coating. Ask trainees if they have questions.

Teaching Tip

See the Teaching Tip at the end of this module.

Classroom

List the 10 types of coatings on the board or an overhead. Ask trainees to name characteristics of each type of coating.

Ask a trainee to distinguish between sagging and wrinkling. Have another trainee name possible causes of each type of coating damage.

conditions because of their thickness and hardness. Grades with various melting points are available to suit pipeline operating temperatures. When working with mastic, always follow the manufacturer's recommendation for the required curing time.

2.10.0 Copolymer

The main purpose of an epoxy-based polymer concrete is to protect the steel pipe under severe handling conditions such as rocky soil. Copolymer concrete coatings are highly abrasion-resistant and provide a smooth surface that allows the pipeline to be pulled under a crossing in a slick bore operation with much less drag resistance. They can also be used as a rock shield.

3.0.0 ◈ TYPES AND CAUSES OF COATING DAMAGE

Coatings can be applied to the piping to increase the resistance between the pipe surface and the soil environment. Care in surface preparation, application, and inspection is the best insurance against coating failure. The following sections describe some of the common failures associated with coatings on buried or submerged pipe.

3.1.0 Sagging

Sagging failures are usually the result of coating thickness problems. Many coatings continue to cure over time. In areas where the coating application was too heavy, they often check or crack, leading to rapid coating failure. Runs and sags in pipe coatings are most often the result of incorrect or poor application practices.

Examples of improperly applied coatings include material that was not brushed out sufficiently to eliminate thick areas, or sprayed materials applied with incorrect spray gun adjustments. Holding the spray gun too close to the surface or not moving it quickly enough over the surface can result in thick applications and/or runs.

Coating sag can also be caused by soil stress where the soil shifts and creates moisture changes or temperature changes that may adversely impact the coating.

3.2.0 Cracking

Cracking failures are often related to the coating formula, and result in premature aging and weathering of the coating material. Cracking is not a surface problem but one in which the cracks at the point of failure extend completely through the coating material to the surface of the pipe.

Cracking occurs due to the varying weather and moisture conditions. Coatings can also experience cracking failure as a result of the contraction and expansion of the pipe or structure being protected. This is more likely to occur when the coating has been applied in a thicker film. The thicker the coating film, the greater the stress within the material as curing takes place.

It is possible to minimize cracking through the correct use of resins, plasticizers, and **pigments**. Reinforcing pigments, fibrous in nature, are effective for reinforcing a coating against cracking failure. The addition of resins or plasticizers also helps to soften the coating, making it more elastic, and less susceptible to the stresses caused by temperature changes, weather, and curing. *Figure 3* shows general cracking of a coating.

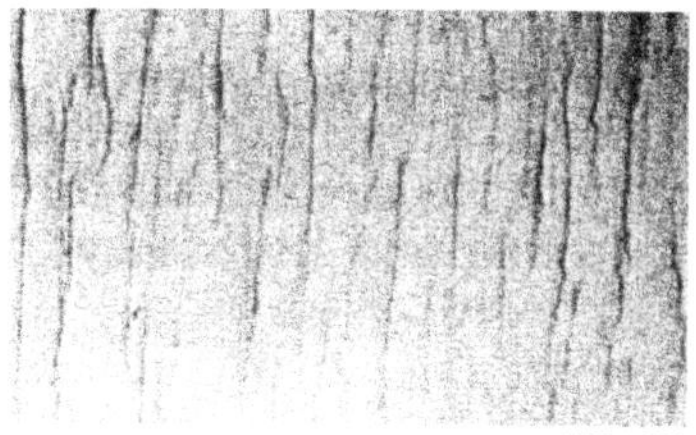

105F03.EPS

Figure 3 ◈ General cracking of a coating.

3.3.0 Wrinkling

Wrinkling, like sagging, is usually associated with coatings that are applied too thickly. As opposed to **shrinkage**, as is the case with checking and cracking, wrinkling is the result of the expansion of a coating. In this case, the coating surface expands more rapidly than the body of the coating during the curing process. Wrinkling is often a problem with oil-based coatings, alkyds, and similar materials.

Many oil-based coating materials contain drying agents. If a coating material contains too much surface drier, wrinkling may occur in spots where the coating layer is thicker. Cobalt driers are commonly used as surface driers. Zinc and lead compounds are classified as the through-drying driers.

Temperature can also impact wrinkling. Coatings that cure without wrinkling at standard application temperatures may wrinkle if the curing is accelerated by heat. An increase in temperature tends to cure the surface too rapidly, often leading to wrinkles in the coating surface.

Instructor's Notes:

Wrinkling is not only detrimental to appearance, but may also reduce the corrosion resistance of a coating. If, however, the coating dries thoroughly and retains its adhesion, the coating should still be effective.

If a coating material seems to have a tendency to wrinkle, apply it in several thin layers so there are no overly thick areas. Eliminating thick areas during the application process will minimize the potential for wrinkling. *Figure 4* shows wrinkling of a coating.

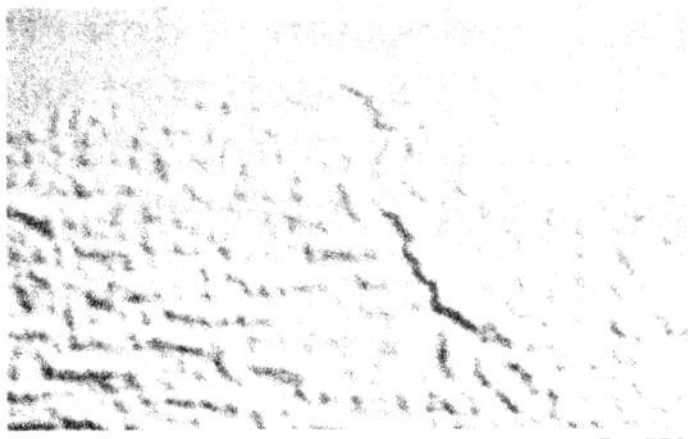

105F04.EPS

Figure 4 ◆ Wrinkling of a coating.

3.4.0 Blistering

Blistering is a common type of coating failure. Blisters, whether large or small, tend to be round and raised from the surface. Blisters might be liquid-filled or dry when they occur. *Figure 5* shows typical hemispherical blistered surface.

Blistering is usually caused by moisture working its way under the surface of the coating in areas where adhesion may break down. This moisture, once it gains an entrance into or under the coating material, can condense and form a liquid-filled blister or form a dry blister from the pressure of the moisture vapor. Blisters typically form in two locations: underneath the coating (separating it from the protected surface) or between coating layers.

Blistering can be caused by hydrogen gas that forms on the pipeline surface under the coating. Gas pressure separates the coating from the pipe surface. It might also occur where surfaces have been corroded and some remaining salts or oxides provide the conditions that support blistering.

Blistering is a serious problem and should be corrected before it can lead to further damage. Localized spots can be treated by removing and re-applying the coating. In the case of general or widespread blistering, the maintenance crew must remove the coating, clean and prepare the surface to accept a new coating, and replace the coating with a suitable material for the location and environment.

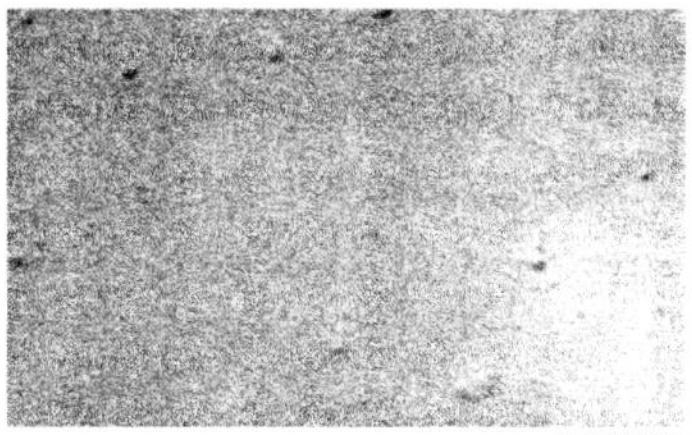

105F05.EPS

Figure 5 ◆ Typical hemispherical blistered surface.

3.5.0 Peeling

When a coating peels (see *Figure 6*) , it is usually the result of an adhesion problem rather than contamination by invading chemicals attacking the coating surface or from the use of an unsuitable coating material. Peeling is often related to expansion or contraction of the coating material to the extent that it loses its surface adhesion.

Peeling can occur when improper surface preparation techniques are used or when some form of contamination is introduced between coating layers as they are applied, causing one or more layers of the coating to peel away. In addition to these failures, coatings that are applied too thickly often develop into a film that is strong enough to overcome the adhesion of the coating to the pipe surface, resulting in a loss of adhesion and eventual widespread peeling.

105F06.EPS

Figure 6 ◆ Peeling of a coating.

3.6.0 Flaking

Flaking (see *Figure 7*) is an adhesion-related failure similar in appearance to peeling. In the case of flaking, the coating material usually becomes hard and brittle from curing too quickly or beyond the desired level of flexibility and, in reaching this inflexible state, shrinks and pulls away from the surface of the pipe. As the coating

Discuss types of coating damage and their causes.

Classroom

Have trainees list the proper response to each type of coating damage. Discuss the consequences of each type of coating damage.

Teaching Tip

See the Teaching Tip at the end of this module.

Homework

Assign review of Sections 4.0.0–6.0.0 for the next session.

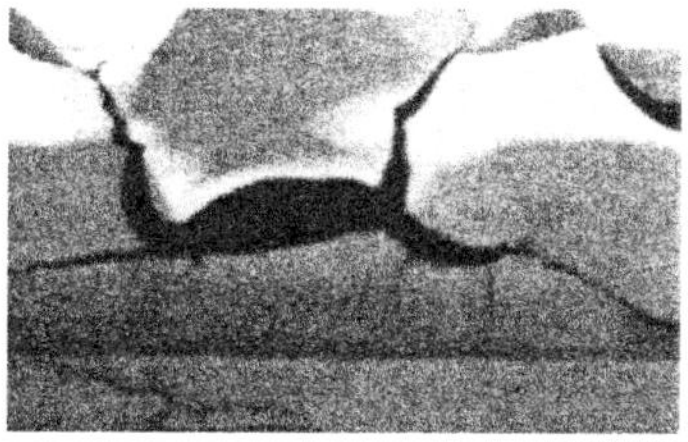

105F07.EPS

Figure 7 ◆ Flaking on a heavy organic coating.

material shrinks, cracks begin to appear in the surface and the further tension caused by the shrinking causes the cracked pieces to curl at the edges and continue to peel, or pull back, away from the pipeline surface.

3.7.0 Intercoat Delamination

Delamination is the term used to describe that condition where the adhesion between coating layers is lost (for example, from blistering) and the layers begin to separate from each other. Coatings often fail in this manner when new coating material is applied over an existing coating after maintenance or repairs have been completed. The old coating material, which may have been in place for an extended period of time, may have started to deteriorate or was not properly prepared to receive the new material. If the new coating fails to adhere completely to the old, delamination may result. *Figure 8* shows delamination of a coal tar epoxy coating on the interior of a large pipe.

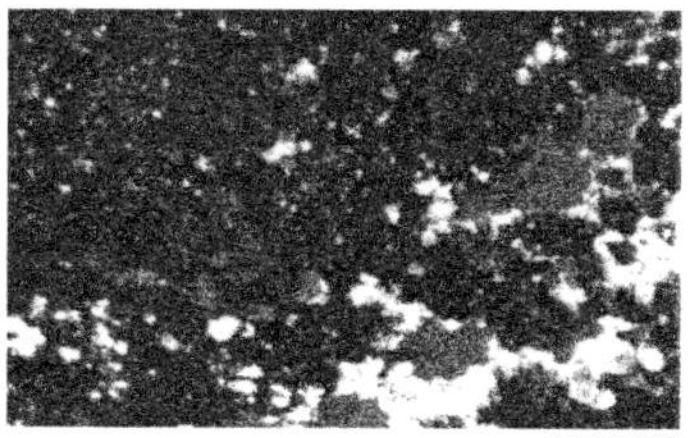

105F08.EPS

Figure 8 ◆ Delamination of a coal tar epoxy coating.

Delamination can be caused by incompatibility of materials, such as when a like material is not used to recover a repair coating that is applied too thick. Chemical invasion from acids or separation of coating layers caused by strong solvents are also causes of delamination.

Localized or spotty delamination can be repaired in the same way as blistering or peeling. Small areas can be treated by removing and reapplying the coating material. When dealing with widespread delamination, the pipeline maintenance crew must remove the coating, clean and prepare the surface to accept a new coating, and replace the coating with a suitable material for the location and environment.

3.8.0 Undercutting

When **rust** forms at a break in the coating material, often caused by damage or other deterioration, it can spread under the coating over the surface of the pipe. This form of corrosion, referred to as undercutting, can also occur without this breakdown in coating material integrity if the coating has been applied over a corroded or contaminated surface.

Undercutting demonstrates how critical careful, thorough pipeline inspections are. It is easy to overlook minor damage to the coating surface, which may lead to undercutting corrosion. *Figure 9* shows an example of the undercutting corrosion that can occur at the location of a minor cut or abrasion on the coated surface.

105F09.EPS

Figure 9 ◆ Type of undercutting due to minor abrasion.

Undercutting corrosion can quickly lead to more widespread corrosion once the protective coating surface has been breached. Corrosion builds up and spreads between the coating material and the pipe surface where typical conditions exist for the formation of an ideal corrosion cell. *Figure 10* shows undercutting beginning at a break in the coating. The coating is actually being turned up at the sides of the break as the corrosion product works underneath.

Damaging levels of **pitting** often result from undercutting. In areas susceptible to serious corrosion, detailed planning and execution are needed to reduce or eliminate undercutting corrosion.

Instructor's Notes:

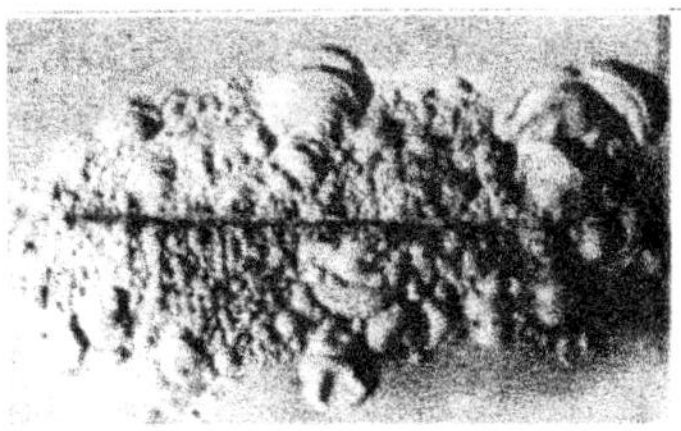

Figure 10 ◆ Undercutting beginning at a break in the coating.

4.0.0 ◆ VISUALLY INSPECT BURIED PIPELINE FOR CORROSION AND MECHANICAL DAMAGE

The repair and proper maintenance of coatings is equally as important as application of the original coating. Repair and maintenance starts with an in-depth inspection of the pipeline to determine the extent of the corrosion or damage that is present. The extent of the coating failure or damage that is found during this inspection process determines to a considerable degree the method as well as the extent of the repair. This inspection should include a visual inspection for corrosion such as pitting, metal loss, and general corrosion on the pipe, as well as for mechanical damage such as **gouges**, scrapes, and **dents**.

The normal operating procedure for visually inspecting a buried pipeline for corrosion and mechanical damage should follow these general steps:

Step 1 Ensure that the areas of the pipeline system have been properly excavated with care not to damage the buried pipeline.

Step 2 Test the atmosphere for any hazardous gases and ensure that it is safe to enter the excavation.

Step 3 Ensure that the trench is safe to enter and that proper safety procedures are in place prior to entering the excavation.

Step 4 Visually inspect the pipe for any signs of soil discoloration, wetness, or fungus growth before cleaning the dirt and contaminants away from the pipe. Take samples of any unusual signs of contamination and document any unusual signs that are observed.

> **NOTE**
> Samples may be taken to the lab for further testing of the contamination that is present.

Step 5 Visually inspect the pipe for signs of pitting to the surface of the pipe.

Step 6 Determine the extent of the pitting to the pipe and document the results per your company specifications.

Step 7 Visually inspect the pipe for signs of metal loss. If signs of metal loss are present, use the appropriate test instrument to determine the extent of the metal loss and document the results per your company specifications.

Step 8 Visually inspect the pipeline that is exposed for any general corrosion problems with special emphasis on pipe connections and welds. Document your observations per your company specifications.

> **NOTE**
> Metal loss could be from general corrosion, long-line corrosion, and biological corrosion. Generally, documentation of your observation should include the type, location, size, and depth of the corrosion.

Step 9 Visually inspect the pipe for mechanical damage such as gouges, scrapes, and dents. If mechanical damage is found, use the appropriate test instrument to determine the extent of the damage and document the results per your company specifications.

Step 10 Investigate any area that adds to or takes away from the appearance of the coating. This may include thin spots or **holidays**.

Step 11 Carefully remove the coating and any by-products that may be under the coating for a sample for later use at a lab for further investigation and testing.

Step 12 Refer to your company policy to determine what action is required for this abnormal condition. If your inspection reveals an unsafe or environmental hazard, notify your supervisor immediately.

Materials

Ensure that you have everything required for the demonstration and testing during this session.

Laboratory

Have trainees practice inspecting a buried pipeline for damage.

Safety

Discuss the dangers of flammable situations in the work area. Ask trainees to explain how to prevent flammable situations.

Safety

Stress the importance of following all fire safety guidelines on the work site.

Laboratory

Have trainees practice visually inspecting the interior of a pipe for corrosion and damage. Stress the danger of fire.

Demonstration

Bring in a mechanical pit gauge. Ask a trainee to explain how to use and read the pit gauge.

Classroom

Ask trainees to explain possible causes of inaccurate mechanical pit gauge readings. Ask trainees if they have any questions.

If the pipe has been opened for maintenance, you must also visually inspect the interior of the pipe for corrosion and damage. Follow these general steps to visually inspect the pipe for internal corrosion and damage:

Step 1 Ensure that the excavation area is still safe to enter.

Step 2 Test the atmosphere for any hazardous gases and ensure that it is safe to work around the open pipe.

> **WARNING!**
> This test should also include any testing for flammable material that may have been spilled when the pipe was opened.

Step 3 Ensure that the pipe is open and repair personnel are ready for your inspection.

> **WARNING!**
> Repair personnel should not perform any work in the area of your inspection that could cause sparks or a flammable situation, such as welding.

Step 4 Obtain an intrinsically safe and/or explosion-proof light source that will provide good illumination of the interior of the pipeline.

Step 5 Inspect the interior of the pipe for any abnormal conditions.

> **NOTE**
> For a liquid pipeline, the bottom of the pipeline is where most corrosion conditions occur. However, for gas pipeline, the corrosion could occur anywhere on the interior surface of the pipe.

Step 6 Visually inspect the interior of the pipe for pitting or tracking from water trailing down the pipeline.

Step 7 Visually check the interface sections of the pipeline for corrosion.

Step 8 Inspect for any deviation from the original pipe and for black scale, black dust, or deposit buildup.

Step 9 Carefully remove deposits and use them as a sample for later testing.

Step 10 Clean under the removed material with a rag to further check the amount of damage that may have occurred.

Step 11 Compare the pipe that was removed with the installed pipe and document your results.

Step 12 Document your findings per your company policy and procedures.

4.1.0 Types of Pit Gauges

There are several different types of pit gauges common to the petroleum industry. These pit gauges are discussed in the following sections.

4.1.1 Mechanical Pit Gauges

One of the most common tools used to check the extent of corrosion to a pipe is the pit gauge. Most pit gauges are easy to use and to carry. These gauges, if properly maintained, can provide accurate measurements.

When corrosion or damage is discovered, the corrosion technician or maintenance team member must determine the remaining wall thickness of the pipe in order to set the priority for repair or replacement of the corroded section of pipe. The common mechanical pit gauge *(Figure 11)* is a small, sturdy tool and is easy to use in making a quick determination of the extent of corrosion damage present. These gauges are usually made of polished stainless steel and can easily be carried in the tool kit. Care must be taken to use these gauges properly and to recognize and understand their limitations.

Inaccurate readings can result if the gauge is not aligned properly, resting on an even surface. Where severely pitted or partially coated pipe and associated structures are involved, these handheld pit gauges will provide the user with only an approximation of pit depth. The depth measurement scale on the mechanical pit gauge is marked in sixteenths of an inch and in mils (thousandths of an inch). Most mechanical pit gauges include useful formulas, tables, and data giving outside diameter (OD), wall thickness, and weight per foot for various sizes of pipe.

4.1.2 Dial Pit Gauges

When pit depth measurements are critical and accuracy is a priority, dial pit gauges, designed for a more accurate measurement of corrosion pit depth, may be used. Most dial pit gauges are

Instructor's Notes:

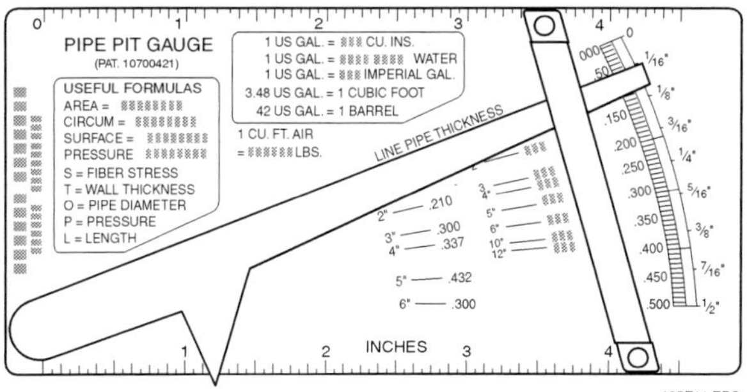

Figure 11 ◆ Mechanical pit gauge.

capable of measuring depth to extremely close tolerances. On the down side, these indicators are more fragile than mechanical gauges and should be handled with care to ensure their repeated accuracy. *Figure 12* shows a dial pit gauge positioned on a pipe.

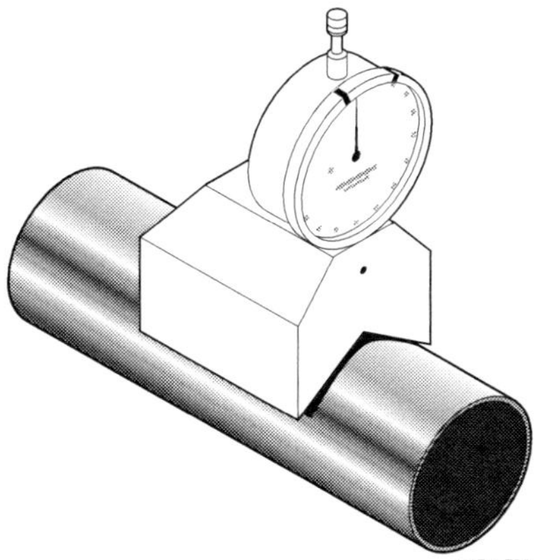

Figure 12 ◆ Dial pit gauge positioned on a pipe.

4.1.3 Micrometer Pit Gauges

A micrometer pit gauge will provide the pipeliner with the most accurate measurement of corrosion pit depth. Some models of manually operated micrometer pit depth indicators measure to an accuracy of +/–0.0002". These measuring tools are somewhat fragile and quite expensive as compared to other indicators.

Micrometer pit gauges should be handled with care to ensure their repeated accuracy and to safeguard the company's investment in these costly tools. Before using a micrometer pit depth indicator, check to see if it has been calibrated and certified as accurate according to your company's tool calibration procedure.

4.2.0 Calibrate Gauge and Check for Worn Point

The accuracy of any pit measurement is only as reliable as the instrument that is used to take the measurement. Before a pit gauge is used for measurement, it must be inspected for any damage and any worn points. During this inspection, always closely inspect the point of the instrument for any worn spots. If a pit gauge does have a worn point, the pit gauge should be replaced. Only use an instrument that is in excellent condition, otherwise your inspection may not be accurate. Prior to using the pit gauge, clean the pit in accordance with your company's approved policy.

> **WARNING!**
> The pipe may still be connected to the system. Make sure you follow your company policies and procedures for work in a hazardous environment.

4.3.0 Use Pit Gauge to Measure Pit Depth

To use a standard pit gauge, align the gauge lengthwise with the pipe to be inspected. The gauge must rest squarely on the pipe surface to provide an accurate reading of the pits. Move the

Audiovisual

Show Transparency 4 (Figure 12). Ask a trainee to explain how to use a dial pit gauge.

Classroom

Bring in examples of the three types of pit gauges. Have trainees familiarize themselves with the gauges. Discuss the advantages and disadvantages of each type of pit gauge.

Safety

Discuss the dangers of working with pipe that is connected to the system. Ask trainees to identify typical procedures for working in a hazardous environment.

Laboratory

Have trainees practice calibrating, inspecting, and cleaning a pit gauge. Ask trainees if they have any questions.

Laboratory

Ask a trainee to demonstrate the proper way to use a pit gauge. Then have trainees practice using a pit gauge to measure pit depth and length.

Classroom

Ask a trainee to explain the differences between the B31G and RSTRENG methods. Demonstrate how to use each method.

Audiovisual

Show Transparency 5 (Figure 14). Have a trainee explain how to interpret the figure.

Classroom

Discuss the types of information required by the inspection report form. Ask trainees to explain why having a reliable history of the pipe is important.

arm on the gauge so that the pointer is at the bottom of the pit. Read and record the corresponding number of the scale of the gauge. Continue this process for a general sampling of the pits to be measured. *Figure 13* shows the proper technique to measure the pit depth.

When using a dial-type pit gauge, the gauge must first be zeroed out. This will calibrate the gauge. Then place the gauge squarely on the pipe surface. Push the pointer into the pit to be measured with the control button on the side of the gauge. This should register a reading on the gauge. Record this measurement. Continue this process until an adequate sample of readings has been recorded.

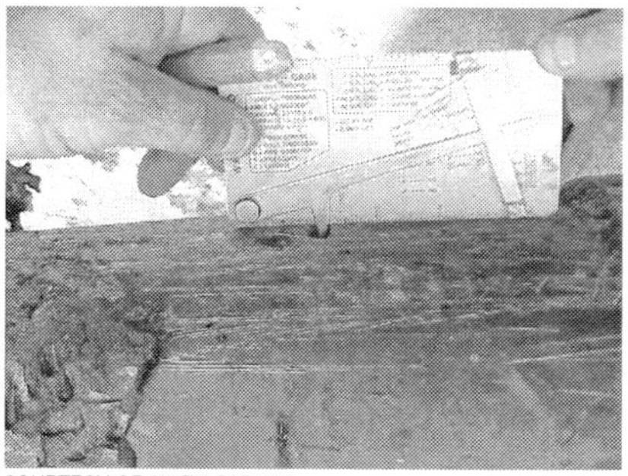

COURTESY OF ALLEN CARLILE

105F13.EPS

Figure 13 ◈ Measuring pit depth.

4.4.0 Use Pit Gauge to Measure Pit Length

With the pointer of the pit gauge in the pit, observe the length of the pit on the bottom scale of the mechanical pit gauge. After this measurement is taken, make sure that the measurement is recorded on the appropriate company forms.

4.5.0 Take Multiple Readings to Gather Remaining Strength (RSTRENG) Data

The B31G and RSTRENG (remaining strength) corrosion assessment methods are used by pipeline operators to evaluate the structural integrity of pipelines with corrosion-based metal loss. The basis for these two calculation approaches has been in existence for more than 25 years. Both methodologies have been validated through burst testing of pipe with corrosion and failures of pipe in service.

While acceptable, the B31G method is recognized to be very conservative in nature. The RSTRENG criterion was developed to offset this conservative characteristic. Pipeline operators are required by *49 CFR 192* and *195* to evaluate the remaining strength in any corroded pipeline, using either ASME/ANSI B31G or RSTRENG to evaluate the remaining strength in any corroded pipeline. The B31G method calculates predicted failure points based on the maximum length and maximum depth of the corrosion defect on the pipe.

The RSTRENG approach combines this method with a more detailed profile of the defect by evaluating a greater number of depth measurements (minimum of three with one being the maximum depth in the defect) taken along the horizontal length of the defect. RSTRENG works equally well with single defects or with several corrosion arrays that interact together. *Figure 14* shows the typical river bottom profile of the corrosion defect as determined through the RSTRENG method.

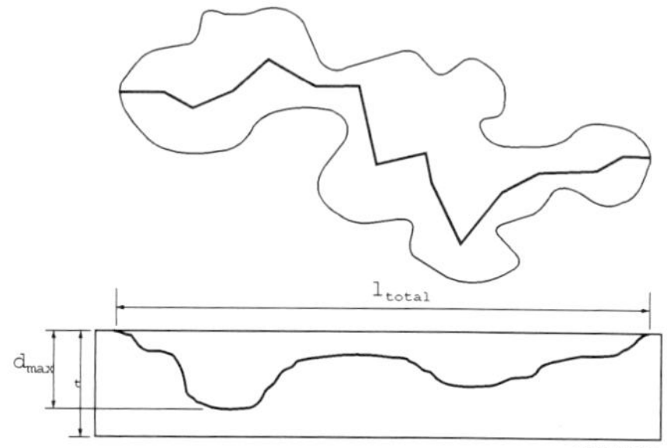

105F14.EPS

Figure 14 ◈ Typical river bottom profile.

5.0.0 ◈ DOCUMENT THE LOCATION OF THE CORROSION OR DAMAGE

The inspection of the pipeline is vital to the safe operation of your pipeline system. The previous section discussed the general requirements for visual inspection of a buried pipeline. This inspection is time-consuming and costly to your company.

The documentation of the inspection fulfills many useful functions, one of which is to maintain a reliable history of the condition of the pipe so that cost-effective repairs can be scheduled before any pipe failure occurs.

Instructor's Notes:

When you inspect a pipe, it is your responsibility to accurately record the results of your observations using the appropriate company procedures so that information will be useful to all company officials. *Figure 15* shows a typical pipeline inspection form. Your company will have a specific form that you are required to use for this inspection; the form that is shown is for general discussion only.

All inspection reports should include a location section. This section is used to describe the exact location of the pipe that is inspected. Whenever possible, try to have a copy of the last inspection report available for comparison, to ensure that the location data is correct. This will help in completing the form and identifying any changes that may have occurred since the last inspection.

A separate section of the form should include the material and condition of the pipe as you observed it during your inspection. Complete this section as the information is asked. Record any other relevant observations. Another section is for your observation of external corrosion. Complete the section as the questions are asked. Normally the last section that you need to complete contains the details of the corrosion that you have observed. When completing this section, be as specific as possible with a drawing, specification numbers, location and depth of corrosion, and any information that you feel is needed to properly document the corrosion problem. Normally, the company form will have specific entries for some of this information.

Now that you have properly inspected the pipe and taken the time to accurately record your findings, the inspection report must be turned in to your company officials in a timely manner. Generally, your company procedure will provide you with the timelines for submission of this report.

6.0.0 ◆ PROCEDURES

The following sections detail basic procedures with which trainees must be familiar when working in the pipeline industry. These are general procedures only. Always consult your supervisor and/or company literature to determine whether your company has more specific policies and procedures in place.

6.1.0 Inspecting for Physical Damage on Buried or Submerged Pipe

The following steps detail a basic procedure of how to inspect for physical damage on buried or submerged pipe. Please note that these steps are very general. Your company may have more specific procedures in place, which you should always follow. If you have any questions, always contact your supervisor. Remember to obtain and properly use the appropriate personal protective equipment for each task.

> **WARNING!**
> Always be alert to the possibility of unexpected leaks from the exposed pipeline such as hazardous liquid or carbon dioxide (CO_2). Safely eliminate all ignition sources and immediately notify the appropriate personnel.

Step 1 Carefully inspect the exposed pipe area for physical/mechanical damage. Note any mechanical damage and/or abnormalities on the pipe surface.

> **CAUTION**
> All corrosion should be thoroughly documented and immediately reported to the person(s) responsible for correction and/or repair.

Step 2 Once you have located damage on the pipe, identify the specific type of damage. That is, classify it as a **scratch**, gouge, dent, or other specific type of damage. This will aid you in deciding upon corrective actions.

> **CAUTION**
> Be alert for pipeline system damage such as dents, gouges, or coating damage. Thoroughly document your findings and notify the person(s) responsible for correction and/or repair.

Step 3 Next, determine what caused the damage. Physical/mechanical damage is commonly caused by equipment during general construction work or farming activities. Also common is damage caused by improper pipeline construction procedures such as pads or skids being left underneath pipelines in the pipeline ditch during construction.

Step 4 Once the cause has been found, you must determine the shape and depth of the physical/mechanical damage. This information is vital in deciding upon the appropriate method of repair.

Safety

Stress the danger of unexpected pipeline leaks. Emphasize the importance of being alert for pipeline damage. Discuss the proper response to leaks and to pipeline damage.

Laboratory

Have trainees practice inspecting a buried pipe for physical damage. Ensure that trainees completely fill out the pipeline inspection form.

PIPELINE INSPECTION REPORT

Frequency: At Time of Inspection

PART A — LOCATION

Reason For Inspection ______________________________

1. Location: Division ______________ County ______________ State ______ Date of Inspection ______________

A) **Mainline**
Alignment Sheet ________ Pipeline Size/Segment ________
Beg. M.P. ______________ End M.P. ______________
Beg. Sta. No. ______________ End Sta. No. ______________

Note: Station numbers shall be measured from a known landmark (Road Crossing, etc.)

B) Company Facility: Pump Station, Tank Farm, etc.
Location Name ______________________
Physical Location of Pipe Inspected ______________________
Milepost ______________________

PART B — MATERIAL/CONDITIONS

2. Piping:
Pipe Size ______________ In. O.D. Nom. W.T. ____________ In. Actual W.T. ____________ Grade ____________

3. External Pipe Coating:

A) Coating
☐ Bare
☐ LAC 450
☐ Taped
☐ Insulated
☐ Other ____________

B) Application
☐ Mill Coated
☐ Yard Coated
☐ Field Coated
☐ Other

C) Coating Thickness
Factory Applied____ mile
Field Applied ____mile

D) Condition of Coating
☐ Good ☐ Fair
☐ Poor

E) Bond
☐ Good
☐ Fair
☐ Poor

F) Cause of Coating Failure
☐ Damage
☐ Disbondment
☐ Defective Application
☐ Decomposition
☐ Other ____________

4. Environment
☐ Clay ☐ Loam ☐ Sand ☐ Gumbo ☐ Caliche ☐ Rock ☐ Other ____________
☐ Wet ☐ Dry ☐ Moist Depth of Cover (Inches) ____________
Soil Resistivity Measurement ______________ (ohm-cm) Measurement Method ______________

5. Coating Repaired ☐ No ☐ Yes
Repair System ______________________________

6. Pipe-to-Soil Potential at Inspected Location (At Grade) ____________ (volts) (At Pipe/Soil Interface) ____________ (volts)

7. Were any Non-Destructive Tests Completed? ☐ No ☐ Yes Type ____________

8. Evidence of: ☐ Corrosion ☐ Defect ☐ Damage ☐ Stress Corrosion Cracking (S.C.C.) ☐ Calcareous Deposits

9. Test Station Added ☐ No ☐ Yes I.D. No. ______________________

NOTES:
1) If S.C.C. or Corrosion is Noted, Complete Below and Back of Report (External Only, See Note 4)
2) If Defect or Damage is Noted, Complete Back of Report
3) If Repairs are Required, AAPL Form 502 "REPAIR REPORT FOR LEAK DAMAGE, CORROSION OR TEST FAILURE" Must be Attached.
4) AAPL Form 508 "INTERNAL CORROSION INSPECTION REPORT" Should be Completed When Internal Corrosion Is Detected.

PART C — EXTERNAL CORROSION

10. General Corrosion Information:

A. DESCRIPTION
☐ Less than 10% of Nom. Wall Thickness, Complete Front of Report
☐ Greater than 10% of Nom. W.T., Complete Front and Back of Report
☐ Pitting ☐ General ☐ Localized
Max Depth ________ (In.) Longitudinal Length ____________ (In.)

B. STATUS
☐ Active
☐ Inactive
☐ Unknown

C. CAUSE
☐ Galvanic
☐ Stray Current
☐ Unknown
☐ Other ________

11. Cathodic Protection: ☐ Impressed ☐ Galvanic

12. Pipe-to-Soil Potential:
a) Last Pipe-to-Soil Measurement at Nearest Point Each Side of Corroded Area
Upstream ____________ (volts) Downstream ____________ (volts)
b) Distance from Corroded Area to Measurement Points:
Upstream ____________ (feet) Downstream ____________ (feet)
c) Date of Measurements: Upstream ____________ Downstream ____________

13. pH of Soil Near Corroded Area ______________ pH of Moisture Under Coating ______________

14. Pipe Surface Temperature if Fluid is Heated ____________ (deg F)

Remarks ______________________________

Distribution: (Original) Corr. Ctrl./Eng (Copy) Division Office (Copy) Other ____________

______________________ ___/___/___ ______________________ ___/___/___
Technician or Inspector *Date* *Supervisor* *Date*

105F15A.EPS

Figure 15 ◈ Typical pipeline inspection form (1 of 2).

Instructor's Notes:

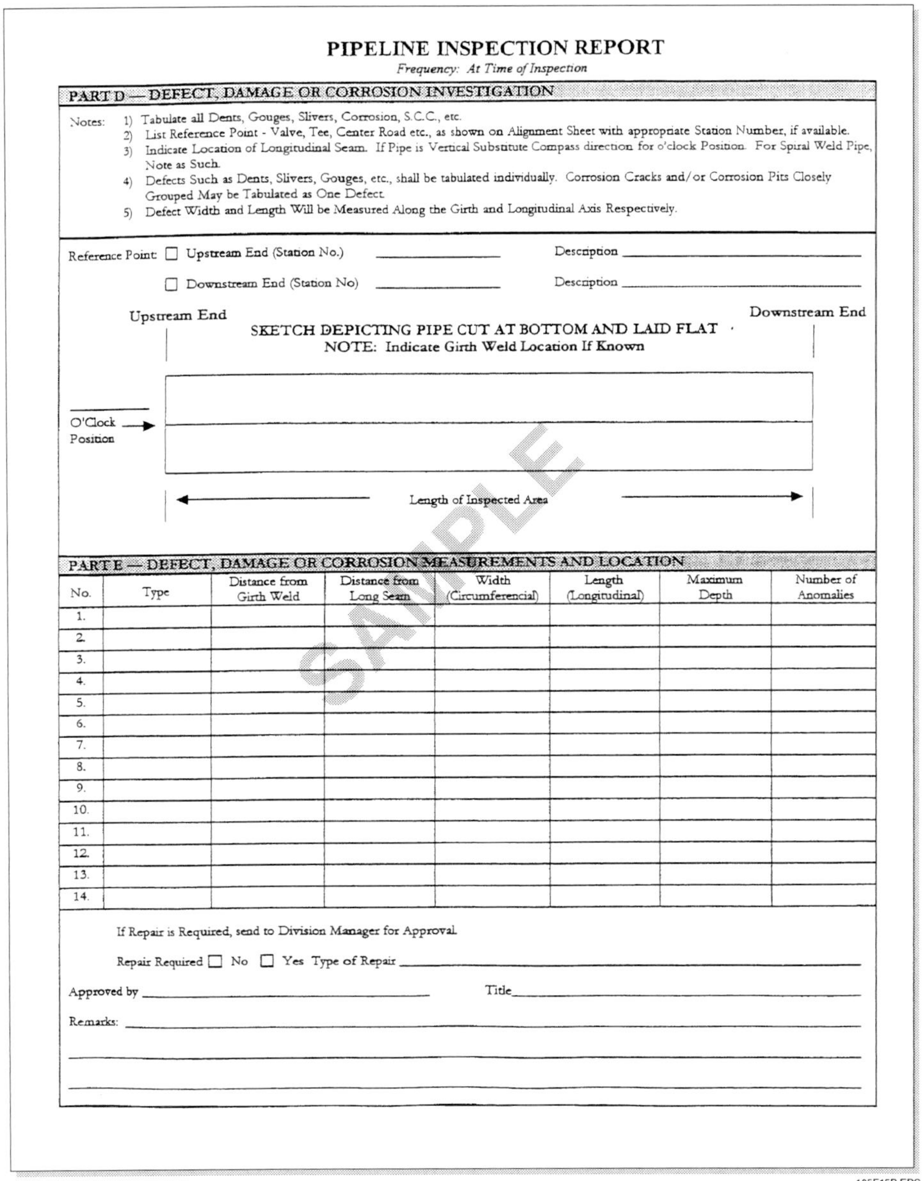

PIPELINE INSPECTION REPORT

Frequency: At Time of Inspection

PART D — DEFECT, DAMAGE OR CORROSION INVESTIGATION

Notes:
1) Tabulate all Dents, Gouges, Slivers, Corrosion, S.C.C., etc.
2) List Reference Point - Valve, Tee, Center Road etc., as shown on Alignment Sheet with appropriate Station Number, if available.
3) Indicate Location of Longitudinal Seam. If Pipe is Vertical Substitute Compass direction for o'clock Position. For Spiral Weld Pipe, Note as Such.
4) Defects Such as Dents, Slivers, Gouges, etc., shall be tabulated individually. Corrosion Cracks and/or Corrosion Pits Closely Grouped May be Tabulated as One Defect.
5) Defect Width and Length Will be Measured Along the Girth and Longitudinal Axis Respectively.

Reference Point: ☐ Upstream End (Station No.) ____________ Description ____________________

☐ Downstream End (Station No) ____________ Description ____________________

Upstream End   Downstream End

SKETCH DEPICTING PIPE CUT AT BOTTOM AND LAID FLAT

NOTE: Indicate Girth Weld Location If Known

O'Clock Position →

← Length of Inspected Area →

PART E — DEFECT, DAMAGE OR CORROSION MEASUREMENTS AND LOCATION

No.	Type	Distance from Girth Weld	Distance from Long Seam	Width (Circumferencial)	Length (Longitudinal)	Maximum Depth	Number of Anomalies
1.							
2.							
3.							
4.							
5.							
6.							
7.							
8.							
9.							
10.							
11.							
12.							
13.							
14.							

If Repair is Required, send to Division Manager for Approval.

Repair Required ☐ No ☐ Yes Type of Repair ____________________

Approved by ____________________ Title ____________________

Remarks: ____________________

105F15B.EPS

Figure 15 ◆ Typical pipeline inspection form (2 of 2).

Safety

Ask trainees to identify the proper response to a fire or explosion. Answer any questions trainees may have.

Performance Testing

Have each trainee perform Performance Task 1 to your satisfaction. Fill out Performance Profile Sheets for each trainee.

Safety

Stress the importance of notifying appropriate personnel about corrosion and pipeline system damage.

Laboratory

Have trainees study the steps for inspecting for external corrosion on a buried pipe. Then have them practice implementing the steps.

Performance Testing

Have each trainee perform Performance Task 2 to your satisfaction. Fill out Performance Profile Sheets for each trainee.

Step 5 Carefully determine the proximity of the dent in relationship to the seam and girth weld, and measure the depth of the dent. This is also important information to have for determining the appropriate method of repair.

Step 6 Thoroughly document your findings and make company notifications as applicable. Proper documentation is critical to future analysis and identification of problem areas.

Step 7 Always be alert for potential abnormal operating conditions. Identify them and determine the appropriate action(s) to be taken.

> **WARNING!**
> In the case of fire, explosion, or ignition of hydrocarbon, evacuate the area immediately and notify the local emergency personnel.

6.2.0 Inspecting for External Corrosion on Buried or Submerged Pipe

The following is a basic guideline of how to inspect for external corrosion on buried or submerged pipe. Please note that these steps are very general. Your company may have more specific procedures in place, which you should always follow. If you have any questions, always contact your supervisor. Remember to obtain and properly use the appropriate personal protective equipment for each task.

> **WARNING!**
> Always be alert to the possibility of unexpected leaks from the exposed pipeline such as hazardous liquid or carbon dioxide (CO_2). Safely eliminate all ignition sources and immediately notify the appropriate personnel.

Step 1 First, you must examine the exposed pipe for any areas of external corrosion.

> **CAUTION**
> All corrosion should be thoroughly documented and immediately reported to the person(s) responsible for correction and/or repair.

Step 2 Specifically locate and identify each area of corrosion on the pipe and determine what type of damage has occurred. Note the particular type of corrosion present, such as isolated, general, pitting, and so on.

> **CAUTION**
> Be alert for pipeline system damage such as dents, gouges, or coating damage. Thoroughly document your findings and notify the person(s) responsible for correction and/or repair.

Step 3 Examine the pipe to determine the extent of corrosion that has occurred. When corrosion is present, you must measure the longitudinal axis of the corroded area(s) to determine the length. Next, measure the depth of the pit(s) using a **pit depth gauge**. Note the location of the corroded area(s) of the pipe (i.e. top, bottom, or sides). Record all information according to company policies and procedures.

Step 4 Thoroughly document your findings and make company notifications as applicable. Proper documentation is critical to future analysis and identification of problem areas.

Step 5 Always be alert for potential abnormal operating conditions. Identify them and determine the appropriate action(s) to be taken.

> **WARNING!**
> In the case of fire, explosion, or ignition of hydrocarbon, evacuate the area immediately and notify the local emergency personnel.

6.3.0 Inspecting the Condition of External Coatings on Buried or Submerged Pipe

The following procedure is a basic guideline for inspecting the condition of external coatings on buried or submerged pipe. Please note that these steps are very general. Your company may have more specific procedures in place, which you should always follow. If you have any questions, always contact your supervisor. Remember to obtain and properly use the appropriate personal protective equipment for each task.

Instructor's Notes:

WARNING!
Always be alert to the possibility of unexpected leaks from the exposed pipeline such as hazardous liquid or carbon dioxide (CO_2). Safely eliminate all ignition sources and immediately notify the appropriate personnel.

Step 1 First, identify areas of the exposed pipe that have coating damage. Recognize any physical abnormalities in the coating that require repair.

CAUTION
All corrosion should be thoroughly documented and immediately reported to the person(s) responsible for correction and/or repair.

Step 2 Next, identify the type of existing pipe coating. You must be able to identify the existing type of coating so that the proper coating selection can be made for repair. This will help ensure, among other things, compatibility, adhesion, mechanical integrity, and proper temperature range.

Step 3 Now identify any flaws or abnormalities in the coating and/or pipe. You must be able to recognize any physical damage to the coating such as peeling, cracking, bubbling, or **disbondment** so that you may make an accurate assessment of the coating damage.

CAUTION
Be alert for pipeline system damage such as dents, gouges, or coating damage. Thoroughly document your findings and notify the person(s) responsible for correction and/or repair.

Step 4 Remove all loose and/or disbonded coating to allow for proper inspection and evaluation of the underlying pipe. Removing defective coating(s) and proper preparation of the pipe surface ensures good adhesion of the repair coating to the pipe.

Step 5 Next, you must identify possible causes of the coating failure. Consider the geographic location of the failed coating. Inspect the environment for clues and causes. Look for signs of poor surface preparation or application, debris in the backfill, third-party encroachments, or the presence of foreign materials.

Step 6 Thoroughly document your findings and make company notifications as applicable. Proper documentation is critical to future analysis and identification of problem areas.

Step 7 Always be alert for potential abnormal operating conditions. Identify them and determine the appropriate action(s) to be taken.

WARNING!
In the case of fire, explosion, or ignition of hydrocarbon, evacuate the area immediately and notify the local emergency personnel.

Summary

Protective coating for buried pipe is a specialty product that represents the most widely used method of corrosion control. The protective coating is designed to give long-term protection under a broad range of corrosive conditions to which the buried pipe is exposed.

There is a variety of common coatings that are used to coat a buried pipe. The most common types of buried pipe coatings that are used today are as follows: coal tar, asphalt, tape, fusion bonded epoxy, extruded polyethylene jacket, concrete, wax, fiberglass, mastic, and copolymer. Each type of coating has a special application, but all the coatings are designed to protect the pipe from its environment.

Coatings for buried pipe are exposed to many different types of environmental conditions found in the soil. The most common failures of buried pipe coatings are sagging, cracking, wrinkling, blistering, peeling, flaking, intercoat delamination, and undercutting.

When inspecting a buried pipe, the corrosion technician needs to be sure that the atmosphere is free of any hazardous gases and that the work site is safe. Prepare the surface for the inspection by removing any foreign matter from the pipe. Inspect the pipe coating for any damage, including pitting. Properly document the results of your inspection, per your company policy, which should include the exact location and extent of any damage that may have been observed.

Safety

Ask trainees to explain safety concerns relevant when inspecting a pipeline.

Laboratory

Have trainees practice inspecting external coatings on buried pipe. Ensure that they wear appropriate personal protective equipment.

Classroom

Go over the Review Questions. Answer any questions trainees may have on materials covered to this point.

Examination

Administer the Module Examination. Be sure to record the results of the Exam on Craft Training Report Form 200 and submit the results to the Training Program Sponsor.

Performance Testing

Have each trainee perform Performance Task 3 to your satisfaction. Fill out Performance Profile Sheets for each trainee.

Ensure that all Performance Tasks have been completed and Performance Profile Sheets for each trainee are filled out. Be sure to record the results of the testing on Craft Training Report Form 200 and submit the results to the Training Program Sponsor.

Review Questions

1. Coal tar pipe coatings have a tendency to _____ when exposed to direct sunlight.
   a. peel
   b. flake
   c. chalk
   d. alligator

2. FBE coatings are applied to _____ pipe.
   a. rusty
   b. tar-coated
   c. cold
   d. hot

3. Wax is normally used with _____.
   a. paint
   b. coal tar
   c. primer
   d. asphalt

4. Copolymer can be used to coat pipe and as a _____.
   a. pipe support
   b. drainage pipe
   c. rock shield
   d. pipe plug

5. Coating cracking is not a _____ problem.
   a. surface
   b. sunlight
   c. soil stress
   d. water exposure

6. Blisters are usually _____ in shape.
   a. square
   b. irregular
   c. round
   d. elongated

7. Undercutting is also called _____.
   a. heat damage
   b. rust
   c. soil stress
   d. mechanical damage

8. If a pipe has been opened for maintenance, always inspect _____.
   a. for welds
   b. for internal corrosion
   c. the pipe supports
   d. the coating

9. The B31G and RSTRENG methods are used to evaluate _____.
   a. average rate of flow
   b. pit depth
   c. repair effectiveness
   d. structural integrity

10. A pipeline inspection report will generally require information about _____.
   a. external corrosion
   b. prior report data
   c. the weather conditions
   d. nearby operations

Instructor's Notes:

GLOSSARY

Glossary

This glossary will help you learn some terms specific to inspecting buried and submerged pipe when exposed. Trade Terms introduced in this module appear in italic text.

**Alligatoring:* Pronounced wide cracking over the surface of a coating, which has the appearance of alligator hide.

Asphalt coating (Somastic): A mastic composition that converts to a solid protective adherent film after application to a surface.

Bacteriological corrosion: This specific type of corrosion is caused by the action of bacteria on various metals (typically SRB and APB bacteria).

Bonding/adhesion: The attachment of a coating to a substrate. Two types of adhesive bonds are chemical and mechanical.

Cigarette wrap: This application method is for cut lengths of sheet coating. The sheet width must be greater than the circumference of the pipe in order to provide an overlap. This method is not commonly used except for heat shrink sleeves.

Coal tar coatings: Coal tar is used as a paint-like coating for petroleum-product pipelines, providing protection from both water and cathodic corrosion (also called coal-tar enamel). Coal tar contains several hundred organic chemicals and may have a felt or mesh outer wrap.

**Coating:* A liquid, liquefiable, or mastic composition that, after application to a surface, is converted into a solid protective, decorative, or functional adherent film.

**Coating disbondment:* See *disbondment*.

Coating overlap: The overlapping of each layer, or application of a tape or cigarette-type coating, to ensure the coating is well bonded to the next layer. This helps prevent separation.

**Coating system:* The complete number and types of coats applied to a substrate in a predetermined order. (When used in a broader sense, surface preparation, pretreatments, dry film thickness, and manner of application are included.)

Dent: This is a depression that produces a reversal in the curvature of the pipe wall yet does not reduce wall thickness.

Dielectric resistance: The opposition that a device or material offers to the flow of direct current, equal to the voltage drop across the element divided by the current through the element.

**Disbondment:* The loss of adhesion between a coating and the substrate.

**Epoxy:* Type of resin formed by the reaction of aliphatic or aromatic polyols (like bisphenol) with epichlorohydrin and characterized by the presence of reactive oxirane end groups.

Extruded coatings (ExtruCoat): A factory-applied coating that is an extruded plastic such as polyethylene.

***Galvanic corrosion:** Accelerated corrosion of a metal because of an electrical contact with a more noble metal or nonmetallic conductor in a corrosive electrolyte.

Gilsonite®: A very pure form of asphalt used as a pipe coating.

Gouge: Surface damage that reduces the pipe wall thickness through displacement or removal of metal.

***Hand tool cleaning:** Removal of loose rust, loose mill scale, and loose paint to degree specified, by hand shipping, scraping, sanding, and wire brushing.

**Holiday:* A discontinuity in a protective coating that exposes unprotected surface to the environment.

Liquid epoxy coatings: A two-part material that, when combined, provides a high-strength low-shrinkage coating.

Petroleum coating products: Rust-prevention products that contain petrolatum.

Phenolic: A thermosetting resin used to coat pipe.

**Pigment:* A solid substance, generally in fine powder form, that is insoluble in the vehicle of a formulated coating material. It is used to impart color or other specific physical or chemical properties to the coating.

Pipe specification: The size, shape, thickness, and coating type of a pipe.

Pit depth gauge: A device that measures the depth of the depressions caused by corrosion.

**Pitting:* Localized corrosion of a metal surface that is confined to a small area and takes the form of cavities called pits.

**Polyester:* Type of resin formed by the condensation of polybasic and monobasic acids with polyhydric alcohols.

Pre-heat: The act of heating a surface and/or material to a desired temperature before the next step in a process.

**Primer:* A coating material intended to be applied as the first coat on an uncoated surface. The coating is specifically formulated to adhere to and protect the surface as well as to produce a suitable surface for subsequent coats. (Also referred to as Prime Coat.)

**Rust:* Corrosion product consisting of various iron oxides and hydrated iron oxides. (This term properly applies only to iron and ferrous alloys.)

Scratch: A thin, shallow cut or mark on the surface.

Shrink: Contraction of a coating that has been heated and then allowed to cool.

Shrink sleeve products: A shrink fit is an extremely tight fit resulting from shrinking the coating component around the structure. Heat is applied around a companion piece and as it is heated, the coating component shrinks around the structure, resulting in a shrink fit.

Specified minimum yield (SMYS): Internal design pressure based on diameter, thickness, and yield strength of the pipe such as 5LX-X60.

Spiral wrap: An application technique for tape wrap coating system. Tape is placed at a 20- to 45-degree angle along the girth of the pipe and overlapped according to the manufacturer's specification.

***Stray current corrosion:** Corrosion resulting from current through paths other than the intended circuit, e.g., by any extraneous current in the earth.

Tape wrap: A wrap made of polyethylene or polyvinylchloride that is applied to the piping over a primer coat.

Instructor's Notes:

MODULE 61105-02—TEACHING TIPS

The following are suggested activities or instructional methods to help you teach the material in this AIG.

General

When you call on someone to answer a question, the rest of the class relaxes or even tunes out because they expect that the question and answer will take place only between you and the trainee you called on. Instead, use this technique to involve more trainees in answering questions and to keep them on their toes.

1. Ask trainees to define a term or explain a concept.
2. After one trainee has answered, ask a trainee seated nearby if the answer is right. Then ask whether a trainee in the back of the room agrees.
3. Ask trainees to explain why they think an answer is right or wrong.
4. Use the session to clear up incorrect ideas and encourage trainees to learn from their mistakes.

Sections 2.0.0–2.10.0

Types of Coating

This Quick Quiz will familiarize trainees with the characteristics of various types of coatings. You will need photocopies of the quiz provided on the following page. Trainees will need pencils. If you allow trainees to use the Trainee Guide, decrease the amount of time you give them to complete the quiz.

1. Make as many photocopies as you will need of the quiz.
2. Give trainees between 5 and 10 minutes to complete the quiz.
3. Go over the answers to the quiz.
4. Ask trainees whether they have questions about the types of coatings.

Answers to Quick Quiz

1. b. wax
2. e. mastic
3. c. tape
4. d. asphalt
5. a. FBE

Section 3.8.0

Types and Causes of Coating Damage

This exercise gives trainees the opportunity to practice identifying types and causes of coating damage. Trainees will need pencils, paper, personal protective equipment, and their Trainee Guides. You will need to provide numbered sections of pipe that evidence each of the 10 types of coating failure. Allow between 30 and 45 minutes for this exercise.

1. Set up workstations at which you have placed numbered sections of pipe that evidence each of the 10 types of coating failure.
2. Have trainees write down the type of coating failure for each pipe section. Also have them list the possible causes for this type of failure.
3. Review trainees' answers for accuracy. Answer any questions trainees may have.

Quick Quiz — *Types of Coatings*

For each description listed, identify the coating the text best describes. Write the corresponding letter in the blank provided.

_____ **1.** This coating is made of a combination of paraffin, plastics, and inhibitors.

_____ **2.** This coating is made with selected sands and other inert materials bound with an insulating compound.

_____ **3.** Usually, this is a plastic film of polyvinyl chloride or polyethylene with an adhesive backing applied to primed pipe surfaces.

_____ **4.** Often, this coating is distilled from petroleum and combined with petroleum resins or inert fillings.

_____ **5.** This is a hard, dense, abrasion-resistant coating applied by placing epoxy powder on a hot surface.

a. FBE
b. wax
c. tape
d. asphalt
e. mastic

MODULE 61105-02—ANSWERS TO REVIEW QUESTIONS

Answer	Section
1. d	2.1.0
2. d	2.4.0
3. c	2.7.0
4. c	2.10.0
5. a	3.2.0
6. c	3.4.0
7. b	3.8.0
8. b	4.0.0
9. d	4.5.0
10. a	5.0.0

CONTREN™ LEARNING SERIES — USER UPDATES

The NCCER makes every effort to keep these textbooks up-to-date and free of technical errors. We appreciate your help in this process. If you have an idea for improving this textbook, or if you find an error, a typographical mistake, or an inaccuracy in NCCER's Contren™ textbooks, please write us, using this form or a photocopy. Be sure to include the exact module number, page number, a detailed description, and the correction, if applicable. Your input will be brought to the attention of the Technical Review Committee. Thank you for your assistance.

Instructors – If you found that additional materials were necessary in order to teach this module effectively, please let us know so that we may include them in the Equipment/Materials list in the Instructor's Guide.

Write: Curriculum Revision and Development Department
National Center for Construction Education and Research
P.O. Box 141104, Gainesville, FL 32614-1104

Fax: 352-334-0932

E-mail: curriculum@nccer.org

Craft ______________________ Module Name ______________________

Copyright Date ______________ Module Number ______________ Page Number(s) ______________

Description
____
____
____
____

(Optional) Correction
____
____
____

(Optional) Your Name and Address
____
____
____

Module 61106-02

Aboveground Pipe Coating and Inspection (CT 7.1, 7.2, 7.3, 7.5 and 13.2)

MODULE OVERVIEW

This module discusses the requirements for preparing, cleaning, and inspecting surfaces using hand and power tools. Trainees will learn how to inspect the surface preparation, select an appropriate coating, and apply a coating to aboveground pipe.

PREREQUISITES

Please refer to the Course Map in the Trainee Module. Prior to training with this module, it is recommended that the trainee shall have successfully completed the following:

Core Curriculum; Pipeline Core, Modules 66101-02 and 66102-02; Pipeline Corrosion Control Level One, Modules 61103-02 through 61105-02

OBJECTIVES

Upon completion of this module, the trainee will be able to:

1. Describe the requirements for surface preparation for aboveground pipe (CT 7.1 and 7.2).
2. Prepare/clean surface for atmospheric coating using hand and power tools (CT 7.2, 7.3, 7.5, and 13.2).
3. Visually inspect the quality of the surface preparation (CT 7.2, 7.3, and 13.2).
4. Inspect the quality of the surface preparation by checking the surface profile (CT 7.2).
5. Properly select a coating to be used.
6. Properly apply a coating to aboveground pipe (CT 7.5).

PERFORMANCE TASKS

Under the the instructor's supervision, the trainee should be able to:

1. Prepare a coating evaluation and prepare the surface of a pipeline system for coating application or repairs (CT 7.1, 7.2, 7.3, 7.5, and 13.2).
2. Visually inspect and verify the quality of the surface preparation by checking the surface profile (CT 7.2).
3. Properly apply corrosion-preventative coating to a pipeline facility exposed to atmospheric conditions (CT 7.5).

NCCER STANDARDIZED CRAFT TRAINING PROGRAM

The National Center for Construction Education and Research (NCCER) provides a standardized national program of accredited craft training. Key features of the program include instructor certification, competency-based training, and performance testing. The program provides trainees, instructors, and companies with a standard form of recognition through a National Craft Training Registry. The program is described in full in the *Guidelines for Accreditation*, published by the NCCER. For more information on standardized craft training, contact the NCCER by writing us at P.O. Box 141104, Gainesville, FL 32614-1104; calling 352-334-0911; or e-mailing info@nccer.org. More information may be found at our Web site, www.nccer.org.

HOW TO USE THIS ANNOTATED INSTRUCTOR'S GUIDE

Each page presents two sections of information. The larger section displays each page exactly as it appears in the Trainee Module. The narrow column ties suggested trainee and instructor actions to each page and provides icons to call your attention to material, safety, audiovisual, or testing requirements. The bottom of each page includes space for your notes.

Teaching Tip

If you see the Teaching Tip icon, that means there is a teaching tip associated with this section. Also refer to any suggested teaching tips at the end of the module.

SAFETY CONSIDERATIONS

Ensure that the trainees are equipped with appropriate personal protective equipment. Emphasize the importance of following all safety precautions and procedures when working with power tools.

PREPARATION

Before teaching this module, you should review the Module Outline, Objectives, Performance Tasks, and the Materials and Equipment List. Be sure to allow ample time to prepare your own training or lesson plan and gather all required equipment and materials.

MATERIALS AND EQUIPMENT LIST

Transparencies

Markers/chalk

Blank acetate sheets

Transparency pens

Pencils and scratch paper

Module Examinations*

Performance Profile Sheets*

Copies of Performance Verifications for Covered Tasks 7 and 13**

Overhead projector and screen

Whiteboard/chalkboard

Appropriate personal protective equipment

Copies of *49 CFR Part 192* (Gas) and/or *49 CFR Part 195* (Liquid)

Copies of your company policy and procedures manual

Copies of *Figure 4* with the callouts covered

Spray equipment with a wand assembly safety latch

Protective clothing

Gloves

Helmet with screen visor

Goggles

A variety of pressure washing nozzles, tips, and accessories, including:
- 0-, 15-, 25-, and 40-degree tips
- Power scrub brushes
- Telescoping extensions
- Horizontal scrubbers
- Rotating water blast nozzles
- Variable tip nozzles
- Rotatable tip nozzles
- Wet abrasive injectors

A variety of common hand tools, including:
- Abrasive pads
- Scrapers
- Putty knives
- Wire brushes
- Chipping hammers

Surfaces that require cleaning

A variety of power tools, including:
- Grinders
- Sanders
- Power files
- Needle guns
- Rotary impact power tools
- Vacuum shrouds

Inspection tools, including:
- Profile comparators
- Profile gauges
- Replica tape

Practice surfaces for preparation and inspection

Sling psychrometers

Electric hydrometers

Surface thermometers

Sample shapes and styles of brushes of various quality

A variety of brushes and rollers

Coatings

Cleaned surfaces for applying coatings

Materials to clean the brushes and roller covers

Roller covers

Brush comb or wire brush

Rags

Brush spinner

Roller spinner

Combination tool

* Located in the Test Booklet.

** Available for download from www.nccer.org.

ADDITIONAL RESOURCES

This module is intended to present thorough resources for task training. The following reference works are suggested for both instructors and motivated trainees interested in further study. These are optional materials for continued education rather than for task training.

API Recommended Practice 651, *CP of Aboveground Petroleum Storage Tanks,* Second Edition (December 1997). Global Engineering Product Code API CERT 651.

Control of Pipeline Corrosion, Second Edition (2001). A. W. Peabody. Houston, TX; NACE International.

Corrosion Prevention by Protective Coatings, 1984. Charles G. Munger. Houston, TX; National Association of Corrosion Engineers (NACE).

TEACHING TIME FOR THIS MODULE

An outline for use in developing your lesson plan is presented below. Note that each Roman numeral in the outline equates to one session of instruction. Each session has a suggested time period of 2½ hours. This includes 10 minutes at the beginning of each session for administrative tasks and one 10-minute break during the session. Approximately 15 hours are suggested to cover *Aboveground Pipe Coating and Inspection* (CT 7.1, 7.2, 7.3, 7.5, and 13.2). You will need to adjust the time required for hands-on activity and testing based on your class size and resources.

Topic	Planned Time
Session I. Metal Substrates, Part One	
A. Surface/Substrate Preparation Agents	________
B. Washing and Cleaning	________
Session II. Metal Substrates, Part Two	
A. Low-Pressure Water Cleaning	________
1. Laboratory	________
B. Water Cleaning Equipment	________
Session III. Metal Substrates, Part Three	
A. Safety Guidelines and Techniques	________
B. Hand Tool Cleaning	________
1. Laboratory	________
Session IV. Metal Substrates, Part Four	
A. Power Tool Cleaning	________
1. Laboratory	________
B. Surface Preparation Standards and Inspection Techniques	________
C. General Preparation Procedures for Specific Surfaces/Substrates	________
1. Performance Testing	________
Session V. Initial Preparation Before Coating Application	
A. Protecting Adjacent Surfaces	________
B. Checking the Surface Condition and Environmental Conditions	________
1. Laboratory	________
C. Properly Selecting and Mixing the Coating	________
D. Brushes	________
1. Laboratory	________
E. Rollers	________
1. Performance Testing	________
Session VI. Applying Coatings with a Brush	
A. Basic Brushing Techniques	________
B. Applying a Coating with a Roller	________
C. Guidelines for Coating Using Brushes and Rollers	________

D. Cleaning and Storing Brushes and Roller Covers
   1. Laboratory
E. Summary
   1. Summarize module
   2. Answer questions
F. Module Examination
   1. Trainees must score 70% or higher to receive recognition from the NCCER.
   2. Record the testing results on Craft Training Report Form 200 and submit the results to the Training Program Sponsor.
G. Performance Testing
   1. Trainees must perform each task to the satisfaction of the instructor to receive recognition from the NCCER.
   2. Record the testing results on Craft Training Report Form 200 and submit the results to the Training Program Sponsor.

Module 61106-02

Aboveground Pipe Coating and Inspection

(CT 7.1, 7.2, 7.3, 7.5, and 13.2)

Instructor's Notes:

Instructor's Notes:

Introduction to the Pipeline Operator Qualification Rule, API, and NCCER

It could be said that petroleum products keep America running. Petroleum products like oil and natural gas are the principal sources of energy in the United States and are used in many manufacturing and chemical processes. Petroleum products are primarily transported by pipeline. There are more than a million miles of liquid and natural gas pipeline in this country, ranging in diameter from two inches to four feet. These pipelines bring crude oil and natural gas from production fields to processing plants and refineries—and ultimately to the consumer.

Pipeline Operator Qualification

Petroleum products are considered hazardous, so pipeline personnel must take extra precautions and follow specific procedures when working with them. To help ensure the safe handling and transportation of petroleum products, the United States Department of Transportation (DOT) mandates safe practices as federal regulations.

One such regulation, the Operator Qualification (OQ) rule, requires pipeline operators to develop a written qualification plan to evaluate their personnel and their contract personnel (see *49 CFR Part 192 Subpart N—Gas,* and *49 CFR Part 195 Subpart G—Liquid*). The evaluation hinges on the operator's ability to perform certain covered tasks and to be able to recognize and respond to abnormal operating conditions that may be encountered while performing these covered tasks.

Covered Tasks

A pipeline performance task is deemed *covered* if the activity passes the following four-part test:

1. Is it performed on a pipeline facility?
2. Is it an operations or maintenance task?
3. Is it performed as a requirement of *49 CFR Part 195 (Liquid)* or *49 CFR Part 192 (Gas)*?
4. Does it affect the operation or integrity of the pipeline facility?

If the answer to all four questions is *yes*, then the activity is considered *covered* and must figure into the evaluation process of pipeline personnel.

American Petroleum Institute

The OQ rule and its related covered tasks provided the catalyst to develop a unified pipeline training and assessment program—so that pipeline operators and owners have ready access to the assessment and training tools they need. This effort was the direct result of the vision and leadership provided by The American Petroleum Institute (API) and API members.

Efforts to unify the industry are nothing new to API. Indeed, the impetus for forming the Institute in 1919 was the need to standardize engineering specifications for drilling and production equipment. Today, the most pressing issues revolve around public perceptions and government policies toward the industry. Speaking with one voice on these issues has become as essential as having interchangeable parts in the field. API's members determine the petroleum industry's positions on public policy as well as the standards that govern its day-to-day operations worldwide.

Consortium on Operator Qualification

API assembled a focus group whose primary purpose was to identify covered tasks and provide compliance recommendations to pipeline operators.

The Consortium on Operator Qualification (COOQ) became the guiding force for the identification and development of covered tasks for the liquid pipeline industry. Their work product, API publication 1161, *Guidance Document for the Qualification of Liquid Pipeline Personnel,* formed the basis for the resulting Pipeliner Training and Assessment Program.

The covered task list created in API publication 1161 has now evolved to include gas covered tasks and is maintained by NCCER. The most current covered task list can be downloaded at www.nccer.org.

National Center for Construction Education and Research

With the guidance and support of API, its members, and the Consortium on Operator Qualification, the National Center for Construction Education and Research (NCCER) was able to design and implement the Pipeliner Training and Assessment Program.

NCCER is a not-for-profit 501(c)(3) education foundation established in 1995 by the world's largest and most progressive construction and maintenance contractors and national contractor associations. NCCER was founded to address the

severe workforce shortage facing the construction and maintenance industries and to develop a standardized training process and curricula. Today, NCCER is supported by hundreds of leading construction and maintenance companies, manufacturers, and national associations and developed more than 35 construction and maintenance titles.

NCCER also maintains a National Registry, which contains the records of trainees and instructors who have received NCCER training. All completion information for every trainee is sent to the NCCER and filed within the Registry. The Registry can then confirm training and skills for workers as they move from company to company, state to state, or even within their own company.

Standardized Curriculum (NCCER's Contren™ Learning Series)

NCCER curricula are developed using the expertise of industry subject matter experts (SMEs). The use of SMEs as an integral part of every new development or revision ensures that the curriculum is current, industry-driven, and indicative of industry practices. The development and/or revision process is facilitated from start to finish by NCCER project managers, who work closely with the SMEs and a select team of technical writer(s) on the development and presentation of the curriculum.

Pipeline: Five Training Paths

The pipeline curriculum focuses on essential skills and competencies within five skill areas:

- Corrosion Control
- Electrical and Instrumentation
- Maintenance
- Mechanical
- Operations (Field and Control Center)

SMEs from the pipeline industry have determined the skills and competencies needed based on API's covered tasks and the consensus of the group/industry. Specific company information is generally not incorporated into the curriculum unless the SMEs determine that the information is the industry standard.

Training Modules and Covered Tasks

The smallest unit of instruction in the curriculum is a *module*. Each curriculum is composed of modules (the equivalent of stand-alone chapters) that are logically sequenced into complete *levels* of instruction. Every module is task-oriented and requires the successful completion of a written (knowledge-based) examination and a performance (hands-on) test. Modules may be individually selected and/or combined for efficient task training. Modules in the pipeline curricula, to the extent possible, will align with API-developed covered tasks. While not all modules will relate to covered tasks, all covered tasks will relate to at least one module.

Benefits of Training to the Pipeline Marketplace

Applying NCCER's standardized training process allows pipeline owners and contractors in the United States to establish the benchmark for acceptable skill and competency levels within the pipeline industry. This process will assist training organizations in developing and/or improving standardized pipeline training programs that effectively match their resources and operating conditions while responding to regulatory demands and to the needs of the industry. The process will also recognize training organizations that meet the minimum standards of competence, warranting regulatory, public, and professional confidence. This in turn will foster national unity within the pipeline industry for the mutual benefit of individual technicians, owners, contractors, and ultimately pipeline-users and society.

Additional Resources

To review the Pipeliner Training and Assessment Program Implementation and Resource Guide and the latest version of pipeline covered tasks, visit NCCER's Web site at www.nccer.org, or contact:

National Center for Construction Education and Research
PO Box 141104
Gainesville, FL 32614-1104
(888) 622-3720
Web site: www.nccer.org

For more information on API, please contact:

American Petroleum Institute
1220 L Street, Northwest
Washington, D.C. 20005
(202) 682-8000
Web site: www.api.org

To order *API Pub 1161*, contact Global Engineering Documents by calling 1-800-854-7149 or visiting their Web site at www.global.his.com.

Additional sources of information on the pipeline operator qualification regulation include:

DOT – Office of Pipeline Safety
www.ops.dot.gov

Transportation Safety Institute
www.tsi.dot.gov

Instructor's Notes:

COURSE MAP

This course map shows all of the modules in the first level of the Pipeline Corrosion Control curriculum. The suggested training order begins at the bottom and proceeds up. Skill levels increase as you advance on the course map. The local Training Program Sponsor may adjust the training order.

PIPELINE CORROSION CONTROL LEVEL ONE

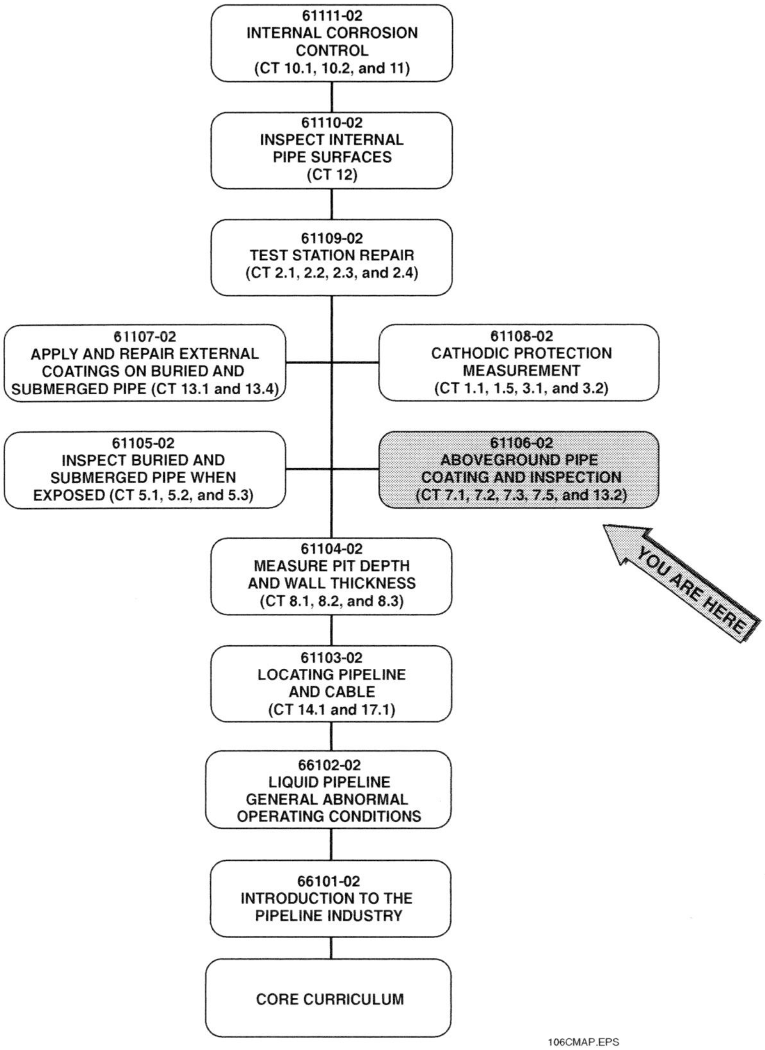

106CMAP.EPS

Homework

Assign reading of Module 61106-02.

Instructor's Notes:

MODULE 61106-02 CONTENTS

MODULE 61106-02 CONTENTS (Continued)

Instructor's Notes:

Figures

Instructor's Notes:

MODULE 61106-02

Aboveground Pipe Coating and Inspection (CT 7.1, 7.2, 7.3, 7.5, and 13.2)

Objectives

When you have completed this module, you will be able to do the following:

1. Describe the requirements for surface preparation for aboveground pipe (CT 7.1 and 7.2).
2. Prepare/clean surface for atmospheric coating using hand and power tools (CT 7.2, 7.3, 7.5, and 13.2).
3. Visually inspect the quality of the surface preparation (CT 7.2, 7.3, and 13.2).
4. Inspect the quality of the surface preparation by checking the surface profile (CT 7.2).
5. Properly select a coating to be used.
6. Properly apply a coating to aboveground pipe (CT 7.5).

Prerequisites

Before you begin this module, it is recommended that you successfully complete the following: Core Curriculum; Pipeline Core, Modules 66101-02 and 66102-02; Pipeline Corrosion Control Level One, Modules 61103-02 through 61105-02.

Required Trainee Materials

1. Pencil and paper
2. Appropriate personal protective equipment

1.0.0 ◆ INTRODUCTION

Applying the finish coat is the least time-consuming part of a pipeline coating job. The most important part, and perhaps the most time-consuming, is the surface/substrate preparation. A coating cannot, and will not, make a bad surface or substrate into a good surface. A finish applied over a poor surface or substrate will not provide a lasting coating. It may temporarily make the surface or substrate look good, but in some cases, it will not even do that.

Since one primary purpose of a coating is protection, you should not invest a large amount of money in coating a surface/substrate that is not prepared properly when new, or has deteriorated to the point where it is not worth protecting. A coating is only as good as the surface or substrate to which it is attached.

Cleaning and preparing of surfaces/substrates is defined in a set of standards developed by **SSPC** (Society for Protective Coatings) and **NACE** International. These standards are referenced by coating manufacturers and in job specifications developed by engineering or architectural firms as universally defined preparation methods and requirements for various coating applications on steel.

SSPC-SP2, Hand Tool Cleaning; SSPC-SP3 and *SSPC-11*, Power Tool Cleaning are covered in this module. Dry abrasive steel blasting standards are covered in *SSPC-SP5/NACE 1, SSPC-SP10/NACE 2, SSPC-SP6/NACE 3*, and *SSPC-SP7/NACE 4* and standard *SSPC-SP12/NACE 5*.

As a result of the Clean Air Act and many other laws and regulations, manufacturers have stepped up research and development for both water-based and oil-based coatings. Coatings have been re-engineered to eliminate or reduce the use of volatile organic compounds (VOCs) and lead. Today's coatings are quite different in their composition and performance than the best coatings of just a few years ago. Improved synthetic resins have yielded better performance, especially in

Materials

Ensure that you have everything required to teach the course. Check the Materials and Equipment List at the front of this module.

Classroom

Explain that terms shown in bold are defined in the Glossary at the back of this module.

Audiovisual

Show Transparency 1, Course Objectives.

Show Transparency 2, Performance Tasks.

Teaching Tip

See the general Teaching Tip at the end of this module.

Classroom

Emphasize the importance of properly preparing the surface/substrate.

Review the standards that regulate the cleaning and preparation of surfaces and substrates.

Explain that, in accordance with the Clean Air Act, many coatings now eliminate or reduce the use of volatile organic compounds.

Classroom

Explain how to determine whether the metal substrate is ferrous or non-ferrous.

Discuss the importance of cleaning the surface/substrate. Review the cleaning agents that may be used. Answer any questions trainees may have.

Review the four types of abrasives that are used in low-pressure water washer systems. Discuss the considerations for selecting the right abrasive for a job.

Have trainees write down the four types of abrasives discussed to this point. Ask trainees to explain the advantages and disadvantages of each and to provide one example of each type of abrasive.

water-based/waterborne coatings. As a result of these efforts to make coatings comply with VOC regulations, many new water-based coatings have been developed that have properties equal to, or better than, oil-based coatings.

2.0.0 ◈ METAL SUBSTRATES

Both ferrous (steel) and non-ferrous (no iron content) metals are used today. If the metal substrate is not known, it may be generally classified using a magnet. A magnet will be attracted to ferrous metals but will have little or no attraction to non-ferrous metals, such as brass. Carefully sanding a hidden portion of a metal to the bare surface may reveal zinc or tin plate on steel or the color and surface characteristics of non-ferrous metals.

2.1.0 Surface/Substrate Preparation Agents

Surface/substrate preparation, the foundation of a good coating job, should always begin with washing away dust, dirt, water-soluble contaminants, oil, grease, and any other residue that may interfere with **adhesion**. Make every effort to clean the surface/substrate as well as possible before coating application, or before using other surface/substrate preparation methods. Cleaning agents include detergents, solvents, and chemical treatments such as alkaline cleaners and acids. In addition, surfaces/substrates can be treated with disinfectants and rust and mildew retardants. This section looks at some of these agents, how they work, and when to use them.

2.1.1 Low-Pressure Washer Abrasives

Abrasives are materials that can be used in low-pressure water washer systems. They come in a range of types, shapes, sizes, and degrees of hardness. Given the multitude of options, selecting the right abrasive for the job becomes a matter of determining which material will give a surface the desired finish, which means the proper degree of cleanliness and roughness, as well as the type of texture needed for the coating. Other factors to consider are the type of surface to be cleaned and its present condition, the consumption rate of the abrasive, its cost, and any environmental regulations that may affect the operation.

There are four types of abrasives: metallic, mineral, by-product, and special purpose. Of these, mineral abrasives and, to a much lesser extent, by-product or special purpose abrasives are used with low-pressure washer, wet-abrasive injectors.

Metallic abrasives are available in a variety of forms. The most common metallic abrasive is cast steel. Others include malleable iron and chilled cast iron. All come in the form of shot and grit and in different degrees of hardness. Abrasive made from wire cut into short pieces is also available. Shot is rounded, which makes it appropriate for cleaning heavy, brittle deposits and for producing a rounded profile or anchor pattern. Grit, which is made by crushing hardened shot pellets, and cut wire both have an irregular, angular shape that effectively removes rust, mill scale, and paint, and leaves a rough profile that feels like abrasive paper. Shot, grit, and cut wire are screened according to size.

Metallic abrasives can be recycled many times before breaking down into a size too small to be reused. The repeated blast cleaning impact causes the abrasive to fracture or flake. Generally, the harder the abrasive, the faster it breaks down. The particles become smaller and smaller until they are eventually filtered out in the recycling process. New abrasive must be added regularly to replace what is lost. The result is a mixture of sizes called a work mix or operating mix of large particles that do the heavy work of **cracking** and breaking up the contaminants and small particles that do the cleanup work of scouring and removing them. Maintaining a properly balanced work mix is important for efficient blast cleaning. This can be done by hand or with a continuous, automated system of replenishing spent abrasives.

Mineral abrasives are commonly used with low-pressure washer, wet-abrasive injectors. Sand has long been used for **abrasive blasting**. In particular, silica sand is readily available, which makes it economical. Its range of angular and spherical shapes gives it good cleaning action. However, it breaks down quickly and produces a large amount of silica dust, which causes a dangerous lung disease known as silicosis when the dust is small enough to be respirable (less than 10 microns).

Wet blasting greatly reduces the dust problem, but dust-laden mist is still a hazard for the operator. Nonsilica or heavy mineral sands used in blasting include magnetite, staurolite, and olivine rutile, which are tough, dense, and smaller than silica sand but contain no dangerous free silica. These abrasives are generally not recycled because of how quickly they break down in blasting. Garnet, a tough abrasive, can be recycled a number of times. Other mineral abrasives, such as zircon and novaculite, are generally used for specialty blasting because of their cost and the finish they leave. Zircon, for example, leaves a smooth, matte finish, and novaculite leaves a satin luster.

Instructor's Notes:

By-product abrasives such as slag, the refuse from smelting or other processes, make good abrasives because of their availability, relatively low cost, and low silica content. Slag's sharp, angular shape gives it good cutting power, and it comes in a full range of sizes. Blasting slags include coal slag (*Figure 1*), copper slag, and nickel slag.

Special purpose sodium bicarbonate abrasives are water-soluble products (one of which is marketed by Church and Dwight Co. through the Sherwin-Williams Co. as ARMEX® Hydroflex® XL) are soft cleaning abrasives designed for use with low-pressure washers.

These abrasives are generally considered non-toxic and non-hazardous. Cleanup may consist of only water flushing of the work site. They are relatively expensive, but their effective flow rate per minute is low compared to other abrasives. These products are usually designed for use with low-pressure washers rated at 2,500 to 4,000 psi using special wet abrasive injection equipment. Use in a conventional wet abrasive injector is not generally recommended.

Since these products are soft abrasives, they are ideal for use on soft or easily damaged surfaces/substrates. They are effective for removal of light coatings. They will not damage seals, bearings, or other moving parts, and are used in industrial plants and food processing operations. When used on steel, the residue of these products will act as a rust inhibitor until rinsed off. Used in low-pressure washers, they will achieve a surface similar to a **brush-off blast cleaned surface**. After rinsing, surfaces should be checked for proper pH with litmus paper. Subsequent coatings must be compatible with sodium bicarbonate.

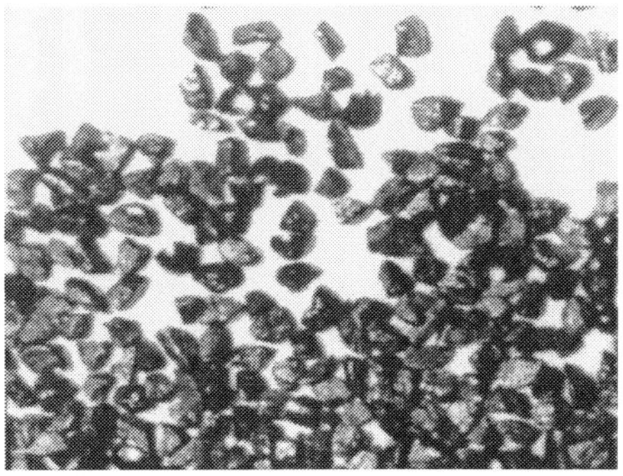

106F01.EPS

Figure 1 ◈ Coal slag.

2.2.0 Washing and Cleaning (*SSPC-SP1* or *ASTM D 4258*)

Cleaning is an indispensable part of surface preparation. Every surface should be thoroughly cleaned so the applied coating can form a strong, firm **bond**. Cleaning methods range from **hand washing** with water to machine blasting with abrasives. A job may require one or more cleaning methods depending on the type of surface, the surface condition, and how clean the surface needs to be for the new coating.

Washing a surface to prepare for a new coating job is a widespread industry practice. In recent years, the number and effectiveness of special cleaning compounds have increased greatly. Cleaning compounds come in varying strengths for different classes of work—some materials are used for cleaning slightly soiled and greasy surfaces, others for heavier duty in offices and public buildings, and still others for industrial plant cleaning.

The following paragraphs are a rundown of common washing and cleaning methods used in surface preparation as defined in *SSPC-SP1, Solvent Cleaning* or American Society for Testing and Materials (ASTM) *D 4258, Surface Cleaning Concrete for Coating*.

These standards define solvent cleaning as the removal of oil, grease, dirt, soil, curing compounds, form compounds, and other soluble contaminants by the use of solvents, emulsions, detergents, or cleaning compounds, applied by hand or by pressure washing with cold or hot water/steam cleaning methods. Solvent cleaning does not remove tightly adherent rust or mill scale unless abrasives are used.

2.2.1 Hydrocarbon-Based Solvent Cleaning

Solvents are liquids capable of dissolving a material. By this definition, water is a solvent. It can be used alone or with soaps or detergents to wash surfaces by hand or, more effectively, with a pressure washer. However, in the coating industry, solvent usually refers to liquids such as denatured alcohol, naphtha, mineral spirits, toluol, kerosene, and others.

These solvents can be used to clean grease, oil, or **waxes** from a solid surface. Heavy deposits of grease, oil, or wax should first be removed by scraping. After washing with a solvent, it is usually necessary to rinse the surface well with a detergent and water to clean away any residue of the contaminant or solvent, which can prevent proper adhesion.

Classroom

Discuss the cleaning methods and the cleaning compounds that may be used for surface preparation.

Review the different classes of work for which cleaning compounds may be used.

Explain that the cleaning methods reviewed in this section are defined in Solvent Cleaning as well as in the American Society for Testing and Materials' Surface Cleaning Concrete for Coating.

Explain how the coating industry defines a solvent. Provide examples of these solvents.

Classroom

Review the hazards associated with hydrocarbon-based solvents. Explain how to safely use these solvents to clean a surface.

Safety

Explain why some ordinances may prohibit the use of solvents.

Audiovisual

Show Transparency 3 (Figure 2). Explain how to properly dispose of solvents. Go over any questions trainees may have.

Safety

Emphasize the importance of following the Occupational Safety and Health Administration's regulations for working with and disposing of solvents.

Check with the coating manufacturer before using solvents in the preparation of surfaces because they may not be compatible with the coating to be applied.

Solvents generally work quickly and are easy to use. They require little equipment. A bucket and a sponge or rag are usually enough. The disadvantages of working with solvents are that many of them are hazardous; they soon become contaminated and must be replaced frequently or else they become ineffective; and they can be expensive.

With few exceptions, hydrocarbon-based solvents are flammable, explosive, and can cause serious health problems if inhaled or absorbed through the skin. Exposure can be especially dangerous to anyone with heart or liver disease. Solvents should always be used with care.

The simple rag or cloth wipe is the most widely used surface preparation device. When used with water, detergent, alkaline cleaner, or solvent, it can clean dirt, grease, oil, mud, markings, and other washable contaminants from all types of surfaces. This simple cleaning method is required by SSPC surface preparation standards before any other surface preparation work.

Rags may be purchased in bulk or recycled from old household items. Cotton fabrics, such as sheets, towels, or diapers, work better than less absorbent synthetic materials. Always start with a clean rag. A dirty rag may spread contamination.

When using solvents, rags become contaminated quickly, so replace them as necessary. Never leave rags soaked in solvent or other flammable material indoors or near any possible source of ignition.

> **CAUTION**
> State and local ordinances may prohibit the use of some solvents in the cleaning and coating process.

Used solvents are considered hazardous wastes. Legal options for disposing of them properly are to pay to have them hauled away and deposited in an appropriate landfill or to recycle them using a solvent recovery system (*Figure 2*).

The distillation process for recycling a solvent first involves evaporating it. The vapor is then condensed back into liquid form by one of several cooling methods. This liquid is the refreshed solvent. The residue is removed and disposed of as hazardous waste. Recycling systems come in various sizes; some are permanent, while others are portable.

> **WARNING!**
> According to the Occupational Safety and Health Administration (OSHA) regulations on disposal of waste materials (*29 CFR 1926.252*), all solvent waste, oily rags, and flammable liquids must be kept in fire-resistant, covered containers until removed from the work site. If rags have become contaminated with a hazardous material such as lead, they must be tested and, if necessary, disposed of as a hazardous waste.
>
> Keep the work area well ventilated. If fresh air is not adequate, provide forced ventilation. Headaches, nausea, dizziness, and skin irritation are signs of overexposure. If they persist, get medical attention.
>
> Wear appropriate protective clothing, including safety glasses, goggles, gloves, and the appropriate National Institute of Occupational Safety and Health (NIOSH)-approved respirator.
>
> Never use flammable solvents near heating units, motors or machines that may spark, electrical outlets, or any possible source of ignition. Note that solvent vapors are heavier than air and can travel undetected along the floor or ground to an ignition source far away from the work area.
>
> Clean up all solvent spills, no matter how small.
>
> Dispose of used solvent and cleaning supplies properly.
>
> Read the Material Safety Data Sheets (MSDS) for information about individual solvents.

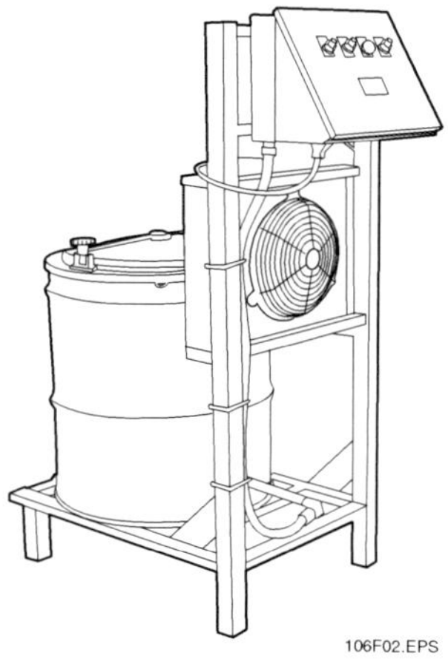

106F02.EPS

Figure 2 ◆ Solvent recovery system.

Instructor's Notes:

In some cases, inadvertent chemical reactions must be controlled during the evaporating process. When heated, some paints emit toxic or volatile gases. In most cases, however, recycling solvents in a distillation system is simple and cost-effective.

Although some solvent is lost in the distillation process, it can be recycled repeatedly, which is economical. Also, it is much less costly to dispose of residue than an entire batch of used solvent. In addition to its other advantages, this process is less damaging to the environment.

2.2.2 Alkaline and Detergent Cleaning

As discussed previously, there are a number of powerful industrial alkaline sodium metasilicate-based cleaners with added detergent and surfactants available. These cleaners, in their concentrated spray-and-wipe forms, are strong enough to remove dried paint, oil, grease, wax, scuff marks, pencil marks, soot, rust stains, and other materials. They are also available in non-foaming versions for dilution and use in pressure washing equipment.

In general, these cleaners should not be allowed to dry on the surface. They should be wiped or rinsed off thoroughly with clean water while still wet, otherwise they may leave a film that is difficult to remove. They are effective, especially when heated to 150°F and above, for removing contaminants from all hard surfaces as well as masonry and concrete.

NOTE
These cleaners should not be used on glass, stainless steel, or aluminum.

WARNING!
Goggles and waterproof gloves should be used when handling these products. In confined areas, respirators should be worn since the fumes from the additives in some of these products are irritating. Manufacturer's instructions and warnings must be followed when using these products.

To remove discoloration caused by industrial gases, wash the surface with a very diluted solution of acetic or muriatic acid, and rinse well with water.

2.3.0 Low-Pressure Water Cleaning

Low-pressure water cleaning uses a low volume of pressurized water to give the water more cleaning power.

Low-pressure water cleaning ranges up to 5,000 psi (352 kg/cm^2). Low-pressure water cleaning equipment can spray water with or without detergents, water with or without abrasives, and steam with or without detergents.

The use of low-pressure water cleaning depends on the surface condition. Its effectiveness depends on the level of pressure and the use of additives. Low-pressure washing or steam cleaning will remove dirt, grime, paint, and heavy soil when solvent cleaning is difficult or impractical.

Injecting an abrasive into the water stream gives pressure washing added power to remove tight, intact coatings and adherent rust and mill scale. It also provides the correct **surface profile**.

NOTE
Always evaluate the blast media for proper dryness and pH.

Low-pressure water cleaning with abrasives is a good alternative when dry blasting is not possible because of environmental restrictions. It does not create as much dust or any sparks, which can be dangerous. Extra caution is required near electrical equipment. Washing with an abrasive requires a second washing to remove surface sediment. It may also require the use of an inhibitor to prevent flash rusting on steel surfaces. One such inhibitor is water containing 0.32% sodium nitrate and 1.28% by weight secondary ammonium phosphate (dibasic). The coating system must be compatible with whatever inhibitor is used.

Low-pressure water, wet-abrasive cleaning can be used to attain a cleaned metal surface similar to an *SSPC-SP7/NACE 4* brush-off blast cleaned surface. A brush-off blast cleaned surface is defined by *SSPC-SP7/NACE 4* as a surface, when viewed without magnification, that is free of all visible oil, grease, dirt, loose mill scale, loose rust, and loose coating. Tightly adherent mill scale, rust, and coating may remain on the surface. Material is considered to be tightly adherent if it cannot be removed by lifting with a dull putty knife.

Working with pressurized water calls for extra safety precautions. Appropriate protective clothing includes a waterproof suit, face shield, goggles, or helmet with visor, hearing protection, gloves, steel-toed shoes, and rubber boots. Operators should be well trained and equipment should be inspected regularly to ensure good working condition.

Classroom

Explain how to use alkaline cleaners and detergents to clean surfaces. Ask a trainee to explain why these cleaners should not be allowed to dry on the surface.

Safety

Discuss the consequences of failing to wear appropriate personal protective equipment when working with alkaline cleaners and detergents.

Homework

Have trainees review Sections 2.3.0–2.3.6 for the next session.

Materials

Ensure that you have everything required for the laboratory during this session.

Classroom

Compare the methods for using low-pressure water cleaning with abrasives with the methods for using low-pressure water, wet-abrasive cleaning.

Discuss the safety precautions associated with low-pressure water cleaning.

Classroom

Review the categories of water cleaning as defined in *SSPC-SP12/NACE 5*.

Refer to Figure 3. Discuss the general guidelines for pressures and water flow.

Audiovisual

Show Transparency 4 (Figure 4). Review the functions of the components in a portable cold water unit.

2.3.1 Pressurized Water Cleaning Equipment

Water cleaning is effective for removing both visible and nonvisible contaminants. As the pressure of the water increases, so does the cleaning ability of the equipment, which falls into one of four categories as defined in *SSPC-SP12/NACE 5, Surface Preparation and Cleaning of Steel and Other Hard Materials by High- and Ultra High-Pressure Water Jetting Prior to Recoating*:

- *Low-pressure water cleaning* – Uses water at pressures less than 5,000 psi (pounds per square inch) or in metric terms, 352 kilograms per square centimeter (352 kg/cm^2).
- *High-pressure water cleaning* – Uses water at pressures from 5,000 to 10,000 psi (352 to 703 kg/cm^2).
- *High-pressure water jetting* – Uses water at pressures from 10,000 to 25,000 psi (703 to 1,758 kg/cm^2).
- *Ultra high-pressure water jetting* – Uses water at pressures greater than 25,000 psi (1,758 kg/cm^2).

SSPC-SP12/NACE 5 covers surface preparation standards for high-pressure and ultra high-pressure water jetting. In this module, only low-pressure water cleaning (less than 5,000 psi [352 kg/cm^2]) will be discussed. The other types will be covered in a later module.

2.3.2 Low-Pressure Water Cleaning Equipment

The type and condition of the substrate being cleaned and the size of the job will determine the type of low-pressure washing equipment required. *Figure 3* is a general guide showing uses for pressures ranging up to 5,000 psi (352 kg/cm^2) and water flow from one to five gallons per minute (gpm) (3.8 to 19 liters/minute) or more.

All units include these basic components:

- A water pump to pressurize water
- A power source for the pump
- An inlet filter and, in some cases, a pressure gauge
- A hydraulic hose and fittings to get the water from the pump to the nozzle
- A wand assembly equipped with a control valve, usually a trigger device to regulate blasting
- Various safety valves

2.3.3 Typical Low-Pressure, Cold Water Washer Equipment

Figure 4 shows a typical portable, gasoline-powered, cold water unit of this pressure class. *Figure 5* shows a large trailer-mounted, diesel-powered unit of this pressure class. The large unit is capable of supporting multiple spray guns. Electric motor-driven units are also available. Externally heated water may also be used with these units.

Several units with on-board hot water or hot water/steam capability will be discussed briefly in later paragraphs. The functions of some of the various components for the cold water unit shown in *Figure 4* are listed here:

- *Engine* – Normally gasoline or electric driven, the engine is coupled directly to the water pump, as shown in *Figure 4*, or by a belt and pulley drive system. The engine supplies the power to rotate the pump.

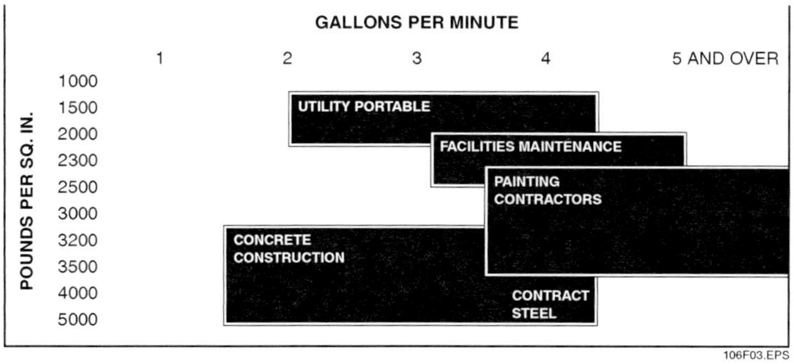

Figure 3 ◆ General guide showing uses for pressures and water flow.

Instructor's Notes:

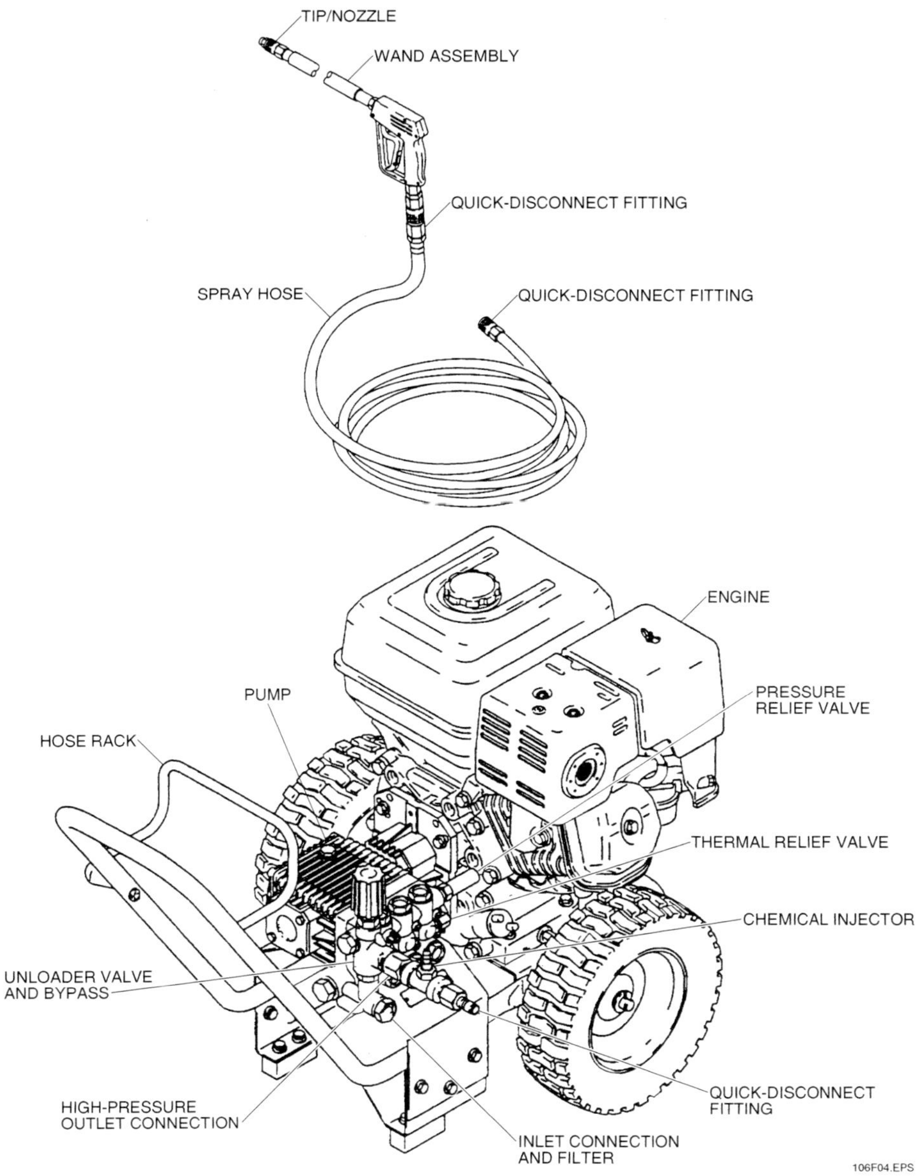

Figure 4 ◈ Portable, gasoline-powered, cold water unit.

Laboratory

Distribute copies of Figure 4 with the callouts covered. Ask trainees to accurately label the cold water unit and explain how each component operates. Ask whether trainees have any questions.

Audiovisual

Show Transparency 5 (Figure 6). Review the options for providing power to on-board heating units.

Classroom

Refer to Figure 7. Compare a portable hot water/steam unit with a trailer-mounted hot water unit.

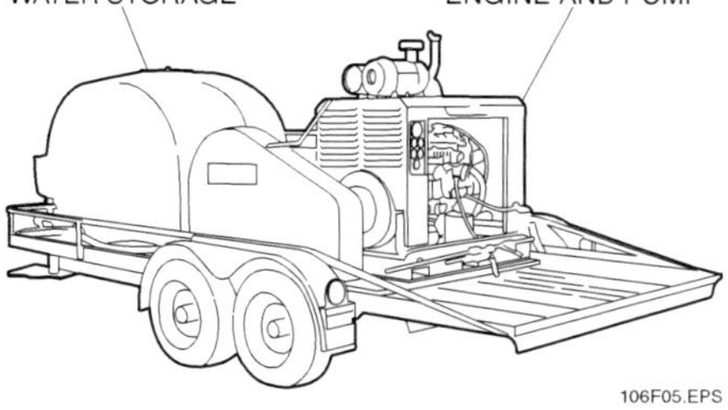

Figure 5 ◈ Large trailer-mounted, diesel-powered unit.

- *Pump* – The water pump is a rotating, multiple-piston, positive displacement pump that provides the rated water flow and pressure for the tips/nozzles designated for use with the equipment.
- *Inlet water connection and filter* – The inlet water connection for these types of portable units contains a fine-mesh screen filter to protect the pump from damage due to small particles. It is usually designed to accept a $\frac{3}{4}$" garden hose.
- *Unloader valve and bypass* – The unloader valve allows water to recirculate from the high-pressure outlet side of the operating pump to the inlet side of the pump at reduced pressure through the bypass when the wand assembly is not in use. This reduces engine load when the equipment is operating but not being used. It also functions to stabilize gun pressure and engine load when the wand assembly is in use. The bypass may be a direct metal connection to the inlet from the valve as shown, or it may be a hose from the valve to a T-connection on the inlet. To provide easier starting, some unloader valves are also designed with an automatic pressure relief valve that prevents pressure buildup until the engine has been started and is running normally.
- *Pressure relief valve* – As a safety feature, a pressure relief valve is used to vent the high-pressure side of the pump to the atmosphere in the event that an unloader valve, thermal relief valve, or some other component malfunction occurs that causes the output pressure to exceed some value in excess of the rated operating pressure of the equipment.
- *Thermal relief valve* – As a pump safety feature, a thermal relief valve is used to vent into the atmosphere any overheated water that is generated when the pump is running in the bypass mode and the unit is accidentally left idle for some period of time. Cool water is introduced through the inlet connection, and the cycle repeats until the equipment is used or shut down. If heated water is being supplied at the inlet, care must be taken not to heat the water to a temperature exceeding the thermal relief valve rating.
- *High-pressure outlet connection* – The high-pressure outlet may be connected to a chemical injector body as shown in *Figure 4*, or directly to the spray hose via a quick-disconnect fitting.
- *Spray hose* – The high-pressure spray hose, rated to withstand the operating pressure of the equipment, transfers the pressurized water from the pump outlet to the spray gun via quick-disconnect fittings.
- *Wand assembly* – The wand assembly controls the pressurized water fed to the nozzle/tip. It is equipped with a trigger guard and a trigger safety latch to prevent accidental discharge of the gun during handling or if the gun is dropped or bumped. Interchangeable wands of various lengths, including telescoping wands, can be used.
- *Nozzle/tip* – Conventional or specialized nozzles and/or tips that are connected to the wand by quick-disconnect fittings apply pressurized water to a surface in various patterns.

2.3.4 *Typical Hot Water or Hot Water/Steam, Low-Pressure Washer Equipment*

Many low-pressure washers are available with on-board heaters that provide hot water or hot water/steam. On-board heating units can be powered by electricity, kerosene, fuel oil, propane, or natural gas. *Figure 6* shows a typical 3,000-psi, portable, gasoline engine-driven, hot water/steam unit with a kerosene/fuel oil-powered heating unit and an electric motor-driven burner. Similar versions with a gasoline- or propane-powered engine and propane-powered burner using an integrally mounted, 30-gallon propane tank are also available.

Figure 7 shows a larger, trailer-mounted, diesel-powered, hot water unit that can support multiple spray guns.

Hot water only units can heat water to temperatures of 200° to 210°F. Hot water/steam units can heat water to temperatures of 210°F or create wet steam at temperatures up to 325°F. The wand assembly (sometimes called a lance when used with these units) is insulated to protect the operator during use. Other than the heating equipment, its associated control and safety devices, and insulated wand assemblies, the units operate the same as cold water, low-pressure washers.

Instructor's Notes:

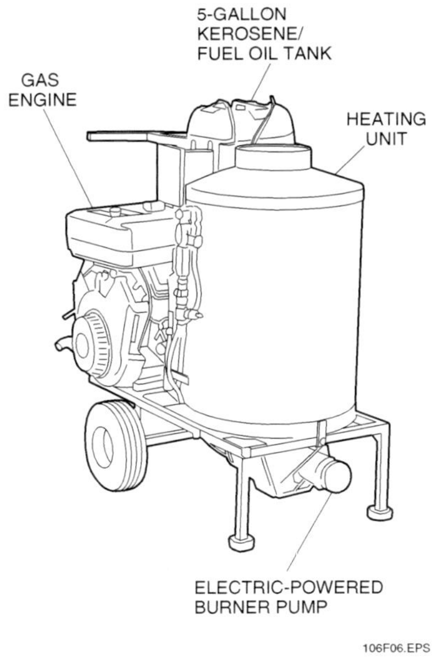

Figure 6 ◈ Typical 3,000-psi, portable, gasoline engine-driven, hot water/steam unit.

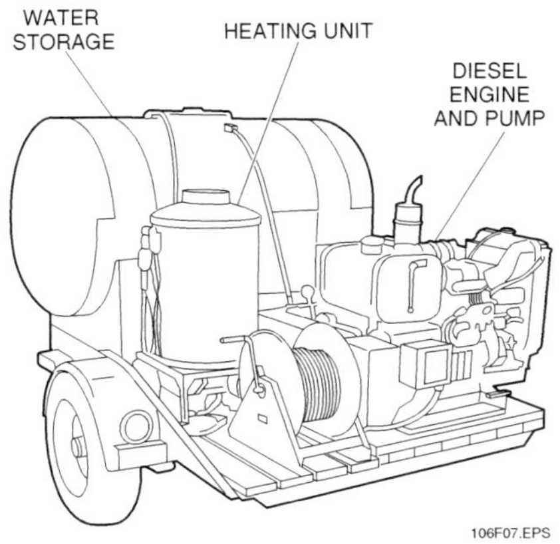

Figure 7 ◈ Larger, trailer-mounted, diesel-powered, hot water unit.

2.3.5 Pressure Washing Nozzles and/or Tips

The nozzles/tips supplied with low-pressure washers are normally provided in up to four different spray angles: 0, 15, 25, and 40 degrees. Different manufacturers have various methods of identifying the tip capacity and spray angle by the use of numbers stamped on the face of the tip. Some may have the capacity preceding the spray angle; others will precede the capacity number with the spray angle and may include a leading zero. For example, in a number such as 904-15, the spray angle is the number 15, representing 15 degrees. In numbers such as 25904 or 025904, the spray angle of 25 degrees is given in the first two or three digits. It will be necessary to refer to the equipment owner's manual to determine how to identify the tip angle.

CAUTION
Specific-capacity tips must be used with specific capacity washers. For example, a large-capacity tip used with a low-capacity unit will cause a large pressure drop and will be ineffective. A small-capacity tip used on a large-capacity unit will cause overheating due to excessive recirculation of the water and in extreme cases may cause an overpressure malfunction.

The following paragraphs provide descriptions and usage guidelines for various tips.

A 0-degree blaster tip provides a concentrated spot of high-pressure water which helps to blast off stubborn material. It also allows the user to stand further away from the surface being cleaned. Its uses include:

- Reaching high areas
- Breaking up large areas of coating to allow stripping
- Removing stubborn contaminants

CAUTION
This tip must be used with care since the concentrated force may erode soft materials or surfaces.

A 15-degree stripper tip provides a narrow fan of concentrated water for removing contaminants. It is held at an angle to the surface and works like a **peeling** tool. Its uses include:

- Loose coating removal from metal
- Oxidation removal
- Cleaning heavy equipment

Classroom

Explain how tip capacity and spray angles are identified on the face of nozzles and tips.

Safety

Discuss problems that could result if a large-capacity tip is used on a small-capacity unit.

Explain how to use the various tips without eroding soft materials or surfaces.

Classroom

Review the usage guidelines for 0-degree blaster tips, 15-degree stripper tips, 25-degree cleaner tips, and 40-degree rinser tips.

Classroom

Explain how to use the various low-pressure water accessories. Answer any questions trainees may have.

Audiovisual

Show Transparency 6 (Figure 8). Discuss the benefits of using a low-pressure flooding tip.

Show Transparency 7 (Figure 12). Explain how a water blast nozzle acts like a sandblaster.

A 25-degree cleaner tip provides a wider fan for pressure washing and rinsing. Its uses include:

- Cleaning dirt from surfaces
- Rinsing stripped areas
- General cleaning before painting

A 40-degree rinser tip provides a wide fan for cleaning and rinsing. Its uses include:

- Light cleaning
- Final rinsing after initial stripping or cleaning with chemicals
- Sweeping horizontal surfaces
- General washdown

2.3.6 Low-Pressure Washer Accessories

Depending on the manufacturer, various accessories can be used with the washer, such as an automatic detergent and cleaning chemical in-line injector with or without a special low-pressure flooding tip (*Figure 8*). Most in-line injector kits provide various metering orifices for use in the injector so that the proper dilution rates for concentrated chemicals can be achieved based on the flow rate of the tip.

Rotary power scrub brushes (*Figure 9*) are used when spray mist is objectionable or hard-to-remove contaminants are encountered.

Also available are telescoping extensions (*Figure 10*), horizontal (floor) scrubbers with rotating nozzles (*Figure 11*), and rotating water blast nozzles (*Figure 12*). Water blast nozzles use a highly concentrated stream (0 degrees) rotating in a circular pattern that acts like a sandblaster. Various other specialty nozzles are available. The variable tip

106F09.EPS

Figure 9 ◈ Rotary power scrub brush.

106F10.EPS

Figure 10 ◈ Telescoping extensions.

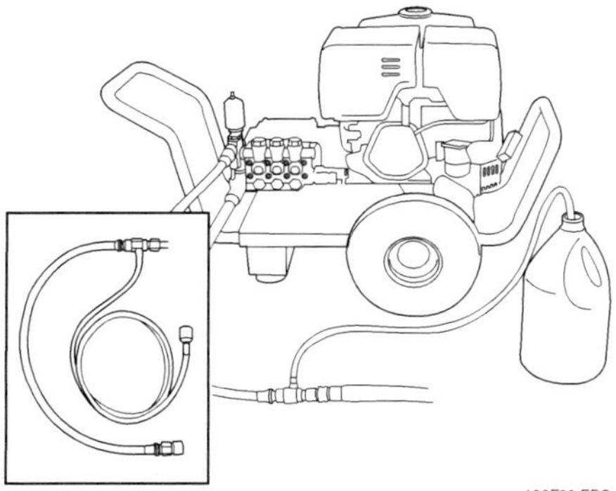

106F08.EPS

Figure 8 ◈ Automatic detergent and cleaning chemical in-line injector with a low-pressure flooding tip.

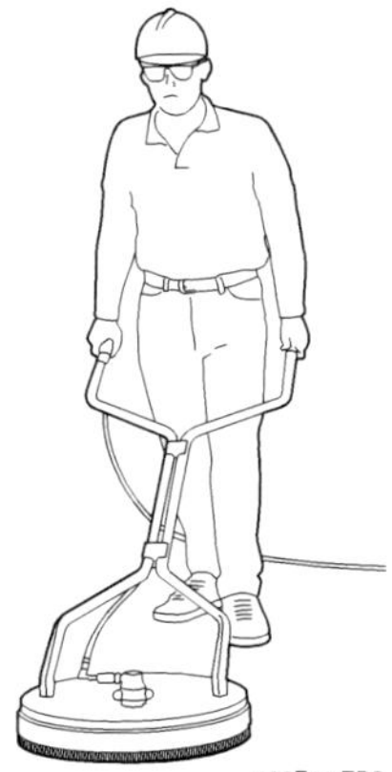

106F11.EPS

Figure 11 ◈ Horizontal scrubber with rotating nozzle.

Instructor's Notes:

nozzle (*Figure 13*) can be switched from a low-pressure, chemical-injector flooding tip to a standard tip by a push/pull action. A twisting action provides spray angles from 0 degrees to 25 degrees. The rotatable tip nozzle (*Figure 14*) allows switching back and forth between two different tips without shutting down and manually changing them.

Wet abrasive injectors (*Figure 15*) are also available for dust-free abrasive blasting. These devices usually have various changeable metering orifices in the abrasive pickup tube so the air/sand ratio can be changed to allow different sand injection rates at the spray tip. Similar special wet abrasive injectors, equipped with special non-clogging tips, are available for use with sodium bicarbonate abrasive injection. They have much lower injection rates (½ to 3 lbs./min.) than conventional wet abrasive injectors, which can inject as much as 5 to 10 pounds per minute.

Wet abrasive blasting with abrasive injection into a water stream is distinguished from another form of wet abrasive blasting in which water is injected into an air and abrasive stream to cut down on dust. In either method, mixing water and abrasive has some drawbacks. Nonsoluble wet abrasive cannot be recycled easily or economically. Soluble sodium bicarbonate abrasive provides a non-toxic, soft abrasive that requires little cleanup after the job is done. However, even though much less abrasive is used, it is relatively expensive compared to sand and is obviously non-recoverable. In either case, the surface must be rinsed with clean water after blasting to remove abrasive residue. Steel surfaces may be subject to flash rusting (rust-back), although the rinse can be treated with a rust inhibitor. The inhibitor must be compatible with the coating, or adhesion problems may develop. Generally, wet abrasive blasting is not the preferred method for obtaining a high-quality surface on steel.

Working with high-pressure water requires extra caution because of the potential risks to workers and the possible damage to nearby property. Operators should be well trained and equipped with appropriate personal protective equipment.

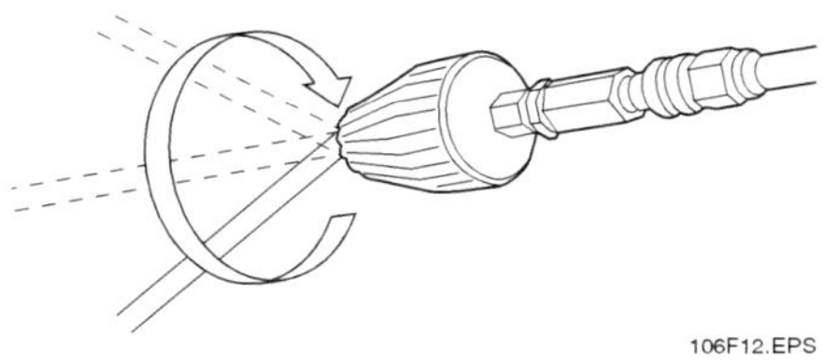

Figure 12 ◈ Rotating water blast nozzle.

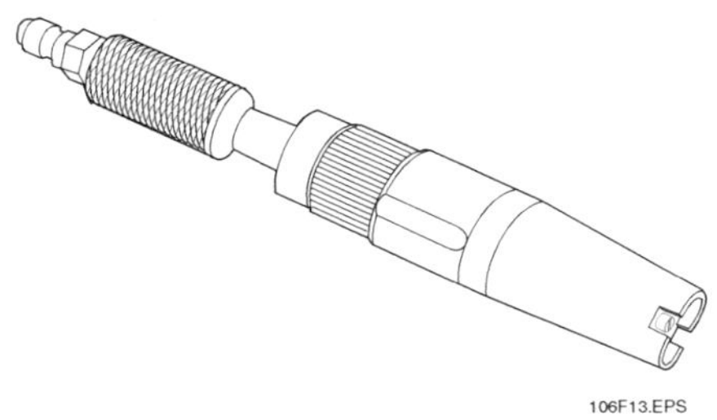

Figure 13 ◈ Variable tip nozzle.

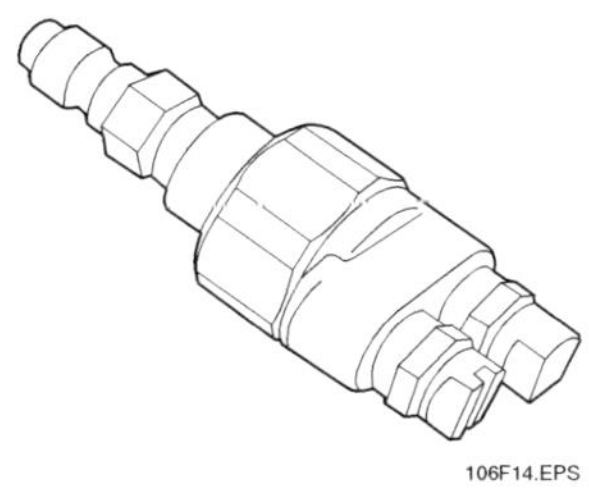

Figure 14 ◈ Rotatable tip nozzle.

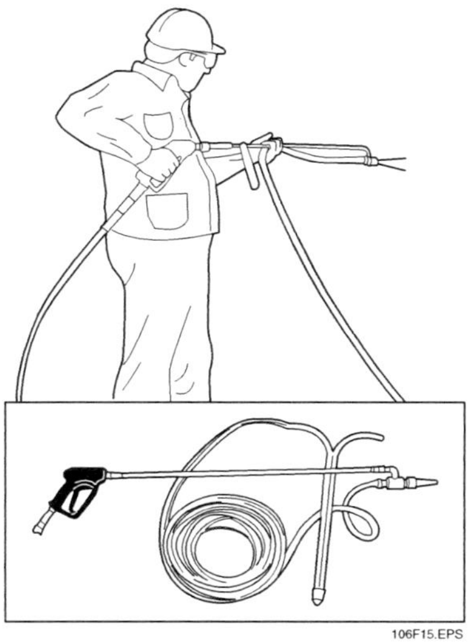

Figure 15 ◈ Wet abrasive injectors.

Audiovisual

Show Transparency 8 (Figure 14). Explain how a rotatable tip nozzle automatically switches between two tips.

Teaching Tip

See the Teaching Tip for Sections 2.3.5–2.3.6 at the end of this module.

Homework

Assign the reading of Sections 2.3.7–2.4.5 for the next session.

Materials

Ensure that you have everything required for the demonstration and laboratory during this session.

Safety

Explain how to prevent injury when working with wet abrasive injectors or chemical injectors.

Classroom

Review the hazards that are associated with using pressure washers or accessories.

Audiovisual

Show Transparency 9 (Figure 16). Ensure that trainees understand how to verify that the trigger safety latch is fully engaged.

> **WARNING!**
> When using wet abrasive injectors or chemical injectors, operators must wear respirators to prevent inhalation of abrasive dust mist or chemical mist. Always follow prescribed safety rules and watch for fatigue.
>
> Fine-mesh screen eye protection that attaches to a helmet must be worn by the operator when using pressurized water equipment.
>
> Be careful when working on a scaffold because using pressurized water equipment can cause problems with balance and footing. For the same reason, keep the work area as clean and unobstructed as possible.
>
> Do not point or aim the water spray at another person.
>
> Equipment should be well maintained and inspected regularly for proper reinforcements and connections. Never operate equipment beyond its recommended limits, and never use water blasting around electrical units.

2.3.7 Safety Guidelines

Review the following safety precautions before attempting to use any pressure washers or accessories.

- *Injection and burn hazard* – Any fluid (particularly fluid containing chemicals or abrasives) injected into the skin from the nozzle/tip, leaks, or ruptured components may look like just a cut, but it is a serious injury. Get emergency surgical treatment immediately. In addition:
  - Do not point the nozzle/tip at any part of your body or at any other person.
  - Do not stop or deflect leaks with any part of your body or with a glove or rag.
  - Do not put your hand or fingers over the nozzle/tip.
  - Make sure high-pressure fluid connections are secure before starting the equipment.
  - Engage the wand assembly trigger safety latch before starting the equipment and whenever you stop spraying, even for a moment. Make sure the safety latch is fully engaged or it will not prevent the wand assembly from being triggered (see *Figure 16*).
  - When using hot water/steam low-pressure washers, use insulated wand assemblies (lances) and gloves to prevent burns.
- *Hazardous fluids* – Improper handling of hazardous fluids can cause serious injury, even death, due to splashing in eyes, ingestion, or bodily contamination.
  - Know the specific hazards of the fluid being used.
  - Store hazardous fluids in approved containers. Dispose of hazardous fluids per local, state, and national guidelines.
  - Wear protective eyewear, gloves, clothing, and a respirator as recommended by OSHA, the fluid manufacturer, or any local regulations.
- *Fuel hazard* – For engine-powered equipment, the fuel used is combustible and, when spilled on a hot surface, can ignite and cause a fire. Do not fill the fuel tank while the engine is running or hot.
- *Exhaust hazard* – Do not operate equipment powered by an engine in a closed building. The exhaust contains poisonous carbon monoxide, which is colorless and odorless.
- *Equipment misuse hazard* – Misuse of the equipment may cause ruptures and result in fluid injection, splashing in the eyes or on the skin, or other serious injuries.

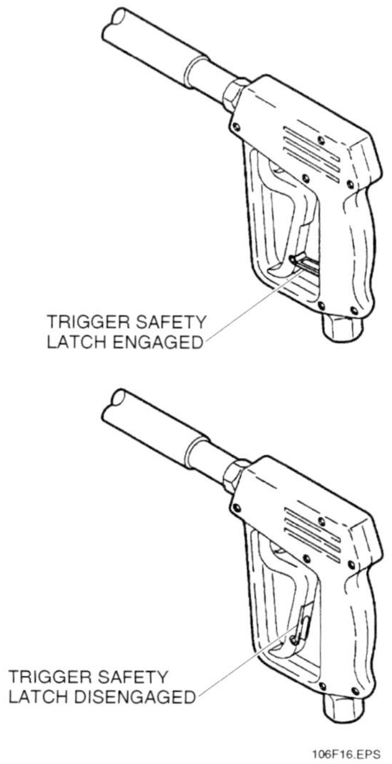

Figure 16 ◆ Trigger safety latch engaged and disengaged.

Instructor's Notes:

- Do not alter or modify any part of the equipment.
- Do not change any factory-set adjustments.
- Do not exceed the maximum working pressure of any component or accessory.
- Do not use any chemicals that are incompatible with the wetted parts as stated in the technical data for the equipment.
- Do not alter the throttle setting.
- Replace worn or damaged parts immediately.
- Check hoses, tubes, and couplings daily. Do not repair high-pressure hoses or couplings; replace the entire hose with one rated for use with the equipment. Make sure fluid hoses have spring guards on both ends to prevent kinks and ruptures.

• *Hypothermia hazard* – When conducting spraying operations in temperatures below 65°F with any amount of wind, a risk of hypothermia exists. Always wear adequate clothing and protective rubber suits, gloves, and a waterproof helmet.

• *Footing and balance hazard* – When conducting spraying operations, the work site will become wet and slick, especially with some types of chemical injection. This can cause difficult footing and balance for the operator.
  - Any scaffolding used must be secure and free of debris and equipment to prevent slipping and tripping.
  - Hoses and other materials must remain clear of your feet at all times.
  - With weight on both feet and feet spaced apart in an open stance, grip the gun and wand securely using both hands before triggering the wand assembly. Be prepared for the reactive kick of the wand assembly when it is triggered.

• *Electrical hazard* – Do not direct spray into or onto electrical boxes or devices. You may be seriously injured or killed by electrocution; damage to the electrical equipment or a fire may occur. Make sure all electrical equipment in the work area is turned off and completely sealed by masking before commencing spraying operations.

2.3.8 Pressure Relief Guidelines

To relieve system pressure:

Step 1 Engage the trigger safety latch.

Step 2 Turn the engine or motor off and disconnect the spark plug wire, if applicable.

Step 3 Shut off the water supply.

Step 4 Disengage the safety latch and trigger the wand assembly to relieve the pressure. Re-engage the safety latch.

> **WARNING!**
> To prevent bodily injury, system pressure must be relieved whenever you are instructed to relieve pressure, stop spraying for more than five minutes, check or service any of the system equipment, or install, clean, or change the nozzle/tip.

If you suspect that the nozzle/tip, wand assembly, or hose is clogged or that the pressure has not been fully relieved, perform the following procedure:

Step 1 Trigger the wand assembly again.

Step 2 Wrap a rag around a pressure hose coupling several times. Using leather gloves, very slowly loosen the coupling to relieve pressure gradually; then loosen it completely.

Step 3 Clear the clog from the hose, wand assembly, or nozzle/tip. Reassemble the equipment, if necessary.

Step 4 Re-engage the wand assembly safety latch.

2.3.9 Start-Up Guidelines

Step 1 Engage the trigger safety latch, then make sure that the motor is off and disconnected or that the engine is disabled.

Step 2 Check that a clean water source is available.

Step 3 Check that the pump inlet screen is in place. Connect all pump inlet and outlet hoses, injector (if required), wand assembly, and nozzle/tip. Check that connections are secure.

Step 4 Check that the correct injector metering orifice is installed (if applicable).

> **WARNING!**
> Fire hazard – Do not refuel a hot engine. Allow the engine to cool for five or more minutes before refueling.

Classroom

Explain how to relieve system pressure from a pressure washer.

Safety

Ask a trainee to explain when system pressure must be relieved.

Classroom

Review the start-up guidelines for a pressure washer. Go over any questions trainees may have.

Safety

Discuss the consequences of refueling a hot engine.

Safety

Ask a trainee to explain why a pump should never be run dry.

Discuss the hazards associated with engine and heater exhaust.

Review problems that could result if equipment is run without spraying it for more than five minutes.

Review the risks of working with spray equipment.

Emphasize the importance of protecting all adjacent surfaces before spraying.

Demonstration

Review the spraying safety guidelines. Demonstrate how to properly disengage a wand assembly safety latch.

Step 5 If an engine is the power source, check that the fuel tank is full and the oil level is correct.

> **CAUTION**
> Never run a pump dry; damage to the pump will occur. Always make sure the water supply is on before starting the equipment.

Step 6 Turn on the water supply and trigger the wand assembly until air is purged from the equipment and water flows from the nozzle/tip. Re-engage the wand assembly safety latch.

Step 7 Connect the motor to the power source or re-enable the engine.

> **WARNING!**
> Exhaust hazard – If the engine or heater is fuel-powered, do not operate the equipment in a closed or confined space.

Step 8 If equipped, start hot water/steam heater.

> **CAUTION**
> Do not allow the equipment to run without spraying for longer than five minutes. Recirculating water may overheat and damage the pump. Reduce this time further if heated water is being used.

Step 9 Start the motor or engine as prescribed in the owner's manual. If necessary for easier starting, have one person trigger the wand assembly while another person starts the equipment. Make sure the wand assembly safety latch is engaged after the equipment has started.

2.3.10 Spraying Safety Guidelines

Do not allow the spray stream to contact any part of the body or another person. Wear protective clothing, gloves, and a helmet with a screen visor or goggles. Always engage the wand assembly safety latch when spraying is stopped, even for a moment. Always be aware of the following hazards:

- *Hypothermia hazard* – Wear warm clothing under a rubber suit and gloves; wear a waterproof helmet.
- *Hazardous fluids and materials* – When working with hazardous fluids or removing hazardous materials, use appropriate respiratory and eye protection.
- *Electrical hazard* – Do not spray directly into or onto electrical boxes or devices. All power to such equipment must be turned off, and the equipment must be masked and sealed.
- *Footing and balance hazard* – Ensure that any scaffolding is secure and that working areas are clear of debris, materials, and hoses. Grip the wand assembly securely and make sure footing is secure before triggering the wand assembly. Be prepared for a reactive kick when the wand assembly is triggered.

> **CAUTION**
> Make sure adjacent surfaces are protected before spraying.

Step 1 Check that adjacent areas have been protected.

Step 2 Check that the electric power to any electrical boxes or devices has been turned off and that the electrical equipment has been masked and sealed.

Step 3 With proper grip and stance, disengage the wand assembly safety latch, trigger the wand assembly, and commence spraying operations.

> **NOTE**
> If wet abrasive blasting is being conducted, follow the manufacturer's instructions for use of the equipment to clean/profile surfaces.

Step 4 Re-engage the safety latch on the wand assembly when the spraying operation is completed or stopped (even for a moment).

> **NOTE**
> If spraying operations will not resume after five minutes or less, turn off the equipment and water supply, then relieve the pressure.

Instructor's Notes:

2.3.11 *Cleaning Techniques*

Use warm or hot water/steam; it will increase the cleaning power of chemicals and/or water. Always apply chemical cleaners to as dry a surface as possible. Wet surfaces dilute the chemicals, reducing their cleaning ability. Apply chemical cleaners from the bottom up and side-to-side on vertical surfaces to avoid streaking. On horizontal surfaces, apply from side-to-side, back-to-front to avoid blowing the chemicals off the surface. Always allow chemicals to work for several minutes before rinsing, but do not let them dry on the surface before rinsing. If necessary, work on small areas at a time. Avoid working on hot surfaces in direct sunlight. Rinse or wash from top down, side-to-side on vertical surfaces, and side-to-side, front-to-back on horizontal surfaces. Using an appropriate fan tip, hold it six to eight inches from the surface at a slight angle (away from uncleaned areas). Use the fan spray like a peeling tool and move it steadily and systematically over the surface. Avoid rapid back and forth motion.

2.3.12 *Nozzle/Tip Changing Guidelines*

Never change nozzles/tips when the system is pressurized. Never place your hands, fingers, or face in front of the nozzle/tip. Keep nozzle/tip directed away from other people.

Step 1 Turn off the equipment and water supply and relieve the pressure.

Step 2 Point the nozzle/tip away from yourself and any other person.

Step 3 Without placing your hand or fingers over the nozzle/tip, pull back the quick-disconnect coupler and push the nozzle/tip out of the coupler with your fingers (*Figure 17*).

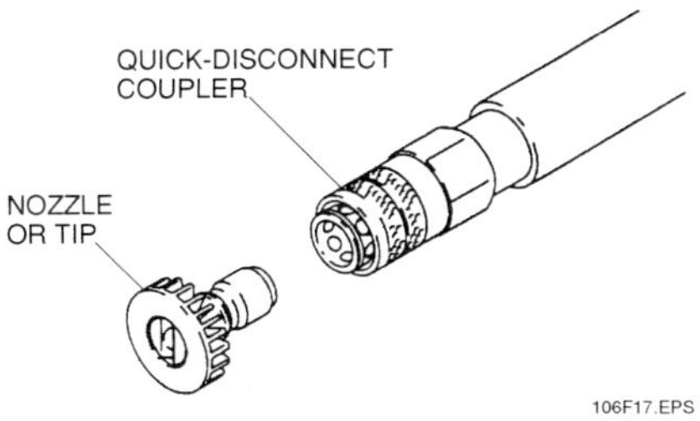

Figure 17 ◆ Quick-disconnect coupler and nozzle.

> **WARNING!**
> Make sure the nozzle/tip is secured in the coupler. Bodily injury or damage to the coupler will occur if the nozzle/tip is blown out of the coupler when system pressure is applied.

Step 4 Select the desired nozzle/tip and insert it in the coupler. Push the coupler forward until it seats. Pull on the nozzle/tip to make sure it is securely engaged.

2.3.13 *Pressure Washer Flushing and Storage*

Ice may damage the pump or heater, if so equipped. If water freezes in the equipment, thaw it at room temperature before attempting to start it. Do not pour hot water on or into the pump; it may crack the pump or its components.

- If the equipment may be exposed to freezing temperatures for a short period of time without being used, drain all the water out of the pump, heater (if equipped), hoses, gun, and wand.
- If the equipment will be stored overnight or longer in freezing temperatures, drain the water from the equipment, flush with a 50% solution of antifreeze, then drain again. Before using the equipment again, flush with fresh, clean water to remove all traces of the antifreeze solution.
- Follow the manufacturer's instructions for long-term storage of equipment and, if applicable, the engine.

2.4.0 Hand Tool Cleaning (*SSPC-SP2*)

Cleaning with hand tools (*Figure 18*) is most effective at removing small amounts of loose contaminants from steel substrates. Wire brushes, sandpaper, nonwoven abrasive pads, scrapers, knives of various shapes and sizes, and chipping hammers are good for cleaning away dirt, loose paint, rust, and mill scale. The best tool to use depends on the type and condition of the surface. One or all of them may be needed.

Hand tools cannot be expected to satisfactorily remove oil and grease, tightly adhered rust, paint, mill scale, or contaminants stuck in pits or crevices. In those cases, other cleaning methods are required.

When using hand tools, it is important not to scar the surface or raise burrs that may later protrude through the coating and provide a place for corrosion to begin.

Classroom

Review proper cleaning techniques. Ask whether trainees have any questions.

Audiovisual

Show Transparency 10 (Figure 17). Explain how to change a nozzle/tip.

Safety

Explain problems that could result if the nozzle/tip is not secured in the coupler.

Classroom

Explain how to properly store pressure washers.

Explain when hand tools are most effective for cleaning. Answer any questions trainees may have.

Classroom

Review the types of common hand tools and the guidelines for working with them. Discuss applications for each.

Audiovisual

Show Transparency 11 (Figure 19). Review the different types, shapes, and sizes of scrapers.

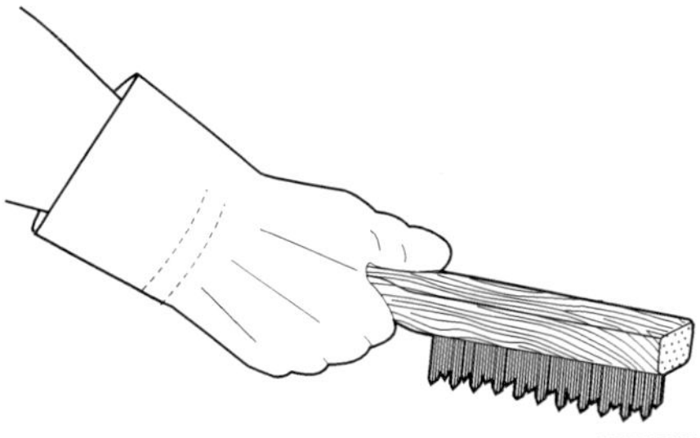

Figure 18 ◈ Cleaning with hand tools.

Before cleaning with hand tools, scrape off any heavy deposits of oil and grease. Then wash the surface with water, detergent, or solvent to remove any remaining dirt, oil, and grease. After cleaning the surface, brush, dust, sweep, or blow away loose debris.

Hand tool cleaning can be done in a confined area or near sensitive equipment since it creates little dust. This method works well when an entire surface does not require cleaning.

For large jobs, it can be slow and costly. Since the quality of cleaning with hand tools may be limited compared to more powerful cleaning methods, coating life expectancy may not be as good. However, some coatings, such as slow-drying, oil-based paints with good wetting ability, or epoxy **mastics**, work well on hand tool-cleaned surfaces, making this an attractive option in certain circumstances.

SSPC-SP2 defines hand tool cleaning as the removal of all loose mill scale, rust, paint, and other loose material (other than adherent). Material is considered adherent if it cannot be removed by lifting with a dull putty knife.

Working safely with hand tools requires the use of safety goggles, gloves, and a NIOSH-approved cartridge respirator to avoid inhaling dust, rust, and material chips. Be careful not to create sparks around flammable materials or volatile vapors.

2.4.1 Hand Tools–General Information

Hand tools are most effective for cleaning small areas. In addition, they are a good choice when power tools are not available, or in confined areas that are inaccessible with power tools. The use of hand tools is cost-effective in maintenance coating programs when only part of the surface needs attention. However, since the user of the tool is the power source, hand-tool cleaning can be slow and tiring work.

Common hand tools include abrasive pads, scrapers, putty knives, wire brushes, and chipping hammers. The best tool to use depends on the job.

Working with hand tools is relatively safe, but it still requires attention to these basic safety guidelines:

- Wear protective glasses or goggles to protect eyes from flying debris or irritating detergents and solvents.
- Keep tools in good condition. Replace or repair any defective equipment.
- Do not leave unused tools on scaffolds, ladders, or overhead work areas where they could fall on someone working below.
- Never throw or drop tools.
- Avoid using tools that may cause sparks near volatile vapors or combustible materials. In some cases, brass tools may be needed to avoid sparks.

Hand tools are harder, rougher, and sharper than your hands, and are used on surfaces that can cause cuts, punctures, and scrapes. Even a splinter of metal or a chip of paint can cause a painful injury. Always wear gloves and goggles when using hand tools and keep alert to the danger of potential injury.

2.4.2 Abrasive Pads, Wet and Dry

Abrasive pads are webs of nylon fibers imbedded with abrasive material. They can be used wet or dry for a variety of jobs, such as removing dried soil, debris, rust, and mill scale from steel. They can also be used to sand surfaces, remove fuzz or nibs between coats, and remove dry spray or to feather the edges of a repair area to disguise the repair.

Nonwoven abrasive pads can be used alone or attached to a holder with a handle. The pads alone are flexible and can be used for contour work. They can be cut to size as needed and rinsed for reuse.

Sanding sponges are flexible and coated with abrasive material—some on one side, some on all sides—for use on curved or contoured surfaces. Some are available for wet or dry sanding. Like nonwoven abrasive pads, they can be rinsed clean for reuse.

2.4.3 Scrapers and Putty Knives

Scrapers are all-purpose tools. They come in a variety of types, shapes, and sizes for different jobs. They are used to remove peeling, flaking, or **blistering** coating, rust, or debris from any type of surface. Different types of scrapers are shown in *Figure 19*.

Instructor's Notes:

Razor-edged scraper knives have a replaceable razor-edged blade that can be resharpened.

Do not use a steel scraper on a rough surface when flammable liquids are present as it may cause a spark and ignite the liquid. As with all hand tools, use scrapers with care and be alert to potential danger and injury.

A putty knife, which has a blade 1" to 2" wide, is used to scrape off loose coating. Sometimes it may be necessary to fabricate a hand tool to clean unusual or irregular surfaces by grinding and shaping putty knives of different styles and sizes to fit a surface profile. For this reason, broken putty knives and other damaged cleaning tools are often saved.

A versatile device called a *combination tool* or a *5-in-1 tool* has a special blade that may be used as a scraper, patching spreader, and paint can opener. It has a sharp point for gouging old, loose paint from crevices or open cracks for patching. Its half-round, cut-out blade is good for cleaning roller covers. Some uses of a combination tool are illustrated in *Figure 20*.

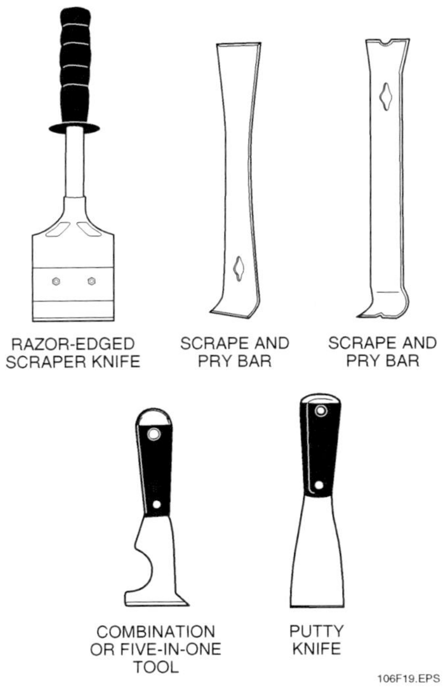

Figure 19 ◆ Different types of scrapers.

2.4.4 Wire Brushes

Wire brushes are useful for cleaning loose rust and other debris from all types of surfaces. There are different types of wire brushes. Within each type, there are variations in size, number and rows of wire, as well as wire stiffness.

One type of brush is the flat block wire brush (*Figure 21*), which has bristles mounted in a flat block of wood like a scrub brush. It is used to clean large, flat areas. Other types of brushes have bristles at one end of a long handle. They include an all-purpose shoe handle brush, a combination wire brush and scraper which provides two ways to remove stubborn contaminants, and a curved handle brush which gives better brushing control because fewer wires touch the surface at a time. The needs of each job should determine the type of brush to use.

2.4.5 Chipping Hammers

A chipping hammer (*Figure 22*) is used to chip away layers of loose rust, coating, and mill scale from steel surfaces. It may be the first tool used to remove such contaminants, followed by other

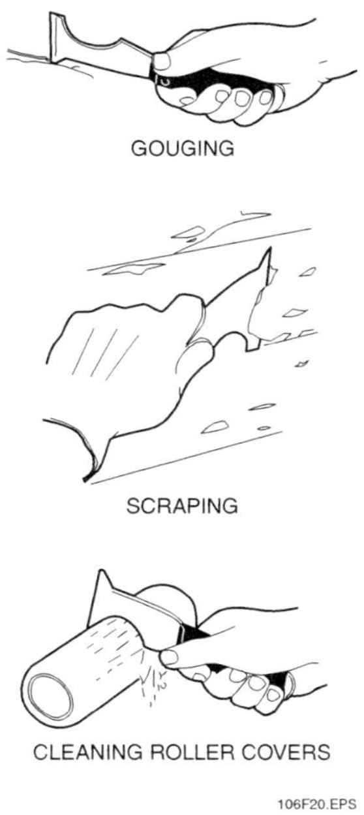

Figure 20 ◆ Some uses of a combination tool.

Audiovisual

Show Transparency 12 (Figure 21). Discuss the different types of wire brushes and their uses.

Laboratory

Set up stations with a variety of common hand tools and several surfaces that require cleaning. Have trainees practice cleaning with hand tools. Discuss the advantages of each tool.

Homework

Have trainees review Sections 2.5.0–2.7.0 for the next session.

Materials

Ensure that you have everything required for the laboratory and testing during this session.

Classroom

Explain when and how power tools are used for surface preparation.

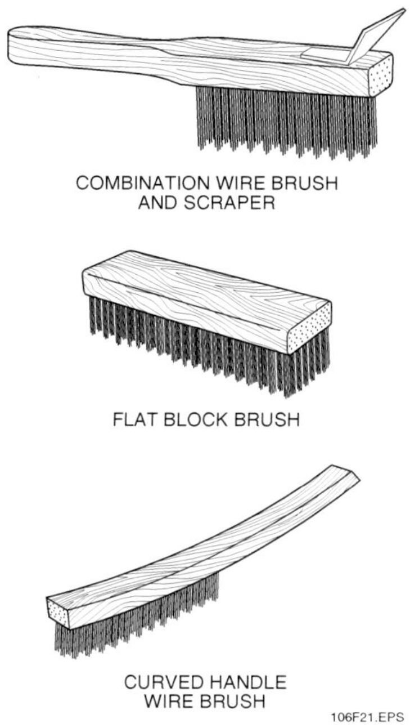

Figure 21 ◆ Wire brushes.

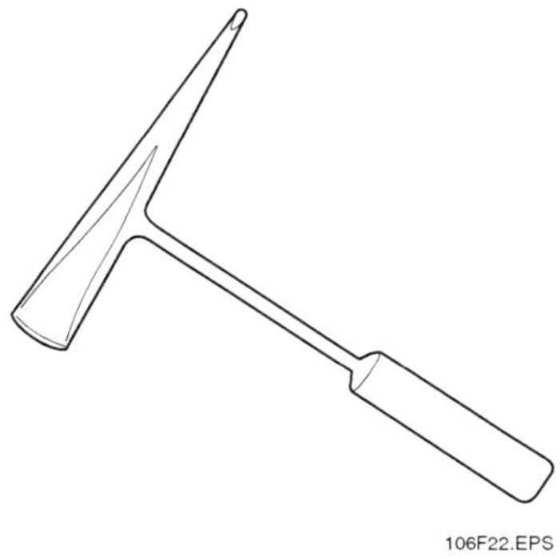

Figure 22 ◆ Chipping hammer.

tools, such as scrapers, nonwoven abrasives, and wire brushes. The heads of chipping hammers come in different configurations. Some have two wedge-shaped faces, one pointed, the other flattened, which make the tool adaptable to different uses. When using a chipping hammer, avoid marking or raising burrs on the surface that may interfere with proper coating.

2.5.0 Power Tool Cleaning (*SSPC-SP3* and *SSPC-SP11*; *ASTM D 4259*)

Power tools are used when hand tools are too slow or ineffective to clean a surface efficiently. Power tools are often selected when abrasive blasting is impractical, is too expensive, or is inappropriate because of environmental factors.

Generally, power tools will remove dirt and hard, brittle material from all surfaces. They also remove loose mill scale or rust, weld spatter, sharp fins and edges, and weld flux from steel.

In addition to removing contaminants, power tools are useful for preparing welds on steel surfaces before blasting or coating, and grinding sharp edges, which may cause paint to draw thin.

As a surface preparation method, power tool cleaning is an important alternative to **abrasive blast cleaning** near sensitive machinery and in populated areas where the dust and debris generated by blasting could be damaging to equipment or dangerous to people. Power tools can also be an effective choice when used with an integrated vacuum system because they limit the amount of hazardous waste that is created compared to conventional abrasive blasting.

In recognition of the increased role of power tool cleaning, SSPC developed a standard, *SSPC-SP11, Power Tool Cleaning to Bare Metal,* to supplement its earlier standard, *SSPC-SP3, Power Tool Cleaning. SSPC-SP3* requires the removal of paint, rust, and mill scale loose enough to be lifted with the blade of a putty knife. *SSPC-SP11* requires removal of all detrimental material except "slight residues of rust and paint in the lower portion of pits." It also requires roughening to a minimum 1-mil surface profile. *SSPC-VIS 3, Visual Standard for Power- and Hand-Tool Cleaned Steel,* was developed to supplement these standards by providing an approximate image of the specified degrees of cleanliness. It shows seven initial conditions and how each of them looks in compliance with SSPC standards for hand tool cleaning, power tool cleaning, and power tool cleaning to bare metal.

Instructor's Notes:

2.5.1 Power Tools–General Information

There are three types of power tools used for surface preparation: impact tools, rotary/vibrating tools, and rotary impact tools.

- Impact tools, such as scaling hammers, needle guns (*Figure 23*), and power chisels, are used to clean steel. They can clean a surface by knocking off the contaminants, including heavy layers of rust, mill scale, thick paint or coating, weld flux, slag, and weld spatter.
- Rotary/vibrating tools are also used on steel. They include sanders, wire brushes, and grinding wheels (*Figure 24*). They brush, grind, or sand away loose rust, paint, loose mill scale, weld flux, and slag. They are also used to prepare welded surfaces.

Figure 23 ◈ Needle gun.

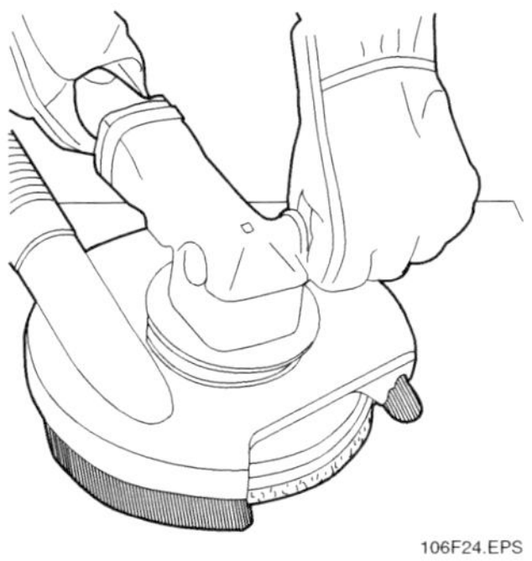

Figure 24 ◈ Grinding wheel.

- Rotary impact tools (*Figure 25*) combine the working features of both rotary and impact tools. They use rotary action to hit a steel surface with an abrasive material. The names of these tools (e.g., rotary peeners, rotary flaps, and rotary hammers) reflect how they work to remove old paint, tight mill scale, rust, weld flux, and slag.

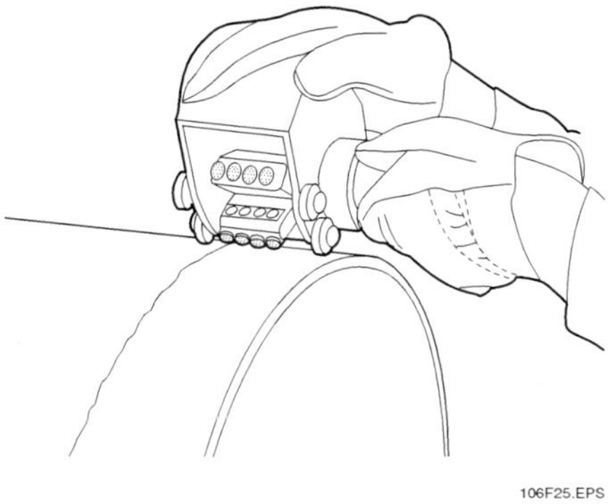

Figure 25 ◈ Rotary impact tool.

As always, start by washing the surface so that oil and grease are not spread around by the action of the power tools.

To create a profile on steel that is suitable for high-performance coatings, follow *SSPC-SP11, Power Tool Cleaning to Bare Metal*. Power tool cleaning to bare metal is defined by *SSPC-SP11* as a surface, when viewed without magnification, that is free of all visible oil, grease, dirt, rust, mill scale, paint, oxides, corrosion products, and other foreign matter. Slight residues of rust or paint may be left in the lower portion of pits if the surface is pitted.

WARNING!
Electric tools must be properly grounded and must never be used in wet areas. Remove any explosive hazard to avoid ignition from sparks and use explosion-proof tools in confined areas.

Safe use of power tools requires good training, regular inspection of equipment, and appropriate personal protective equipment, including hearing protection and a respirator, especially if removing any respirable debris.

Classroom

Review the general guidelines for using power tools. Ask whether trainees have any questions.

Audiovisual

Show Transparency 13 (Figure 24). Explain how to use a grinding wheel.

Safety

Emphasize the importance of properly grounding electric tools.

Safety

Explain how to safely remove lead-based materials.

Classroom

Explain how grinders and sanders are used to remove contaminants from steel.

Audiovisual

Show Transparency 14 (Figure 26). Review the two common types of attachments for grinding and sanding tools.

Classroom

Refer to Figure 27. Explain how pneumatic-powered orbital sanders prevent scratching and gouging.

Audiovisual

Show Transparency 15 (Figure 28). Explain how to attach stiff wire bristles to a rotary cleaning tool.

> **WARNING!**
> Removing lead-based materials requires strict and careful compliance with the regulations designed to protect the worker and work area from the dangers of contamination.

2.5.2 Grinders and Sanders

Grinding and sanding tools use rotary or vibrating action to remove contaminants from steel.

Two common types of attachments for grinding and sanding tools are made of nonwoven abrasive and coated abrasive material. Nonwoven abrasive is a synthetic fiber web treated with abrasive grit in different grades and densities. It comes in the form of wheels, cups, or discs, which wear evenly to provide a constant rate of cleaning. Coated abrasive is made of abrasive material bonded to a backing material. It also comes in various forms, including sheets, rolls, discs, belts, and flaps.

There are several distinctions between these types of abrasives. One is that nonwoven abrasives are less susceptible to loading or filling up with dust and debris. Another distinction is that coated abrasives can remove surface contaminants as well as base metal, which is useful for weld grinding.

Nonwoven and coated abrasive attachments can be mounted on either straight, in-line rotary tools or vertical, right-angle rotary tools (*Figure 26*). Wheel-shaped attachments work well on straight rotary tools; discs and cups work well on vertical tools.

Pneumatic- or electric-powered disc sanders and grinders have a revolving, flat, circular abrasive which can be used to aggressively remove heavy coatings and other contaminants. They work well on metal.

Pneumatic-powered orbital sanders (*Figure 27*) can be used for coarse to very fine sanding, depending on the particular tools and the abrasive used. If used properly, they can sand without scratching or gouging a surface. The flat abrasive pad of an orbital sander rotates at high speed while moving in a small, circular motion.

Belt sanders have a continuous belt of abrasive that moves in one direction. The width of the abrasive belt varies, depending on the size of the tool. Belt sanders can only remove limited layers without gouging a surface.

Pneumatic (or electric) half-sheet sanders, palm sanders, and die grinders are all finishing tools. They are vibrating sanders with either a straight-line or oscillating action. Half-sheet sanders have flat, rectangular pads to which a half sheet of sandpaper is attached. They work well on large, flat surfaces. Palm sanders, so named because they fit comfortably in the palm of the hand, are good for fine finishing of metal. They work well on both flat and rounded surfaces.

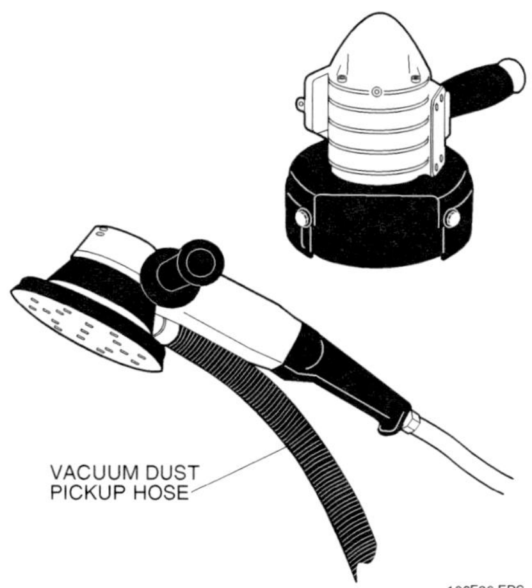

106F26.EPS

Figure 26 ◆ Nonwoven and coated abrasive attachment.

106F27.EPS

Figure 27 ◆ Pneumatic-powered orbital sander.

2.5.3 Power Wire Brushes

Stiff wire bristles attached to a rotary cleaning tool (*Figure 28*) will brush away loose mill scale, loose rust, weld flux, slag, dirt, and old paint or coating from steel.

Bristles come knotted or crimped and in the shape of a wheel or a cup. They attach to either a straight rotary tool with a rotating axis in line with

Instructor's Notes:

the handle, or a vertical rotary tool with an axis perpendicular to the handle. The tool to use depends on the job and shape of the bristles that would work best. Wheel bristles are usually used on straight rotary tools. Cup bristles are normally used on vertical rotary tools.

Operating wire brushes at excessive speed or holding them in one spot for too long may burnish the surface and leave it too smooth and polished to be coated effectively.

2.5.4 Power Files

Power files are a substitute for tedious hand filing to smooth or remove burrs from metal surfaces. The power advantage is the regulated length of each stroke and the number of strokes per minute, which help to achieve a uniform finish. It is also useful in doing intricate work that requires precision. The files, which are either straight or curved, can be used to reach into confined areas.

2.5.5 Needle Guns

A needle gun (*Figure 29*) is a type of compressed air-powered impact tool. It has a bundle of long, cylindrical steel rods mounted in front of a piston that strikes them against a surface. Because each needle hits the surface individually, a needle gun is good for cleaning irregular areas, such as corners (*Figure 30*) and crevices or around bolt and rivet heads. It works best on loose or brittle material.

Other types of impact tools include chipping and scaling hammers, piston scalers, and power chisels (*Figure 31*). All work on the same principle as the needle gun. These tools will remove heavy rust, mill scale, thick paint, weld flux, and slag.

Figure 28 ◈ Stiff wire bristles attached to rotary cleaning tool.

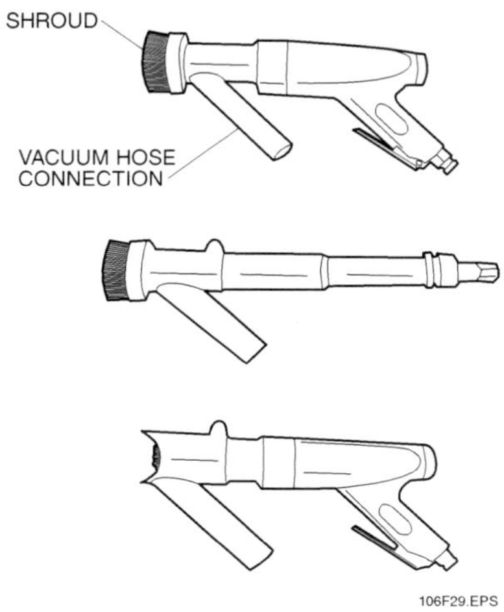

Figure 29 ◈ Needle gun.

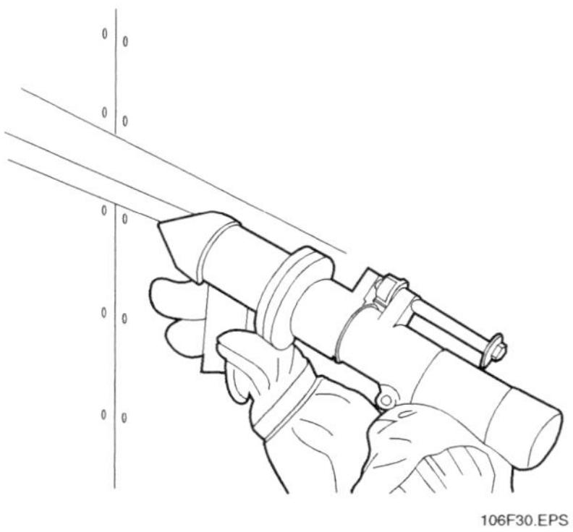

Figure 30 ◈ Needle gun cleaning a corner.

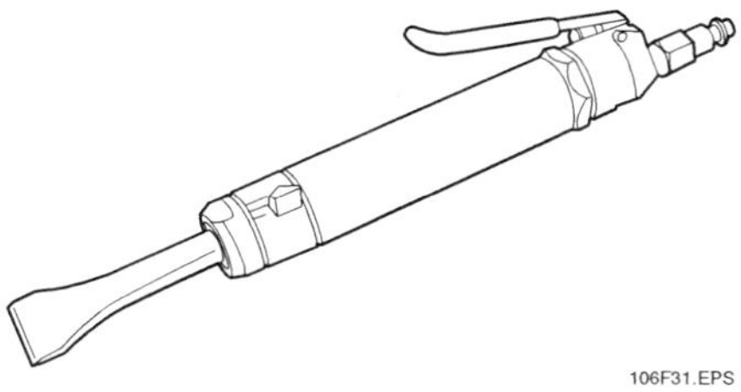

Figure 31 ◈ Power chisel.

Classroom

Discuss the advantages of using power files.

Audiovisual

Show Transparency 16 (Figure 29). Explain how needle guns use compressed air.

Audiovisual

Show Transparency 17 (Figure 32). Explain how rotary impact power tools remove heavy coatings and contaminants.

Classroom

Refer to Figures 33 through 35. Discuss the three common types of abrasive hubs.

Discuss the role of a shroud that is attached to a power tool.

Audiovisual

Show Transparency 18 (Figure 37). Review the benefits of using an integrated vacuum system.

2.5.6 Rotary Impact Power Tools

A rotary impact tool, such as the one shown in *Figure 32*, uses rotating action to throw different types of fixed abrasives against a steel surface to remove heavy coatings and contaminants.

Three common types of abrasive hubs are:

- Cutter bundles or stars (*Figure 33*) are hardened, star-shaped steel washers that rotate on spindles around the tool's axis.
- Rotary hammers (*Figure 34*) are free-swinging steel plates attached to the tool's rotating hub.
- Rotary flaps or peeners (*Figure 35*) are round carbide cleats attached to flexible flaps mounted in the rotating hub.

In addition to hand-held models, peening equipment is also available in upright models for descaling floors.

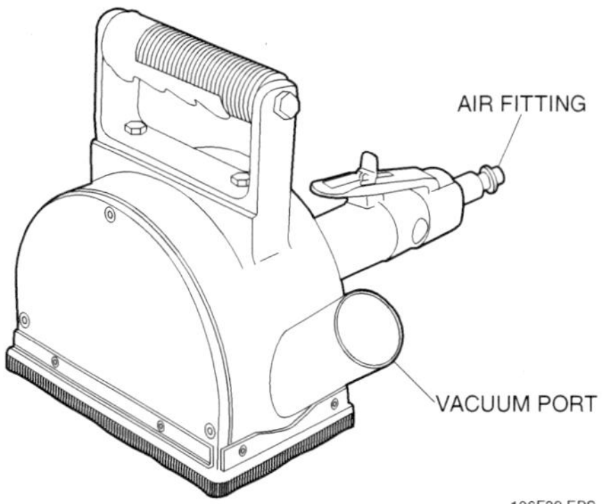

Figure 32 ◈ Rotary impact tool.

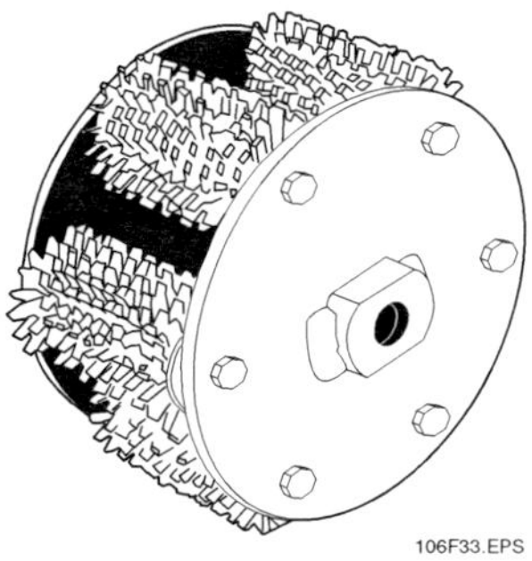

Figure 33 ◈ Cutter bundles or stars.

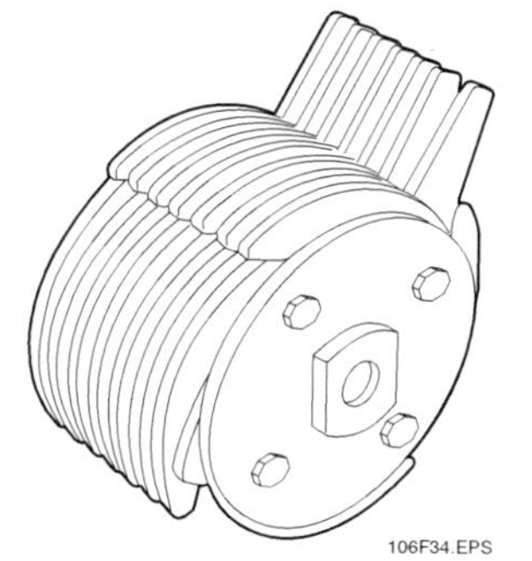

Figure 34 ◈ Rotary hammer.

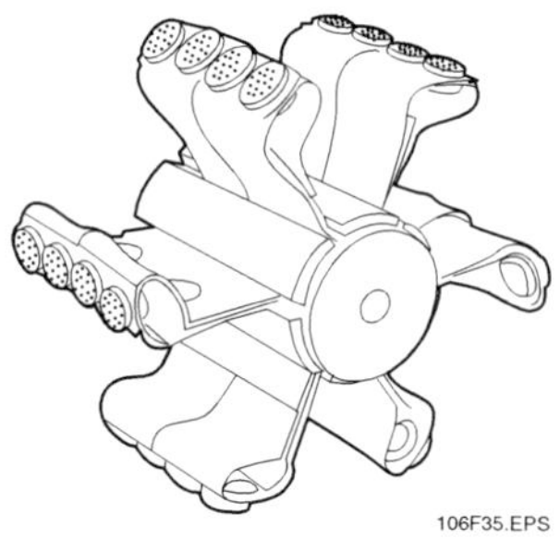

Figure 35 ◈ Rotary flaps or peeners.

2.5.7 Vacuum Shrouds

Some power tools are equipped with a vacuum system to capture dust, debris, and other airborne contaminants at the surface and prevent them from escaping into the environment. The head of the tool is enclosed in a shroud or casing (*Figure 36*), which is attached to a vacuum hose. A variation of this is a sander with holes in and around the sanding pad that draw dust into a vacuum hose.

In some cases, the vacuum system is tool-triggered; that is, the vacuum system starts and stops with the tool. Other tools allow independent operation of the vacuum system.

Some vacuum systems are equipped with a high-efficiency particulate air (HEPA) filter (*Figure 37*) which is designed to remove 99.97% of all particles above 0.3 micron. To put that size into perspective, 25 microns equal 1 mil, and 1,000 mils equal 1 inch.

The benefits of using power tools with an integrated vacuum system are that by removing dust and debris, the workplace is safer and cleaner, and

Instructor's Notes:

the worker can see better. Dust is not only a health risk, it is also an environmental hazard. Under certain conditions, it can even be combustible. Also, tools equipped with a vacuum system reduce the need for containment systems and protective gear for workers. However, the tools with which these shrouds are used are often very loud and require adequate hearing protection.

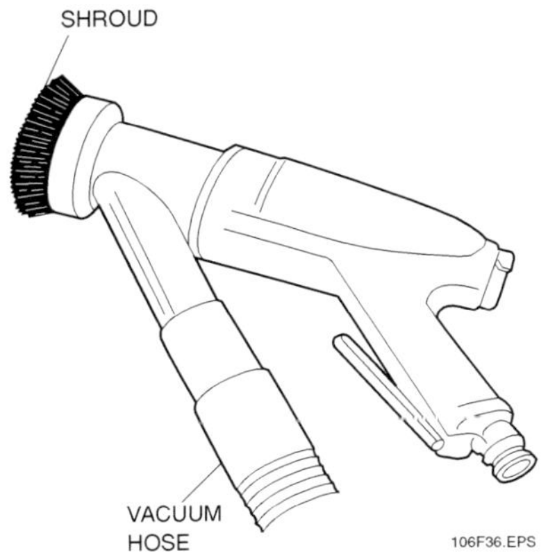

Figure 36 ◈ Shroud.

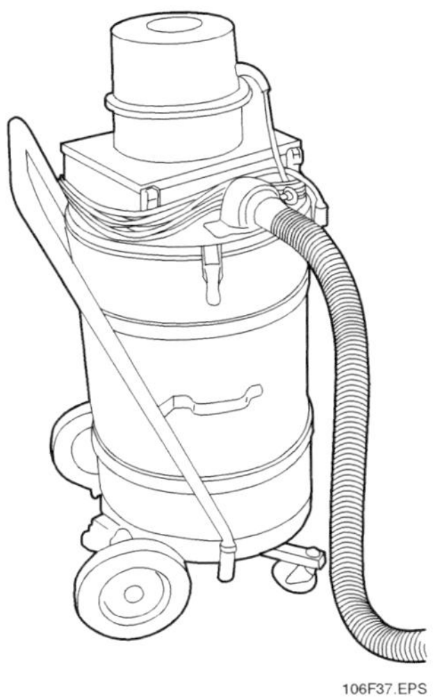

Figure 37 ◈ Vacuum system equipped with a HEPA filter.

2.6.0 Surface Preparation Standards and Inspection Techniques

The quality of the surface preparation has a direct bearing on how well a coating performs and how long it lasts. It is believed that the majority of all premature coating failures are due, all or in part, to faulty surface preparation and/or coating application. That being the case, it is important to determine how well a surface has been prepared and to catch any deficiencies in the quality of the preparation before they are covered with a coating.

Standards and inspection techniques for surface preparation have been established to help protect against premature coating failures. Generally, the more industrial the application, the more standards there are likely to be. And the more standards there are, the more formal the inspection process is.

The following is a brief overview of some generalized standards for surface preparation.

When preparing steel surfaces:

- Slag, flux, weld spatter, and rough edges should be removed.
- The surface should be washed with solvents to remove grease, dirt, oil, debris, and other loose contaminants (*SSPC-SP1*).
- The surface should be cleaned of all rust, dirt, mill scale, old paint or coating, or other tightly adhered contaminants according to the specification requirements, using one or more of the following: hand tools, power tools, wet or dry abrasive blast machines, and high-pressure or ultra high-pressure water jetting machines with or without abrasives (*SSPC-SP2, -SP3, -SP5, -SP6, -SP7, -SP10, -SP11, -SP12,* and *-SP13*).
- The surface should be profiled with power tools or abrasives to produce the specified anchor pattern.

The two primary areas of concern to inspectors in the surface preparation of steel are cleanliness and profile. SSPC and NACE have developed joint standards for evaluating four degrees of abrasive blast cleaning. Starting at the lowest degree, they are brush-off blast cleaning (*SSPC-SP7/NACE 4*), **commercial blast cleaning** (*SSPC-SP6/NACE 3*), **near-white blast cleaning** (*SSPC-SP10/NACE 2*), and **white metal blast cleaning** (*SSPC-SP5/NACE 1*). These standards are explained in more detail later in your training.

As a guide for comparing surface cleanliness to these standards, SSPC has developed reference photographs to illustrate each grade. These visual standards (*SSPC-VIS 1*) are for steel that has not previously been coated.

Laboratory

Set up several stations with grinders, sanders, power files, needle guns, rotary impact power tools, and vacuum shrouds. Ask trainees to identify each tool and explain how each is used to remove contaminants.

Classroom

Review the standards and inspection techniques for surface preparation.

Discuss the four degrees of abrasive blast cleaning. Answer any questions trainees may have.

Audiovisual

Show Transparency 19 (Figure 39). Review the tools that inspectors use to check surface profiles.

Show Transparency 20 (Figure 41). Discuss the criteria for checking adhesion when evaluating existing coatings.

Classroom

Ensure that trainees understand that surface type dictates how to best prepare it for coating.

SSPC also has visual standards (*SSPC-VIS 3*) for hand and power tool cleaning (*SSPC-SP2, -SP3, and -SP11*) of steel.

To check surface profile, inspectors use tools, such as a profile comparator (*Figure 38*), which helps them to make a visual analysis of the surface compared to established standards; a profile gauge (*Figure 39*), which measures the depth of the profile; or replica tape (*Figure 40*), which makes an impression of the profile. These tools and how they are used are covered in more detail in another module.

Other standards for surface preparation of steel concern the cleanliness of air used for blast cleaning, the quality of abrasives, and ambient weather conditions during blasting. During blasting and coating operations, the temperature of the steel must be at least 5°F above the dew point to prevent condensation, which causes flash rusting. The steel temperature and dew point can be measured with two simple tools: a surface temperature gauge and a sling psychrometer.

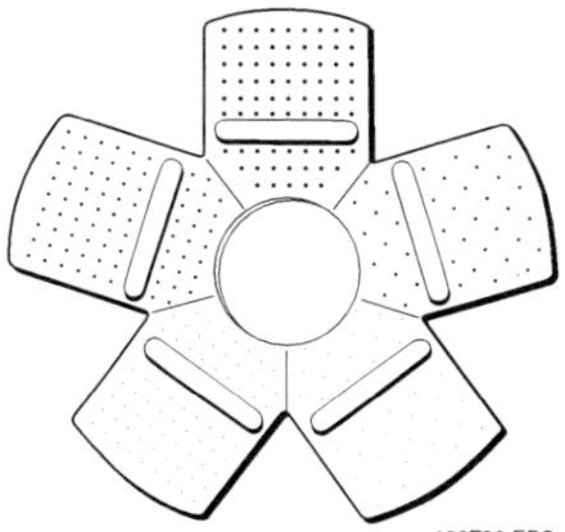

Figure 38 ◈ Profile comparator.

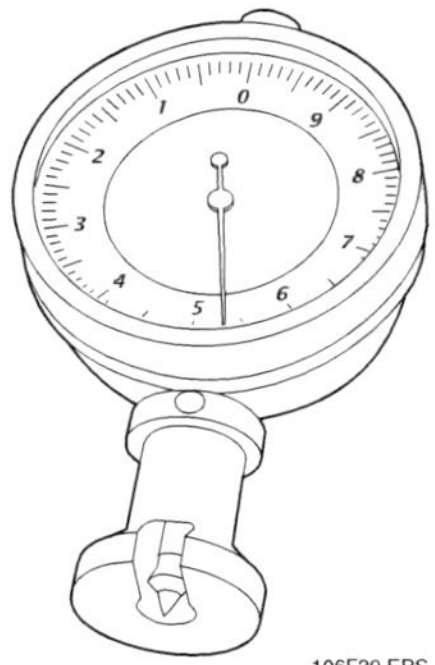

Figure 39 ◈ Profile gauge.

In evaluating existing coatings, inspectors check the quality of adhesion, the film thickness, and the condition of the substrate. Adhesion can be evaluated visually or with the aid of adhesion testing equipment (*Figure 41*) or by the use of a tape test (*ASTM D 3359*). The condition of the substrate can be checked by removing a little bit of coating and visually inspecting it. Among the problems that inspectors look for are evidence of mill scale or rust on steel. A variety of gauges are available to measure film thickness, which indicates the quantity of the coating and how well it was applied.

2.7.0 General Preparation Procedures for Specific Surfaces/Substrates

The goal of surface preparation is to create proper conditions for a coating to adhere and form a strong, durable, attractive barrier against contamination and deterioration. Each surface type has features that warrant individual consideration of how best to prepare it for coating. Following are general procedures for surface preparation of steel and other metals. Depending on the circumstances encountered, the steps of each procedure may not have to be accomplished completely or in the order given. In any case, the specific instructions of the coating manufacturer should be followed in lieu of these procedures, where applicable.

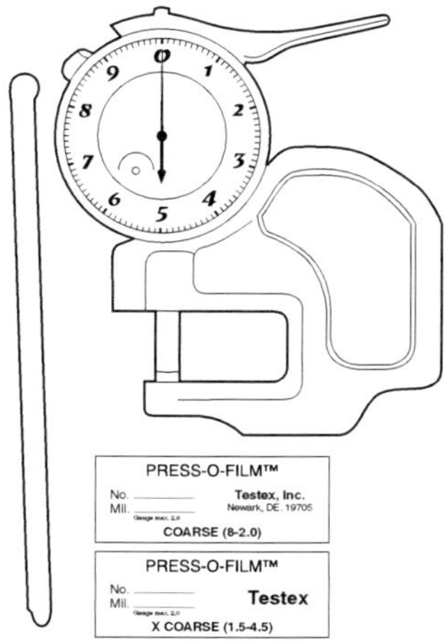

Figure 40 ◈ Replica tape.

Instructor's Notes:

Instructions such as clean, fill, repair, etc., in the following procedures assume that the reader is familiar with the tools, materials, and methods previously covered. Remember, the goal of surface/substrate preparation is to have the surface/substrate clean, dry, and dull.

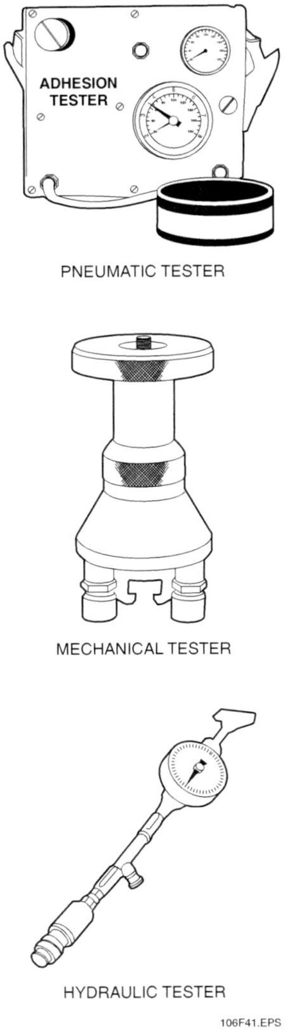

Figure 41 ◆ Adhesion testing equipment.

3.0.0 ◆ INITIAL PREPARATION BEFORE COATING APPLICATION

Before coatings can be applied to a surface/substrate using a brush, roller, or the spray method, initial preparation tasks must be accomplished. These tasks include:

- Protecting or masking the adjacent surfaces
- Checking and establishing the proper surface and environmental conditions
- Selecting and mixing the proper coating

3.1.0 Protecting Adjacent Surfaces

Before spray coating any surface, always ensure that proper containment is in place and that all adjacent surfaces have been properly protected with tarps and/or masked against spills. When spraying, a hand-held shield can be used to help contain **over spray** or bounceback. For detailed surface protection guidelines, refer to the containment/ventilation methods discussed in a previous module.

3.2.0 Checking the Surface Condition and Environmental Conditions

Before coating any surface, it must be clean, dry, and otherwise properly prepared. For guidelines on preparing the different kinds of surfaces, refer to the manufacturer's recommendations for the type of coating being used and the type of substrate to which it will be applied.

Another important check to make before coating involves the current environmental conditions, including the air and surface temperature, humidity, dew point, and wind speed. For example, the application data sheet for a particular coat may state that the surface temperature should be 65°F to 85°F, the ambient temperature 60°F to 85°F, and the relative humidity 35% to 85%. Applying coating outside these ranges can result in failure. High winds can result in a variety of problems for outdoor painting unless containment is used. Wind can blow dust and dirt onto the wet surface; even a light wind (as little as 3 to 5 mph) will blow the over spray around and can cause damage. Coating steel normally requires a surface temperature at least 5°F above the dew point (the temperature at which the water vapor in the air condenses). Condensation or surface moisture can interfere with adhesion or proper curing of the coating.

Surface temperature of a substrate can be measured with a thermometer (see *Figure 42*) in direct contact with the substrate. Air conditions can be

Teaching Tip

See the Teaching Tip for Sections 2.4.0–2.5.0at the end of this module.

Performance Testing

Have each trainee perform Performance Task 1 to your satisfaction. Fill out a Performance Profile Sheet for each trainee.

Homework

Assign the reading of Sections 3.0.0–3.5.6 for the next session.

Materials

Ensure that you have everything required for the laboratories, demonstration, and testing during this session.

Classroom

Ensure that trainees understand how to protect adjacent surfaces.

Review the guidelines for checking surface and environmental conditions before coating any surface.

Audiovisual

Show Transparency 21 (Figure 42). Explain how to measure the surface temperature of a substrate.

Laboratory

Set up stations with sling psychrometers, electric hydrometers, and surface thermometers. Ask trainees to practice using the instruments to measure environmental conditions.

Classroom

Review factors for properly selecting and mixing the coating. Go over any questions trainees may have.

Demonstration

Bring in a typical brush. Identify and define the parts of the brush.

checked using a sling psychrometer (refer to *Figure 42*) which contains two thermometers, one to measure the dry air temperature (dry bulb) and the other to measure the wet air temperature (wet bulb). The measured wet and dry bulb temperatures can be used to find the percent of relative humidity (RH) and dew point of the measured air. This is done by using a built-in chart on the psychrometer or a separate psychrometric chart or table. An electronic hygrometer (refer to Figure *42*) can also be used to measure the air temperatures and give a direct reading of the relative humidity. Some are also capable of giving a direct reading of the dew point.

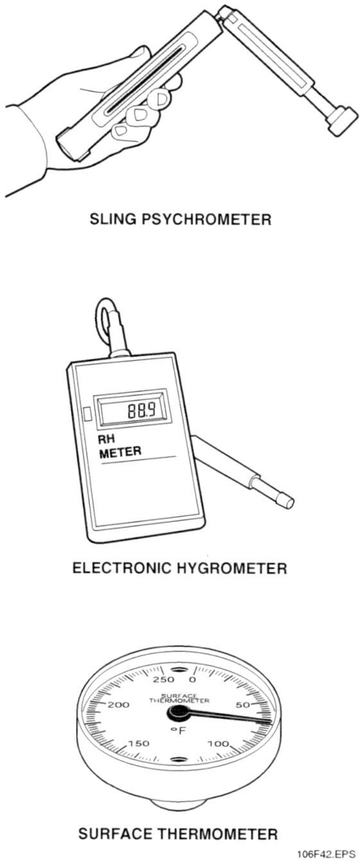

Figure 42 ◆ Sling psychrometer, electronic hygrometer, and surface thermometer.

3.3.0 Properly Selecting and Mixing the Coating

The quality of a finished spraying application depends on the quality of the coating, the equipment, and the operator. The coating should be selected based on how well it can protect a specific surface. It should be the type of coating recommended by the specifications and/or manufacturer for use on the surface to be protected. If applying multiple coats, such as a **primer**, intermediate, and/or final coat to a surface, all coatings should be compatible. Equally as important is the preparation of the coating before use. To ensure uniform application, follow the instructions on the product label and the application sheet to mix the paint and thin it, if necessary. It must be thoroughly mixed to achieve uniform color and consistency. Mixing also determines the durability of the coat. Poor adhesion, uneven color, inadequate film thickness, poor hiding ability, cracking, and other failures can all result from incomplete mixing. The process for preparation can involve:

- Manual or mechanical stirring
- Boxing
- Straining
- Thinning
- Catalyzing

3.4.0 Brushes

All brushes consist of bristles, a heel, a ferrule, and a handle (*Figure 43*). The portion of the bristles nearest the ferrule is called the *butt end*; the lower part of the brush is its *heel*; and the tip of the brush is called the *flag end*. The bristles of the brush are held in position at the lower end of the ferrule by

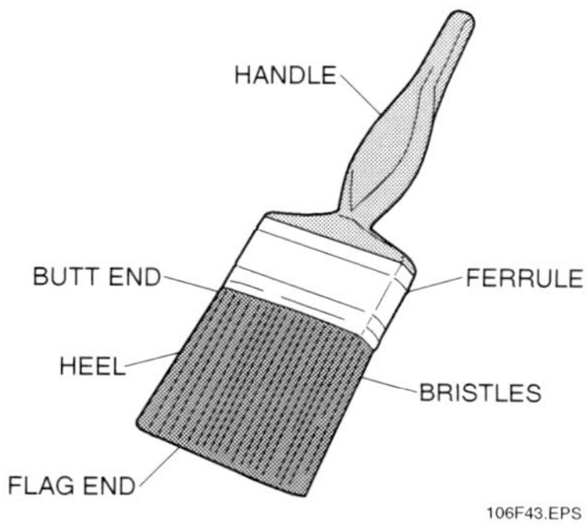

Figure 43 ◆ Brush parts.

Instructor's Notes:

an epoxy adhesive. Normally, the bristles are separated by wooden or aluminum spacer plugs. These spacer plugs provide support for the bristles and help keep them from fanning out when a coating of paint is applied.

3.4.1 Natural and Synthetic Bristle Brushes

Either natural or synthetic bristles can be used in brushes. Hogs' hair bristles are used in good brushes made with natural bristles. There are two benefits of using hogs' hair bristles over other animal hairs: they are tapered and their tips are ragged or flagged (*Figure 44*), a natural feature that helps the bristles to hold more paint and to spread it evenly onto a surface. Flagging also helps the brush to keep its shape. Synthetic bristles, also called filaments, can be made from nylon, polyester, or a blend of these materials. Synthetic fibers can be made in three basic shapes: solid, cross-sectional, and hollow. The solid shape is the easiest to clean and lasts the longest. The cross-sectional filament is an economical choice; the hollow shape is the least desirable. Synthetic filaments have no natural flags, but are usually tapered and flagged in the manufacturing process to closely match the features of hogs' hair bristles.

> **NOTE**
> Many new reduced-solvent coatings are heavier and resist flowing with flag-tipped brushes. For this reason, certain synthetic brush tips are made with sanded and tapered bristles to make the coating flow more easily.

A general rule of thumb is that brushes made with natural bristles are used only with oil or **alkyd** coatings, while those made with synthetic bristles are used with water-based coatings. This is because natural bristles can be softened by the water in water-based coatings, causing the natural bristles to puff up and become unruly. Another reason is that natural bristles can tolerate the harsh solvents used to clean them, while some synthetic brushes will deteriorate in the solvents. In recent years, brush manufacturers have developed synthetic bristle brushes designed for use with both oil-based and water-based coatings.

3.4.2 Types of Brushes

Selection of a brush is determined by the type of finish being applied, the size and shape of the area being covered, and the quality and speed of application desired. Many types of brushes are made, each with a specific purpose and/or performance benefit.

3.4.3 Guidelines for Selecting Quality Brushes

Brush quality is determined both by materials and construction. Only high-quality brushes should be used because they hold more, leave a better finish, and do not shed their bristles. They also last longer and generally are easier and faster. Price and feel are both factors in selecting a quality brush. Normally, the higher the price, the better the brush. Other guidelines for testing the quality of a brush are:

Audiovisual

Show Transparency 22 (Figure 44). Review the different types of bristles and bristle tips.

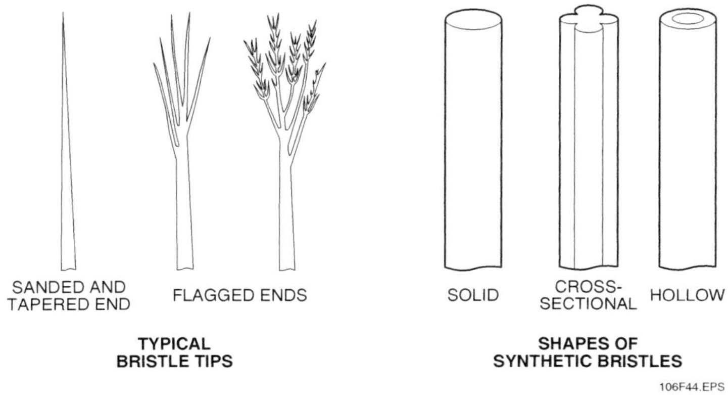

Figure 44 ◆ Typical bristle tips and shapes of synthetic bristles.

Classroom

Review the guidelines for selecting the correct brush for each job.

Laboratory

Bring in a variety of shapes and sizes of brushes. Have trainees inspect the brushes and make determinations regarding the quality of the brush.

Classroom

Review the classifications of and uses for rollers. Ask whether trainees have any questions.

Audiovisual

Show Transparency 23 (Figure 45). Emphasize the goal of dip rolling and explain how roller design affects how the roller can be used.

- Pull gently on the bristles. An inferior brush will leave several bristles in your palm, whereas a good brush will lose only one or two.
- Bounce and wiggle the bristles in your palm. A good brush has bristles that feel soft and springy, and bounce back into shape quickly when let go.
- Make sure the bristles are thick and plentiful. Good brushes will have bristles that are at least 2" long. The wider the brush, the longer the bristles. For example, a brush that is 1" or 2" wide should have bristles about 3" long; a brush that is 3" or 4" wide should have bristles about 4" long. The longer the bristles and the thicker their mass, the more the brush can hold.
- Good brushes have a substantial handle that is usually made of lacquered beech, birch, or maple wood. It should be balanced, comfortable, and shaped to fit the hand.
- The metal ferrule that holds the handle should be made of a rust-resistant metal and fastened to the handle with brads, screws, or rivets. Make sure it does not rock when the bristles are wiggled back and forth. Poor brushes usually have larger spacer plugs and, therefore, have fewer bristles than better brushes.

3.5.0 Rollers

Rollers are used to apply coatings to large, regular surfaces. They transfer and spread a coating onto a surface by rolling it with a fabric-covered roller cover. Rollers can be classified by the method of supplying a coating to the roller cover and/or the function they serve. These classifications are:

- Dip rollers
- Self-feeding rollers
- Special-purpose rollers

3.5.1 Dip Rollers

Dip rollers (*Figure 45*) are the most common type of roller. They are so named because the coating is supplied to the roller cover by dipping it into a roller tray or a five-gallon can. Dip rollers consist of a handle, a frame, and a cage or cylinder which holds the roller cover. Normally, the cage has nylon end caps or bearings that help it to rotate smoothly. The diameter and width of the roller cover is selected to match the diameter and width of the cage on which it is to be mounted.

Roller covers consist an of inner cylinder, called the core, which is commonly made of baked phenolic fiberboard to which a fabric is bonded. The fabric holds the coating much as a sponge holds water. Standard-size roller covers have a core with an inside diameter of 1½". Common widths range from 3" to 18". However, 9" is the most popular width because it is easy to control, is readily available, and fits easily into a tray or five-gallon pail for dipping and cleaning. Narrower roller covers are used for trimming and similar applications. Wider rollers are used to cover large, flat areas.

> **NOTE**
> The use of wider rollers for overhead work is impractical because they are heavy when filled with a coating and are clumsy to use in close or busy areas.

The roller cover fabric can be made of knitted or woven natural or synthetic fibers. Natural fibers include lamb's wool, sheepskin, and mohair (goat hair). Synthetic fibers include polyester, acrylic, and nylon. Special-purpose covers are also made of foam for use on smooth or textured surfaces. Covers with non-shedding naps and phenolic cores are made for use with industrial coatings.

The length of the roller cover fibers, called the *nap*, ranges from ⅛" to 1½" long. The nap is important because it determines the kinds of surfaces on which the cover can be used and its overall performance on those surfaces. For practical purposes, naps can be classified as short, medium, and long. Short naps down to ⅛" provide a smooth finish on very flat and smooth surfaces, such as metal. Medium naps ranging between ¼" and ¾" are typically used on walls and ceilings made of concrete or cement block, etc. Long naps between 1" and 1½" are suitable for use on rough surfaces where the coating must be worked into crevices, pits, and other irregularities.

The goal in rolling is to put as much coating on the surface as quickly as possible without leaving roller marks, craters, bubbles, or other surface imperfections. A roller cover with a long nap will leave waves and craters on a smooth surface, while a nap that is too short holds relatively little coating, so more effort goes into frequent reloading of the roller and keeping a wet edge. For this reason, manufacturers label their roller cover packages to show the types of coatings and surfaces for which a roller cover is designed. When choosing a roller cover, always follow the manufacturer's recommendations.

Instructor's Notes:

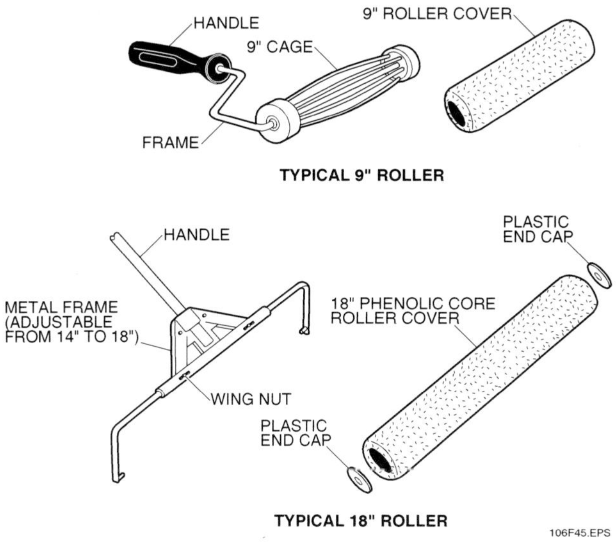

Figure 45 ◈ Dip rollers.

3.5.2 Dip Roller Accessories

Common accessories used with dip rollers include trays, grids, and extension poles (*Figure 46*). A tray is used to hold the coating supply for the roller. It has a well for the coating and a ramp that leads into the well. As the roller is moved back and forth over the roughened surface of the ramp, it works the coating into the roller cover and removes excess coating from the cover before it is applied to the surface. Quality trays are made from sturdy metal or plastic. In general, the use of a deeper pan is best because it allows more work to be done between refills.

A grid or mesh screen is used when rolling from a five-gallon pail. The grid is used to work the coating into the roller cover and remove excess coating from the cover before it is applied to the surface. Grids or screens hook over the rim of a five-gallon pail that is about half-full, allowing the grid to hang from the top without being immersed in the coating.

3.5.3 Pressure-Fed Rollers

Pressure-fed rollers have an intermittent supply of coating applied to the roller cover, eliminating the need for dipping the roller into a pail or pan. They are the most efficient types of rollers for use on very large rolling jobs, such as the exterior of a tank.

A production-type, pressure-fed roller is usually an attachment (*Figure 47*) that can be used with an airless sprayer pump operating at 3,000 pounds per square inch (psi) or less. The roller can be attached directly to the gun, to an in-line valve, or through a fixed or telescoping extension, as shown in *Figure 47*. The coating, controlled by the spray gun or valve, flows through the inside of the tubular frame of the roller and into the core of the roller cover. The core is perforated to allow the coating to pass into the fabric of the cover. Roller attachment widths of 12", 9", and 3" are generally available.

Also shown in *Figure 47* is a non-pressurized roller with airless spray tips, all attached to a wand used with an airless spray gun. This assembly is used to apply a coating and back-roll the coating in one operation. Assemblies with roller widths up to 18 inches or more are available.

3.5.4 Special-Purpose Rollers

There are many kinds of rollers that have been developed for special purposes (*Figure 48*). Common types include:

Classroom

Refer to Figure 46. Review the different types of dip roller accessories.

Audiovisual

Show Transparency 24 (Figure 47). Explain how pressure-fed rollers can be used with an airless sprayer pump.

Classroom

Discuss the various types of special purpose rollers and their applications.

Audiovisual

Show Transparency 25 (Figure 49). Explain how mitts and squeegees are used to apply coating.

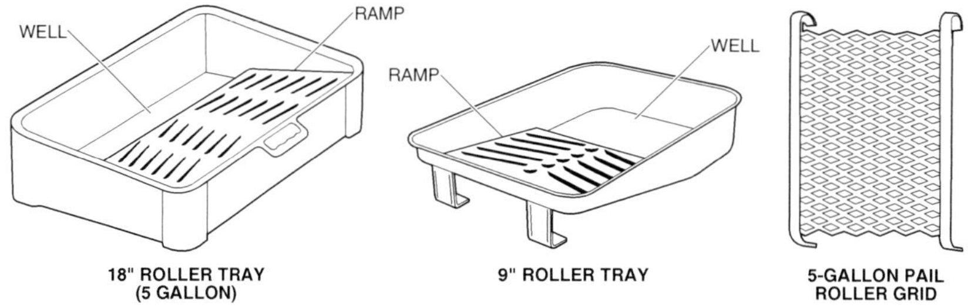

Figure 46 ◈ Common accessories used with dip rollers.

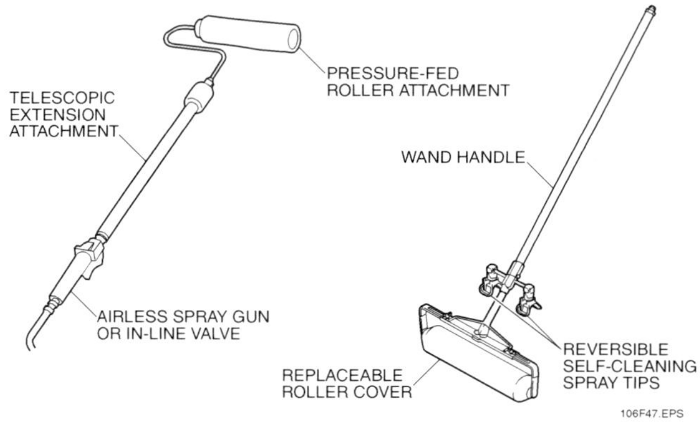

Figure 47 ◈ Production-type, pressure-fed roller.

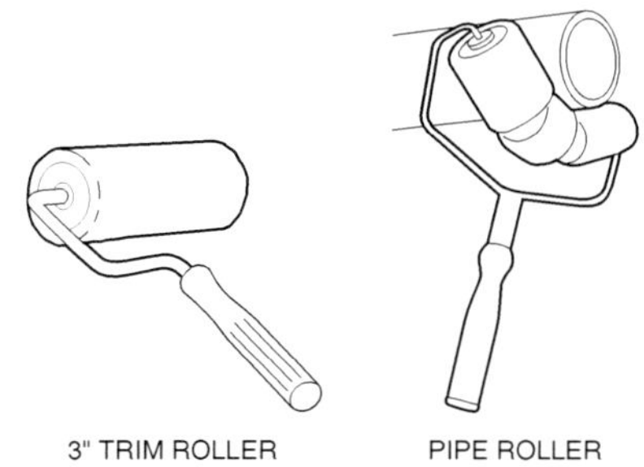

Figure 48 ◈ Special-purpose rollers.

- *Trim rollers* – Small rollers of various types, shapes, and sizes designed for covering trim, corners, edges, and other narrow surfaces.
- *Pipe rollers* – A roller unit that has a spring spindle with a number of narrow rollers that move under pressure to conform to contoured surfaces, such as pipes.

3.5.5 Mitts

Mitts (*Figure 49*) are useful when coating pipe and similar surfaces that are too small for rolling, too irregular for brushing, or impractical for spraying. Mitts are fabric mitts worn on the hand. They can be made of natural and synthetic fibers and with long or short naps for rough or smooth surfaces. Mitts are usually shaped in either a mitten

Instructor's Notes:

(with thumb) or square-end style. When worn, the mitt is dipped into the coating, then used to spread the coating by rubbing the mitt over the surface. Protective gloves should always be worn under a mitt to prevent the coating absorbed by the mitt from getting on your skin.

3.5.6 Squeegees

A toothed, rubber squeegee (*Figure 49*) may be used to spread a measured amount of coating per square foot on a flat, horizontal surface/substrate, such as the floor of a tank. After the coating is spread over an area, the coating is back-rolled with a roller to achieve the final uniform coating thickness and finish.

4.0.0 ◆ APPLYING COATINGS WITH A BRUSH

Applying coatings with a brush is typically done when coating small areas, irregular surfaces, or when cutting in or stripe coating. Cutting in is the term used for coating inside corners and around trim edges with a brush or trim roller before applying the coating with a roller. It also means coating a clean edge, such as needed at an adjacent surface.

Stripe coating is usually applied to welded areas or welded edges. It is accomplished by applying a coat of primer to the weld area or edge that extends out from the weld area or edge by 1" or more. Stripe coating is done before the first coat of primer is applied and results in a thicker overall coating over welded areas and edges.

Choose the correct brush for the coating being used. Use natural bristles for oil-based coatings and synthetic bristles for water-based coatings. As a rule, choose the largest brush that is suitable for the job. Typically, a 4" to 6" brush is used for large surfaces. Make sure the brush is free of dust, dirt, loose bristles, or other particles that may show up on the surface.

4.1.0 Basic Brushing Techniques

When using a brush, work from a partially filled container (bucket or pot). When dipping the brush into the container, the bristles should be dipped no more than halfway (*Figure 50*). Dipping deeper than halfway can cause coating to build up in the heel of the brush, and it also increases the chances of dripping and splattering. To set the coating in the brush and remove the excess coating, gently tap or slap one side of the brush against the inside of the container. Do not remove excess coating by wiping the bristles against the rim of the container because this removes too much coating from the brush and may damage some types of bristles.

The method for holding the brush is one of personal preference. Obviously, the job will be easier if the brush is held comfortably. *Figure 51* shows three common methods. Apply the coating with two or three light strokes, then spread it smoothly and evenly using long, even, parallel strokes. Whenever possible, use the tip and flat sides of the brush and not the side edges. Use just enough pressure to bend the bristles slightly.

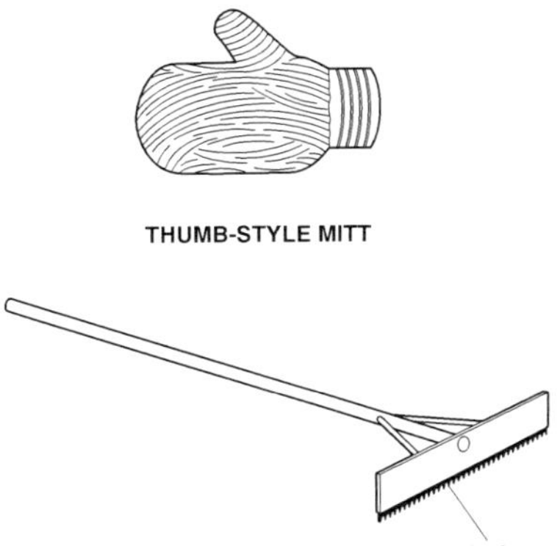

Figure 49 ◆ Paint mitts and toothed, rubber squeegee.

Figure 50 ◆ Dipping brush in container.

Teaching Tip

See the Teaching Tip for Section 3.0.0 at the end of this module.

Performance Testing

Have each trainee perform Performance Task 2 to your satisfaction. Fill out a Performance Profile Sheet for each trainee.

Homework

Have trainees review Sections 4.0.0–4.4.3 and answer to Review Questions for the next session.

Materials

Ensure that you have everything required for the demonstrations, laboratory, and testing during this session.

Classroom

Explain when coating should be applied with a brush.

Demonstration

Review basic brushing techniques. Demonstrate the common methods of holding a brush.

Audiovisual

Show Transparency 26 (Figure 52). Explain how to feather using a brush.

Classroom

Explain how to apply a coating with a roller. Answer any questions trainees may have.

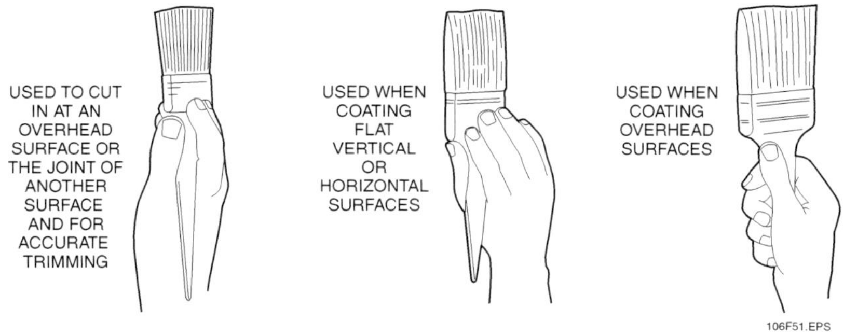

Figure 51 ◆ Common methods of holding a brush.

Finish off the leading edge of the coating by gradually pulling the bristles of the brush off the painted surface (*Figure 52*). This is called feathering. Whenever possible, move vertically because up and down strokes are less tiring. On a vertical surface, like a tank wall, move from top to bottom, and do not stop until you reach a natural stopping point such as a joint. As you move, keep a wet edge. This means working from an uncoated area into a previously coated, still-wet area. To avoid lap marks, feather the edges. When the surface is covered and while the coating is still wet, the final step is to lay it off, which means to go over the entire section you have just painted using light, quick strokes all in one direction in order to smooth brush marks, laps, and joints, and to cover any thin or missed spots.

NOTE

Glossy coatings show brush marks more than flat coatings. When using glossy coatings, work in smaller areas near the wet edge and lay it off frequently.

4.2.0 Applying a Coating with a Roller

Rolling is typically done when coating large, flat surfaces such as tanks.

The rolling procedure begins by following the manufacturer's recommendations to choose the correct roller cover for the type of coating being used and the correct nap for the type of surface involved. Make sure the roller cover is free of loose nap, dust, dirt, and any other particles that may show up on the surface. Before use, a new roller cover should always be rubbed down by wringing the cover in your bare hands vigorously to remove all loose nap. The roller, roller tray or pail, and roller grid should also be checked and wiped clean to prevent any dirt, sand, dirty solvent, or previously used coating from mixing with the new coating.

When using a dip-type roller, fill the tray with the coating material until the well is full and about one-third of the ramp is covered. Do not overfill the tray. Dip but do not submerge the roller into the coating, then roll it back and forth over the roughened surface of the ramp several times until the cover is well coated. If using a five-gallon pail with a roller grid or mesh screen, fill the pail about halfway. Dip the roller into the coating, and roll it back and forth over the roller grid

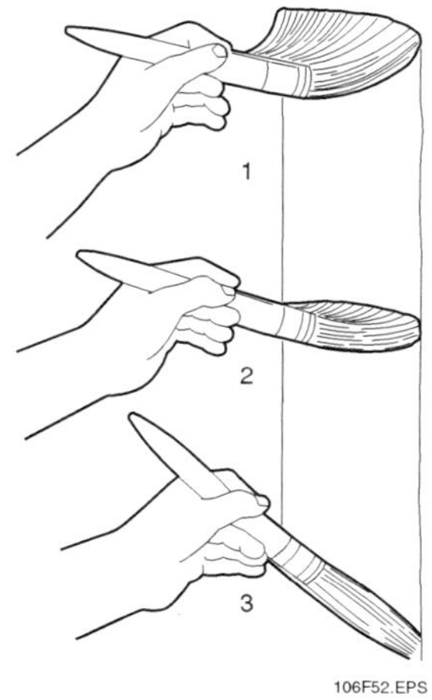

Figure 52 ◆ Feathering.

Instructor's Notes:

or mesh screen until the cover is well coated. When done, the roller cover should be full of coating material, but not dripping. If using a pressure-fed roller, fill it with the coating and prepare it for use according to the manufacturer's instructions. A useful way to break in and wet the roller cover fibers is to roll out the first load of coating onto craft paper, scrap drywall, or another disposable surface. This also helps to remove any air bubbles that may be trapped in the roller cover.

Apply the coating from a fully loaded roller to the surface in a zigzag pattern like the letters N or M (*Figure 53*). The first stroke should be in the upward or forward direction. This helps to prevent dripping and splattering. Be careful not to roll too fast because this also helps to avoid splattering. Use an even pressure on the roller to achieve an even coating. Without lifting the roller from the surface, cross roll to spread the coating. Then finish with light roller strokes all in the same direction but at a right angle to the cross roll. As the roller nears the end of the area being coating, feather the coating by slowly pulling the spinning roller off the surface.

When coating a surface, work from the wet edge to prevent lapping. For example, when coating a tank wall, work from left to right, keeping the leading edge of the coating wet by working from top to bottom across the surface. Do not stop until you reach a natural break. If you cannot go from top to bottom because of scaffolding, make sure that the edges are not allowed to dry before the scaffolding is moved because this will most likely cause lap marks.

4.3.0 Guidelines for Coating Using Brushes and Rollers

Before starting to coat an interior or exterior area or surface, always make sure that all adjacent surfaces have been properly masked against unwanted coating or spills. The coatings being used should be those recommended by the specification and/or manufacturer for use on the surfaces to be coated. If applying multiple coats such as a primer, intermediate, and/or final coat to a surface, all coatings must be compatible. This section describes coating surfaces using brushes and rollers. Methods for coating surfaces using a sprayer are covered later in this module.

Begin coating flat, vertical surfaces at the top and work down. Work from top to bottom and (usually) horizontally from left to right. Coat horizontal panels in horizontal strips; coat vertical panels in vertical strips. You should try to work from one natural break point to the next. Where there are no natural breaks, work across the width of the surface, coating only as much length as will allow you to keep blending the coating before it dries. Some other general considerations are:

- Follow the coating manufacturer's directions concerning temperatures. Be cautious of sudden drops in temperature. If outdoors, work only when the weather moderates to a fairly even temperature.
- If working with oil-based coatings, do not apply after a rain. After an extended wet period, allow a few days of good drying weather. If you must work on a damp surface, make sure to use appropriate moisture-tolerant coatings such as epoxies and alcohol-based coatings.
- If outdoors, do not work in strong winds; fresh coatings can collect leaves, dirt and dust particles, and insects.
- If outdoors, do not start working too early in the morning. Give the sun a chance to dry the surface. Do not work in the direct rays of hot summer sun because coatings do not spread well on a hot surface, and can dry too fast. Ideally, you should work in areas protected by containment and with moisture control. This allows the surface to be free of moisture and also allows the coating to dry properly.

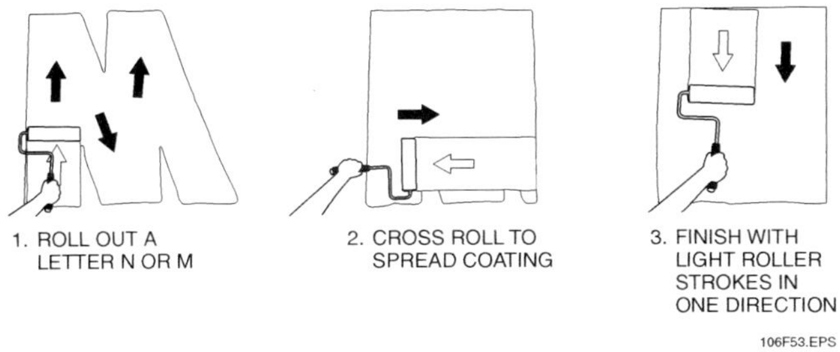

Figure 53 ◆ Zigzag pattern.

Audiovisual

Show Transparency 27 (Figure 53). Review the techniques for applying coating.

Classroom

Discuss the general considerations for coating using brushes and rollers.

Teaching Tip

See the Teaching Tip for Section 4.0.0 at the end of this module.

Classroom

Emphasize the importance of properly cleaning and storing brushes and roller covers.

Safety

Emphasize the importance of following the instructions on the manufacturer's material safety data sheet.

Discuss the consequences of mixing solvent waste with nonliquid or coating solid materials.

Demonstration

Ask a trainee to demonstrate how to properly clean a brush used in oil-based or specialty coatings. Have trainees observe the technique and identify what is done correctly and/or incorrectly.

Audiovisual

Show Transparency 28 (Figure 55). Explain how to use a brush spinner.

4.4.0 Cleaning and Storing Brushes and Roller Covers

If a brush or roller is to be used again, it must be well cleaned immediately after use, before the coating dries.

4.4.1 Brushes

Brushes used with oil-based or specialty coatings must be cleaned with an appropriate solvent. Most brushes used with water-based coatings can be cleaned with warm, soapy water after being flushed well to remove any remaining wet or dry hardened coating. This applies to water-based coatings that have a polyvinyl acetate (PVA) vehicle or an acrylic resin base with no additional resins.

Always refer to the coating manufacturer's cleanup directions because some water-based coatings contain emulsified-oil components that water alone will not clean. These coatings have PVA or an acrylic resin base and an additional resin base of linseed alkyd, polyester, or other emulsified resins. As the coating dries, the additional emulsified resins undergo chemical changes that will leave a crust on the bristles if the brush is not cleaned properly. Normally, a brush used with these water-based coatings must first be cleaned with a solvent such as lacquer thinner or paint thinner before it can be cleaned with soap and water.

Clean brushes used in oil-based or specialty coatings according to the following guidelines.

WARNING!

To prevent injury, always refer to and follow the instructions given on the coating and/or solvent manufacturer's MSDSs and product labels concerning the use of appropriate personal protective equipment and the handling of coatings and related solvents.

CAUTION

Collect used solvent waste in a container with a lid. Once waste solvent is no longer usable, pour it into a suitable waste solvent barrel. Do not mix other items such as used roller covers, rags, stir sticks, or any nonliquid or coating solid material in with the waste solvent. Clean, dry, and dispose of these items as dry waste. Dispose of dirty solvents and dry waste in accordance with all federal, state, and local laws.

Step 1 Use a rag to wipe off as much coating material as you can from the brush and ferrule.

Step 2 Let the brush soak in a bucket of solvent for a few minutes. Next, use a brush comb or wire brush (*Figure 54*) to comb the flat sides and edges of the brush. Comb from the ferrule toward the end of the bristles, forcing the coating out. Repeat this process while dipping the brush into the solvent to loosen any coating trapped inside the bristles near the ferrule.

Step 3 Hold the brush upside down, and bend the bristles back and forth, forcing solvent into the upper heel and ferrule of the brush. Repeat this action several times while pouring solvent into the bristles. Note that dirty solvent can often be used for the early rinses but clean solvent must be used for the later rinses. Continue the process until clear solvent runs out of the brush.

Step 4 Use a brush spinner (*Figure 55*) to remove the solvent and dry the brush. Make sure to hold the spinner inside a paint bucket or other suitable container while pumping it to prevent getting solvent on yourself, others, and the surrounding area.

106F54.EPS

Figure 54 ◆ Using a brush comb to comb flat sides and edges of the brush.

Instructor's Notes:

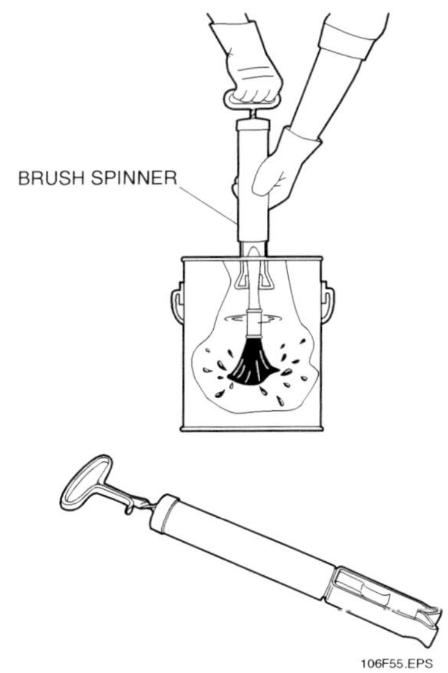

Figure 55 ◆ Using a brush spinner.

Step 5 Use a clean brush comb or wire brush to return the bristles to their original shape, with the bristles straight and parallel.

CAUTION
Make sure to dispose of all unusable latex and water-based coating and any wash water in accordance with all federal, state, and local laws that apply to your area. Do not dispose of this material in a storm sewer or in the ground.

The procedure for cleaning brushes used with water-based coatings is basically the same as those used with oil-based or specialty coatings. The difference is that most water-based coatings are cleaned using soap and water rather than solvents.

4.4.2 Cleaning Rollers

If a roller cover is to be used again, it must be cleaned immediately after use, before the coating sets or dries. Roller covers used with oil-based or specialty coatings are cleaned with an appropriate solvent. Those used with water-based coatings are cleaned with soap and water. Clean roller covers according to the following guidelines:

Step 1 Without removing the roller cover from the roller, use the semi-circular groove of a painter's tool (*Figure 56*) or the edge of a putty knife to scrape the wet coating from the roller cover into the pan or five-gallon pail.

Step 2 Remove the roller cover from the roller and soak it in solvent or water, as applicable. Use a curved painter's tool to scrape the cover again after it has been soaked.

Step 3 Use a roller spinner (*Figure 57*) to clean out as much solvent or water from the cover as possible. Make sure to hold the spinner inside a paint bucket or other suitable container while pumping it to prevent getting solvent on yourself, others, and the surrounding area.

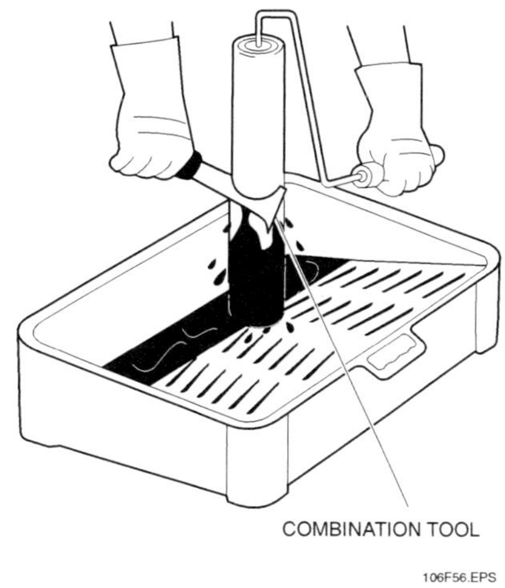

Figure 56 ◆ Using a combination tool.

Step 4 Repeat Steps 1 through 3 as necessary until the cover is clean.

Step 5 Stand the cover on end and allow it to dry thoroughly. Do not allow anything to lean against the cover as it dries.

Step 6 Clean, rinse, and dry the roller, tray, and other accessories.

Step 7 Make sure to dispose of waste coating, solvent, and/or waste water in accordance with all federal, state, and local laws and regulations.

Safety

Explain problems that can result if unusable latex and water-based coatings are disposed of in a storm sewer or on the ground.

Audiovisual

Show Transparency 29 (Figure 57). Discuss the advantage of using a roller spinner.

Laboratory

Ask a trainee to demonstrate how to properly clean a roller using a combination tool. Ask additional trainees to repeat the procedure until the cover is clean.

![Audiovisual]

Audiovisual

Show Transparency 30 (Figure 58). Explain how to properly store rollers and brushes.

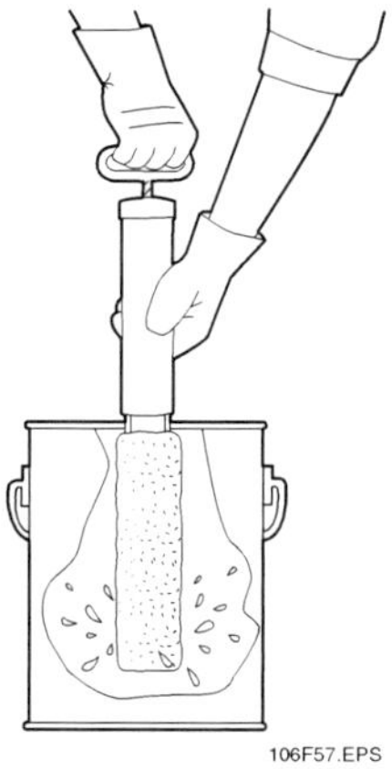

Figure 57 ◈ Roller spinner.

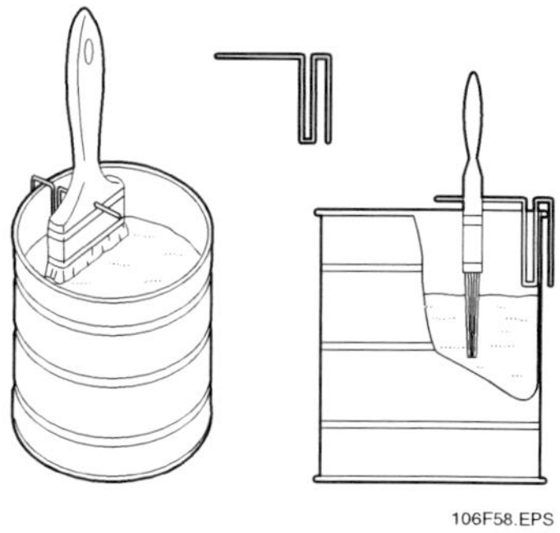

Figure 58 ◈ Proper storage of a brush.

4.4.3 Storage of Brushes and Rollers

Brushes and rollers can be stored wet or dry, usually depending on how soon they are going to be reused. After cleaning and drying, brushes and rollers can be wrapped in their original cover, brown kraft paper, aluminum foil, or wax paper. This helps to keep the brush or roller in shape. After wrapping, brushes can be stored either flat or by hanging. Brushes that are going to be reused soon are often suspended in a solvent base (not in water because water rots and warps the handles). While suspended in the solvent, they must not sit on their bristles as this can cause the bristles to lose their shape. Brushes can be suspended wet or dry in a can by placing a rod through a hole drilled near the upper part of the ferrule and laying the rod across the top of the can (*Figure 58*). Special brush containers, called brush keepers, are also available. They allow several brushes to be suspended for storage in the container at the same time.

Overnight storage of roller covers can be done by first stripping the paint out of the roller cover, then removing it from the frame. Following this, the cover is wrapped in plastic or aluminum foil and left to float in a container of solvent or water.

Summary

Protective coating for aboveground pipe is a specialty product that represents the most widely used method of corrosion control. The protective coating is designed to give long-term protection under a broad range of corrosive conditions to which the aboveground pipe is exposed.

This module covered the proper surface preparation, inspection, and coating application methods for aboveground pipe. Using low-pressure cleaning and washing were discussed as well as hand and power tool cleaning.

As explained in this module, the most common application methods are by brush, roller, and spray. All three methods should be considered when applying coating. The method must suit the situation in which the coating is going to be applied and the type of coating previously used on the pipe. The proper application method must provide the pipe with a uniform film of coating material in order to protect the aboveground pipe.

Instructor's Notes:

Review Questions

1. The least time-consuming part of a pipeline coating job is _____.
   a. cleaning the pipe surface
   b. preparing the surface/substrate for coating
   c. applying primer
   d. applying the finish coat

2. The most important part of a pipeline coating job is _____.
   a. cleaning the pipe surface
   b. preparing the surface/substrate for coating
   c. applying primer
   d. applying the finish coat

3. Non-ferrous metal substrates that are used on pipelines contain _____.
   a. steel
   b. iron
   c. no iron
   d. hardened steel

4. Abrasive materials not typically used with low-pressure washer, wet-abrasive injectors are known as _____ abrasives.
   a. metallic
   b. mineral
   c. by-product
   d. special purpose

5. _____ is a mineral abrasive which may be repeatedly recycled.
   a. Garnet
   b. Magnetite
   c. Staurolite
   d. Olivine rutile

6. Slag is the refuse from smelting or other processes and is a _____ abrasive.
   a. mineral
   b. by-product
   c. metal
   d. special purpose sodium bicarbonate

7. The most widely used surface preparation device is the _____.
   a. pressure washer
   b. scrape blade
   c. sander
   d. rag or cloth wipe

8. The distillation process for recycling a solvent first involves _____ the solvent.
   a. evaporating
   b. heating
   c. separating
   d. condensing

9. To remove discoloration caused by industrial gases, wash the surface with a very diluted solution of _____.
   a. vinegar and water
   b. lemon juice and water
   c. acetic or muriatic acid and water
   d. detergent and water

10. Low-pressure water cleaning ranges up to _____ psi.
    a. 1,000
    b. 3,000
    c. 5,000
    d. 7,000

11. High-pressure water jetting uses water at pressures from _____ psi.
    a. 3,000 to 5,000
    b. 5,000 to 10,000
    c. 5,000 to 20,000
    d. 10,000 to 25,000

12. A valve that allows water to recirculate from the high-pressure outlet side of the operating pump to the inlet side of the pump at reduced pressure through the bypass when the wand assembly is *not* in use is a(n) _____ valve.
    a. unloader
    b. pressure relief
    c. thermal relief
    d. vent

13. A blaster tip that provides a concentrated spot of high-pressure water which helps to blast off stubborn material is a _____ blaster tip.
    a. 0-degree
    b. 15-degree
    c. 25-degree
    d. 40-degree

Classroom

Go over the Review Questions for this module. Answer any questions trainees may have.

Examination

Administer the Module Examination. Be sure to record the results of the Exam on Craft Training Report Form 200 and submit the results to the Training Program Sponsor.

Performance Testing

Have each trainee perform Performance Task 3 to your satisfaction. Fill out a Performance Profile Sheet for each trainee.

Ensure that all Performance Tests have been completed and Performance Profile Sheets for each trainee are filled out. Be sure to record the results of the testing on Craft Training Report Form 200 and submit the results to the Training Program Sponsor.

14. A stripper tip that provides a narrow fan of concentrated water for removing coating and oxidation is a _____ stripper tip.

a. 0-degree
b. 15-degree
c. 25-degree
d. 40-degree

15. Heavy equipment cleaning and oxidation removal are uses for a _____ stripper.

a. 0-degree
b. 15-degree
c. 25-degree
d. 40-degree

16. A _____ rinser tip is useful for general wash-down and sweeping horizontal surfaces.

a. 0-degree
b. 15-degree
c. 25-degree
d. 40-degree

17. Always apply chemical cleaners to as _____ a surface as possible.

a. wet
b. dry
c. dirty
d. clean

18. A versatile device called a combination tool or a _____ is a special tool used by painters.

a. six-in-one tool
b. five-in-one tool
c. four-in-one tool
d. three-in-one tool

19. A useful tool for cleaning loose rust and other debris from all types of surfaces is called a _____.

a. chipping hammer
b. hammer and chisel
c. putty knife
d. wire brush

20. Impact tools such as _____ are used to clean steel and concrete substrates.

a. power chisels
b. grinding wheels
c. sanders
d. power file

21. Pneumatic- or electric-powered _____ sanders and grinders have a revolving flat, circular abrasive that can be used to aggressively remove heavy coatings and other contaminants.

a. disc
b. orbital
c. belt
d. half-sheet

22. Pneumatic-powered _____ sanders can be used for coarse to very fine sanding, depending on the particular tools and the abrasive used.

a. disc
b. orbital
c. belt
d. half-sheet

23. The portion of the bristles on a brush nearest the ferrule is called the _____.

a. bottom edge
b. heel
c. flag end
d. butt end

24. The lower part of the brush is called the _____.

a. bottom edge
b. heel
c. flag end
d. butt end

25. The tip of the brush is called the _____.

a. bottom edge
b. heel
c. flag end
d. butt end

Instructor's Notes:

Glossary

This glossary will help you learn some terms specific to aboveground pipe coating and inspection. Trade Terms introduced in this module appear in italic text.

***Abrasive blast cleaning:** Cleaning and roughening of a surface produced by the high-velocity impact of an abrasive that is propelled by the discharge of pressurized fluid from a blast nozzle or by a mechanical device such as a centrifugal blasting wheel. [Also referred to as Abrasive Blasting.]

Alkyd: A type of resin formed by the reaction of polyhydric alcohols and polybasic acids, part of which is derived from saturated or unsaturated oils and fats.

Blister: A roughly circular or elliptic unbonded area between plies of a laminated material; usually caused by trapped moisture.

Bond/adhesion: The attachment of a coating to a substrate. Adhesion is the key to coating, effectively allowing the coating to become part of the substrate. The three types of adhesive bonds are chemical, polar, and mechanical.

***Brush-off blast cleaned surface:** A brush-off blast cleaned surface, when viewed without magnification, shall be free of all visible oil, grease, dirt, dust, loose mill scale, loose rust, and loose coating. Tightly adherent mill scale, rust, and coating may remain on the surface. Mill scale, rust and coating are considered tightly adherent if they cannot be removed by lifting with a dull putty knife. [See *NACE No.4/SSPC-SP7.*]

Cigarette wrap: Application technique for cut lengths of sheet coating. The width of the sheet must exceed the circumference of the pipe to provide an overwrap. The sheet is placed longitudinally and wrapped no more the twice around the pipe. Ending seams should be alternated along the pipe so they are not continuous. Each sheet should overlap the previous. This method is not commonly used, with the exception of heat shrink sleeves.

Coal tar: A tar distilled from bituminous coal used to coat piping. Usually has a paper wrap over the coal tar.

***Commercial blast cleaned surface:** A commercial blast cleaned surface, when viewed without magnification, shall be free of all visible oil, grease, dust, dirt, mill scale, rust, coating, oxides, corrosion products, and other foreign matter. Random staining shall be limited to no more than 33% of each unit area (approximately 58 $cm.^2$ [9.0 $in.^2$]) of surface and may consist of light shadows, slight streaks, or minor discolorations caused by stains of rust, stains of mill scale, or stains of previously applied coating. [See *NACE No.3/SSPC-SP 6.*]

Confined space: A space that: 1) Is large enough and so configured that an employee can bodily enter and perform assigned work; and 2) Has limited or restricted means for entry or exit (e.g. tanks, vessels, vaults, etc.); and 3) Is not designed for continuous employee occupancy.

* Term published in NACE *Glossary of Corrosion-Related Terms*. Reproduced with permission from NACE International, Houston, TX. For a copy of the complete *Glossary of Corrosion-Related Terms*, contact NACE International, 1440 South Creek Dr., Houston, TX 77084. The Glossary is revised periodically, so information in an outdated version may not be accurate.

GLOSSARY (Continued)

**Cracking:* Breaks in a coating that extend through to the substrate.

Dry film thickness (DFT): The thickness of a coating after it has completely dried or cured, usually measured in mils (1 mil = 0.001 inch).

Fusion bond: Application of a powder coating to a heated pipe by electrostatic methods. The powder gels and flows with the heat, then cures and hardens during cooling. The process creates a tight physical bond between the coating and the metal.

Hand cleaning: The process of removing burrs, slag, rust, mill scale, and other contaminants on steel pipe. Burrs or slag may be removed by grinding. Abrasive blasting is the preferred method for recoating.

Hand power tool cleaning: Removal of any loose rust, loose mill scale, and loose paint by using a chipping power tool, a power sander, or a power grinder.

**Hand tool cleaning:* Removal of loose rust, loose mill scale, and loose paint to degree specified, by hand chipping, scraping, sanding, and wire brushing. [*See SSPC-SP2.*]

Hand wash: Removal of oil, dirt, soil, grease, and other contaminants by cleaning with solvent and detergents, etc.

***Holiday:** A discontinuity in a protective coating that exposes unprotected surface to the environment.

Mastic: Asphalt-like mixture used to coat piping.

NACE: National Association of Corrosion Engineers.

**Near-white blast cleaned surface:* A near-white blast cleaned surface, when viewed without magnification, shall be free of all visible oil, grease, dust, dirt, mill scale, rust, coating, oxides, corrosion products, and other foreign matter. Random staining shall be limited to not more than 5% of each unit area of surface (approximately 58 cm.^{2} [9.0 in.^{2}]), and may consist of light shadows, slight streaks, or minor discolorations caused by stains of rust, stains of mill scale, or stains of previously applied coating. [See *NACE No. 2/SSPC-SP 10.*]

Orange peel: Hills and valleys in a coating resembling the rind of an orange.

Over spray: Dry, flat, pebbly surface resulting from paint particles falling outside the spray pattern.

Overlap: An amount of coating covering the previous row of coating in a row-to-row sequence.

Peeling: Pulling a layer of material away from another layer, breaking one row of bonds at a time.

Permit-required confined space: A confined space that has one or more of the following characteristics: 1) Contains or has a potential to contain a hazardous atmosphere; 2) Contains a material that has the potential for engulfing an entrant; 3) Has an internal configuration such that an entrant could be trapped or asphyxiated by inwardly converging walls or by a floor which slopes downward and tapers to a smaller cross-section; or 4) Contains any other recognized serious safety or health hazard.

Instructor's Notes:

GLOSSARY (Continued)

Pre-heat: Pre-heating of the pipeline metal prior to application of certain coatings. Heat shrink sleeves and several other coatings require the pipe to be pre-heated prior to applying the coating. Care should be taken not to exceed specific temperatures as it can damage the coating. Epoxy coatings require specific temperature ranges to properly cure.

******Primer:*** A coating material intended to be applied as the first coat on an uncoated surface. The coating is specifically formulated to adhere to and protect the surface as well as to produce a suitable surface for subsequent coats. [Also referred to as Prime Coat.]

Sags: Coating defects, they are also called runs or curtains caused from excess flow of paint.

Shrink sleeve: A conformable wrap-around sleeve designed to protect belowground and submerged metal surfaces from corrosion and electrolysis. The sleeve consists of two overlapping pieces or a single tube surrounding the pipe. Heat is applied to the sleeves until an adhesive material in the sleeve bonds the sleeve to the pipe and joins the overlapping sections to form a tight bond.

SSPC: Society for Protective Coatings.

Surface profile: Graphic representation of surface condition.

Wax: The wax present in petroleum crudes primarily consists of paraffin hydrocarbons known as paraffin wax and naphthenic hydrocarbons.

Wet Film Thickness: The thickness of a fresh coating before it dries, usually measured in mils (1 mil = 0.001 inch).

White metal blast cleaned surface*: A white metal blast cleaned surface, when viewed without magnification, shall be free of all visible oil, grease, dust, dirt, mill scale, rust, coating, oxides, corrosion products, and other foreign matter. [See *NACE No. 1/SSPC-SP 5.*]

Wrinkling: A coating defect appearing as a rough, crinkled surface.

REFERENCES

Additional Resources

This module is intended to present thorough resources for task training. The following reference works are suggested for further study. These are optional materials for continued education rather than for task training.

API Recommended Practice 651, *CP of Aboveground Petroleum Storage Tanks*, Second Edition (December 1997). Global Engineering Product Code API CERT 651.

Control of Pipeline Corrosion, Second Edition (2001). A. W. Peabody. Houston, TX; NACE International.

Corrosion Prevention by Protective Coatings, 1984. Charles G. Munger. Houston, TX; National Association of Corrosion Engineers (NACE).

Instructor's Notes:

MODULE 61106-02—TEACHING TIPS

The following are suggested activities or instructional methods to help you teach the material in this AIG.

General

When you call on someone to answer a question, the rest of the class relaxes or even tunes out because they expect that the question and answer will take place only between you and the trainee you called on. Instead, use this technique to involve more trainees in answering questions and to keep them on their toes.

1. Ask trainees to define a term or explain a concept.
2. After one trainee has answered, ask a trainee seated nearby if the answer is right. Then ask whether a trainee in the back of the room agrees.
3. Ask trainees to explain why they think an answer is right or wrong.
4. Use the session to clear up incorrect ideas and encourage trainees to learn from their mistakes.

Sections 2.3.5–2.3.6

Pressure Washing Nozzles, Tips, and Accessories

Trainees have reviewed the various pressure washing nozzles, tips, and accessories. This exercise will familiarize trainees with these items and their uses. For this exercise, you will need 0-, 15-, 25-, and 40-degree tips; power scrub brushes; telescoping extensions; horizontal scrubbers; rotating water blast, variable tip, and rotatable tip nozzles; and wet abrasive injectors. Trainees will need their Trainee Guides, appropriate personal protective equipment, and pencils and paper to take notes for discussion. Allow 30 minutes for this exercise.

1. Set up several stations with the various pressure washing nozzles, tips, and accessories.
2. Number, but do not label, each of the items. Ask trainees to visit each station and identify each of the nozzles, tips, and accessories.
3. After trainees have correctly identified the items, have trainees briefly explain how each is used and the advantages and/or disadvantages of each.
4. Discuss safety considerations when working with pressure washing equipment.
5. Answer any questions trainees may have about pressure washing.

Sections 2.4.0–2.5.0

Hand and Power Tool Cleaning

This exercise will give trainees the opportunity to practice using a variety of hand and power tools. For this exercise, you will need a variety of hand and power tools and surfaces to practice preparation and inspection techniques. Trainees will need their Trainee Guides, appropriate personal protective equipment, and pencils and paper to take notes for discussion. Allow 45 minutes for this exercise.

1. Set up several stations with a variety of hand and power tools and surfaces to practice preparation and inspection techniques.
2. Have trainees practice using the hand and power tools to clean the surfaces you provide. Emphasize safety precautions.
3. Have trainees inspect the surface preparations according to SSPC visual standards by using profile comparators, profile gauges, and replica tape.

5. Answer any questions trainees may have about hand and power tool cleaning.

Section 3.0.0 *Initial Preparation Before Coating Application*

This exercise will emphasize the importance of properly selecting the coating and the materials to apply the coating. For this exercise, trainees will need their Trainee Guides and pencils and paper to take notes for discussion. Allow 20 minutes for this exercise.

1. Present a variety of scenarios for applying coatings to surfaces/substrates.
2. Ask trainees to discuss the scenarios and recommend the appropriate coating and application tools.
3. Discuss the advantages of the materials and equipment that the trainees recommend.
4. Provide additional recommendations and suggestions, as warranted.
5. Answer any questions trainees may have about initial preparation.

Section 4.0.0 *Applying Coatings with a Brush*

This exercise will give trainees practice applying coatings with brushes and rollers. For this exercise, you will need a variety of brushes and rollers, coatings, and clean surfaces to apply the coatings. Trainees will need their Trainee Guides and appropriate personal protective equipment. Allow 30 minutes for this exercise.

1. Set up several stations with brushes, rollers, coatings, and clean surfaces.
2. Ask trainees to select a brush or roller and practice applying the coatings to the surfaces.
3. Have trainees try as many different types of brushes and rollers as time permits.
4. Review trainees' techniques and make recommendations for applying coating.
5. Ask trainees to appropriately clean the brushes and roller covers. Answer any questions trainees may have about applying coatings.

MODULE 61106-02 — ANSWERS TO REVIEW QUESTIONS

Answer	Section
1. d	1.0.0
2. b	1.0.0
3. c	2.0.0
4. a	2.1.1
5. a	2.1.1
6. b	2.1.1
7. d	2.2.1
8. a	2.2.1
9. c	2.2.2
10. c	2.3.0
11. d	2.3.1
12. a	2.3.3
13. a	2.3.5
14. b	2.3.5
15. b	2.3.5
16. d	2.3.5
17. b	2.3.11
18. b	2.4.3
19. d	2.4.4
20. a	2.5.1
21. a	2.5.2
22. b	2.5.2
23. d	3.4.0
24. b	3.4.0
25. c	3.4.0

CONTREN™ LEARNING SERIES — USER UPDATES

The NCCER makes every effort to keep these textbooks up-to-date and free of technical errors. We appreciate your help in this process. If you have an idea for improving this textbook, or if you find an error, a typographical mistake, or an inaccuracy in NCCER's Contren™ textbooks, please write us, using this form or a photocopy. Be sure to include the exact module number, page number, a detailed description, and the correction, if applicable. Your input will be brought to the attention of the Technical Review Committee. Thank you for your assistance.

Instructors – If you found that additional materials were necessary in order to teach this module effectively, please let us know so that we may include them in the Equipment/Materials list in the Instructor's Guide.

Write: Curriculum Revision and Development Department
National Center for Construction Education and Research
P.O. Box 141104, Gainesville, FL 32614-1104

Fax: 352-334-0932

E-mail: curriculum@nccer.org

Craft ____________________ Module Name ____________________

Copyright Date __________ Module Number __________ Page Number(s) __________

Description

____

____

____

(Optional) Correction

____

____

(Optional) Your Name and Address

____

____

Module 61107-02

Apply and Repair External Coatings on Buried and Submerged Pipe

(CT 13.1 and 13.4)

Apply and Repair External Coatings on Buried and Submerged Pipe Annotated Instructor's Guide

Module 61107-02

MODULE OVERVIEW

This module explains how to select coatings for buried and submerged piping. Trainees will learn how to remove coating using chemical strippers or blasting and to apply coating using brushes, rollers, and conventional spray systems.

PREREQUISITES

Please refer to the Course Map in the Trainee Module. Prior to training with this module, it is recommended that the trainee shall have successfully completed the following:

Core Curriculum; Pipeline Core, Modules 66101-02 and 66102-02; Pipeline Corrosion Control Level One, Modules 61103-02 through 61106-02

OBJECTIVES

Upon completion of this module, the trainee will be able to:

1. Select a coating for buried or submerged piping.
2. Remove coating from buried or submerged pipe using chemical strippers or blasting systems (CT 13.1 and 13.4).
3. Apply coatings with a brush and roller.
4. Apply coatings with a conventional spray system.

PERFORMANCE TASKS

Under the the instructor's supervision, the trainee should be able to:

1. Remove pipe coating from buried or submerged piping in the area of defect (CT 13.1).
2. Clean and prepare pipe to accept coating repair (CT 13.1).
3. Properly prepare and apply different types of coatings according to the manufacturer's instructions (CT 13.4).

NCCER STANDARDIZED CRAFT TRAINING PROGRAM

The National Center for Construction Education and Research (NCCER) provides a standardized national program of accredited craft training. Key features of the program include instructor certification, competency-based training, and performance testing. The program provides trainees, instructors, and companies with a standard form of recognition through a National Craft Training Registry. The program is described in full in the *Guidelines for Accreditation*, published by the NCCER. For more information on standardized craft training, contact the NCCER by writing us at P.O. Box 141104, Gainesville, FL 32614-1104; calling 352-334-0911; or e-mailing info@nccer.org. More information may be found at our Web site, www.nccer.org.

HOW TO USE THIS ANNOTATED INSTRUCTOR'S GUIDE

Each page presents two sections of information. The larger section displays each page exactly as it appears in the Trainee Module. The narrow column ties suggested trainee and instructor actions to each page and provides icons to call your attention to material, safety, audiovisual, or testing requirements. The bottom of each page includes space for your notes.

If you see the Teaching Tip icon, that means there is a teaching tip associated with this section. Also refer to any suggested teaching tips at the end of the module.

SAFETY CONSIDERATIONS

Ensure that the trainees are equipped with appropriate personal protective equipment. Emphasize the importance of following all safety precautions and procedures when working with power tools.

PREPARATION

Before teaching this module, you should review the Module Outline, Objectives, Performance Tasks, and the Materials and Equipment List. Be sure to allow ample time to prepare your own training or lesson plan and gather all required equipment and materials.

MATERIALS AND EQUIPMENT LIST

Transparencies

Markers/chalk

Blank acetate sheets

Transparency pens

Pencils and scratch paper

Module Examinations*

Performance Profile Sheets*

Copies of Quick Quiz**

Copies of Performance Verifications for Covered Task 13†

Overhead projector and screen

Whiteboard/chalkboard

Appropriate personal protective equipment

Copies of *49 CFR Part 192* (Gas) and/or *49 CFR Part 195* (Liquid)

Copies of your company policy and procedures manual

Manufacturers' product information data sheets for coating

SSPC-VIS 1 photographs that illustrate each grade of steel cleanliness

A variety of steel surface samples

A variety of surface profiling instruments, including:
Profile comparator
Profile gauge
Replica tape

A variety of brushes

Containers filled with coating

Practice surfaces for applying coating with a brush

A variety of rollers

Trays filled with coating

Practice surfaces for applying coating with a roller

A variety of practice surfaces

Conventional and airless spray systems

Coatings

Respirators

* Located in the Test Booklet.
** Located at the end of this module.
† Available for download at www.nccer.org.

ADDITIONAL RESOURCES

This module is intended to present thorough resources for task training. The following reference works are suggested for both instructors and motivated trainees interested in further study. These are optional materials for continued education rather than for task training.

API Recommended Practice 651, *CP of Aboveground Petroleum Storage Tanks,* Second Edition (December 1997). Global Engineering Product Code API CERT 651.

Control of Pipeline Corrosion, Second Edition (2001). A. W. Peabody. Houston, TX; NACE International.

Corrosion Prevention by Protective Coatings, 1984. Charles G. Munger. Houston, TX; National Association of Corrosion Engineers (NACE).

TEACHING TIME FOR THIS MODULE

An outline for use in developing your lesson plan is presented below. Note that each Roman numeral in the outline equates to one session of instruction. Each session has a suggested time period of 2½ hours. This includes 10 minutes at the beginning of each session for administrative tasks and one 10-minute break during the session. Approximately 10 hours are suggested to cover *Apply and Repair External Coatings on Buried and Submerged Pipe* (CT 13.1 and 13.4). You will need to adjust the time required for hands-on activity and testing based on your class size and resources.

Topic	Planned Time
Session I. Selecting a Coating	
A. Testing and Inspecting Coatings	
B. Preparation Standards, Profiling, and Inspection Techniques	
1. Laboratory	
C. Blast Surface Profiling and Measurement	
1. Laboratory	
Session II. Blast Cleaning Media	
A. Types of Abrasives	
B. Properties of Abrasives	
1. Laboratory	
C. Blast Cleaning Safety Guidelines	
1. Laboratory	
Session III. Chemical Strippers; Surface Preparation Technologies	
A. Nonflammable Removers	
B. Flammable Removers	
C. Alkali Removers	
1. Laboratory	
D. Other Surface Preparation Technologies	
1. Performance Testing	
Session IV. Coating Application	
A. Applying a Coating with a Brush	
1. Laboratory	
B. Applying a Coating with a Roller	
1. Laboratory	
C. Spraying with a Conventional Spray System	
D. Airless Spray Systems	
E. Summary	
1. Summarize module	
2. Answer questions	
F. Module Examination	
1. Trainees must score 70% or higher to receive recognition from the NCCER.	
2. Record the testing results on Craft Training Report Form 200 and submit the results to the Training Program Sponsor.	
G. Performance Testing	
1. Trainees must perform each task to the satisfaction of the instructor to receive recognition from the NCCER.	
2. Record the testing results on Craft Training Report Form 200 and submit the results to the Training Program Sponsor.	

Module 61107-02

Apply and Repair External Coatings on Buried and Submerged Pipe

(CT 13.1 and 13.4)

Instructor's Notes:

Instructor's Notes:

Introduction to the Pipeline Operator Qualification Rule, API, and NCCER

It could be said that petroleum products keep America running. Petroleum products like oil and natural gas are the principal sources of energy in the United States and are used in many manufacturing and chemical processes. Petroleum products are primarily transported by pipeline. There are more than a million miles of liquid and natural gas pipeline in this country, ranging in diameter from two inches to four feet. These pipelines bring crude oil and natural gas from production fields to processing plants and refineries—and ultimately to the consumer.

Pipeline Operator Qualification

Petroleum products are considered hazardous, so pipeline personnel must take extra precautions and follow specific procedures when working with them. To help ensure the safe handling and transportation of petroleum products, the United States Department of Transportation (DOT) mandates safe practices as federal regulations.

One such regulation, the Operator Qualification (OQ) rule, requires pipeline operators to develop a written qualification plan to evaluate their personnel and their contract personnel (see *49 CFR Part 192 Subpart N—Gas*, and *49 CFR Part 195 Subpart G—Liquid*). The evaluation hinges on the operator's ability to perform certain covered tasks and to be able to recognize and respond to abnormal operating conditions that may be encountered while performing these covered tasks.

Covered Tasks

A pipeline performance task is deemed *covered* if the activity passes the following four-part test:

1. Is it performed on a pipeline facility?
2. Is it an operations or maintenance task?
3. Is it performed as a requirement of *49 CFR Part 195 (Liquid)* or *49 CFR Part 192 (Gas)*?
4. Does it affect the operation or integrity of the pipeline facility?

If the answer to all four questions is *yes*, then the activity is considered *covered* and must figure into the evaluation process of pipeline personnel.

American Petroleum Institute

The OQ rule and its related covered tasks provided the catalyst to develop a unified pipeline training and assessment program—so that pipeline operators and owners have ready access to the assessment and training tools they need. This effort was the direct result of the vision and leadership provided by The American Petroleum Institute (API) and API members.

Efforts to unify the industry are nothing new to API. Indeed, the impetus for forming the Institute in 1919 was the need to standardize engineering specifications for drilling and production equipment. Today, the most pressing issues revolve around public perceptions and government policies toward the industry. Speaking with one voice on these issues has become as essential as having interchangeable parts in the field. API's members determine the petroleum industry's positions on public policy as well as the standards that govern its day-to-day operations worldwide.

Consortium on Operator Qualification

API assembled a focus group whose primary purpose was to identify covered tasks and provide compliance recommendations to pipeline operators.

The Consortium on Operator Qualification (COOQ) became the guiding force for the identification and development of covered tasks for the liquid pipeline industry. Their work product, API publication 1161, *Guidance Document for the Qualification of Liquid Pipeline Personnel*, formed the basis for the resulting Pipeliner Training and Assessment Program.

The covered task list created in API publication 1161 has now evolved to include gas covered tasks and is maintained by NCCER. The most current covered task list can be downloaded at www.nccer.org.

National Center for Construction Education and Research

With the guidance and support of API, its members, and the Consortium on Operator Qualification, the National Center for Construction Education and Research (NCCER) was able to design and implement the Pipeliner Training and Assessment Program.

NCCER is a not-for-profit 501(c)(3) education foundation established in 1995 by the world's largest and most progressive construction and maintenance contractors and national contractor associations. NCCER was founded to address the

severe workforce shortage facing the construction and maintenance industries and to develop a standardized training process and curricula. Today, NCCER is supported by hundreds of leading construction and maintenance companies, manufacturers, and national associations and developed more than 35 construction and maintenance titles.

NCCER also maintains a National Registry, which contains the records of trainees and instructors who have received NCCER training. All completion information for every trainee is sent to the NCCER and filed within the Registry. The Registry can then confirm training and skills for workers as they move from company to company, state to state, or even within their own company.

Standardized Curriculum (NCCER's Contren™ Learning Series)

NCCER curricula are developed using the expertise of industry subject matter experts (SMEs). The use of SMEs as an integral part of every new development or revision ensures that the curriculum is current, industry-driven, and indicative of industry practices. The development and/or revision process is facilitated from start to finish by NCCER project managers, who work closely with the SMEs and a select team of technical writer(s) on the development and presentation of the curriculum.

Pipeline: Five Training Paths

The pipeline curriculum focuses on essential skills and competencies within five skill areas:

- Corrosion Control
- Electrical and Instrumentation
- Maintenance
- Mechanical
- Operations (Field and Control Center)

SMEs from the pipeline industry have determined the skills and competencies needed based on API's covered tasks and the consensus of the group/industry. Specific company information is generally not incorporated into the curriculum unless the SMEs determine that the information is the industry standard.

Training Modules and Covered Tasks

The smallest unit of instruction in the curriculum is a *module*. Each curriculum is composed of modules (the equivalent of stand-alone chapters) that are logically sequenced into complete *levels* of instruction. Every module is task-oriented and requires the successful completion of a written (knowledge-based) examination and a performance (hands-on) test. Modules may be individually selected and/or combined for efficient task training. Modules in the pipeline curricula, to the extent possible, will align with API-developed covered tasks. While not all modules will relate to covered tasks, all covered tasks will relate to at least one module.

Benefits of Training to the Pipeline Marketplace

Applying NCCER's standardized training process allows pipeline owners and contractors in the United States to establish the benchmark for acceptable skill and competency levels within the pipeline industry. This process will assist training organizations in developing and/or improving standardized pipeline training programs that effectively match their resources and operating conditions while responding to regulatory demands and to the needs of the industry. The process will also recognize training organizations that meet the minimum standards of competence, warranting regulatory, public, and professional confidence. This in turn will foster national unity within the pipeline industry for the mutual benefit of individual technicians, owners, contractors, and ultimately pipeline-users and society.

Additional Resources

To review the Pipeliner Training and Assessment Program Implementation and Resource Guide and the latest version of pipeline covered tasks, visit NCCER's Web site at www.nccer.org, or contact:

National Center for Construction Education and Research
PO Box 141104
Gainesville, FL 32614-1104
(888) 622-3720
Web site: www.nccer.org

For more information on API, please contact:

American Petroleum Institute
1220 L Street, Northwest
Washington, D.C. 20005
(202) 682-8000
Web site: www.api.org

To order *API Pub 1161*, contact Global Engineering Documents by calling 1-800-854-7149 or visiting their Web site at www.global.his.com.

Additional sources of information on the pipeline operator qualification regulation include:

DOT – Office of Pipeline Safety
www.ops.dot.gov

Transportation Safety Institute
www.tsi.dot.gov

Instructor's Notes:

COURSE MAP

This course map shows all of the modules in the first level of the Pipeline Corrosion Control curriculum. The suggested training order begins at the bottom and proceeds up. Skill levels increase as you advance on the course map. The local Training Program Sponsor may adjust the training order.

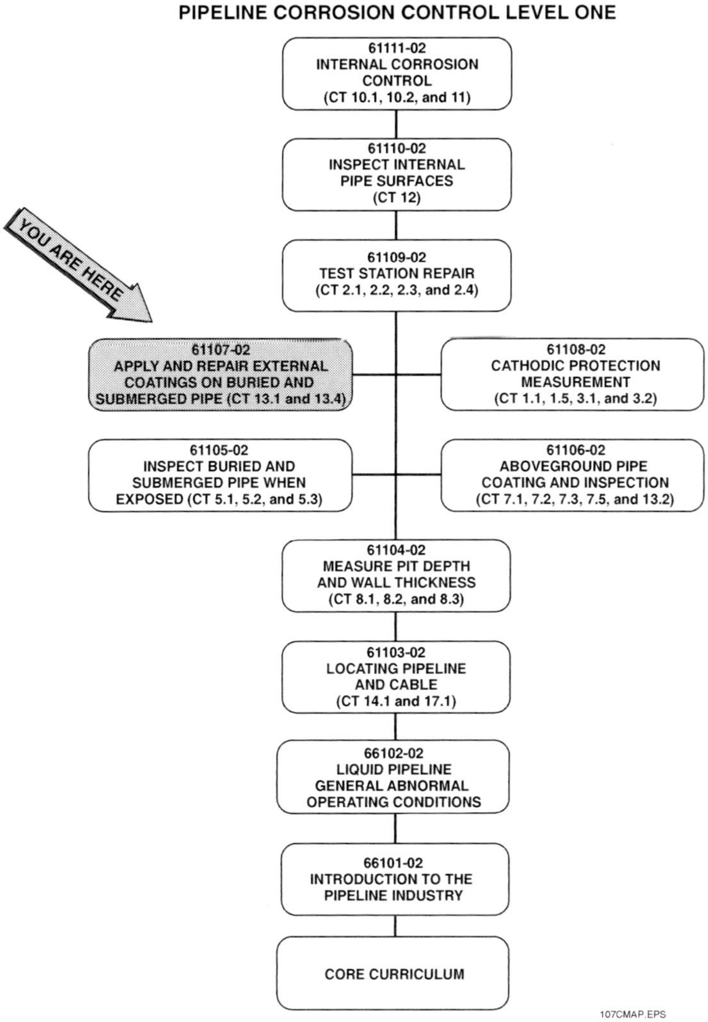

Homework

Assign reading of Module 61107-02.

Instructor's Notes:

MODULE 61107-02 CONTENTS

MODULE 61107-02 CONTENTS (Continued)

Instructor's Notes:

Figures

Tables

Instructor's Notes:

MODULE 61107-02

Apply and Repair External Coatings on Buried and Submerged Pipe (CT 13.1 and 13.4)

Objectives

When you have completed this module, you will be able to do the following:

1. Select a coating for buried or submerged piping.
2. Remove coating from buried or submerged pipe using chemical strippers or blasting systems. (CT 13.1 and 13.4).
3. Apply coatings with a brush and roller.
4. Apply coatings with a conventional spray system.

Prerequisites

Before you begin this module, it is recommended that you successfully complete the following: Core Curriculum; Pipeline Core, Modules 66101-02 and 66102-02; Pipeline Corrosion Control Level One, Modules 61103-02 through 61106-02.

Required Trainee Materials

1. Pencil and paper
2. Appropriate personal protective equipment

1.0.0 ◆ INTRODUCTION

The maintenance and repair of coatings for buried pipe is one of the biggest challenges that faces today's corrosion control individual. This is particularly true when dealing with the critical nature of the substance that is being transported through today's pipelines. Since the coating represents the first line of defense in the protection of the pipelines, it must be kept in the best possible condition for the longest period to guarantee proper protection over the intended life span of the piping. Coatings require constant maintenance and repair to ensure their proper function and maximize their life span. This module explains some of the activities involved in maintaining and repairing a buried or submerged pipeline coating.

2.0.0 ◆ SELECTING A COATING

Coatings and coating systems are selected for specific reasons and job requirements. The coating must be compatible with the substrate and the environmental conditions. In general, coating selection is based on the following considerations:

- *Type of substrate and exposure* – The type of substrate affects the chemical compatibility, thermal compatibility, flexing, and adhesion of the coating applied to it. A reaction between the coating and the substrate could be harmful if it lessens adhesion or leads to bleeding. Exposure is also a highly important consideration. Coatings should always be selected with regard to the service environment.
- *Operating conditions/upset conditions* – The specific operating conditions to which a coating is subjected are a key factor in coating selection. For example, if a coating is to be applied to a structure operating at a high temperature and exposed to acid fumes, the coating selected is limited to one that can withstand high temperatures and also has good resistance to acid.

  Upset conditions are changes in the operation that could damage or destroy a coating. For example, a tank properly protected on its interior and containing an acid or alkali might be coated on its exterior with a mildly resistant coating. As long as the contents stay in the tank,

Materials

Ensure that you have everything required to teach the course. Check the Materials and Equipment List at the front of this module.

Classroom

Explain that terms shown in bold are defined in the Glossary at the back of this module.

Audiovisual

Show Transparency 1, Course Objectives.

Show Transparency 2, Performance Tasks.

Teaching Tip

See the general Teaching Tip at the end of this module.

Classroom

Explain that the coating is the pipeline's first line of defense against corrosion and contamination.

Discuss the considerations for selecting an appropriate coating.

Classroom

Explain how to select a coating that is highly resistant to a tank's contents.

Emphasize the importance of selecting a coating of the proper thickness.

Distribute copies of a manufacturer's data sheets for a variety of coating products. Have trainees review the material's data and identify information you specify.

Discuss the consequences of coating failures. Review the considerations for inspecting and testing the protective qualities of a coating.

no harm may come to the coating. However, if the tank were to overflow, the exterior coating could fail. If there is a possibility of this occurring, then the coating selected for the exterior of the vat should be highly resistant to the vat's contents.

- *Ambient conditions during application* – The anticipated ambient conditions that exist during the application of a coating must be considered. Temperature, relative humidity, and dew point can all affect how well a coating cures. The coating selected must be one that can be successfully applied in the anticipated ambient conditions.
- *Government and environmental regulations* – In all cases, the coating selected must comply with the prevailing federal, state, and local laws. In many locations, government regulations may restrict the use of certain techniques and/or materials such as abrasive blasting or solvent emission.
- *Budget* – The amount of money allocated for the purchase of coating materials for the job can influence the selection of the coating.
- *During operation or shutdown* – If the coating must be applied when a unit is in operation, various factors may influence the coating selection. Some of these factors include metal temperature during operation, application methods permissible during operation, plant personnel who will be present during operation, and safety.
- *Time constraints* – The time allowed to work on and/or to complete the job may affect the coating selection.
- *Surface preparation* – Generally, new construction projects specify the degree of surface preparation and allow for, or require, a complete coating system. With maintenance work, surface preparation efforts usually tend to be minimal, and the coating may be only touched up. Coatings compatible with those to be maintained must be chosen, and a coating that can tolerate a lesser degree of surface preparation must also be selected.

In order to determine the individual coating to use, *always* check the manufacturer's recommendations and the job requirements. Job requirements are usually expressed as one or more ASTM International standard(s). The ASTM standards used to select these types of coatings usually refer to coating thickness, moisture resistance, fire resistance, specific chemical resistance, and abrasion resistance. Always resolve any coating issues before starting the job. For example, if the exposure level and substrate require a coating 20 mils thick, but the intended epoxy coating is supposed to be no more than 7 mils thick, an alternative coating or system must be selected. Coating failures can be life-threatening in these situations.

Selection of a correct coating is also aided by referring to and comparing the manufacturer's product information data sheets for each coating product being considered. *Figure 1* shows an example of the typical product data sheets used by one manufacturer. The method and layout used by different manufacturers varies, but all contain basically the same information. As shown, it includes the following information:

- Product description
- Recommended uses
- Product characteristics
- Performance characteristics
- Recommended coating systems
- Surface preparation
- Color availability and tinting
- Application
- Ordering information
- Safety precautions

2.1.0 Testing and Inspecting Coatings

Industrial coatings are chosen and applied in jobs with special requirements. Coating failures in these situations usually result in more than just staining or discoloration of the substrate. They can result in loss of structure, loss of contained resources, spills, accidents, other hazards, and even loss of life. It is especially important to ensure proper surface preparation and application. The protective qualities of a coating can be verified with inspections and tests. Each inspection or test verifies a particular property, such as minimum chemical, corrosion, and fire resistance/retardance. Control and checking begin with determining the environmental conditions of the job, especially if multicomponent coatings are used, because relative humidity and temperature affect their application. Once applied, wet-film thickness can be measured. Once the coating is dry, check the dry-film thickness, film continuity, and gloss. The inspection and test results are often expressed as Society for Protective Coatings (SSPC) or ASTM standards. The coating is often required to meet the ASTM standards in the specification requirements.

Instructor's Notes:

Sherwin Williams

Industrial and Marine Coatings

4.66

POLAR-POXY™

LOW TEMPERATURE CURE EPOXY

PART A	B62W700	WHITE
PART A	B62S701	ALUMINUM
PART B	B62V702	HARDENER

PRODUCT INFORMATION

PRODUCT DESCRIPTION

POLAR-POXY LOW TEMPERATURE EPOXY is a high performance, interior/exterior, self-priming, high build, VOC compliant, epoxy amine adduct epoxy, formulated to be applied at temperatures as low as 20°F. Specifically designed as a one coat primer or finish coat where conventional coatings cannot be used due to cold weather conditions.

- Will wet and penetrate firm rust and tight crevices
- Chemical and moisture resistant
- Suitable for use in USDA inspected facilities

RECOMMENDED USES

For use over prepared steel surfaces such as:

- Structural steel
- Industrial equipment
- Railings
- Tank exteriors
- Bridges
- Catwalks
- Piping
- Water towers

Do not use in hot weather applications over 80°F.
Not intended for architectural applications.

PRODUCT CHARACTERISTICS

Finish: Low Sheen

Color: White and Aluminum

Volume Solids: 67% ± 2%, mixed

Weight Solids: 80% ± 2%, mixed

VOC (EPA Method 24):
White, mixed — Unreduced: 280 g/L; 2.33 lb/gal
Aluminum, mixed — Unreduced: 316 g/L; 2.63 lb/gal

Mix Ratio: 2 components, 1:1 by volume

Recommended Spreading Rate per coat:
Wet mils: 5.75 - 8.75
Dry mils: 5.0 - 7.0
Coverage: 160 - 220 sq ft/gal approximate

NOTE: Brush or roll application may require multiple coats to achieve maximum film thickness and uniformity of appearance.

Drying Schedule @ 7 mils wet @ 50% RH:

	@ 35°F	@ 50°F	@ 75°F
To touch:	12 hours	6 hours	4 hours
To handle:	48 hours	12 hours	6 hours
To recoat:			
minimum:	24 hours	24 hours	18 hours
maximum:	6 months	6 months	6 months
Pot Life:	8 hours	5 hours	3 hours

If maximum recoat time is exceeded, abrade surface before recoating.
Drying time is temperature, humidity, and film thickness dependent.

Sweat-in-time: 15 minutes, @ 50°F, 50% RH

Shelf Life: 24 months, unopened, at 77°F

Flash Point: mixed — White: 87°F Seta Flash; Aluminum: 81°F Seta Flash

Reducer/Clean Up: Reducer #54, R7K54

PERFORMANCE CHARACTERISTICS

System Tested: (unless otherwise indicated)
Substrate: Steel
Surface Preparation: SSPC-SP6
Finish: 1 ct. Polar-Poxy @ 6 mils dft

Abrasion Resistance:
Method: ASTM D4060, CS17 wheel, 1000 cycles, 1 kg load
Result: 80 mg loss

Adhesion:
Method: ASTM D3359 Method B
Result: 5B

Direct Impact Resistance: (on cold rolled steel)
Method: Fed. Standard 141, Method 6226
Result: 24 in. lbs. minimum

Dry Heat Resistance:
Method: ASTM D2485
Result: 175°F, intermittent 200°F

Exterior Durability:
Method: 1 year at 45° south
Result: Chalks, discolors (untopcoated)

Flame Spread Rating:
Method: ASTM-E84 Tunnel Test
Result: Class A on noncombustible surfaces

Flexibility:
Method: ASTM D522, 180° bend, 1/8" mandrel
Result: Passes

Moisture Condensation Resistance:
Method: ASTM D4585, 100°F, 2000 hours
Result: Passes, no evidence of rusting, blistering, cracking, or adhesion loss

Pencil Hardness:
Method: ASTM D3363
Result: 7H

Prohesion:
Method: ASTM G85, Annex A5, 2500 hours
Result: Rating 10 per ASTM D610 for rusting; Rating 10 per ASTM D714 for blistering

Salt Fog Resistance:
Method: ASTM B117, 1000 hours
Result: No surface or pinpoint rusting, softening, or delamination; no more than 1/8" creepage at scribe.

Epoxy coatings may darken or yellow following application and curing.

SAMPLE

107F01A.EPS

Figure 1 ◆ Typical product data sheet (1 of 2).

Sherwin Williams

Industrial and Marine Coatings

4.66

POLAR-POXY™

LOW TEMPERATURE CURE EPOXY

PART A	B62W700	WHITE
PART A	B62S701	ALUMINUM
PART B	B62V702	HARDENER

PRODUCT INFORMATION

RECOMMENDED SYSTEMS

Steel:
1-2 cts. Polar-Poxy @ 5.0 - 7.0 mils dft/ct

Steel, acrylic topcoats:
1-2 cts. Polar-Poxy @ 5.0 - 7.0 mils dft/ct
1-2 cts. DTM Acrylic Finish @ 2.5 - 4.0 mils dft/ct
or Metalatex Semi-Gloss

Steel, epoxy topcoats:
1-2 cts. Polar-Poxy @ 5.0 - 7.0 mils dft/ct
1-2 cts. Sher-Tile Epoxy @ 4.0 - 8.0 mils dft/ct
or Heavy Duty Epoxy
or Macropoxy
or Water Based Catalyzed Epoxy

Steel, polyester epoxy:
1-2 cts. Polar-Poxy @ 5.0 - 7.0 mils dft/ct
1-2 cts. Armor-Tile HS @ 2.5 - 4.0 mils dft/ct

Steel, polyurethane topcoats:
1-2 cts. Polar-Poxy @ 5.0 - 7.0 mils dft/ct
1-2 cts. Hi-Solids Polyurethane @ 3.0 - 4.0 mils dft/ct
or Corothane II Polyurethane

Concrete:
2 cts. Polar-Poxy @ 5.0 - 7.0 mils dft/ct

Check minimum application temperatures of topcoats prior to use.

The systems listed above are representative of the product's use. Other systems may be appropriate.

SURFACE PREPARATION

Surface must be clean, dry, and in sound condition. Remove all oil, dust, grease, dirt, loose rust, and other foreign material to ensure adequate adhesion.

Refer to product Application Bulletin for detailed surface preparation information.

Minimum recommended surface preparation:
Iron & Steel: SSPC-SP2
Concrete: Cured, clean, sound, dry

COLOR AVAILABILITY / TINTING

Do not tint. Not controlled for tint strength.

Colors: White and Aluminum

APPLICATION

Temperature: 20°F minimum, 80°F maximum (air and surface; material must be mixed at 50°F minimum) At least 5° above dew point
Relative humidity: 85% maximum

Refer to product Application Bulletin for detailed application information.

ORDERING INFORMATION

Packaging:
Parts A & B: 5 gallon containers

Weight per gallon:
White: 12.78 ± 0.3 lb, mixed
Aluminum: 11.88 ± 0.3 lb, mixed

SAFETY PRECAUTIONS

Refer to the MSDS sheet before use.

Published technical data and instructions are subject to change without notice. Contact your Sherwin-Williams representative for additional technical data and instructions.

SAMPLE

107F01B.EPS

Figure 1 ◈ Typical product data sheet (2 of 2).

Instructor's Notes:

2.2.0 Preparation Standards, Profiling, and Inspection Techniques

The quality of the surface preparation has a direct bearing on how well a coating performs and how long it lasts. It is important to determine how well a surface has been prepared and to catch any problems before they are covered with a coating.

Standards and inspection techniques for surface preparation have been established to help protect against premature coating failures. Generally, the greatest number of standards and the most formal inspection processes are applied to pipeline applications.

The following is a brief overview of steel surface preparation.

- Steel surfaces:
  - Slag, flux, weld spatter, and rough edges should be removed.
  - The surface should be washed with solvents to remove grease, dirt, oil, debris, and other loose contaminants.
  - The surface should be cleaned of all rust, dirt, **mill scale**, old paint, or other tightly adhered contaminants to the degree required by the specifications. Use one or more of the following: hand tools, power tools, wet or dry abrasive blast machines.
  - The surface should be profiled with abrasives to produce the specified anchor pattern.

The two primary areas of concern to inspectors in the surface preparation of steel are cleanliness and profile. SSPC and the National Association of Corrosion Engineers (NACE) have developed joint standards for evaluating abrasive blast cleaning.

2.2.1 Blast Surface Cleanliness Standards and Production Rates

As a guide for comparing surface cleanliness, SSPC has developed reference photographs (*SSPC-VIS 1*) to illustrate each grade. These visual standards are for steel that has not previously been painted.

Abrasive blast cleaning can be used to achieve various degrees of surface cleanliness to meet job specifications. These specifications usually define the minimum level of cleaning required by referencing and supplementing the standards (SSPC/NACE) used for steel finishing. They are based on dry abrasive blasting. If wet abrasive blasting is being used by mutual agreement between the parties specifying the work and the parties performing the work, then inspectors must be made aware of the agreement. In any case, inspectors should be made aware of the abrasive used and the appearance of the final blasted surface by a test demonstration before the job commences. The test demonstration area can be coated with clear urethane and used as a reference standard for cleaning and profiling by both blast operators and inspectors.

> **NOTE**
> Once steel surfaces are cleaned and bare metal is exposed, flash-rusting of the surface (rust-back) may begin almost immediately. If the primer coating cannot be applied within a short period of time, a wash of the bare surfaces using a rust inhibitor may be required. One such inhibitor is water that contains 0.32% sodium nitrate and 1.28% (by weight) secondary ammonium phosphate (dibasic). In any event, the coating system must be compatible with whatever inhibitor is used.

2.3.0 Blast Surface Profiling and Measurement

With many high-performance coatings, profiles on new surfaces or on previously coated surfaces that were not profiled are required to provide an anchor pattern for subsequent application.

In most cases, the project specifications will dictate the amount of profiling required. If not, the coating manufacturer's data sheets will provide this information. *Table 1* is a guide for obtaining various depths of profiling using a variety of dry mineral and ferrous abrasives.

The data in the table should be used only for approximating the abrasive size required to obtain a specified anchor pattern. The resulting depth of anchor pattern varies with the method used for measuring depths as well as with a number of other variables (for example, type and hardness of steel, thickness of mill scale, and degree of cleaning specified). This information can be used for centrifugal wheel blasting as well as pressure blasting. Pressure blasting should be done using 90- to 100-psi nozzle pressure. The depth of anchor pattern referred to in *Table 1* is an average and not a minimum or maximum depth obtainable.

To check the surface profile, use tools such as a profile comparator (see *Figure 2*) to compare the surface to established standards; a profile gauge (see *Figure 3*) to measure the depth of the profile;

Classroom

Discuss the standards for inspecting a steel surface preparation. Answer any questions trainees may have.

Laboratory

Display the *SSPC-VIS 1* photographs that illustrate each grade of steel cleanliness. Have trainees inspect the photographs and identify the standards that are illustrated in the photographs.

Classroom

Explain how anchor patterns are used for coating applications on previously coated surfaces.

Refer to Table 1. Ensure that trainees understand how to interpret the data in the table. Emphasize that the information should be used only for approximate sizing.

Audiovisual

Show Transparency 3 (Figure 2). Explain how a profile comparator compares the surface to established standards.

Audiovisual

Show Transparency 4 (Figure 4). Explain how replica tape makes an impression of the profile.

Laboratory

Set up stations with a variety of surface profiling instruments. Ask trainees to identify each instrument and explain how it is used to check the surface profile. Ask whether trainees have any questions.

Teaching Tip

See the Teaching Tip for Section 2.0.0 at the end of this module.

Homework

Have trainees review Sections 3.0.0–3.3.8 for the next session.

Table 1 Obtaining Various Depths of Profiling

Profile	Abrasive Required
1 mil	30/50 mesh silica sand
	G-80 steel grit
	S-110 steel shot*
	80-mesh garnet
	100-grit aluminum oxide
	Coal slag 3060
1.5 mil	16/35 mesh silica sand
	G-50 steel grit
	S-170 steel shot*
	36-mesh garnet
	60-grit aluminum oxide
	Coal slag 3060
2 mil	16/35 mesh silica sand
	G-40 steel grit
	S-230 steel shot*
	36-mesh garnet
	36-grit aluminum oxide
	Coal slag 2040
2.5 mil	8/35 mesh silica sand
	G-40 steel grit
	S-280 steel shot*
	16-mesh garnet
	24-grit aluminum oxide
	Coal slag 2040
3–4 mil	8/20 mesh silica sand
	G-25 steel grit
	S-330 or 390 steel shot*
	16-mesh garnet
	16-grit aluminum oxide
	Coal slag 1240

* Steel shot alone will not give a good angular anchor pattern and should be used in combination with steel grit for best results.

107T01.EPS

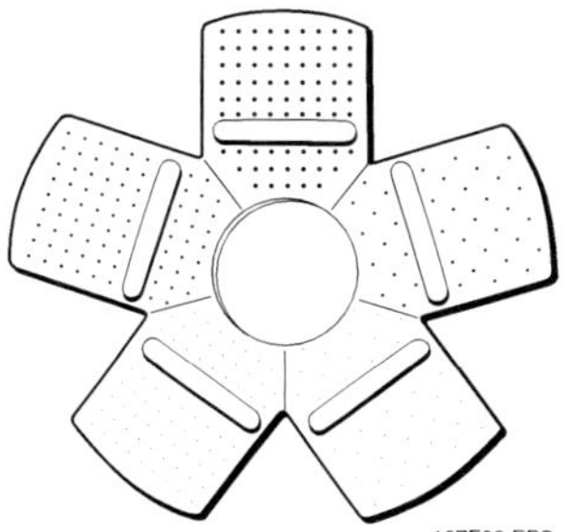

107F02.EPS

Figure 2 ◆ Profile comparator.

or a replica tape (see *Figure 4*) to make an impression of the profile. How clean and profiled a surface needs to be depends on the type of structure, its location, and the job specifications. Again, inspectors and blast operators should be made aware of the appearance of the final blasted and profiled surface by a test demonstration before the job commences.

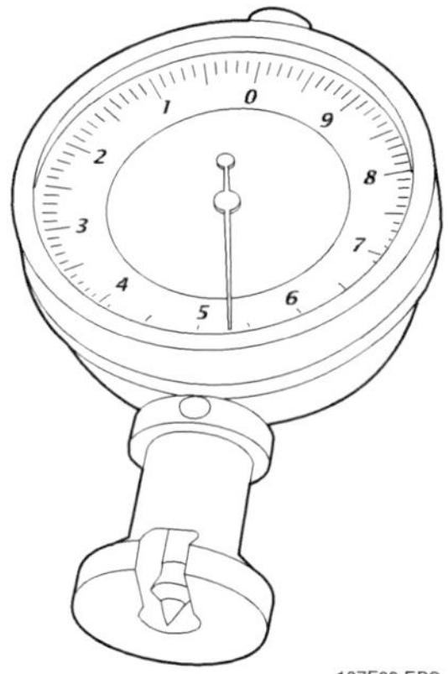

107F03.EPS

Figure 3 ◆ Profile gauge.

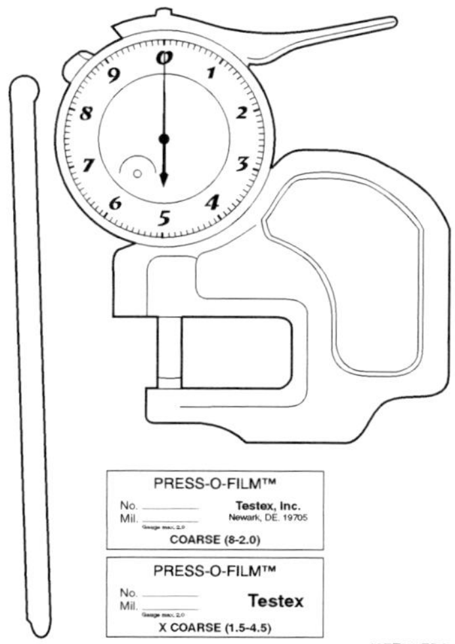

107F04.EPS

Figure 4 ◆ Replica tape.

Instructor's Notes:

3.0.0 ◈ BLAST CLEANING MEDIA

Abrasive, the actual cleaning material of the blast cleaning process, comes in a range of types, shapes, sizes, and degrees of hardness. Given the multitude of options, selecting the right abrasive for the job becomes a matter of determining which material will give a surface the desired finish, which means the proper degree of cleanliness and roughness as well as the type of texture needed for coating. Other factors to consider in choosing an abrasive are the type of surface to be blasted and its present condition, the consumption rate of the abrasive, its cost, and any environmental regulations that may affect the blast cleaning operation.

3.1.0 Types of Abrasives

Certain characteristics of an abrasive media influence the productivity of the operator. Work performed by the media is controlled by its size and density (weight), and more importantly, by its impact velocity.

Some projects may use a specification to identify the abrasive required. Federal government agencies, the Society of Automotive Engineers (SAE), and others have abrasive specifications applicable to their operations.

There are four types of abrasives: metallic, mineral, by-product, and special-purpose.

3.1.1 Metallic Abrasives

The most common metallic abrasive is cast steel. Others include malleable iron and chilled cast iron. All come in the form of shot and grit (see *Figure 5*) and in different degrees of hardness. Abrasive made from wire cut into short pieces is also available. Shot is rounded, which makes it appropriate for cleaning heavy, brittle deposits and for producing a rounded profile or anchor pattern. Grit, which is made by crushing hardened shot pellets, and cut wire both have an irregular, angular shape that effectively removes rust, mill scale, and paint, and leaves a rough profile that feels like abrasive paper. Shot, grit, and cut wire are screened according to size.

Metallic abrasives can be recycled many times before breaking down into a size too small to be reused. The repeated blast cleaning impact causes the abrasive to fracture or flake. Generally, the harder the abrasive, the faster it breaks down. The particles become smaller and smaller until they are eventually filtered out in the recycling process. New abrasive must be added regularly to replace what is lost. The result is a mixture of sizes called a *work mix* or *operating mix* of large and small particles. Large particles do the heavy work of cracking and breaking up the contaminants; small particles do the cleanup work of scouring and removing them. Maintaining a properly balanced work mix is important for efficient blast cleaning. This can be done by hand or with a continuous, automated system of replenishing spent abrasives.

3.1.2 Mineral Abrasives

From the beginning, sand has been used for abrasive blast cleaning. Silica sand in particular is readily available, which makes it economical. Its range of angular and spherical shapes gives it good cleaning action. However, it breaks down quickly and produces a large amount of silica dust, which causes a dangerous lung disease known as *silicosis* when the dust is small enough to require respiratory protection (10 microns or less). Non-silica or heavy mineral sands used in blast cleaning include magnetite, staurolite, and olivine rutile, which are tough, dense, and smaller than silica sand but contain no dangerous free silica. These abrasives are generally not recycled because they quickly break down in blast cleaning. On the other hand, garnet, a tough abrasive, can be recycled several times. Other mineral abrasives, such as zircon and novaculite, are generally used for specialty blast cleaning because of their cost and the finish they leave. Zircon, for example, gives a smooth, matte finish, and novaculite leaves a satin luster.

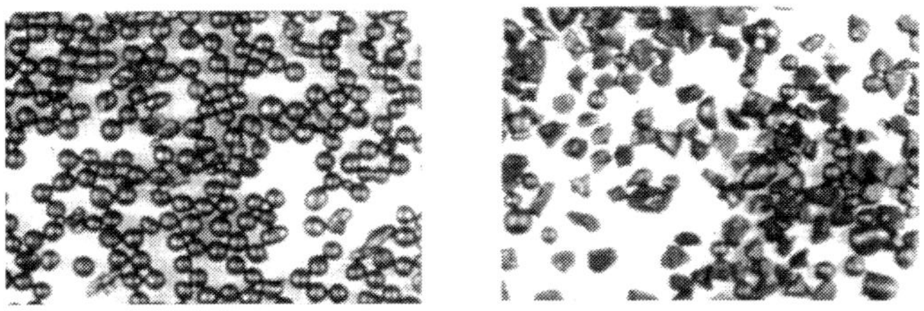

107F05.EPS

Figure 5 ◈ Shot and grit.

Materials

Ensure that you have everything required for laboratories during this session.

Classroom

Review the factors for selecting an appropriate abrasive for a job.

Discuss the four types of abrasives and their properties.

Compare the finishes that are achieved by using the various abrasives. Go over any questions trainees may have.

Teaching Tip

See the Teaching Tip for Section 3.1.0 at the end of this module. Administer the Quick Quiz.

Classroom

Refer to Table 2. Discuss the benefits, usage properties, and physical properties of various abrasives.

Laboratory

On the whiteboard/ chalkboard, list the benefits and properties of various abrasives. Ask trainees to identify the abrasive based on the information you provide.

Classroom

Emphasize the importance of following the manufacturer's safety recommendations when blast cleaning.

Review the safety guidelines for working with blast nozzles.

3.1.3 By-Product Abrasives

Slag, the refuse from smelting or other processes, makes a good abrasive because of its availability, relatively low cost, and low silica content. Its sharp, angular shape gives it good cutting power, and it comes in a full range of sizes. Blast cleaning slags include coal slag (see *Figure 6*), copper slag, and nickel slag. By-product abrasives are not recycled because they break down too rapidly.

107F06.EPS

Figure 6 ◆ Coal slag.

3.1.4 Special-Purpose Abrasives

There are specialty abrasives made with walnut shells or peach pits which will remove light deposits without damaging the surface. Corncob and sodium bicarbonate abrasives can also be used for this purpose, or for removing surface grease. Other special abrasives include sponge, silicon carbide, aluminum oxide, cut glass, and plastic or glass and ceramic beads. Plastic and glass abrasives will clean small, delicate surfaces. Aluminum oxide and silicon carbide are tough, hard, aggressive abrasives that can be recycled. Aluminum oxide works on stainless steel, and silicon carbide does specialty etching. A polyurethane sponge, sometimes incorporating various abrasives, can be used to remove oil and other contaminants while also cleaning and roughening a surface for painting. Most of these abrasives can be recycled a limited number of times.

3.2.0 Properties of Abrasives

Shape, size, density, and hardness are the other factors to consider in selecting the proper abrasive material. As noted, abrasives come in spherical and angular shapes, as well as irregular shapes that have features of both. Rounded or spherical abrasive particles clean by hitting the surface; sharp or angular abrasives clean by cutting or gouging it. Large abrasive particles remove thick, heavy contaminants, while small ones work on the fine, residual debris. A mix of sizes is often best. Generally, the denser the abrasive, the more effective it is, and the harder it is, the deeper it will clean the surface. All abrasives should be kept clean and dry before blast cleaning.

Table 2 presents a condensed summary of the properties for a number of different types of abrasive media that, with the exception of sand and sodium bicarbonate, can be recycled a varying number of times.

3.3.0 Blast Cleaning Safety Guidelines

Blast cleaning produces large amounts of dust, debris, chips, and spent abrasives, making safety a primary concern. You should always follow the manufacturer's safety recommendations and any applicable company safety guidelines. Be sure to comply with environmental regulations when containing and disposing of the waste.

3.3.1 Nozzle Blast Hazard

The abrasive/air jet from a blast nozzle can cause serious physical injury including injection of abrasive particles, rust inhibitors if used, and other contaminants into the body. If this occurs, get emergency surgical treatment immediately. In addition:

- Do not point a blast nozzle at any part of your body or at any other person.
- Do not put your hand or fingers in front of the blast nozzle.
- Make sure the nozzle and hose connections are secure and safety-pinned or wired before applying pressure to the hoses and nozzle. In addition, make sure the hose connections have whip checks installed.
- Make sure the dead-man control handle is functioning correctly and that the safety latch engages properly before commencing blast cleaning operations. If you stop blast cleaning, even for a moment, ensure that the safety latch is fully engaged or it will not prevent the dead-man control from being accidentally triggered if the hose and nozzle are dropped or bumped.
- Make sure that you wear the proper blast suit, air-fed helmet, and hearing and respiratory protection when entering or working in a blast cleaning area. Make sure the inner safety lens of the helmet is in place and is not damaged.
- Check the hoses, couplings, nozzles, and gaskets daily for signs of damage, leaks, and wear. Do not attempt to repair damaged, leaking, or worn components; replace them.

Instructor's Notes:

Table 2 Properties of Abrasives

Abrasive	Benefits				Usage Properties					Physical Properties			
	Finishing	Cleaning and Removal	Peening	Etching	Average Cleaning Speed	Reusability	Dust Level(s)	Chance of Metal Removal	Typical Blast Pressures (psi)	Hardness MOH Scale (Rockwell C)	Bulk Density (ft.³)	Mesh Sizes	Angular (◊) Spherical (○)
Glass Beads	X	X	X		MED	HI—LO	LO—HI	Very LO	20—55	5.5	100	30—440	○
Ceramic Shot	X	X	X		MED	HI	LOW	Very LO	20—90	7 (57—63)	150	8—46	○
S.S. Cut Wire	X	X	X	X	MED	HI	LOW	MED—HI	20—90	6—7.5 (35—55)	280	20—62	◊
Steel Shot	X	X	X		MED	Very HI	LOW	Very LO	20—90	6—7.5 (20—66)	280	8—200	○
Steel Grit	X	X		X	MED—HI	Very HI	LOW	MED	20—90	6—7.5 (40—66)	230	10—325	◊
Aluminum Oxide	X	X		X	HI	MED—HI	LO—MED	MED—HI	20—90	8—9	125	12—325	◊
Silicon Carbide	X	X		X	Very HI	LO—MED	MED—HI	MED—HI	20—90	9	95	36—220	◊
Crushed Glass	X	X		X	HIGH	LO—MED	MED—HI	LO—MED	20—50	5.5	100	30—400	◊
Plastic Media		X		X	MED—HI	MED	MED	Very LO	20—60	3—4	45—60	12—80	○
Wheat Starch		X			MED	MED	MED	Very LO	20—35	3	40	12—100	◊
Walnut Shell		X			LO—HI	LOW	MED—HI	Very LO	10—40	3	40—80	Many	◊
Corn Cob		X			LO—MED	LOW	MED—HI	Very LO	20—60	1	30	8—40	◊
Sodium Bicarb.[1]		X			MED—HI	NR[3]	HI	Very LO	20—90	3	62	54—150	◊
Silica Sand[2]	X	X		X	HI	NR[3]	HI	HI	30—90	7.5	100	12—100	◊

NOTES:
Above information is intended as a general reference guide.
[1] Sodium bicarbonate must be treated with a flow agent.
[2] Do not use silica sand in a hand cabinet.
[3] Not recommended.

107T02.EPS

Classroom

Ask a trainee to explain how to ensure proper footing and balance.

Explain how to avoid respiratory injury when working with hazardous materials.

Discuss the consequences of failing to follow proper safety recommendations when working with carbon monoxide.

3.3.2 Hypothermia Hazard

When conducting outdoor blast cleaning operations in temperatures below 60°F with any amount of wind, a risk of hypothermia exists, especially with wet abrasive blast cleaning. Always wear adequate protective clothing, head covering, and gloves.

3.3.3 Footing and Balance Hazard

When conducting dry blast cleaning operations, vision may become temporarily impaired due to dust which could obscure hazardous footing conditions. When wet abrasive blast cleaning, footing surfaces could become slick. Either of these conditions can cause difficult footing and/or balance problems for the operator. In addition:

- A proper fall protection system should be used with any scaffolding or staging.
- Hoses and other materials must remain clear of your feet at all times.
- With weight on both feet and feet spaced apart in an open stance, grip both the control handle and the nozzle before triggering the control handle to commence blast cleaning. Be prepared for the reactive force from the nozzle which will occur shortly after the control handle is triggered.

3.3.4 Hazardous Materials

Dust from disintegrating coatings and/or abrasives during blast cleaning operations must never be inhaled. In fact, any dust (including dust from abrasive handling, loading, and site cleanup operations) must never be inhaled. Severe, incurable health problems and/or eventual death may occur. To avoid respiratory injury:

- Always wear a National Institutes of Occupational Safety and Health (NIOSH)-approved air-fed helmet or hood before, during, and after blast cleaning, until cleanup is completed and the air has been tested and found clear of toxic dust.
- Always clean, maintain, and inspect your helmet, breathing hose, air control device, and breathing air filter daily. Replace any damaged components with approved parts; otherwise, the NIOSH approval of the equipment will be voided.
- Always use dry blast media that is prepared for that purpose and is properly sized and free of harmful substances such as free silica, cyanide, arsenic, or lead. Check Material Safety Data Sheets (MSDS) for presence of toxic or harmful substances.
- Surfaces to be blasted must be tested for toxic substances before blast cleaning commences. If toxic, additional protective measures for operators and bystanders must be taken as required by OSHA/NIOSH.
- If blast cleaning operations are conducted in a confined area, proper sealing of the area and separate, adequate, negative pressure ventilation (in addition to any abrasive recovery equipment) with appropriate filtering must be in place to allow removal of airborne toxic dust from the blast cleaning area and to prevent leakage of the dust to other areas.

3.3.5 Carbon Monoxide Hazard

Breathing air compressors can and do produce deadly, colorless, odorless carbon monoxide gas that can be fatal to blast operators. In addition, carbon monoxide gas from the exhaust of other engine-driven equipment can be sucked in by breathing air compressors or breathing air pumps and transmitted to operators. To avoid death caused by carbon monoxide:

- Make sure the breathing air is not supplied from an oil-lubricated, piston-type compressor.
- Make sure the breathing air compressors are checked and maintained strictly according to the manufacturer's instructions and schedule, and document the actions in an equipment log as required by OSHA.
- Calibrate and perform maintenance on any carbon dioxide (CO_2) monitors or converters used for breathing air, strictly following the manufacturer's instructions and schedule, and document the actions in an equipment log as required by OSHA. These devices, when used in the field, usually require more frequent calibration/maintenance. If monitors or converters are not used, frequent periodic testing of breathing air must be accomplished and documented as required by OSHA.
- Outdoors, if at all possible, locate all compressors or air pumps used to supply breathing air away from, and upwind of, any other engine-operated equipment and the dust created by the blast cleaning operation.
- Do not operate engine-powered equipment in a confined area without adequate ventilation to remove exhaust fumes and carbon monoxide (CO) from the area.

Instructor's Notes:

3.3.6 Fire/Explosion Hazard

The fuel used in gasoline- or diesel-engine powered equipment is combustible. When spilled on a hot surface, it can ignite and cause a fire or explosion. Unless the fuel tanks are located remotely from the equipment, do not fill fuel tanks while the engine is running or hot. Make sure that the blast machine and any conductive work surfaces are grounded and that the blast hose is of the static-suppressing or grounded type. Electrical static discharges from the blast nozzle to the work surface may cause a flash fire or explosion if flammable vapor or dust is present and if a discharge occurs. When working in confined areas such as tanks, make sure all volatile fumes and dust are removed by ventilation. Sparks created by abrasive striking a surface may cause a flash fire or explosion.

3.3.7 Electrical Hazard

If wet blast attachments or injectors are used, all electricity to motors, electrical boxes, or electric devices must be turned off and locked out, or the motors, boxes, or devices must be completely water sealed by materials resistant to the abrasive blast jet. This prevents accidental death from electrocution or fires caused by shorts in the electrical equipment.

3.3.8 Equipment Misuse Hazard

Misuse, unauthorized repair, alteration, or modification of the equipment may cause violent ruptures or explosions that result in serious physical injury or death. In addition, unauthorized repairs, alterations, or modifications may violate American Society of Mechanical Engineers (ASME), OSHA, or NIOSH certifications or regulations.

- Do not alter or modify any part of the equipment.
- Do not change any factory-set adjustments.
- Do not exceed the maximum rated pressure of any equipment or component.
- Do not repair ASME-coded pressure vessels or equipment without re-certifying the vessels or equipment.
- Do not alter the throttle settings of any engine-operated equipment unless recommended or allowed by manufacturer's instructions.
- Do not perform unauthorized repairs of equipment. Replace worn or damaged parts immediately with new, recommended, and approved parts.
- Do not allow the blast nozzles to be dropped, roughly handled, or used as hammers or signaling devices. The inner lining may crack, causing premature failure of the nozzle.
- Do not allow the blast hoses or air hoses to be curved with sharp bends or to be driven over repeatedly by vehicular traffic. Premature wear or damage can occur, which may result in an unexpected rupture of the hose.

4.0.0 ◆ CHEMICAL STRIPPERS

In some situations in which minimal substrate damage is desired, chemical stripping may be the only surface preparation method available to accomplish the task. Chemical strippers are suitable for removing most oil-based, thermoplastic, and water-based coatings. An example of a coating removal operation is shown in *Figure 7*. However, it is very difficult and labor intensive, and sometimes impossible, to remove thermoset materials, such as epoxies, urethanes, polyesters, and other high-performance coating materials with conventional removers. Specialty strippers are available that will be effective in removing many of these surfaces.

There are three general classes of removers, or strippers, that are available in either thin or heavy-body viscosities. The classes are nonflammable, flammable, and alkali removers. Some removers are blends of different types.

107F07.EPS

Figure 7 ◆ Coating removal operation.

Classroom

Explain problems that can result from equipment misuse and from unauthorized alterations, repairs, or modifications. Ask whether trainees have any questions.

Laboratory

Ask trainees to number a sheet of paper from 1 through 8. Have trainees identify each of the hazards discussed in this section and to explain how to avoid the associated dangers. Review and discuss trainees' answers.

Homework

Assign the reading of Sections 4.0.0–5.5.0 for the next session.

Materials

Ensure that you have everything required for laboratory and testing during this session.

Classroom

Review the general classes of removers and strippers.

Classroom

Discuss the characteristics of methylene chloride. Review the disadvantages of using nonflammable removers instead of methylene chloride.

Safety

Emphasize the importance of wearing appropriate personnel protective equipment when working with methylene chloride.

Classroom

Discuss effective uses for flammable removers. Explain why flammable removers should be used carefully in confined spaces.

Explain why alkali removers are effective only on oleoresinous coatings. Answer any questions trainees may have.

Safety

Explain problems that could result if an adjacent surface is exposed to alkali.

4.1.0 Nonflammable Removers

These strippers traditionally have been based on methylene chloride, a strong, nonflammable solvent being withdrawn from the market because of health and environmental concerns. Other, more environmentally acceptable solvents are being formulated to replace methylene chloride in nonflammable removers.

> **WARNING!**
> OSHA has drastically reduced the personal exposure limits for methylene chloride, which in turn requires the use of extensive personnel protection gear and ventilation equipment. Before using methylene chloride-based products, make sure that you thoroughly understand the exposure limits and equipment required by OSHA *29 CFR Parts 1910, 1915*, and *1926*.

Since methylene chloride is still available to some extent, its characteristics are worth noting. This material works well on surfaces coated with oil-based coatings, as well as on certain thermoplastic acrylics and waterborne coatings. Thick, gel-like formulations enable methylene chloride to remain in contact with a vertical surface for an extended period of time, enhancing its stripping ability. Scrubbing with a stiff bristle brush can accelerate the removal process.

After the coating has been removed, methylene chloride stripper should be washed thoroughly from the surface using water and, if necessary, detergent. While methylene chloride is not flammable, it is toxic, and appropriate personal protective equipment, such as gloves and respirators, must be used when working with it. Also, it is hazardous after use and must be disposed of in accordance with appropriate environmental regulations.

Nonflammable removers developed with replacement ingredients for methylene chloride are also effective strippers, but they generally do not work as fast, and contractors have had to adjust to the longer time required for these products to work.

4.2.0 Flammable Removers

These strippers are based on a variety of solvents or solvent blends, most commonly aliphatic, aromatic, and ester-type solvents. Ketone and other oxygenated solvents are sometimes added in small amounts to enhance the solvent action. These removers are similar to thinners, although they are designed specifically to remove coatings. They are particularly effective on vinyls, acrylics, and synthetic resin-based thermoplastic coatings.

Because they are flammable, these removers are dangerous in **confined spaces** and around electrical equipment, open sparks, or flames. Furthermore, adequate ventilation is required in the working space to prevent explosive concentrations of solvent vapors. Rags used with flammable removers must be stored or disposed of properly after use to prevent possible spontaneous combustion. Used or contaminated solvents are considered hazardous waste because of their flammability, and must also be disposed of properly in accordance with applicable environmental regulations.

4.3.0 Alkali Removers

Alkali strippers are made of varying concentrations of caustic soda (sodium hydroxide), potassium hydroxide, and other highly alkaline materials, often combined with solvents and detergents. Alkali strippers are effective only on oleoresinous coatings, and a considerable amount of time, often measured in hours, is necessary before the alkali attacks and breaks down the coating resin. However, these materials can be formulated so that they do not dry out. As a result, they are effective outdoors, where long contact times between the stripper and the coating are allowable.

> **CAUTION**
> When using these removers, extra caution is needed to protect any adjacent surface that might be subject to alkali attack.

Alkali strippers are sometimes available as a saturated cloth with a polyethylene backing that can be applied by hand as a thick layer over the coating surface to be removed. The plastic backing prevents evaporation of the removal ingredients. After a specified period of time, the stripper is removed and disposed of, and the surface is washed and neutralized. After drying and other cleaning, as appropriate, the surface can be refinished.

Alkali strippers are sometimes used to remove hazardous coatings such as those containing lead. Although the removal process is slow, expensive, and labor-intensive, it reduces or eliminates the need for containment facilities that are necessary for most mechanical means of hazardous material removal. However, even though this type of stripper removes most of the hazardous coating, a thin residual layer may remain on the surface. This residual layer must also be removed. Options include using a solvent-based remover, washing with water, blast cleaning, or power tool cleaning.

Instructor's Notes:

Mechanical cleaning may require the use of shrouded power tools or possibly even containment for collection of debris. All activities that disturb hazardous materials such as lead must be performed in accordance with all OSHA safety regulations.

> **WARNING!**
> Many chemical strippers are very strong and may give off highly hazardous and toxic vapors. Always wear appropriate personal protective gear and follow prescribed safety procedures.

5.0.0 ◆ OTHER SURFACE PREPARATION TECHNOLOGIES

Many technologies have been introduced for specialized surface cleaning and preparation. A few of these technologies are relatively new and are still being developed. Typical versions of some of these systems are briefly described in the following paragraphs.

5.1.0 Sodium-Bicarbonate Blasting Systems

These systems, or variations of these systems, were developed by various manufacturers. A version of a sodium-bicarbonate abrasive blasting system is shown in *Figure 8* and diagrammed in *Figure 9*. This particular equipment, known as the *Accustrip System*®, is available in various machine sizes (SX line) marketed by Church and Dwight Co. The machine pictured is equipped with a 3 cubic-foot pressurized abrasive tank that can hold enough media for approximately two hours of blasting before refilling is required.

These blast machines function either as dry abrasive or water-wetted abrasive blast machines. On these units, the pressure tank is operated at a slightly higher pressure than the blast hose. The 2- to 4-pound pressure differential between the tank and the blast hose is used to help media flow through the media valve.

The blast system illustrated is specifically designed to use a specially formulated, free-flowing, sodium-bicarbonate abrasive called

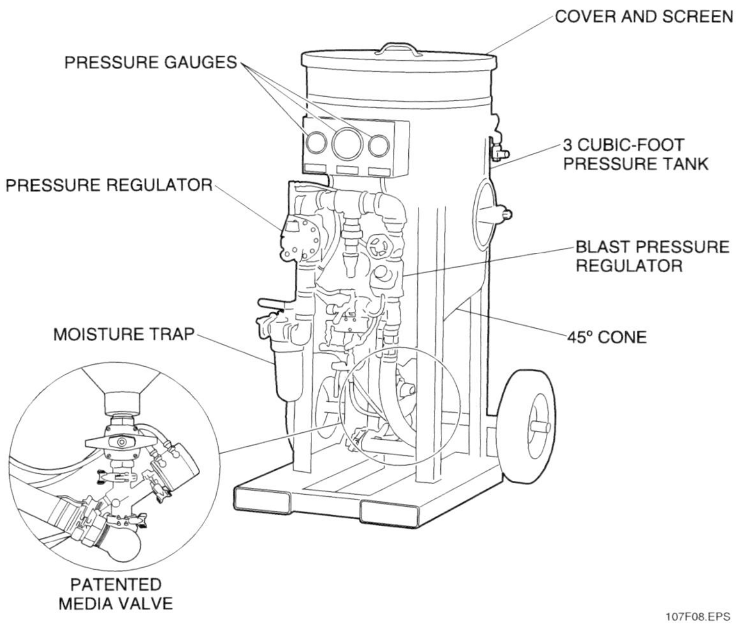

Figure 8 ◆ Version of a sodium-bicarbonate abrasive blasting system.

Safety

Ask a trainee to explain the safety precautions that should be taken when working with chemical strippers.

Laboratory

On the whiteboard/chalkboard, list the coatings that trainees may encounter when working on surfaces. Ask trainees to review the list and determine which remover or stripper would be most effective to remove the coatings.

Audiovisual

Show Transparency 5 (Figure 8). Explain how a sodium-bicarbonate blasting system is used for specialized surface cleaning.

Audiovisual

Show Transparency 6 (Figure 9). Review the diagram of a sodium-bicarbonate abrasive blasting system.

Classroom

Discuss the blast media that can be used with sodium-bicarbonate abrasive blasting systems.

Audiovisual

Show Transparency 7 (Figure 10). Review the types of blast media for which water-injected nozzles are used.

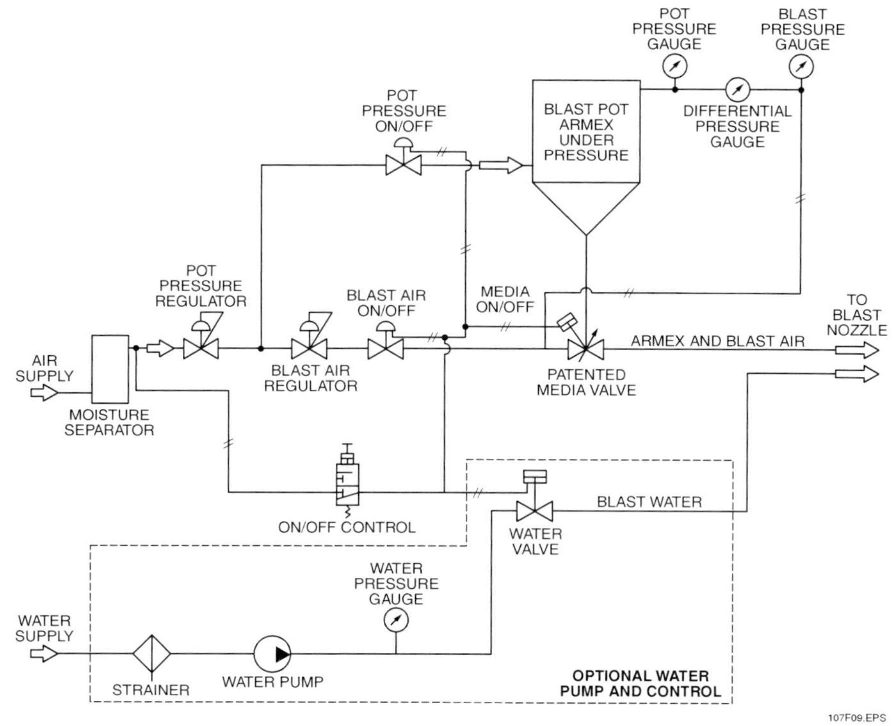

Figure 9 ◈ Diagrammed sodium-bicarbonate abrasive blasting system.

Armex® Blast Media. This nonrecyclable media can be delivered through specially designed, water-injected nozzles (see *Figure 10*) which are used for dust suppression and are available in a number of styles and sizes. These systems operate at blasting pressures ranging from 35 to 120 pounds per square inch (psi) (241 to 827 kiloPascals [kPa]) with air consumption ranging from 31 to 478 cubic feet per minute (CFM).

Blast media are available in a variety of grit sizes and formulations that enable blast cleaning from brush-off to white-metal quality cleaning. The media is soft and nonetching (except on very soft substrates) and is considered nontoxic. It is effective for removing paint, fats, grease, oil, various stains, rust, and other contaminants. As well, it will not damage seals, bearings, and other moving parts. When used on steel, formulations that include 20% to 30% hard abrasive for profiling purposes are available. The residue of these products will act as somewhat of a rust inhibitor until rinsed off. After rinsing, surfaces should be checked for proper pH with litmus paper. Subsequent coatings used should be determined to be compatible with sodium bicarbonate.

Since it is a water-wetted, soft, nontoxic media, operator protective gear is less stringent than with conventional blasting systems, unless a toxic coating is being removed or the formulations containing a hard abrasive are used. Usually, only NIOSH-approved goggles, a dust/mist mask, and hearing protection along with some form of waterproof or water-repellent clothing are required. When the job is complete, water-flussh cleanup using available sewers and drains is generally all that is required (unless toxic coatings or hard abrasives are involved) along with some insoluble residue cleanup which can usually be land-filled. The medium is relatively expensive, but its consumption rate is only about 100 pounds per hour at most.

Instructor's Notes:

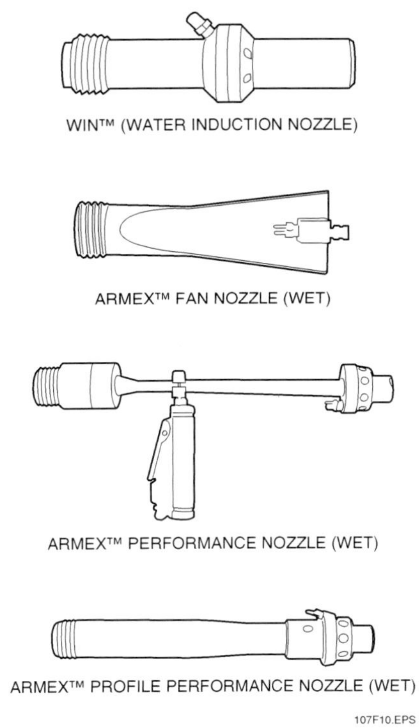

Figure 10 ◆ Water-injected nozzles.

5.2.0 Sponge-Jet® Blasting System

This very specialized blasting system is manufactured by Sponge-Jet, Inc. When dust control is a major issue and either conventional or wet blasting is restricted and the only approach seems to be hand-finishing, this process may be an effective solution. It can be useful in reducing airborne particles in situations involving toxic coating removal such as asbestos when abatement procedures must be used.

Figure 11 shows a Sponge-Jet® blasting system. It is a dry-process system that uses particles of a patented, synthetic, open-cell polymer (urethane) sponge media that may be chemically bonded with various grit materials. The following media are available from Sponge Jet, Inc. without an abrasive or with various abrasives for different degrees of cleaning or profiling:

- Sponge only (cleaning of oil or grease)
- Sponge with **bonded** plastic chips (non-profiling)
- Sponge with bonded Starblast® (up to 1-mil profile)
- Sponge with bonded garnet (2- or 3-mil profile)
- Sponge with bonded aluminum oxide (2- or 3-mil profile)
- Sponge with bonded steel grit (2- to 3-mil profile)

The lower blast pressures used in this system, plus the cushioning provided by the sponge material, help prevent breakdown of both the abrasive and the sponge material.

When the sponge particles impact a surface under air pressure, they flatten out slightly, dissipating the force of the blast pressure. Then they contract and rebound with very little force. As the particles flatten out, the abrasive bonded to the sponge impacts and scrubs the surface. As the sponge particles contract, the abrasive again scrubs the surface and the open cells capture much of the surface contaminant or coating dislodged by the impact or the scrubbing action of the abrasive.

This micro-containment process reduces airborne dust. Compared to dry sand blasting, in which most of the abrasive disintegrates, the dust levels can be reduced by as much as 94%. Since the sponge media captures much of the dust and there is very little media rebound force, operator protective gear can be minimal (except for toxic coating removal). Under normal conditions (non-toxic coating removal), all that may be required are NIOSH-approved goggles, a dust mask, and hearing protection, along with normal-weight work clothes.

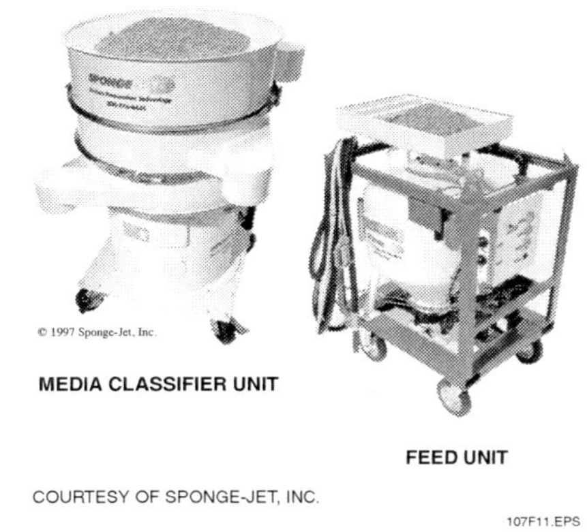

Figure 11 ◆ Sponge-Jet® blasting system.

Classroom

Refer to Figure 11. Discuss the media that may be used with a Sponge-Jet® blasting system.

Explain how the Sponge-Jet® system operates. Go over any questions trainees may have.

Classroom

Review the special features of the pressurized tank that is used in the Sponge-Jet® system.

Explain how carbon dioxide blasting systems are used to remove surface coatings.

The various types of sponge media available for the system can be used to attain anything from a brush-off to white-metal blast-cleaned surface with no profile or profiles up to 3 mil. While the media is relatively expensive, the use of the media classifier unit will allow the media to be cleaned and reused from five to six times, depending on the application. It should be noted, however, that at the completion of any project involving toxic coating removal, any dust and debris as well as any leftover used media normally has to be discarded as hazardous waste.

Even though the blast machine is a pressurized tank system, its special features require some explanation. The feed unit contains a hopper with a pneumatic vibrator that is used to load either a 33-gallon or a 93-gallon pressurized abrasive tank, depending on the size of the unit. Once the tank is pressurized by activation of the dead-man switch at the nozzle handle, a pneumatic-powered actuator inside the tank begins to rotate 90 degrees back and forth to ensure that the media does not bridge or clog within the tank. At the same time, media at the bottom of the tank is transferred by a pneumatic-powered, variable-speed auger unit in precise amounts to the blast hose air stream. The auger unit performs the same function as an adjustable metering valve in a conventional, pressurized, gravity-fed blast machine.

The media classifier unit is an adjustable-speed, electric-powered, media-recycling device normally located next to the feed unit. This unit removes spent undersized sponge, debris, dust, contaminants, and fines from expended sponge particles and allows them to be returned (recycled) to the feed unit. During normal blasting operations, expended sponge media is collected periodically by hand or by vacuum equipment from simple 6-mil plastic containment used on floors with or without an enclosure. The expended media is then transferred to a hopper at the top of the classifier. An adjustable-speed, two-stage, rotary-actuated, vibratory screen system is used to separate and clean the sponge particles. The upper first-stage screen with large holes separates oversize debris and ejects it out the top port. The media, along with spent undersized media, dust, contaminants, and fines, falls through to the lower second-stage screen. This lower screen, equipped with small holes, vibrates and bounces the media to clean it of dust, fines, and contaminants as well as to separate it from spent undersized media. The cleaned, correctly sized media is ejected from the middle port and manually returned to the feed unit for reuse. All spent undersized media, dust, contaminants, and fines that passed through the second-stage screen are ejected as waste from the lower port. Lowering the speed of the unit increases the cleaning ability somewhat and lowers the output of the unit. Increasing the speed increases the output but lowers the cleaning ability.

An optional washer unit is also available to process sponge media contaminated with grease, oils, inks, etc. This unit is a portable, electric-powered, adjustable-speed, closed-cycle device that centrifugally launders, spin dries, and unloads the media. On-board electric heaters are used to preheat and maintain a water/detergent solution to temperatures up to 170°F. During use, the wash water is collected, filtered, and reused within the washer unit.

After the water is heated, the washer is started and the contaminated media is loaded into a rotating wash/spin drum through a hopper equipped with a vibrator. Up to 300 lbs. of media can be loaded at one time into the drum. After the adjustable automatic wash cycle is complete, an adjustable automatic high-speed spin cycle is used to dry the media. When the spin cycle is complete, the drum rotation is reversed by the operator to unload the cleaned media from the drum for reuse in the feed unit.

5.3.0 Carbon Dioxide (CO_2) Blasting Systems

In recent years, a nonabrasive cryogenic cleaning and stripping system has been developed. This system uses liquid carbon dioxide to form nonabrasive dry-ice (frozen CO_2) pellets about the size of a grain of rice. The pelletized dry ice is propelled by high-pressure air (100 to 300 psi [690 to 2,069 kPa]) through blast hoses to special nozzles. Sprayed at high velocities (600 to 1,000 ft./sec.) against the surface of a substrate, the extremely cold (–108°F), pelletized dry ice, which disintegrates on impact, quickly draws heat from the surface coatings and/or contaminates and causes their temperature to drop far below zero in relation to the temperature of the substrate. This thermal shock, which causes the surface coatings/contaminants to quickly contract in relation to the substrate, literally fractures the bond between the surface coating/contaminants and the substrate. In addition, the very low temperature causes the surface coating/contaminants to become very brittle. The concentration and velocity of the disintegrating pellets breaks and then scrubs the residue of frozen surface coatings and/or contaminants off the substrate. Heat absorbed by the disintegrating dry ice causes the dry ice to quickly sublimate (evaporate) directly to a gaseous state, leaving only the freeze-dried

Instructor's Notes:

residue of the surface coatings/contaminants removed from the substrate to clean up. Because carbon dioxide is nonflammable, nonconductive, nontoxic, and does not create a secondary waste stream, the FDA, USDA, and EPA have approved it for a variety of uses in the electrical, manufacturing, and petro-chemical industries.

Carbon dioxide blasting systems are primarily used for maintenance purposes and are particularly effective on coatings and contaminants that react quickly to a thermal difference between themselves and a substrate. Coatings and contaminants can be very effectively removed by the systems. However, laminate coatings such as FRP or graphite/epoxy cannot be removed by carbon dioxide blasting. With the exception of very thin or dead-soft aluminum, no peening, stress cracking, profiling, or damage to a substrate occurs.

Figure 12 is an illustration of a carbon dioxide blasting system manufactured by Cold Jet, Inc. This system consists of a dry ice pelletizer unit, a delivery unit, and special nozzles. The units are equipped with programmable microprocessors that allow precise control of pellet density, flow velocities, thermal temperature, and air pressure for different tasks. In addition to the system, a 300 psi (2,069 kPa), 350 CFM source of compressed air plus large storage tank(s) or tank truck(s) for the liquid carbon dioxide are required. At a minimum, an 18 kilowatts (kW), three-phase, 480 VAC power generator or power source is also required. The pelletizer and delivery units can be equipped with large casters or skids, or they can be set up inside a large cargo trailer for mobile use.

The largest mobile pelletizer unit available from Cold Jet, Inc. can produce 165 to 1,500 lbs. of dry ice pellets per hour while consuming 360 to 2,900 lbs. of liquid carbon dioxide per hour. The pellets, ranging in diameter from 0.125" to 0.438", are formed using interchangeable die plates. The unit, which weighs 3,500 lbs. dry, can be operated independently or in coordination with a companion

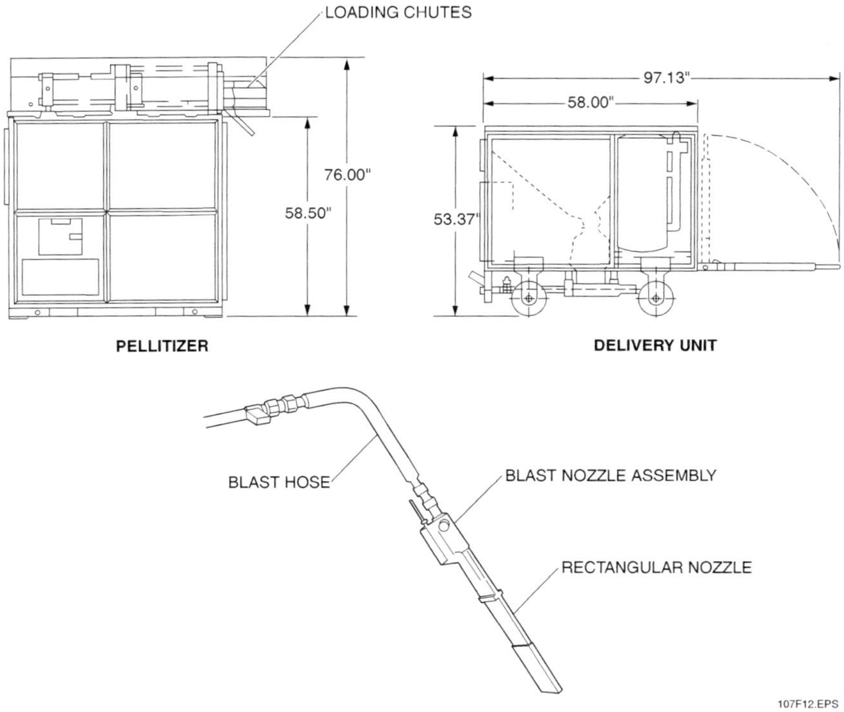

Figure 12 ◈ Carbon dioxide blasting system.

Classroom

Discuss suitable applications for carbon dioxide blasting systems.

Audiovisual

Show Transparency 8 (Figure 12). Review the components of a carbon dioxide blasting system.

Classroom

Review the various blast nozzles that can be used with a carbon dioxide blasting system.

Audiovisual

Show Transparency 9 (Figure 13). Review the three actions that are used when cleaning with ice blasting systems.

Classroom

Review the contaminants or coatings for which laser ablation systems are used.

delivery unit. Dry ice pellets are supplied to the hopper of the delivery unit by twin pellet-loading chutes on top of the pelletizer unit. The unit requires 30 amperes (A) of three-phase, 480 VAC power for operation.

The delivery unit, which can be located up to 200' from the work area, weighs 2,250 lbs. dry. Its hopper can hold up to 640 lbs. of dry ice pellets. An optional internal dryer for the incoming blast air can add another 1,000 lbs. to the weight of the unit. Using a special feed device designed for low-temperature operation, this unit can deliver up to 1,000 lbs. (453 kg) of dry ice pellets per hour through standard blast hoses to the special blast nozzle assembly at air pressures up to 300 psi (2,069 kPa) and with air flow rates up to 350 CFM. The unit requires 15 amperes (A) of single-phase, 120 VAC power for operation.

The special blast nozzle assembly uses various sizes of interchangeable, rectangular nozzles for different applications. The rectangular nozzles create an even spray pattern and eliminate the hot spot normally created by straight-bore round nozzles.

Because the blasting operation is very noisy, adequate hearing protection must be worn by the operators. Unless toxic coatings are being removed, normal dust filtering respirators and goggles can be used if the blasting is in the open. In containment or confined spaces, monitoring is required and extra ventilation may be required to keep the carbon dioxide levels suitable for breathing; otherwise, air-supplied blasting helmets must be used.

5.4.0 Ice Blasting Systems

These specialized systems are primarily nonabrasive, nondetergent cleaning systems used to remove surface contaminants from a sound surface without damage to the surface. When used after water-soluble softeners or strippers have been sprayed onto a surface, they can effectively remove certain coatings from a substrate without damaging it. Cleaning with ice blasting is based on three actions:

- Ice particles striking the surface at high velocity displace contamination before a phase change (melting) to water.
- The blast air causes the ice particles to scrub the surface during the phase change to water.
- The water from the melted ice particles along with the blast air rinses and carries away the surface contaminants.

These systems create a very limited secondary waste stream; typically, only about two to seven gallons of water per hour are produced. They are dustless, therefore minimal containment is needed for hazardous materials. They require no separate media source because the ice particles are generated as required from a water supply. If necessary, the units can be operated with a siphon injection unit at the blast nozzle for inhibitor application to reduce flash rusting.

Figure 13 is an illustration of an ice blasting system manufactured by Universal Ice Blast, Inc. This unit is equipped with casters for mobility and can continuously generate up to 300 lbs. of ice particles per hour. The system weighs 1,400 lbs. and requires inputs of 0.4 U.S. gallon of water per minute and 300 CFM of air at 120 psi (827 kPa) from outside sources. In addition, a 15 kW, three-phase, 480 VAC external power source or generator is required for the internal, air-cooled, R404 refrigeration system used to produce the ice particles.

The special lightweight, low-thrust blast nozzle uses blast air to siphon the ice particles through an insulated hose to the nozzle where the particles are combined with the blast air and applied to a surface. When blasting is temporarily stopped, ice is still siphoned through the nozzle by a stream of idling air because ice is continuously produced by the unit when operating.

Unless toxic materials are being removed or blasting is taking place within a confined space, waterproof suits with soft-head shrouds and dust respirators, along with face shields, are all that is usually required for personal protection. Otherwise, air-supplied blasting helmets must be used.

5.5.0 Laser Ablation Systems

These very sophisticated and specialized systems are primarily used to remove precise layers of contaminants or coatings from a surface. This technology uses extremely short repetitive pulses of intense monochromatic light from a pulsed laser (light amplification by stimulated emission

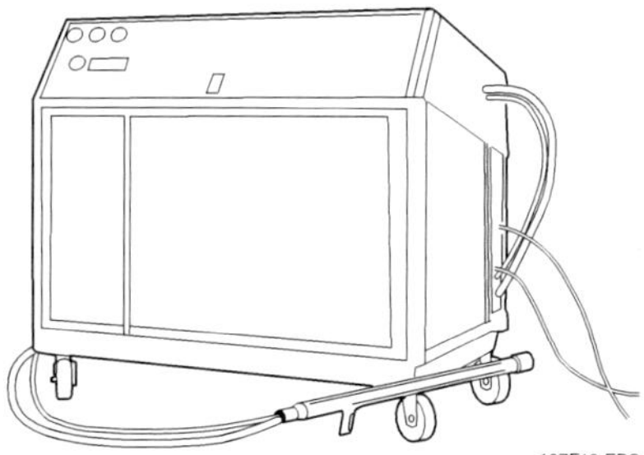

Figure 13 ◆ Ice blasting system.

Instructor's Notes:

of radiation) device to instantly heat and vaporize (ablate) coatings or contaminants. Thin layers at a time can be instantly transformed from a solid to a gas-and-particulate mixture which literally explodes off the layer or substrate below. The laser pulse durations are so short and the beam scanning dwell times can be so closely regulated that the heat used for vaporizing the target coating/contaminant layer is not able to propagate inward to the next layer or the substrate.

Laser ablation is a noncontact technology. No chemicals or blasting media are required and no secondary waste streams are generated. Many traditional and leading-edge coatings (organic and inorganic) are reduced to their fundamental molecular components. Organic coatings are typically reduced to pigment particulate, water vapor, carbon dioxide, and trace amounts of other gases. For hydrocarbon-based coatings, the waste volume is reduced by up to 80%. This minimal waste volume is a particular benefit when removing hazardous coating where the cost of disposal is expensive. These systems can be used to remove rust, adhesives, and other hazardous material contaminants on substrates.

Figure 14 is an illustration of a mobile, pulsed, carbon-dioxide laser cleaning system manufactured by F2 Associates, Inc. These systems are also available as self-contained, either fixed or mobile systems mounted on robotic platforms for remote-control use in hazardous areas. The system pictured is mounted on casters and equipped with an extendable, hard-optics beam-delivery device that consists of an overhead support arm and jointed, rigid, transmission tubes that conduct the laser beam to a hand-held beam-delivery device. The beam-delivery device is equipped with a vacuum system and comes with a set of shrouds to conform to various surface geometries. The vacuum system, equipped with a HEPA filter, removes the gases and particulate matter being stripped by the laser beam. The beam-delivery device is equipped with stand-off safety sensors which stop the laser beam if the delivery device is more than a specified distance from a surface. Decoating parameters such as scan speed, or pulses-per-second, and energy-per-pulse are manually inputted to the carbon-dioxide laser power and control unit. In some cases, especially the robotic units, real-time spectral sensors can also be incorporated in the delivery device to provide process feedback information to control the laser scan and pulse rate for selective layer decoating or decontamination and to maximize cleaning efficiency.

6.0.0 ◆ COATING APPLICATION

There are a number of methods of applying coating after the surface has been properly prepared. The application method selected depends on the type of coating, the environment in which the work will take place, the accessibility of the surface, and the proximity of other structures. The more common methods of application are brush, roller, conventional spray system, and airless spray system.

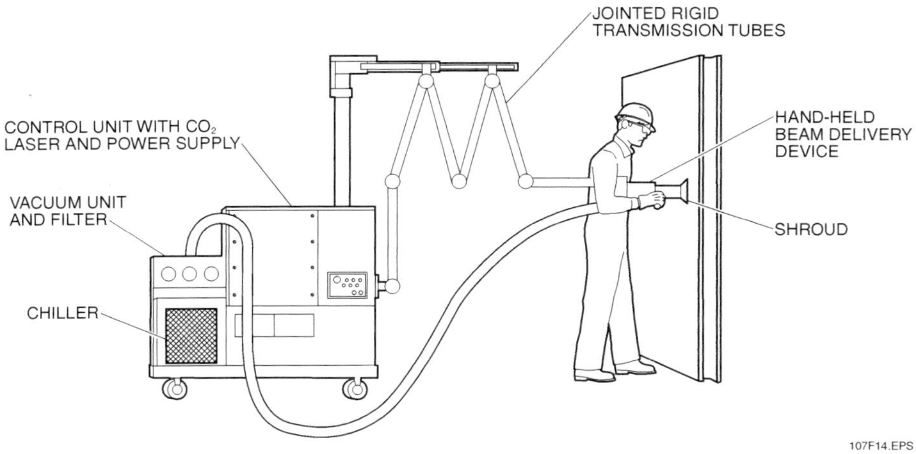

Figure 14 ◆ Mobile, pulsed, carbon-dioxide laser cleaning system.

Audiovisual

Show Transparency 10 (Figure 14). Explain how a carbon-dioxide laser cleaning system operates. Ask whether trainees have any questions.

Teaching Tip

See the Teaching Tip for Section 5.0.0 at the end of this module.

Performance Testing

Have each trainee perform Performance Tasks 1 and 2 to your satisfaction. Fill out a Performance Profile Sheet for each trainee.

Homework

Have trainees review Sections 6.0.0–6.4.0 and answer the Review Questions for the next session.

Materials

Ensure that you have everything required for the laboratories and testing during this session.

Classroom

Review techniques that are used when applying coatings with a brush.

Explain how to select the appropriate brush for the coating materials that are being applied.

Laboratory

Set up several stations with a variety of brushes, containers filled with coating, and practice surfaces. Ask trainees to practice cutting in, stripe coating, and feathering techniques. Observe and assess trainees' work.

6.1.0 Applying Coatings with a Brush

Applying coatings with a brush is typically done when covering small areas, irregular surfaces, or when cutting in or stripe coating. *Cutting in* is the term used for covering inside corners and around trim edges with a brush or trim roller before applying the coating to a flat or large, open surface with a roller. It also means leaving a clean edge, such as needed at an adjacent surface.

Stripe coating is usually applied to welded areas or welded edges. It is accomplished by applying a coat of **primer** to the weld area or edge that extends out from the weld area or edge by 1" or more. Stripe coating is done before the first coat of primer is applied and results in a thicker overall coating over welded areas and edges.

Figure 15 ◆ Dipping bristles no more than halfway.

Coating begins by choosing the correct brush for the coating material being used. Use natural bristles for oil-based coatings and synthetic bristles for water-based coatings. As a rule, choose the largest brush that is suitable for the job. Typically, a 4" to 6" brush is used for large surfaces. Make sure the brush is free of dust, dirt, loose bristles, or other particles that may show up on the surface.

When working with a brush, work from a partially filled container (bucket or pot). When dipping the brush into the container, the bristles should be dipped no more than halfway (see *Figure 15*). Dipping deeper than halfway can cause coating to build up in the heel of the brush, and it also increases the chances of dripping and splattering. To set the coating in the brush and remove the excess coating, gently tap or slap one side of the brush against the inside of the container. Do not remove excess coating by wiping the bristles against the rim of the container because this removes too much coating from the brush and may damage some types of bristles.

The method for holding the brush is one of personal preference. Obviously, the work is easier if the brush is held comfortably. *Figure 16* shows three common methods. Apply the coating with two or three light strokes, then spread it smoothly and evenly using long, even, parallel strokes. Whenever possible, use the tip and flat sides of the brush and not the side edges. Use just enough pressure to bend the bristles slightly.

Finish off the leading edge of the coating by gradually pulling the bristles of the brush off the painted surface (see *Figure 17*). This is called feathering. Whenever possible, use vertical strokes because they are less tiring. On a vertical surface, work from top to bottom, and do not stop until you reach a natural stopping point such as trim or a joint. As you work, keep a wet edge. This means

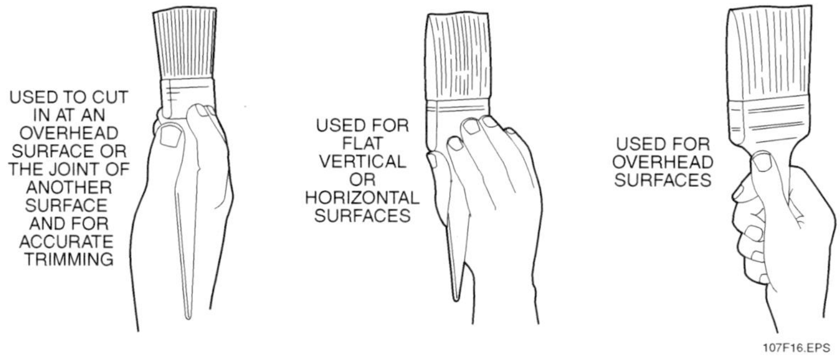

Figure 16 ◆ Common methods of holding the brush.

Instructor's Notes:

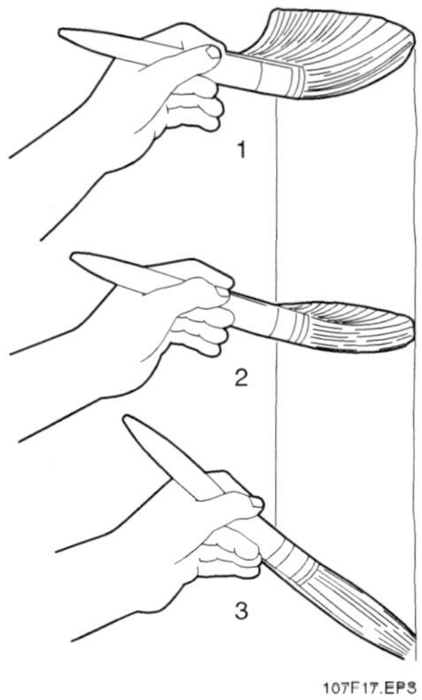

Figure 17 ◆ Feathering.

working from an uncoated area into a previously coated, still-wet area. To avoid lap marks, feather the edges. When the surface is covered and while the coating is still wet, the final step is to lay it off, which means to go over the entire section you have just coated using light, quick strokes all in one direction in order to smooth brush marks, laps, and joints, and to cover any thin or missed spots.

> **NOTE**
> Glossy coatings show brush marks more than flat coatings. When using glossy coatings, work in smaller areas near the wet edge and lay it off frequently.

6.2.0 Applying a Coating with a Roller

Rolling is typically done when coating large, flat surfaces such as tanks. These surfaces can be coated much faster with a roller than with a brush. As previously mentioned, cutting in corners and/or trim must be done with a brush or special-purpose roller before beginning to roll the surface. Cutting in should be done as the rolling progresses in order to maintain a wet edge to prevent lapping. Because rolling can be done faster than coating with a brush or trim roller, it is common for two or more workers to do the cutting in while a third worker is doing the rolling.

The rolling procedure begins by following the manufacturer's recommendations to choose the correct roller cover for the type of coating being used and the correct nap for the type of surface being coated. Make sure the roller cover is free of loose nap, dust, dirt, and any other particles that may show up on the surface. Before use, a new roller cover should always be rubbed down by wringing the cover in your bare hands vigorously to remove all loose nap. The roller, paint tray or pail, and roller grid should also be checked and wiped clean to prevent any dirt, sand, dirty solvent, or previously used coating from mixing with the new coating.

When using a dip-type roller, fill the tray with the coating material until the well is full and about one-third of the ramp is covered. Do not overfill the tray. Dip but do not submerge the roller into the coating, then roll it back and forth over the roughened surface of the ramp several times until the cover is well coated. If using a five-gallon pail with a roller grid or mesh screen, fill the pail about halfway. Dip the roller into the coating, and roll it back and forth over the roller grid or mesh screen until the cover is well coated. When done, the roller cover should be full of coating material, but not dripping. If using a pressure-fed roller, fill it with the coating and prepare it for use according to the manufacturer's instructions. A useful way to break in and wet the roller cover fibers is to roll out the first load of coating onto craft paper, scrap drywall, or another disposable surface. This also helps to remove any air bubbles that may be trapped in the roller cover.

Apply the coating from a fully loaded roller to the surface being coated in a zigzag pattern like the letter N or M (see *Figure 18*). The first stroke should be in the upward or forward direction. This helps to prevent dripping and splattering. Be careful not to roll too fast because this also helps to avoid splattering. Use an even pressure on the roller to achieve an even coating. Without lifting the roller from the surface, cross roll to spread the coating. Then finish with light roller strokes all in the same direction but at a right angle to the cross roll. As the roller nears the end of the area being coating, feather the coating by slowly pulling the spinning roller off the surface.

When coating a surface, work from the wet edge to prevent lapping. For example, when coating an 8' high surface, a right-handed painter should work from left to right, keeping the leading edge of the coating wet by working from top to bottom across the surface. Do not stop until reaching a natural break, such as a joint.

Classroom

Discuss the advantages of applying coating with a roller.

Review the procedures to accurately apply coating with a roller. Explain how to apply the coating to the roller.

Laboratory

Set up several stations with a variety of rollers, trays filled with coating, and practice surfaces. Ask trainees to practice loading the roller, applying coating in a zigzag pattern, and achieving an even coating. Observe and assess trainees' work.

Classroom

Emphasize the importance of following the manufacturer's guidelines when operating and maintaining conventional spray systems. Go over any questions trainees may have.

Audiovisual

Show Transparency 11 (Figure 19). Explain how to properly use a respirator.

6.3.0 Spraying with a Conventional Spray System

There are many different kinds of conventional spray equipment and spray system configurations. Regardless of the type, all should be operated and maintained as directed in the manufacturer's operation and service manuals. General guidelines that apply to the operation and maintenance of most conventional spray systems are described in the following paragraphs.

6.3.1 Personal Protective Equipment/Safety Guidelines

When using any type of spray system, always use the proper respirator (see *Figure 19*). Make sure that the respirator is approved by NIOSH and certified for use with the material you are spraying. Also, before attempting any spraying, ensure that the respirator has a good face seal by performing both positive and negative fit checks. Use of protective hand cream and/or gloves and eye protec-

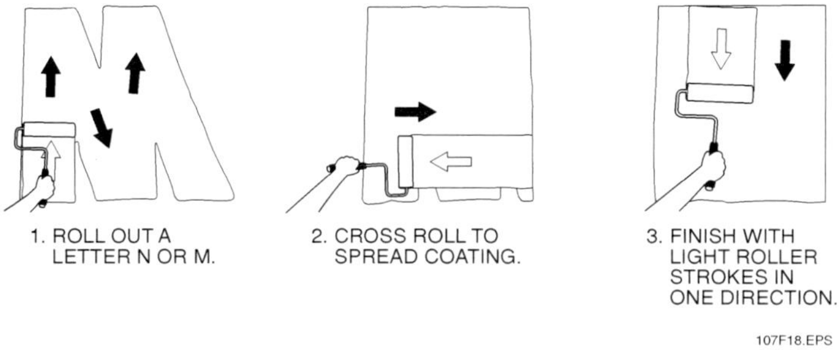

Figure 18 ◈ Zigzag pattern.

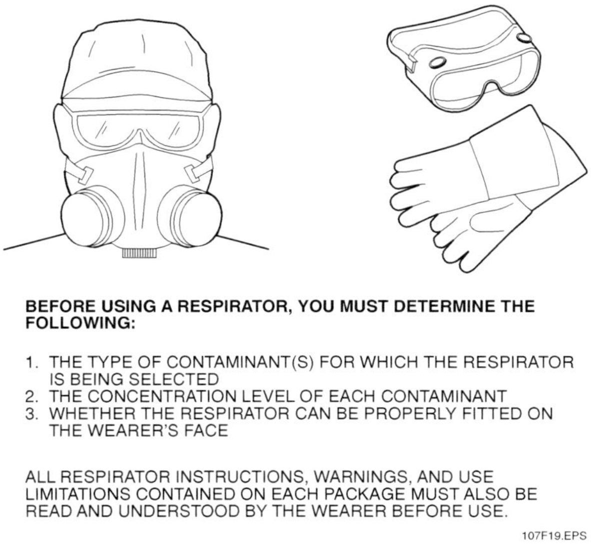

Figure 19 ◈ Proper respirator.

Instructor's Notes:

tion is also important. Goggles offer good protection because of their wrap-around design. Because overspray will obscure your goggles while spraying, you should consider keeping one or two extra pairs of goggles handy to exchange as necessary. Clean them often using non-abrasive cleaner. If in doubt about which respirator, gloves, or face/eye protection to use, refer to the Material Safety Data Sheet for the material being sprayed.

Whenever operating or maintaining a conventional spray system, comply with the following general safety guidelines. The following information applies to spray equipment.

- Ensure that there is adequate ventilation before starting. Mechanical ventilation must be provided when spraying in enclosed areas. Ventilation equipment must be located so that it will not circulate contaminated air back into the spraying area.
- The spray area must be free of open flames, sparks, and other ignition sources.
- Locate appropriate fire extinguishers in or near the spraying area. Make sure all personnel know where they are located.
- Use explosion-proof portable lamps when required. Check with your supervisor.
- Appropriate warning signs must be posted in hazardous areas.
- The spray area must be free of combustible materials and residue.
- OSHA requires that an operator's instruction manual be at the job site whenever spray equipment is in use. The manual should be available from your supervisor.
- Check the equipment carefully before use. Ensure that the electrical cords, plugs, air and fluid hoses, couplings, and spray tips are in good condition. Remove any damaged equipment from service.
- Check the spray equipment's safety shut-off valve and pressure gauge(s) to be sure that they are in good condition prior to each use. Equipment with a damaged or missing safety shut-off valve or pressure gauge must be removed from service.
- Close the safety shut-off valve when spraying operations are not being done.
- Do not exceed the equipment manufacturer's recommended safe working pressure. If the equipment is faulty and the safe working pressure cannot be maintained, immediately remove it from service for repairs or replacement.
- Before repairing or servicing the equipment, disconnect the power source and completely relieve the pressure from the system.
- Clean up overspray because it can be flammable.

WARNING!
Halogenated hydrocarbon solvents, or fluids containing such solvents, can cause an explosion when used in a pressurized spray system containing aluminum or galvanized components. Cleaning agents, coatings, and paints may contain halogenated hydrocarbon solvents. If in doubt about the material you intend to use, always check the MSDS.

Adhere to the following guidelines when using compressors:

- Never operate a compressor unless it is equipped with a pressure relief valve and pressure gauges.
- Operate and lubricate compressors in accordance with manufacturers' recommendations.
- Inspect all safety devices to ensure that they are in proper working order before using the compressor.
- Before any repair work is done on the pressure system of the compressor, bleed off the pressure and lock out the system.
- Obey all signs posted on compressors. They are there to warn you of important safety hazards, including the automatic starting feature of the compressor, if so equipped.
- Do not aim compressor air at yourself or toward others.
- Use safety cables, also known as whip checks, on couplings of high-pressure hose lines where a connection failure could occur.
- The compressed air pressure of spray pots must be checked and must not exceed the safe working pressure of the container.

Consider the following when using gasoline engine-driven compressors.

- Shut the compressor down while filling the fuel tank. Do not refuel a hot engine.
- Always ground the fuel nozzle against the filler neck to avoid sparks.
- Handle all flammable material according to the procedures on the container.
- Refueling operations should be done outdoors. If you must refuel the compressor inside, make sure that the area is well ventilated.
- Never smoke while handling fuel. The fumes in an empty fuel container are explosive.
- Avoid spilling fuel. If a spill occurs, clean it up immediately.

Safety

Review the safety considerations for using spray equipment.

Discuss the hazards associated with halogenated hydrocarbon solvents.

Classroom

Review the guidelines for using compressors and gasoline engine-driven compressors.

Classroom

Explain how to select the appropriate spray gun nozzles, tips, and valves to achieve required results.

Refer to Figure 20. Review the various fan spray patterns.

Audiovisual

Show Transparency 12 (Figure 21). Explain how airless spray systems control the flow of fluid.

Show Transparency 13 (Figure 22). Review the basic components of an airless spray system and explain how it functions.

Classroom

Discuss suitable applications for airless spray systems.

6.3.2 Selecting the Conventional Spray Gun Nozzle Combination And Fan Spray

To obtain the required results, and to achieve the desired fan pattern and application rate, the spray gun air cap, fluid tip, and needle valve (needle) must be correctly selected for use with the material being sprayed. Because they all affect the spray pattern, the air cap, fluid tip, and needle must be selected as a matched set, commonly called the nozzle combination. Generally, the larger the fluid tip orifice, the more paint is discharged and the faster the application. However, if the fluid tip orifice is too large, control over the material discharge can be lost. If the fluid tip orifice is too small, the discharge velocity will be too high. This means that the air coming from the air cap will not atomize the paint properly and the pattern will be distorted.

To select the proper nozzle combination, refer to the gun manufacturer's specifications for the particular spray gun being used and choose a nozzle combination recommended by the manufacturer for use with the type of material being sprayed. Most coating manufacturers also give cap/tip recommendations for using each of their products. For example, to conventionally spray almost all interior and exterior latex and alkyd coatings, one coating manufacturer recommends using a general-purpose cap/tip combination that has an orifice inside diameter (ID) of 0.042" (flow rate up to 20 oz./min.) and a typical pattern size of 10" to 12".

Once the correct cap/tip combination is installed in the spray gun, the fan spray is adjusted using the pattern control on the gun. Typically, turning it clockwise until it is fully closed produces a round spray; turning it counterclockwise widens the spray into a fan shape. The fan spray pattern can be oriented horizontally, vertically (see *Figure 20*), or anywhere through 360° by loosening the spray gun retaining ring and rotating the nozzle or air cap to the required position.

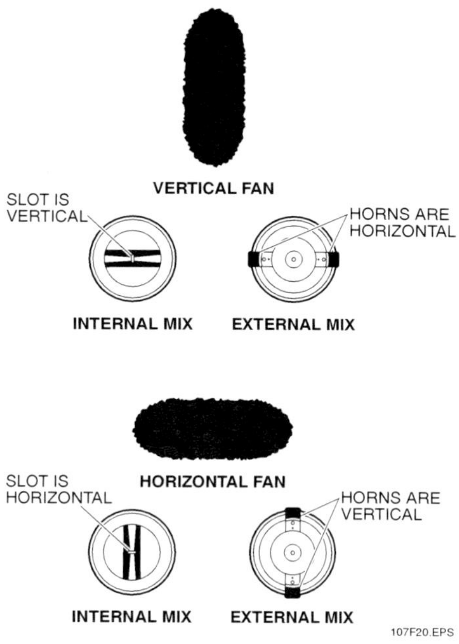

Figure 20 ◈ Fan spray patterns.

6.4.0 Airless Spray Systems

Airless spray systems are so named because air is not used to atomize the paint as with conventional air spray systems. Instead, fluid is pumped under high pressure through a small, specially shaped opening or hole, or orifice, in the spray gun tip (see *Figure 21*). The tip controls the flow of the fluid and creates the spray pattern. As the fluid passes through the orifice at a high pressure (typically 2,000 to 3,000 psi), it separates into small droplets, resulting in a very fine spray mist. The

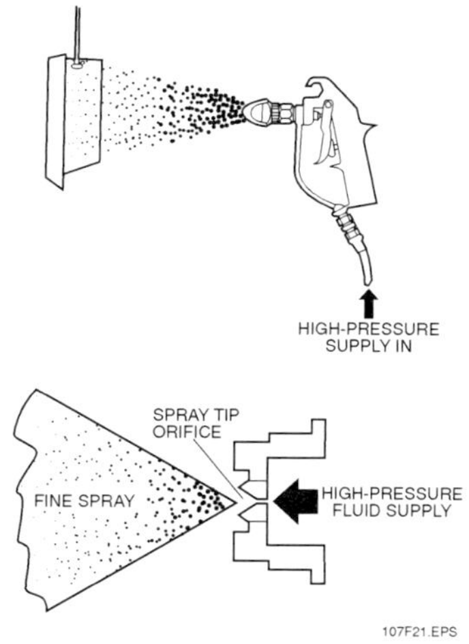

Figure 21 ◈ Spray tip orifice.

Instructor's Notes:

process is similar to the way a garden hose works. The velocity with which the spray leaves the nozzle gives it enough momentum to reach the surface. The higher the viscosity of the coating material, the higher the pressure required to atomize it.

Airless systems (especially the heavy-duty, air-powered units) can be used to spray almost any type of coating. There are many types of airless spray systems. However, all consist of the same basic components (see *Figure 22*) including a spray gun, fluid pump and pump power source, and fluid hose, couplings, and fittings.

The heavy-duty units can readily support multiple guns operating simultaneously. Airless systems can provide complete and uniform coverage because they apply only the coating material, not air, to the surface. They can coat surfaces evenly because there is little air turbulence to deflect the coating material or cause it to rebound. Airless spraying is best suited for jobs where medium to large areas must be covered quickly. Its strong points are its simplicity and speed. Airless systems can achieve up to 50% more coverage per day on large, regular surfaces than conventional or HVLP air spray systems. Because there is little overspray, typical airless systems have a transfer efficiency that ranges between 50% and 60%. The disadvantages associated with airless spray systems include safety hazards due to its high operating pressures, a greater chance of clogging problems, moderate overspray, and limited methods of controlling the spray pattern.

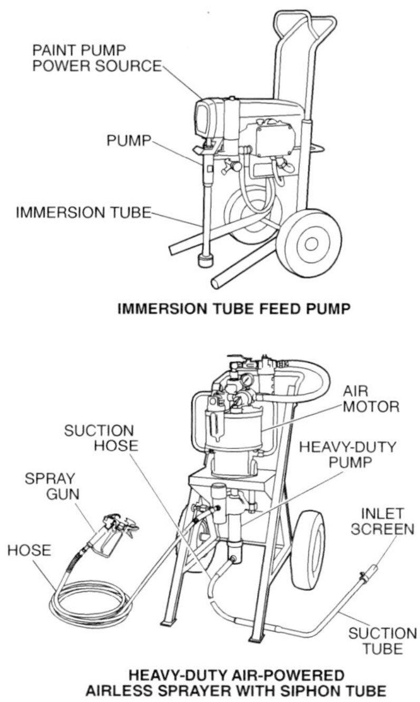

Figure 22 ◈ Airless spray system components.

Teaching Tip

See the Teaching Tip for Sections 6.3.0–6.4.0 at the end of this module.

Classroom

Go over the Review Questions for this module. Answer any questions trainees may have.

Examination

Administer the Module Examination. Be sure to record the results of the Exam on Craft Training Report Form 200 and submit the results to the Training Program Sponsor.

Performance Testing

Have each trainee perform Performance Task 3 to your satisfaction. Fill out a Performance Profile Sheet for each trainee.

Ensure that all Performance Tests have been completed and Performance Profile Sheets for each trainee are filled out. Be sure to record the results of the testing on Craft Training Report Form 200 and submit the results to the Training Program Sponsor.

Summary

Coatings must be in good condition to protect the buried or submerged pipeline, and it is essential to maintain the coatings. Pipeline individuals must be able to select the proper coating for piping, whether the piping has never been coated or is being repaired. The coating must be compatible with the substrate, any existing coating, and the environment in which it will be used; it must also meet all applicable specifications.

Existing coatings must be tested and inspected for damage, defects, and deterioration. When the pipeline individual finds areas that need repair, the piping surface must be prepared according to the specification and the manufacturer's instructions in order for the coating to adhere properly and remain intact for the expected life span.

Abrasives are used in surface preparation. The pipeline individual must be able to select the proper one for the substrate and the profile. Abrasives create dust and debris and must be used safely.

Once the coating has been selected and the surface prepared, the pipeline individual must be able to apply the coating properly. Different types of coatings require different methods of application, such as brushing, rolling, and spraying.

When all of these tasks are done correctly, the coating will provide an effective, long-lasting defense against corrosion.

Review Questions

1. Curing of a coating can be affected by the temperature, the relative humidity, and the _____.
   a. water table
   b. dew point
   c. soil type
   d. time allowed to complete the job

2. Job requirements for coatings are usually expressed as one or more _____ standards.
   a. NIOSH
   b. OSHA
   c. SSPC
   d. ASTM

3. Coating inspection and test results are often expressed as _____ standards.
   a. SAE
   b. OSHA
   c. SSPC
   d. NIOSH

4. To remove grease, dirt, oil, debris, and other loose contaminants, the substrate surface should be washed with _____.
   a. enzyme-containing detergents
   b. mild soap and water
   c. kerosene
   d. a solvent

5. If a clean, exposed steel surface cannot be primed immediately, the surface should be _____ to prevent flash-rusting.
   a. sanded again with a specialty abrasive
   b. washed with a rust inhibitor
   c. treated with kerosene
   d. covered with a protective film

6. In most cases, the amount of profiling required on a substrate surface is detailed _____.
   a. in the project specification sheets
   b. by the supervisor
   c. by the vendor
   d. in the SSPC standards

7. The most common metallic abrasive is _____.
   a. slag
   b. malleable iron
   c. chilled cast iron
   d. cast steel

8. To remove light deposits without damaging the surface of the substrate, abrasives made with _____ can be used.
   a. corncobs
   b. slag
   c. garnet
   d. zircon

9. Strippers made of varying concentrations of potassium hydroxide, sodium hydroxide, and other materials are called _____ strippers.
   a. acid
   b. alkali
   c. flammable
   d. oil-based

10. When covering small areas, irregular surfaces, or when cutting in, the coating is usually applied with a(n) _____.
    a. edge roller
    b. airless sprayer
    c. air brush
    d. brush

Instructor's Notes:

GLOSSARY

Glossary

This glossary will help you learn some terms specific to applying and repairing coatings on buried and submerged pipe. Trade Terms introduced in this module appear in italic text.

Bond/Adhesion: The attachment of a coating to a substrate. Adhesion is the key to coating, effectively allowing the coating to become part of the substrate. The three types of adhesive bonds are chemical, polar, and mechanical.

Cigarette wrap: Application technique for cut lengths of sheet coating. The width of the sheet must exceed the circumference of the pipe to provide an overwrap. The sheet is placed longitudinally and wrapped no more the twice around the pipe. Ending seams should be alternated along the pipe so they are not continuous. Each sheet should overlap the previous. This method is not commonly used, with the exception of heat shrink sleeves.

Coal tar: A tar distilled from bituminous coal used to coat piping. Usually has a paper wrap over the coal tar.

Confined space: A space that: 1) Is large enough and so configured that an employee can bodily enter and perform assigned work; and 2) Has limited or restricted means for entry or exit (e.g. tanks, vessels, vaults, etc.); and 3) Is not designed for continuous employee occupancy.

**Cracking:* Breaks in a coating that extend through to the substrate.

Fusion bond: Application of a powder coating to a heated pipe by electrostatic methods. The powder gels and flows with the heat, then cures and hardens during cooling. The process creates a tight physical bond between the coating and the metal.

Hand cleaning: The process of removing burrs, slag, rust, mill scale, and other contaminants on steel pipe. Burrs or slag may be removed by grinding. Abrasive blasting is the preferred method for recoating.

***Holiday:** A discontinuity in a protective coating that exposes unprotected surface to the environment.

Mastic: Asphalt-like mixture used to coat piping.

Mill scale: The oxide layer formed during hot fabrication or heat treatment of metals.

Overlap: An amount of coating covering the previous row of coating in a row-to-row sequence.

Permit-required confined space: A confined space that has one or more of the following characteristics: 1) Contains or has a potential to contain a hazardous atmosphere; 2) Contains a material that has the potential for engulfing an entrant; 3) Has an internal configuration such that an entrant could be trapped or asphyxiated by inwardly converging walls or by a floor which slopes downward and tapers to a smaller cross-section; or 4) Contains any other recognized serious safety or health hazard.

Pre-heat: Pre-heating of the pipeline metal prior to application of certain coatings. Heat shrink sleeves and several other coatings require the pipe to be pre-heated prior to applying the coating. Care should be taken not to exceed specific temperatures as it can damage the coating. Epoxy coatings require specific temperature ranges to properly cure.

******Primer:* A coating material intended to be applied as the first coat on an uncoated surface. The coating is specifically formulated to adhere to and protect the surface as well as to produce a suitable surface for subsequent coats. [Also referred to as Prime Coat.]

Shrink sleeve: A conformable wrap-around sleeve designed to protect belowground and submerged metal surfaces from corrosion and electrolysis. The sleeve consists of two overlapping pieces or a single tube surrounding the pipe. Heat is applied to the sleeves until an adhesive material in the sleeve bonds the sleeve to the pipe and joins the overlapping sections to form a tight bond.

Wax: The wax present in petroleum crudes primarily consists of paraffin hydrocarbons known as paraffin wax and naphthenic hydrocarbons.

* Term published in NACE *Glossary of Corrosion-Related Terms.* Reproduced with permission from NACE International, Houston, TX. For a copy of the complete *Glossary of Corrosion-Related Terms*, contact NACE International, 1440 South Creek Dr., Houston, TX 77084. The Glossary is revised periodically, so information in an outdated version may not be accurate.

Instructor's Notes:

REFERENCES

Additional Resources

This module is intended to present thorough resources for task training. The following reference works are suggested for further study. These are optional materials for continued education rather than for task training.

API Recommended Practice 651, *CP of Aboveground Petroleum Storage Tanks*, Second Edition (December 1997). Global Engineering Product Code API CERT 651.

Control of Pipeline Corrosion, Second Edition (2001). A. W. Peabody. Houston, TX; NACE International.

Corrosion Prevention by Protective Coatings, 1984. Charles G. Munger. Houston, TX; National Association of Corrosion Engineers (NACE).

MODULE 61107-02—TEACHING TIPS

The following are suggested activities or instructional methods to help you teach the material in this AIG.

General

When you call on someone to answer a question, the rest of the class relaxes or even tunes out because they expect that the question and answer will take place only between you and the trainee you called on. Instead, use this technique to involve more trainees in answering questions and to keep them on their toes.

1. Ask trainees to define a term or explain a concept.
2. After one trainee has answered, ask a trainee seated nearby if the answer is right. Then ask whether a trainee in the back of the room agrees.
3. Ask trainees to explain why they think an answer is right or wrong.
4. Use the session to clear up incorrect ideas and encourage trainees to learn from their mistakes.

Section 2.0.0 *Selecting a Coating*

This exercise will give trainees the opportunity to select coatings and determine surface grades. For this exercise, you will need a variety of steel surface samples, the *SSPC-VIS 1* reference photographs, and manufacturers' product information data sheets. Trainees will need their Trainee Guides and pencils and paper to take notes for discussion. Allow 30 minutes for this exercise.

1. Display the illustrations of the surface grades.
2. Set up a station for each of the steel surface samples. Include a list of information describing the sample at each station, including:
   - Type of substrate and exposure
   - Operating conditions
   - Upset conditions
   - Ambient conditions
   - Regulations
   - Budget and time constraints
   - Surface preparation
3. Ask trainees to visit each station and determine the surface's cleanliness, according to the SSPC visual standards.
4. Next, have trainees select the correct coating based on the information you provided. Have trainees refer to manufacturers' product information data sheets, as needed.
5. Review and discuss trainees' answers. Go over any questions trainees may have about selecting a coating.

Section 3.1.0 *Types of Abrasives*

This Quick Quiz will familiarize trainees with the types of abrasives. You will need photocopies of the quiz provided on the following page. Trainees will need pencils. If you allow trainees to use the Trainee Guide, decrease the amount of time you give them to complete the quiz.

1. Make as many photocopies as you will need of the quiz.
2. Give trainees between 5 and 10 minutes to complete the quiz.
3. Go over the answers to the quiz.
4. Review the trainees' examples for each type of abrasive.
5. Ask trainees whether they have questions about types of abrasives.

Quick Quiz — *Types of Abrasives*

For each description listed, identify the type of abrasive that the text best describes. Write the corresponding letter in the blank provided. Ask trainees to provide an example of each abrasive on the line following the question.

_____ **1.** These come in a range of angular and spherical shapes, which gives them good cleaning action; however, they may break down quickly.

Example: ______________________________

_____ **2.** These can remove surface grease as well as light deposits without damaging the surface.

Example: ______________________________

_____ **3.** These come in the form of shot and grit and in different degrees of hardness; they can be recycled many times.

Example: ______________________________

_____ **4.** These have a sharp, angular shape, which gives them good cutting power; they have a relatively low cost.

Example: ______________________________

_____ **5.** These can clean small, delicate surfaces and can be recycled a limited number of times.

Example: ______________________________

a. Metallic abrasives
b. Mineral abrasives
c. By-product abrasives
d. Special-purpose abrasives
e. Plastic and glass abrasives

Section 3.1.0 *Answers to Quick Quiz*

1. b. Mineral abrasives
2. d. Special-purpose abrasives
3. a. Metallic abrasives
4. c. By-product abrasives
5. e. Plastic and glass abrasives

Section 5.0.0 *Other Surface Preparation Technologies*

This exercise will reinforce what trainees have learned about a variety of surface preparation technologies. For this exercise, you will need an overhead projector, sheets of acetate, and transparency markers. Trainees will need their Trainee Guides and pencils and paper to take notes for discussion. Allow 30 minutes for this exercise.

1. On a sheet of acetate, write several scenarios that require the use of a specialized surface cleaning system.
2. Ask trainees to review each scenario and to determine which system would be most suitable for each application.
3. Have trainees explain why they selected each system and discuss the relevant safety considerations for using each system.
4. Review and discuss trainees' answers. Provide recommendations for using these systems based on your own experience with surface cleaning and preparation.
5. Answer any questions trainees may have about surface preparation technologies.

Sections 6.3.0–6.4.0 *Spraying with Conventional and Airless Spray Systems*

This exercise will give trainees the opportunity to practice using conventional and airless spray systems. For this exercise, you will need a variety of practice surfaces, conventional and airless spray systems, and coating. Trainees will need appropriate personnel protective equipment and their Trainee Guides. Allow 30 minutes for this exercise.

1. Set up an outdoor location where trainees can practice using the spray systems without risking damage to an adjacent surface.
2. Have trainees practice using a respirator. Ensure that trainees understand how to properly select and wear a respirator.
3. Ask a trainee to demonstrate how to use the conventional and/or airless spray system. Have trainees observe the procedure.
4. Ask trainees to comment on what was done correctly and incorrectly. Make recommendations for effective spraying.
5. Answer any questions trainees may have about spraying with conventional and airless spray systems.

MODULE 61107-02 — ANSWERS TO REVIEW QUESTIONS

Answer	Section
1. b	2.0.0
2. d	2.0.0
3. c	2.1.0
4. d	2.2.0
5. b	2.2.1
6. a	2.3.0
7. d	3.1.1
8. a	3.1.4
9. b	4.3.0
10. d	6.1.0

CONTREN™ LEARNING SERIES — USER UPDATES

The NCCER makes every effort to keep these textbooks up-to-date and free of technical errors. We appreciate your help in this process. If you have an idea for improving this textbook, or if you find an error, a typographical mistake, or an inaccuracy in NCCER's Contren™ textbooks, please write us, using this form or a photocopy. Be sure to include the exact module number, page number, a detailed description, and the correction, if applicable. Your input will be brought to the attention of the Technical Review Committee. Thank you for your assistance.

Instructors – If you found that additional materials were necessary in order to teach this module effectively, please let us know so that we may include them in the Equipment/Materials list in the Instructor's Guide.

Write: Curriculum Revision and Development Department
National Center for Construction Education and Research
P.O. Box 141104, Gainesville, FL 32614-1104

Fax: 352-334-0932

E-mail: curriculum@nccer.org

Craft ______________________ Module Name ______________________

Copyright Date ______________ Module Number ______________ Page Number(s) ______________

Description
____
____
____
____

(Optional) Correction
____
____
____

(Optional) Your Name and Address
____
____
____

Module 61108-02

Cathodic Protection Measurement

(CT 1.1, 1.5, 3.1, and 3.2)

MODULE OVERVIEW

This module familiarizes trainees with corrosion monitoring and various types of cathodic protection (CP). Trainees will also learn how to measure CP components and inspect CP systems.

PREREQUISITES

Please refer to the Course Map in the Trainee Module. Prior to training with this module, it is recommended that the trainee shall have successfully completed the following:

Core Curriculum; Pipeline Core, Modules 66101-02 and 66102-02; Pipeline Corrosion Control Level One, Modules 61103-02 through 61107-02

OBJECTIVES

Upon completion of this module, the trainee will be able to:

1. Explain the basic theory of cathodic protection (CP).
2. Define galvanic and impressed current CP and explain the advantages and limitations of each system.
3. Demonstrate how to use instrumentation and meters to monitor CP systems (CT 1.1).
4. Measure structure-to-soil potentials (CT 1.1).
5. Inspect and perform electrical tests of bonds.
6. Obtain voltage and current output readings from a rectifier (CT 3.1).
7. Check proper operation of a rectifier (CT 3.2).
8. Interpret and test isolation devices (CT 1.5).

PERFORMANCE TASKS

Under the supervision of the instructor, the trainee should be able to:

1. Properly locate the reference electrode relative to the structure (CT 1.1).
2. Measure structure-to-soil potential (C 1.1).
3. Obtain a voltage and current output reading from a rectifier (CT 3.1).
4. Check for proper operation of a rectifier (CT 3.2).
   - Determine voltage.
   - Determine current.
   - Obtain the shunt ratio.
   - Examine rectifier for abnormal defects.
5. Inspect and test isolation devices (CT 1.5).

NCCER STANDARDIZED CRAFT TRAINING PROGRAM

The National Center for Construction Education and Research (NCCER) provides a standardized national program of accredited craft training. Key features of the program include instructor certification, competency-based training, and performance testing. The program provides trainees, instructors, and companies with a standard form of recognition through a National Craft Training Registry. The program is described in full in the *Guidelines for Accreditation*, published by the NCCER. For more information on standardized craft training, contact the NCCER by writing us at P.O. Box 141104, Gainesville, FL 32614-1104; calling 352-334-0911; or e-mailing info@nccer.org. More information may be found at our Web site, www.nccer.org.

HOW TO USE THIS ANNOTATED INSTRUCTOR'S GUIDE

Each page presents two sections of information. The larger section displays each page exactly as it appears in the Trainee Module. The narrow column ties suggested trainee and instructor actions to each page and provides icons to call your attention to material, safety, audiovisual, or testing requirements. The bottom of each page includes space for your notes.

If you see the Teaching Tip icon, that means there is a teaching tip associated with this section. Also refer to any suggested teaching tips at the end of the module.

SAFETY CONSIDERATIONS

Ensure that the trainees are equipped with appropriate personal protective equipment. Emphasize the importance of following all safety precautions and procedures when working with power tools.

PREPARATION

Before teaching this module, you should review the Module Outline, Objectives, Performance Tasks, and the Materials and Equipment List. Be sure to allow ample time to prepare your own training or lesson plan and gather all required equipment and materials.

MATERIALS AND EQUIPMENT LIST

Overhead projector and screen

Transparencies

Blank acetate sheets

Whiteboard/chalkboard

Transparency pens

Markers/chalk

Pencils and scratch paper

Appropriate personal protective equipment

Copies of your company's policy and procedures manual

Copies of *49 CFR 192* and *195*

Copies of *API 650-653*

Test equipment for measuring structure-to-soil electrolyte potential

Forms for recording soil potential

Digital voltmeter

Forms for recording rectifier readings

Ammeters

Access to a rectifier with and without a meter installed and with and without a shunt

Copies of NACE *Standard RP-01-69*

Samples of piping with various coating defects

Several types of voltmeters with various features

Ammeters, various types of shunts, and access to measurable circuits

Soil resistance meters, soil resistance pins, selector boxes, and soil boxes

Variety of copper sulfate electrodes

Multimeters and audio and visual continuity testers

Multimeters with damaged and undamaged test leads, probes, and plugs

Nonmetallic abrasive pads, distilled water, and copper sulfate crystals

Insulation materials and access to existing flanges that need to be insulated

Access to functioning and defective insulated flanges and appropriate measuring equipment

Isolation surge protectors and polarization cells

Module Examinations*

Performance Profile Sheets*

Copies of the Performance Verifications for Covered Tasks 1.1, 1.5, 3.1, and 3.2**

* Located in the Test Booklet.

** Available for download at www.nccer.org

TEACHING TIME FOR THIS MODULE

An outline for use in developing your lesson plan is presented below. Note that each Roman numeral in the outline equates to one session of instruction. Each session has a suggested time period of 2½ hours. This includes 10 minutes at the beginning of each session for administrative tasks and one 10-minute break during the session. Approximately 7½ hours are suggested to cover *Cathodic Protection Measurement (CT 1.1, 1.5, 3.1, and 3.2)*. You will need to adjust the time required for hands-on activity and testing based on your class size and resources.

Topic	Planned Time
Session I. Common Types of Monitoring	
A. Introduction	________
B. Identifying Types of Structures	________
C. Measuring Structure-to-Soil Electrolyte Potential	________
D. Rectifiers	________
1. Performance Testing	________
E. Reading Bond Drains	________
Session II. Corrosion and Cathodic Protection	
A. Basic Theory of CP	________
B. CP with Galvanic Anodes	________
C. CP with Impressed Current	________
D. Criteria for Cathodic Protection	________
E. Planning CP Systems	________
1. Laboratory	________
F. Effects of CP	________
Session III. Instrumentation, Inspection, and Fitting	
A. Instrumentation and Meters	________
1. Checking CP with Instrumentation and Meters	________
2. Recognizing Electrical Discontinuity	________
3. Laboratories	________
B. Visual Half-Cell Inspection	________
1. Laboratory	________
C. Electrically Insulated Fittings	________
1. Insulated Flanges	________
2. Testing Insulated Conductive Fittings	________
3. Laboratory	________
4. Grounding Cells	________
5. Isolation Surge Protectors and Polarization Cells	________
D. Summary	
1. Summarize module	________
2. Answer questions	________
E. Module Examination	________
1. Trainees must score 70% or higher to receive recognition from the NCCER.	
2. Record the testing results on Craft Training Report Form 200 and submit the results to the Training Program Sponsor.	
F. Performance Testing	________
1. Trainees must perform each task to the satisfaction of the instructor to receive recognition from the NCCER.	
2. Record the testing results on Craft Training Report Form 200 and submit the results to the Training Program Sponsor.	

Module 61108-02

Cathodic Protection Measurement

(CT 1.1, 1.5, 3.1, and 3.2)

Instructor's Notes:

Instructor's Notes:

Introduction to the Pipeline Operator Qualification Rule, API, and NCCER

It could be said that petroleum products keep America running. Petroleum products like oil and natural gas are the principal sources of energy in the United States and are used in many manufacturing and chemical processes. Petroleum products are primarily transported by pipeline. There are more than a million miles of liquid and natural gas pipeline in this country, ranging in diameter from two inches to four feet. These pipelines bring crude oil and natural gas from production fields to processing plants and refineries—and ultimately to the consumer.

Pipeline Operator Qualification

Petroleum products are considered hazardous, so pipeline personnel must take extra precautions and follow specific procedures when working with them. To help ensure the safe handling and transportation of petroleum products, the United States Department of Transportation (DOT) mandates safe practices as federal regulations.

One such regulation, the Operator Qualification (OQ) rule, requires pipeline operators to develop a written qualification plan to evaluate their personnel and their contract personnel (see *49 CFR Part 192 Subpart N—Gas*, and *49 CFR Part 195 Subpart G—Liquid*). The evaluation hinges on the operator's ability to perform certain covered tasks and to be able to recognize and respond to abnormal operating conditions that may be encountered while performing these covered tasks.

Covered Tasks

A pipeline performance task is deemed *covered* if the activity passes the following four-part test:

1. Is it performed on a pipeline facility?
2. Is it an operations or maintenance task?
3. Is it performed as a requirement of *49 CFR Part 195 (Liquid)* or *49 CFR Part 192 (Gas)*?
4. Does it affect the operation or integrity of the pipeline facility?

If the answer to all four questions is *yes*, then the activity is considered *covered* and must figure into the evaluation process of pipeline personnel.

American Petroleum Institute

The OQ rule and its related covered tasks provided the catalyst to develop a unified pipeline training and assessment program—so that pipeline operators and owners have ready access to the assessment and training tools they need. This effort was the direct result of the vision and leadership provided by The American Petroleum Institute (API) and API members.

Efforts to unify the industry are nothing new to API. Indeed, the impetus for forming the Institute in 1919 was the need to standardize engineering specifications for drilling and production equipment. Today, the most pressing issues revolve around public perceptions and government policies toward the industry. Speaking with one voice on these issues has become as essential as having interchangeable parts in the field. API's members determine the petroleum industry's positions on public policy as well as the standards that govern its day-to-day operations worldwide.

Consortium on Operator Qualification

API assembled a focus group whose primary purpose was to identify covered tasks and provide compliance recommendations to pipeline operators.

The Consortium on Operator Qualification (COOQ) became the guiding force for the identification and development of covered tasks for the liquid pipeline industry. Their work product, API publication 1161, *Guidance Document for the Qualification of Liquid Pipeline Personnel*, formed the basis for the resulting Pipeliner Training and Assessment Program.

The covered task list created in API publication 1161 has now evolved to include gas covered tasks and is maintained by NCCER. The most current covered task list can be downloaded at www.nccer.org.

National Center for Construction Education and Research

With the guidance and support of API, its members, and the Consortium on Operator Qualification, the National Center for Construction Education and Research (NCCER) was able to design and implement the Pipeliner Training and Assessment Program.

NCCER is a not-for-profit 501(c)(3) education foundation established in 1995 by the world's largest and most progressive construction and maintenance contractors and national contractor associations. NCCER was founded to address the

severe workforce shortage facing the construction and maintenance industries and to develop a standardized training process and curricula. Today, NCCER is supported by hundreds of leading construction and maintenance companies, manufacturers, and national associations and developed more than 35 construction and maintenance titles.

NCCER also maintains a National Registry, which contains the records of trainees and instructors who have received NCCER training. All completion information for every trainee is sent to the NCCER and filed within the Registry. The Registry can then confirm training and skills for workers as they move from company to company, state to state, or even within their own company.

Standardized Curriculum (NCCER's Contren™ Learning Series)

NCCER curricula are developed using the expertise of industry subject matter experts (SMEs). The use of SMEs as an integral part of every new development or revision ensures that the curriculum is current, industry-driven, and indicative of industry practices. The development and/or revision process is facilitated from start to finish by NCCER project managers, who work closely with the SMEs and a select team of technical writer(s) on the development and presentation of the curriculum.

Pipeline: Five Training Paths

The pipeline curriculum focuses on essential skills and competencies within five skill areas:

- Corrosion Control
- Electrical and Instrumentation
- Maintenance
- Mechanical
- Operations (Field and Control Center)

SMEs from the pipeline industry have determined the skills and competencies needed based on API's covered tasks and the consensus of the group/industry. Specific company information is generally not incorporated into the curriculum unless the SMEs determine that the information is the industry standard.

Training Modules and Covered Tasks

The smallest unit of instruction in the curriculum is a *module*. Each curriculum is composed of modules (the equivalent of stand-alone chapters) that are logically sequenced into complete *levels* of instruction. Every module is task-oriented and requires the successful completion of a written (knowledge-based) examination and a performance (hands-on) test. Modules may be individually selected and/or combined for efficient task training. Modules in the pipeline curricula, to the extent possible, will align with API-developed covered tasks. While not all modules will relate to covered tasks, all covered tasks will relate to at least one module.

Benefits of Training to the Pipeline Marketplace

Applying NCCER's standardized training process allows pipeline owners and contractors in the United States to establish the benchmark for acceptable skill and competency levels within the pipeline industry. This process will assist training organizations in developing and/or improving standardized pipeline training programs that effectively match their resources and operating conditions while responding to regulatory demands and to the needs of the industry. The process will also recognize training organizations that meet the minimum standards of competence, warranting regulatory, public, and professional confidence. This in turn will foster national unity within the pipeline industry for the mutual benefit of individual technicians, owners, contractors, and ultimately pipeline-users and society.

Additional Resources

To review the Pipeliner Training and Assessment Program Implementation and Resource Guide and the latest version of pipeline covered tasks, visit NCCER's Web site at www.nccer.org, or contact:

National Center for Construction Education and Research
PO Box 141104
Gainesville, FL 32614-1104
(888) 622-3720
Web site: www.nccer.org

For more information on API, please contact:

American Petroleum Institute
1220 L Street, Northwest
Washington, D.C. 20005
(202) 682-8000
Web site: www.api.org

To order *API Pub 1161*, contact Global Engineering Documents by calling 1-800-854-7149 or visiting their Web site at www.global.his.com.

Additional sources of information on the pipeline operator qualification regulation include:

DOT – Office of Pipeline Safety
www.ops.dot.gov

Transportation Safety Institute
www.tsi.dot.gov

Instructor's Notes:

COURSE MAP

This course map shows all of the modules in the first level of the Pipeline Corrosion Control curriculum. The suggested training order begins at the bottom and proceeds up. Skill levels increase as you advance on the course map. The local Training Program Sponsor may adjust the training order.

PIPELINE CORROSION CONTROL LEVEL ONE

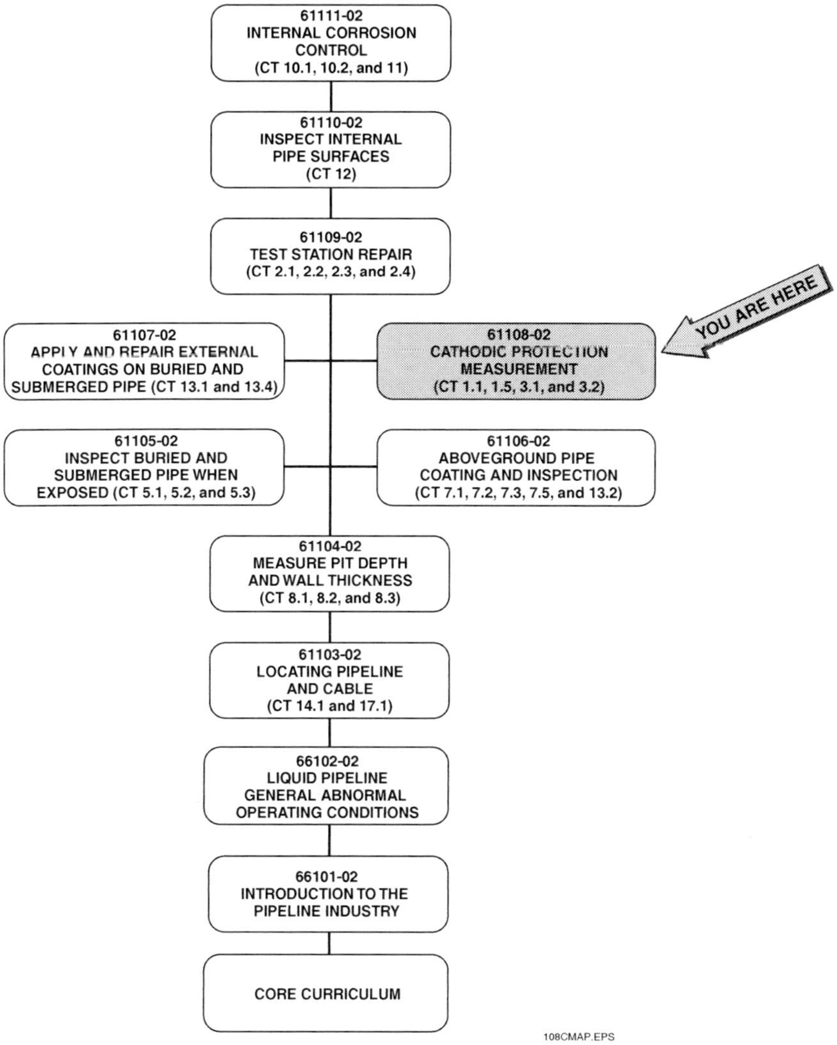

108CMAP.EPS

Homework

Assign reading of Module 61108-02.

Instructor's Notes:

MODULE 61108-02 CONTENTS

MODULE 61108-02 CONTENTS (Continued)

Instructor's Notes:

Figures

Tables

Instructor's Notes:

MODULE 61108-02

Cathodic Protection Measurement (CT 1.1, 1.5, 3.1, and 3.2)

Objectives

When you have completed this module, you will be able to do the following:

1. Explain the basic theory of cathodic protection (CP).
2. Define galvanic and impressed current CP and explain the advantages and limitations of each system.
3. Demonstrate how to use instrumentation and meters to monitor CP systems (CT 1.1).
4. Measure structure-to-soil potentials (CT 1.1).
5. Inspect and perform electrical tests of bonds.
6. Obtain voltage and current output readings from a rectifier (CT 3.1).
7. Check proper operation of a rectifier (CT 3.2).
8. Interpret and test isolation devices (CT 1.5).

Prerequisites

Before you begin this module, it is recommended that you successfully complete the following: Core Curriculum; Pipeline Core, Modules 66101-02 and 66102-02; Pipeline Corrosion Control Level 1, Modules 61103-02 through 61107-02.

Required Trainee Materials

1. Pencil and paper
2. Appropriate personal protective equipment

1.0.0 ◆ INTRODUCTION

Corrosion is a natural process that, given the right conditions, can attack any metal or alloy. Most pipeline workers are familiar with corrosion but do not have specific knowledge of the various types of corrosion. Everyone is surely familiar with rust, the most common form of corrosion, which results in metal loss. Most corrosion failures are not caused by rust or general corrosion but by localized attack.

Each year corrosion costs the pipeline industry millions of dollars in material damage and downtime. Corrosion can cause catastrophic failures with potentially severe consequences for the environment and serious injury or even death to personnel. It is necessary today to be able to identify potential problems and risks through a corrosion management program. An entire industry has built up around this need.

Federal guidelines have been implemented to support safety of both personnel and the environment. The complete text of the corrosion control requirements can be found in *49 CFR 192*, and *49 CFR 195*. Companies are required to have written guidelines and procedures for most corrosion-related activities.

2.0.0 ◆ COMMON TYPES OF MONITORING

Corrosion workers should be aware of the effects and the testing techniques associated with cathodic corrosion protection. This module describes some

Materials

Ensure that you have everything required to teach the course. Check the Materials and Equipment List at the front of this module.

Classroom

Explain that terms shown in bold are defined in the Glossary at the back of this module.

Audiovisual

Show Transparency 1, Course Objectives.

Show Transparency 2, Performance Tasks

Teaching Tip

See the general Teaching Tip at the end of this module.

Classroom

Ask trainees to name possible consequences of corrosion.

Hand out copies of *49 CFR 192* and *195.* Review the contents with the trainees.

Classroom

Have a trainee explain the meaning of soil potential.

Ask trainees to name the components of both a pipeline and a pipeline facility. Write their answers on the board or an overhead.

Ask a trainee to explain the problems that can occur with casings. Ask another to discuss how to maintain the cathodic protection of pipelines.

Distribute copies of *API 650–653* (see text for edition numbers). Review contents with trainees. Answer any questions trainees may have.

of the most common activities required of a corrosion worker for:

- Identifying types of structures
- Measuring **structure-to-soil potential**
- Monitoring **shorted casings**
- Reading **rectifiers** with meters installed
- Reading rectifiers without meters installed
- Checking visual indicators to verify operation of a rectifier
- Reading bond drains

2.1.0 Identifying Types of Structures

Soil potentials are commonly measured for the following types of structures:

- Pipelines
- Casing
- Tank bottoms
- **Foreign structures**
- Well casings
- Underground storage tanks

2.1.1 Pipelines

The terms pipeline or pipeline system refer to all parts of a pipeline facility through which a hazardous liquid is transported. Pipeline systems include pipe, valves, and other components connected to line pipe, pumps, and other associated assemblies.

The pipeline facility encompasses piping and equipment used in production, extraction, recovery, lifting, stabilization, separation, or treating of petroleum. This includes piping between the production facilities and refineries.

2.1.2 Casings

Casings are pipes that contain a carrier pipeline. Casings are typically used when a pipeline passes under railroad rights-of-way and under some roads to protect the pipeline from high load stresses. State and local regulations identify specific mandatory casing locations.

Problems created by casings result from electrical contact between the coated pipeline and the bare casing. Contact with casings can interfere with the **cathodic protection** (CP) of pipelines, so transport pipes must be provided with sufficiently high, mechanically stable spacers to exclude electrical contacts. Designers and planners should use only the type of casing seals that prevent ground water from entering into the ring-like space, called the annulus, between the casing and the pipeline. If the casings are in contact with the carrier pipe, CP can be lost over wide stretches by electrical contact with the carrier pipe. Improper installation of a pipeline within a casing can be a cause of contact.

Proper installation and maintenance are essential to eliminate contact between the carrier pipe and the casing. On cathodically protected pipelines, casings should be used only where required by law or where the pipeline cannot be laid at crossing points without risk of damage.

2.1.3 Tank Bottoms

An aboveground storage tank is any tank or other container that is aboveground, partially buried, bunkered, or in a subterranean vault. This includes floating fuel systems. The tank bottom and any portion that is buried or submerged in **electrolyte** must be protected.

With CP, accepted criteria must be established. The potential is commonly measured using a copper-copper sulfate **reference electrode** in contact with the electrolyte. **Voltage** drops other than those across the structure-to-electrolyte boundary must be considered for valid interpretation of the voltage measurement.

The following American Petroleum Institute (API) references may be applicable:

- API 650 *Welded Steel Tanks for Oil Storage*, 10th/1998
- API 651 *CP of Aboveground Storage Tanks*, 2nd/1997
- API 652 *Lining of Aboveground Storage Tank Bottoms*, 2nd/1997
- API 653 *Tank Inspection, Repair, Alteration, and Reconstruction*, 2nd/1996

2.1.4 Foreign Structures

A foreign structure is any structure or pipeline that is not owned by your company. When steel structures contact the earth or foreign metallic structures, corrosion can occur. Although the corrosive process may take years, if **interference** is undetected, its effects can be significant and may result in metal loss.

Foreign metal structures typically include:

- Water lines
- Telephone lines
- Electric power lines
- Traffic control cables and conduits
- Steam lines
- Pipe cables
- Street light cables

Instructor's Notes:

- Conduits
- Light standards
- Electric ground rods
- Sewage lines
- Miscellaneous manholes and drainpipes

Additionally, junk metal, such as pipe and rails from abandoned facilities, are sometimes present. Other metal structures also include reinforcing rod or wire mesh in concrete.

When installing CP systems, notify owners of nearby pipelines and utilities. Pipeline and utility companies with buried metallic structures close to a petroleum terminal's impressed-current ground-bed must be notified so that interference testing can be cooperatively conducted.

When a foreign structure is near enough, **current** takes an unplanned path. If another metal structure or pipeline is in the current's path, or if the foreign structure offers a path of lower resistance back to the current's source, the protective current will be picked up. The foreign structure receives some CP by picking up this stray current, but when the current leaves the foreign structure, it takes metal into solution and will eventually cause corrosion of the foreign structure.

Pipeline mapping systems help gas utilities and pipeline companies to locate contacts with foreign metallic structures. The use of mapping systems significantly reduces the cost of troubleshooting CP pipeline systems.

2.1.5 Well Casings

Wells drilled in oil and natural gas fields are fitted with casings for protection and stabilization. Cement is poured into the annular space between the outer pipe, or casing, and the surrounding rock from bottom to top, to seal the deposits and to keep fresh and salt water separate. The cement filling is not uniform in thickness and may have areas with no cement. The pipeline well casings are subject to various types of corrosion, depending on the surrounding soil content and the degree of protection provided by the cement. Boreholes are often located close to one another, so they may be affected by the CP from other wells. In large oil fields and reservoirs, CP for all the wells is appropriate.

2.1.6 Underground Storage Tanks

An underground storage tank (UST) or underground storage tank system refers to one or a combination of stationary devices designed to contain an accumulation of petroleum. This includes underground pipes connected to the devices. Referring to storage tanks as underground storage tanks typically means that the volume, including the volume of underground pipes, is 10% or more beneath the surface of the ground.

2.2.0 Measuring Structure-to-Soil Electrolyte Potential

Measurement of potential between a pipeline or structure and a reference electrode is the most frequent measurement in pipeline corrosion testing. This section explains measurements using the copper sulfate electrode.

The essential parts of a copper sulfate electrode are:

- Copper rod to which the instrument is connected
- Saturated solution of copper sulfate in contact with the rod
- Porous membrane in contact with the soil
- Surplus crystals of copper sulfate in the solution

Be sure that the test equipment is in proper working condition. Place the electrode near the structure to be monitored. The porous plug, with cap removed, should be in firm contact with the moist earth. This may require digging if the surface is dry. If the area is extremely dry, moisten the ground around the electrode with fresh water to obtain good contact. Readings through concrete or asphalt may not be accurate. Keep grass or weeds (particularly when wet) away from the area with exposed electrode terminals because they may affect the observed potential.

Connecting the electrode to the negative terminal of a digital voltmeter and connecting the positive instrument terminal to the pipeline, should result in a negative reading. *Figure 1* shows voltmeter measurement of **pipe-to-soil** potential.

When taking the readings, notice the **polarity** and the magnitude. Once the readings are completed, record the results on the company form designated for that purpose.

> **NOTE**
> Always consult company procedures for the proper polarity connections. Company procedures should indicate which positive and which negative lead to connect to a particular source to obtain a correct reading.

Classroom

Have a trainee explain what happens when foreign structures are located too close to cathodic protection systems.

Have trainees write out three corrosion concerns relating to well casings. As a class, discuss how each corrosion problem can be avoided.

Ask a trainee to identify the percentage of volume at which a tank is considered to be underground.

Audiovisual

Show Transparency 3 (Figure 1). Discuss how to use the voltmeter to take a reading.

Teaching Tip

See the Teaching Tip at the end of this module.

Safety

Stress the danger of contact with an energized rectifier. Discuss proper safety precautions.

Audiovisual

Show Transparency 4 (Figure 2). Discuss the operation of CP rectifiers.

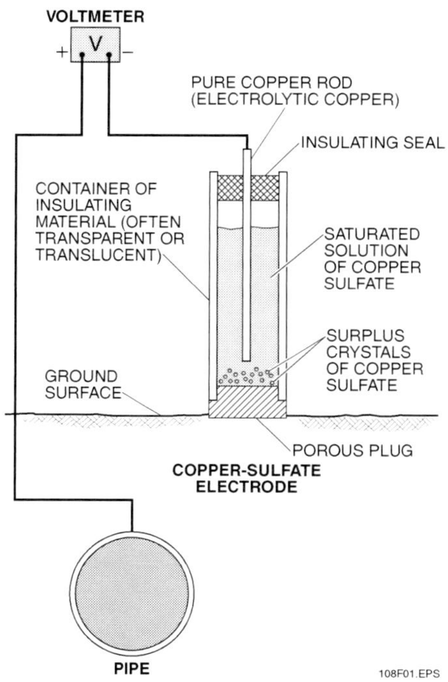

Figure 1 ◆ Voltmeter measurement of pipe-to-soil potential.

2.3.0 Rectifiers

The major components of a typical rectifier include:

- *Transformer* – Steps down AC line voltage to low voltage AC on the secondary
- *Tap arrangement* – Allows selection of a DC output voltage range
- *Rectifying element* – Rectifies AC current to a DC output
- *Housing* – Permits mounting the rectifier at the point of use
- *AC circuit breakers* – Circuit protection devices
- *DC output meters* – Monitor rectifier output, voltage and current
- ***Shunts*** – Resistors of a known value

A typical rectifying unit has high resistance to current flow in one direction and low resistance in the other. Current can flow on only one route through the rectifying element for a given direction of current flow in the transformer secondary winding. This flow runs out through the positive terminal to the ground bed and back from the pipeline connection to the negative terminal.

> **WARNING!**
> Contact with the components of an energized rectifier can cause serious injury or death. Follow all safety precautions and be extremely careful when working on or near a rectifier.

2.3.1 Reading Rectifiers With Meters Installed

Rectifier control panels may contain an **ammeter** and a DC voltmeter. To verify the operation of the rectifier, turn off the power to the rectifier at the main circuit **breaker** and then restore the power to the rectifier at the circuit breaker. Observe the ammeter and the voltmeter; both should show a deflection and the status of the rectifier. If either meter fails to show a deflection, the rectifier is not operating correctly. Use a digital voltmeter to verify proper rectifier operation. Visually check for:

- Arcing
- Broken or missing gauges
- Smells
- Tripped breakers
- Bad mounting brackets
- Burned wires
- Animals/insects

2.3.2 Reading and Recording Volts and Amps

After verifying that the rectifier is operating, observe the reading on the DC voltmeter and the ammeter and record the values on the appropriate forms. Compare the values of the current readings with those of the previous reading. *Figure 2* shows the schematic diagram for CP rectifiers.

2.4.0 Reading Rectifiers Without Meters Installed

Some rectifiers do not have a voltmeter or ammeter installed in the control panel. The corrosion control tester must take and record these measurements with a voltmeter and/or ammeter.

On the face of the control panel, the test terminal should be marked with (–) for negative and (+) for positive. They may also be color-coded: black for negative and red for positive. Verify that the correct polarity is being used for the readings. To read

Instructor's Notes:

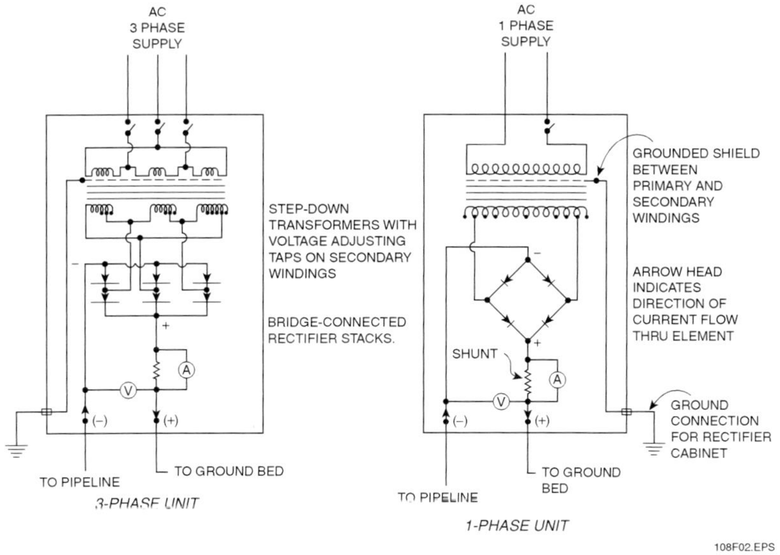

Figure 2 ◆ Rectifier schematic diagrams.

direct-current (DC) voltage, connect the positive lead from the meter to the positive terminal of the rectifier. Connect the negative lead to the negative terminal.

2.4.1 Reading Current Without a Shunt

If a shunt is not present, use a clamp-on ammeter or break the circuit and put an ammeter in line. This allows you take a direct reading that does not require conversion.

2.4.2 Reading DC Amperage

To read amperage with a shunt present, set the voltmeter to the millivolt (mV) scale and record the positive reading. To convert this reading to an amperage value, use the formula illustrated in the example that follows.

Assume that the shunt has a millivolt rating of 50 and amperage rating of 30. For the purposes of this exercise, assume the millivolt reading taken from the electrode was 22.2 mV.

To convert millivolts to amps, take the reading from the shunt and multiply it by the ratio of the shunt. The result is the amperage output of the unit. The shunt ratio factor is equal to the amperage rating divided by the millivolt rating of the shunt.

The millivolt reading (22.2) times the ampere rating, divided by the millivolt rating of the shunt (30÷50 or 0.6), is equal to 22.2 times 0.6, or 13.32 amps. *Table 1* shows shunt types.

2.5.0 Reading Bond Drains

This procedure is similar to reading rectifiers without meters. For a **bond** with a shunt, use a voltmeter. For a bond without a shunt, use an ammeter. It is critical that polarity be understood when reading a bond.

3.0.0 ◆ CORROSION AND CATHODIC PROTECTION (CP)

Four conditions must be present for corrosion to occur. There must be an **anode**, a **cathode**, an electrolyte, and a metallic path. The four conditions can involve two separate structures or can be present on one structure. If any one of the four conditions is eliminated, the corrosion can be prevented.

In anodic areas, current flows from the steel in the structure into the surrounding electrolyte (soil and water), causing the structure to corrode. At the cathodic areas, current flows from the electrolyte onto the structure surface, and corrosion is eliminated.

Teaching Tip

See the Teaching Tip at the end of this module.

Audiovisual

Have trainees compare and contrast the procedures for reading rectifiers with and without meters and shunts.

Performance Testing

Have each trainee perform Performance Tasks 1 and 2 to your satisfaction. Fill out a Performance Profile Sheet for each task for each trainee.

Classroom

Ask a trainee to explain how to read a bond drain with and without a shunt. Ask another trainee to explain why polarity must be understood when reading a bond.

Homework

Have trainees review Section 3.0.0 for the next session.

Materials

Ensure that you have everything required for demonstrations and teaching during this session.

Classroom

Ask a trainee to name the four conditions required for corrosion to occur. Have another trainee explain the corrosion process.

Audiovisual

Show Transparency 5 (Figure 3). Ask a trainee to explain how a CP system works. Have another trainee name the two basic methods of CP.

Table 1 Shunt Types

Type	Shunt Rating		Shunt Valve	Shunt Factor
Holloway	**Amps**	**MV**	**Ohms**	**A/mV**
RS	5	50	.01	0.1
SS	25	25	.001	1.0
SO	50	50	.001	1.0
SW or CP	1	50	.05	.02
SW or CP	2	50	.025	.04
SW or CP	3	50	.017	.06
SW or CP	4	50	.0125	.08
SW or CP	5	50	.01	0.1
SW or CP	10	50	.005	0.2
SW	15	50	.0033	0.3
SW	20	50	.0025	0.4
SW	25	50	.002	0.5
SW	30	50	.0017	0.6
SW	50	50	.001	1.0
SW	60	50	.008	1.2
SW	75	50	.0067	1.5
SW	100	50	.0005	2.0
J. B.				
Agra-Mesa	5	50	.01	0.1
Cott or MCM				
Red (MCM)	0.1	100	.1	.01
Red (Cott)	0.5	50	.1	.01
Yellow	5.0	50	.01	0.1
Orange	25	25	.001	1.0

108T01.EPS

The positive area, the anode, is the one that corrodes and loses metal. The metal that is lost is directly proportional to the amount of current flow. For example, 1 ampere of direct current discharging into a soil electrolyte (the **ionic** current flow path) can remove approximately 20 pounds of steel in one year. Rates vary with different metals; normal corrosion currents are usually given in thousandths of amperes, or milliamperes. This information can be used to calculate the amount of protection needed.

Protective coatings are one method of corrosion protection. They prevent contact between the structure and the electrolyte. However, these coatings are not perfect: they often have holidays, or voids, that allow current to flow from the structure to the electrolyte.

CP is used to prevent corrosion in metals exposed to an electrolyte such as soil or water. CP is directed at reducing the potential difference between the anodic and cathodic areas of the structure to zero.

3.1.0 Basic Theory of CP

Cathodic protection is the use of direct current (DC) electricity, provided through an impressed or **galvanic** external source, to oppose the discharge of corrosion current from anodic areas. To mitigate corrosion, a CP system induces current onto the structure from the electrolyte.

The rate of corrosion can be reduced if all the exposed metal on the surface of a pipeline can be made to collect current. To make the surface of the structure collect current, direct current is forced onto all surfaces of the pipeline. This direct current shifts the potential of the pipeline in the active (negative) direction, resulting in a reduction in the corrosion rate of the metal. When the amount of applied direct current is adjusted properly, it overpowers the corrosion current that is discharging from the anodic areas on the pipeline and generates a net current flow on the pipe surface at these points. The entire surface then becomes a cathode, and the corrosion rate is thereby reduced. A significant activity of a CP tester is to determine the actual level of CP needed to reduce the corrosion rate to an acceptable level.

When current is forced to flow into the pipe at areas that were previously discharging current, the voltage provided by the CP system must be greater than the voltage coming from the corrosion cells.

There are two basic methods of CP. The first method is by use of a galvanic anode. The second common type of CP is called impressed current CP.

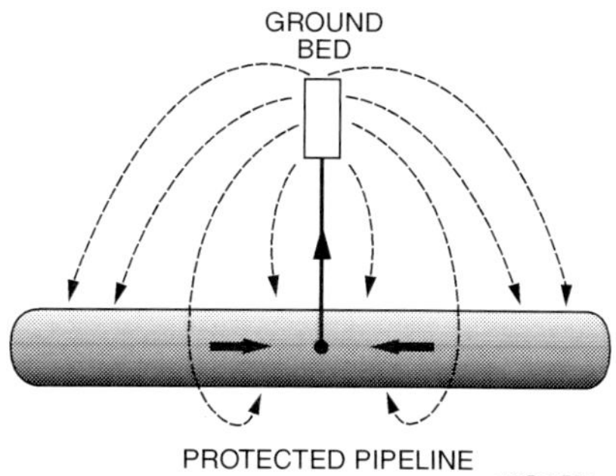

108F03.EPS

Figure 3 ◆ Diagram of galvanic anode circuit.

Instructor's Notes:

3.2.0 CP With Galvanic Anodes

Galvanic anode systems are simple and reliable. In a galvanic system, another metallic element that has a higher natural electromotive force is introduced; it must be strong enough to counteract the corrosion cell that is usually present on the structure. The new material must be anodic to the structure. A working galvanic anode of zinc or magnesium buried in the earth and connected to the pipeline will discharge current and protect the pipeline.

This element, which is called a sacrificial metal, is connected through a metallic connector or wire to the structure to be protected. CP makes use of the potential, or voltage, difference between the anodic and the cathodic areas of the structure by causing electrical current to flow onto the structure from the second anodic area, which is produced by the sacrificial metal. The sacrificial metal then attracts the corrosive action. Magnesium, zinc, and aluminum are common sacrificial or galvanic anodes.

This type of galvanic CP relies on the natural **electrical potential** between the two metals to cause the CP current to flow. Since the driving voltage is limited to the very small potential difference existing between the metals, and the current output is relatively low, this type of CP is normally associated with structures requiring small current. The galvanic anodes used in soil applications are usually made of a magnesium alloy.

For the CP system to work, current must be discharged from an earth connection, or ground bed (*Figure 3*). The purpose of the ground bed is to discharge current. In the process of discharging current, the anodes in the ground bed are consumed by corrosion. The rate of corrosion consumption is described in pounds per ampere-year.

3.2.1 Galvanic Anode CP Application

Galvanic anodes are normally used where relatively small amounts of currents are needed and in soils with relatively low resistance.

Even when a structure is protected through coating, some defects or holidays occur. Galvanic anode CP is used to protect these small areas of exposed steel rather than the entire surface of an uncoated structure. CP of a bare structure would require much more energy than would the protection of a coated structure.

There are also instances where galvanic anodes are placed at anodic areas on bare or poorly coated structures. This practice is called hot spotting and could include repair locations. This approach is often taken if the structure is poorly coated or there is much bare metal.

At the discharge point on a foreign line, under the right conditions, galvanic anodes may be used to return the interfering current in cases of cathodic interference. In an impressed current system, galvanic anodes may be used if electrical shielding gets in the way of effective current distribution from a remote point. Galvanic anodes may be an economical choice where conditions make it difficult to install impressed current systems without creating problems with stray current interference.

3.2.2 Advantages and Limitations of Galvanic Anodes

Among the advantages of using a galvanic anode CP system is that no external power source is needed. This type of system is easy to install, needs only minimum right-of-way or easement, and has low maintenance requirements. The small current output generates little or no stray current interference, and it is easy to add more anodes.

The limitations of galvanic anodes are the low driving voltage or current output; they may not be effective in a highly resistive environment. If the structure is bare or poorly coated, a large number of anodes may be required, and spent anodes may be difficult or expensive to replace. Because of lower efficiency, the cost per unit ampere is higher than that of impressed current. Efficiency varies with the type of alloy and the environment in which it is used.

3.3.0 CP With Impressed Current

The second widely used type of CP is impressed current. In impressed current CP, an external direct current source such as a rectifier or battery is used. An anode material is placed in the electrolyte with the structure to be protected. The anode is then made more positive than the structure by connecting both the anode and the structure to the direct current supply. Any conductive material can be used as an impressed current anode, but since corrosion takes place at the anode, materials with very low consumption rates are most desirable.

Impressed current CP systems (see *Figure 4*) can supply greater protection than the limited driving voltage of galvanic anodes. Direct current voltage from an outside source is impressed on the circuit between the protected structure and the anodes. The most common sources of impressed power are the CP rectifier and the DC power supply, but solar power and thermoelectric generators may also be used.

The rectifiers used in impressed current are devices that convert alternating current electric

Classroom

Discuss the process by which galvanic anode systems work. Ask a trainee to name the unit for the rate of corrosion consumption.

Ask a trainee to define hot spotting. Ask another trainee to name several applications of galvanic anodes.

Have trainees list at least three advantages and three disadvantages of galvanic anodes. Review their answers and compile a master list on the board or an overhead.

Audiovisual

Show Transparency 6 (Figure 4). Have trainees identify four sources of impressed power. Ask a trainee to identify the two that are most common.

Classroom

Describe three scenarios in which CP is needed. Have trainees decide whether galvanic anodes or impressed current CP should be used. Call on trainees to explain their choices.

Distribute copies of the latest revision of NACE Standard *RP-01-69*. Review the criteria for cathodic protection.

Ask a trainee to explain what NACE Standard *RP-01-69, Section 6* requires. Discuss how special conditions such as shielding can affect the requirement.

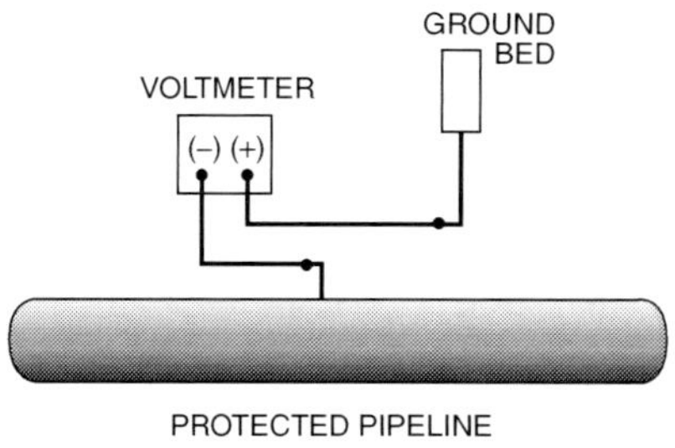

Figure 4 ◆ Diagram of impressed current circuit.

power to low-voltage direct current power. The current output of these rectifiers is usually adjustable over a range of 10 to 50 volts, but greater ranges are available. The output voltage of the unit is usually adjustable. CP rectifiers are available in output capacities from one ampere to many hundreds of amperes.

3.3.1 Impressed Current CP Application

Impressed current CP is used where large amounts of current are required, such as for bare or poorly coated structures. It can be used in electrolytic **resistivity** and can overcome stray current or cathodic interference problems. Impressed current CP is an economical way to protect structures with dissipated galvanic anodes and for a number of different types of structures.

Except for those with active anode surfaces, all electrical systems used in impressed current applications must be completely sealed with dielectric insulating materials.

3.3.2 Advantages and Limitations of Impressed Current CP

Impressed current systems are flexible and capable of handling a wide range of voltage and current outputs. They are effective in high-resistivity environments and in protecting uncoated or poorly coated structures. With impressed current, the anode consumption is less than with galvanic anodes, and a single installation can satisfy high current requirements.

Among the disadvantages or limitations of impressed current systems are greater complexity than galvanic systems and greater maintenance requirements. They require an external power source, making the overall power supply costs higher than those of galvanic anodes. With impressed current CP, there is also a greater risk of causing stray current interference. Overprotection, a risk with impressed current, causes coating damage and hydrogen embrittlement.

3.4.0 Criteria for Cathodic Protection

Criteria have been developed to ensure that structures are fully protected from corrosion. These criterion were developed by the National Association of Corrosion Engineers (NACE) and are listed in *Standard RP-01-69* titled, *Recommended Practice — Control of External Corrosion on Underground or Submerged Metallic Piping Systems*. These criteria are also contained in *Section I, Corrosion, of Part 192 (Transportation of Natural and Other Gas by Pipeline: Minimum Federal Safety Standards), Title 49 of the Code of Federal Regulations*. Corrosion testers should always be sure to use the latest revision of these standards.

All underground steel piping systems must be protected against corrosion by a cathodic protection system. The system will create a cathodic potential that meets a negative (cathodic) voltage of at least 0.85 volt, with reference to a saturated copper/copper sulfate half cell or other requirements as described in Appendix D of *CFR Title 49, Part 192. RE: 192.463*.

3.4.1 NACE Standard

NACE *Standard RP0169-96*, Section 6, requires a negative (cathodic) voltage of at least 0.85 volt as measured between the structure surface and a saturated copper-copper sulfate reference electrode contacting the electrolyte. Determination of this voltage is to be made with the protective current applied.

Special conditions such as elevated temperatures, disbonded coatings, thermal insulating coatings, shielding, bacterial attack, and unusual contaminants in the electrolyte are not covered.

3.4.2 Basis for Criteria

These criteria are based on the fact that when a pipeline is under CP, direct current flows from the conducting environment onto the pipeline, forcing the pipeline to assume a negative electrical polarity with respect to the environment. The question is how negative the pipe should be to serve as an indication that full protection has been attained. The value of –0.85 volt is used for steel pipe as measured using a standard copper-copper sulfate reference electrode to contact the environment. If other types of reference electrodes are used, the values will be different.

Instructor's Notes:

Potentials on steel pipe that are less negative than –0.85 volt to **copper-copper sulfate electrode (CSE)** indicate less than full CP. Potentials more negative than –0.85 volt to copper-copper sulfate electrode may indicate excessive energy use because more negative potentials are not necessary once corrosion is stopped at –0.85 volt. However, to maintain the minimum of –0.85 volt at locations some distance from the drainage points, it is usually necessary to have higher negative potentials at drainage points.

3.4.3 Level of Protection

Determining the correct level of protection is very important; overprotection can cause damage in most coated structures. When CP potentials are more negative than necessary on coated pipelines, gaseous hydrogen is generated at coating defects in the steel surface. Hydrogen bubbles may cause the paint to lift around defects, increasing the current requirements for CP. This effect is most likely to occur in low-resistance environments. Some coatings are more resistant to this effect than others, but none are immune.

Excessive amounts of cathodic current may produce a defect called cathodic **disbondment** on coated piping or equipment. Once the cathodic current exceeds approximately –1.17 volts with a copper-copper sulfate reference electrode, the area starts to become overprotected. Once the current reaches approximately –1.22 volts, hydrogen can evolve in the form of gas bubbles on the surface of the pipe, causing disbondment of the coating. Negative potentials are best kept in the area of –0.85 to –1.15 volts with a copper-copper sulfate reference electrode.

3.5.0 Planning CP Systems

The two basic considerations when planning a CP system are determining the amount of protection needed overall and then the type of system, galvanic or impressed current, that would fit the situation. In a galvanic CP system, the layout of the anodes must be carefully planned to ensure that the needed protection is provided; in an impressed current system, the external power source must be considered. Another factor to consider is the proximity of any other underground structures that could produce stray currents. Stray currents can weaken the protection provided by the CP system and may even reverse its effects, increasing the rate of corrosion.

Soil type and conditions must be considered. Perhaps drainage could be improved, reducing the potential for corrosion and decreasing the amount of protection needed. Environmental conditions, local weather, and temperature all play a part.

Another important consideration in planning a CP system is access for testing, repairs, and replacement.

3.5.1 Effect of Coating on CP

When the piping underlying the coating fails prematurely because of corrosion, the cause is often holidays in the coating. Properly applied coating on newer piping has relatively few defects, approximately 1%. But coating defects do occur, and they can be very difficult to locate, especially when the piping is buried. To prevent these costly failures and the repairs and downtime that inevitably follow, regular coating inspections should be performed, beginning early in the life of the coating. *Figure 5* shows some typical coating defects.

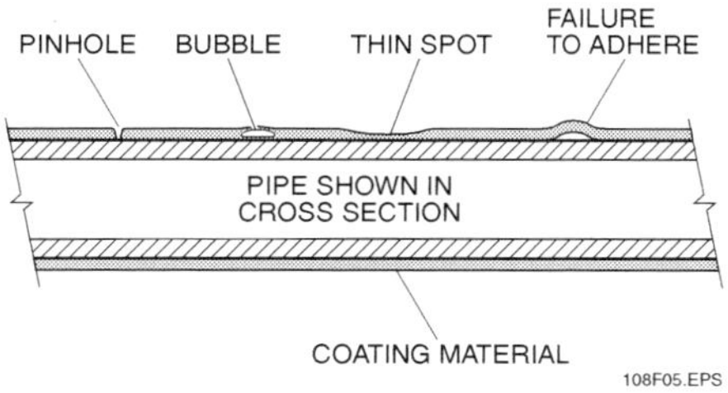

Figure 5 ◈ Typical coating defects.

For buried piping, sophisticated holiday detectors can be used. In locations where coating defects exist, measurable voltage gradients can be detected at ground level. The larger the defect, the greater the current flow.

Although the type of coating may vary with the physical environment, the situation of the piping, and the piping material, the most widely used coating in the United States is fusion bonded epoxy (FBE). It is also the most prevalent pipeline coating in Saudi Arabia, the United Kingdom, and Canada. FBE provides very effective protection against corrosion, especially when combined with CP. CP helps to overcome deficiencies in the coating, which are usually related to the application process or to the organic components of the coating material.

Classroom

Have trainees explain the consequences of overprotection and underprotection. Answer any questions trainees may have.

Teaching Tip

See the Teaching Tip at the end of this module.

Classroom

Have trainees explain the considerations for planning a CP system.

Audiovisual

Show Transparency 7 (Figure 5). Review the typical coating defects on piping and the possible causes of each.

Laboratory

At several stations, place samples of piping with coating defects. Have trainees find the defects and identify each type.

Classroom

Ask a trainee to name the most widely used pipe coating in the United States. Have another trainee explain why CP is combined with fusion bonded epoxy.

Ask a trainee to explain the difference between FBE and Powercrete®. Ask another trainee to identify the advantages of Powercrete®.

Classroom

Have trainees compare and contrast remote and close ground beds.

Audiovisual

Show Transparency 8 (Figure 6). Ask a trainee how to identify whether the proximity of structures has produced a significant shield.

In the FBE process, metallic iron is converted to ferrous **ions** and electrons. Problems with FBE arise because these properties can be removed by the presence of certain chemical substances such as oxygen, chloride, and water, so it does need to be combined with another form of protection. CP is very effective when used with FBE coatings.

In some circumstances, another layer of coating is used over the FBE, especially where the piping is exposed or is subject to extreme conditions. Casings are used to avoid impediments such as open trenches and road and railroad crossings. In many situations in which casing is used, the carrier pipe inside the casing is unprotected, especially with a CP system. Concrete coatings have been used, and although effective, the coatings were heavy and abrasive, making the piping hard to pull. Other coatings were tried over FBE, but when the piping was pulled through rough areas, the coating sometimes peeled or sustained heavy damage.

Finally an epoxy-based polymer concrete was developed that could be applied directly to the FBE. This coating, called Powercrete®, has many advantages over plain concrete. It provides a smooth outer shell with good toughness and abrasion resistance. The thickness of the Powercrete® coating can vary, depending on several factors.

3.5.2 Remote Versus Close Ground Beds

In a CP system, the pipeline is negative and the earth is positive. One of the factors in determining the degree of protection is the potential difference between the pipeline and the earth. The potential can be controlled by increasing either the positive current in the earth or the negative current in the pipeline.

Remote ground beds are those in which the anodes that form the ground bed produce a negative current that produces voltage drops from point to point in the earth. Remote ground beds are used to protect large sections of piping. These voltage drops spread outward from the ground bed. The difference in voltage is greatest close to the ground bed and lessens with distance until it becomes insignificant: the point of insignificance is referred to as remote earth. If a circle were drawn with the ground bed at its center, extending to remote earth, it would show the area of influence of the ground bed.

In a close ground bed, only the area close to the anodes is protected. An area of influence surrounds both the ground bed and the piping. As with the remote ground bed, a voltage drop is produced by the current that flows from the anodes in the ground bed to the piping. The area of influence surrounds only the piping close to the ground bed.

If the area of influence includes both the piping and the ground bed with no uncovered area in between, the ground bed is considered close. If there is an area in between that is not covered, the ground bed is considered remote.

Earth is considered an infinite conductor, and in a remote ground bed, the current between areas of influence flows into the earth and to the pipeline.

3.5.3 Electrical Shielding and CP

An electrical shield can be defined as any barrier that will prevent or divert the flow of CP current from soil or water away from a pipeline for which protection is intended. This electrical shielding can be of two types. One may result from a nonmetallic insulating barrier that prevents current flow. The other involves diversion of current to other metallic structures.

3.5.4 Shielding in Congested Areas

When underground metal structures are close together, an electrical shield may be produced that will lower the protective effect of the CP system. Pumping stations and tank farms are subject to this type of shielding. The severity of the shielding effect varies with conditions at each location. Uncoated metallic structures, especially those in direct contact, tend to increase the effects. Uncoated piping would be severely affected. Even grounding systems and rebar in buried concrete can affect the current flow. Areas with only well-coated piping may not be affected.

In these areas, a remote CP system is unlikely to provide adequate protection. A close ground bed, using either a galvanic or an impressed current system, may provide the needed protection.

Measuring the potential between the earth and the structures will show whether the proximity of the structures has produced a significant shield. The shield will be strongest near the center of the congested area. Full protection is –0.85 volt, according to CSE guidelines (see *Figure 6*). If the reading is as low as –0.7 volts, there is a problem. According to Peabody in *Control of Pipeline Corrosion*:

> The remote ground bed and suitably sized rectifier may change the potential of the entire structure sufficiently to give an indication of full protection when measured with respect to a remote reference electrode.

Instructor's Notes:

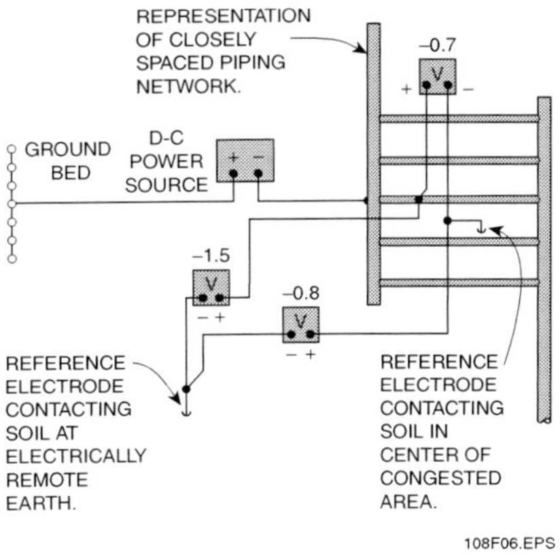

Figure 6 ◆ Electrical shielding in congested areas.

3.6.0 Effects of CP

Underground metallic structures in the vicinity of the CP system may suffer from stray current effects, increasing the rate of corrosion. Designers and planners for the CP system and those for any neighboring structures must be kept informed, both of existing structures and any changes that may be considered. Corrosion testers should always note unusual changes and check for other structures or systems; it may be necessary to alter the system.

3.6.1 Other Structures

All buried steel will corrode unless an effort is made to stop or prevent the corrosion. When steel comes into contact with soil or water—the electrolyte—it immediately starts to corrode. Stray currents, those on an unintended path, cause even more problems. Stray currents occur when two or more operators install CP systems without consulting each other, or possibly when an older structure exists that an operator is unaware of. It should be company policy to notify all operators in an area when a CP system is planned or when changes to an existing system are planned.

Stray current damage to another underground structure is usually referred to as interference. Because of the higher current, impressed current CP systems are usually connected to this type of corrosion damage. Galvanic CP systems are also capable of producing corrosion damage, but because of their lower current, it is far less likely.

3.6.2 Bonds

In cases in which the current pickup by the foreign pipeline is not too great, a metallic **bond** can be installed between the two lines. However, severe cases make it necessary to abandon the ground bed when an adequate bond between the two systems circulates so much of the rectifier current through the foreign structure that little is left for CP of the pipeline for which it was intended. This emphasizes the need for care in selecting CP installation sites so that such situations may be avoided.

4.0.0 ◆ INSTRUMENTATION AND METERS

Corrosion testers use a variety of instruments and meters in their work. Some of the more typical items are described in this section, but corrosion testers should refer to manufacturers' instructions and manuals for specific information. To obtain valid results, the corrosion tester must select the proper instrument for the task and use it correctly.

4.1.0 Checking CP With Instrumentation and Meters

Measuring voltage, such as between a pipeline and a reference electrode, is probably the most commonly performed task in corrosion testing. There are several types of voltmeters. With a digital voltmeter, an unknown voltage is compared with an internally generated analog voltage, and the result is displayed as a number. An analog voltmeter displays results through a needle moving over a scale.

4.1.1 Voltmeters

In a voltmeter, the meter coil has a fixed resistance, and therefore when current flows through the coil, a voltage drop will be developed across this resistance. According to **Ohm's law**, the voltage drop will be directly proportional to the amount of current flowing through the coil. In addition, the amount of current flowing through the coil is directly proportional to the amount of voltage applied to it. By calibrating the meter scale in units of voltage instead of current, the voltage in a circuit can be measured.

Classroom

Ask trainees to name two causes of stray current. Answer any questions trainees may have.

Discuss how a metallic bond between two pipelines can prevent current damage.

Homework

Assign the review of Sections 4.0.0–6.0.0 and the completion of the Review Questions for the next session.

Materials

Ensure that you have everything required for demonstrations and teaching during this session.

Classroom

Ask a trainee to distinguish between a digital voltmeter and an analog voltmeter.

Demonstration

Provide a variety of voltmeters with different features. Call on trainees to identify the features of each voltmeter and its applicability to various jobs.

Demonstrate the proper way to use a voltmeter.

Laboratory

Have trainees practice taking measurements with various types of voltmeters.

Voltmeters must be in parallel with the circuit component being measured. On the higher ranges, the amount of current flowing through the meter is much lower due to its very high total resistance. However, an inaccurate reading will result if the voltmeter is placed in series rather than in parallel with a circuit component. When connecting a DC voltmeter, always observe the proper polarity. The negative lead of the meter must be connected to the negative or low potential end of the component, and the positive lead to the positive or high potential end of the component. If you connect a voltmeter to the component with the opposing polarities, the meter coil will move to the left and the pointer could be bent. In an AC circuit, the voltage constantly reverses polarity, so there is no need to observe polarity when connecting the voltmeter to a component in an AC circuit.

To make a voltmeter capable of measuring multiple ranges, you need several multiplier resistors that are switch-selectable for the different ranges that are required. Some multi-range voltmeters have only one range of values marked on the scale, and the scale reading must be multiplied by the range switch setting to obtain the correct voltage.

Other voltmeters have different ranges on the scale for each setting of the range switch. When using one of these voltmeters, be sure that you read the set of values that corresponds to the position of the range switch. See *Figure 7* for a typical voltmeter.

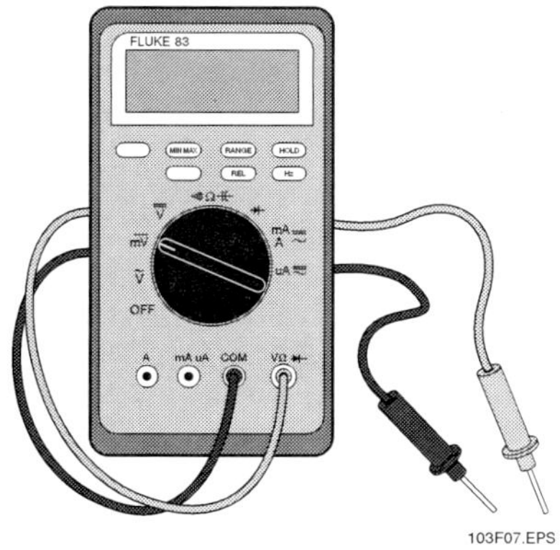

Figure 7 ◈ Voltmeter.

Some digital voltmeters are autoranging. These types of voltmeters do not have a range switch. The internal construction of the meter will select the proper resistance for the current being detected. But when using a voltmeter that is not autoranging, always start with the highest voltage setting and work down until the indication reads somewhere between half-scale and three-quarters scale. This will give a more accurate reading and will prevent damage to the voltmeter.

With regard to meter accuracy, many meters offer a 3½- or 4½-digit resolution. This refers to the number of significant digits that will be used to display an accurate voltage value. For CP measurements, where a voltage of, for instance, –0.972 volt is being measured, a more accurate measurement of –0.9721 may not be relevant to a field determination of protection. The significance of extreme precision must be viewed in conjunction with the necessity of such accuracy.

The selection of voltmeters should also include an evaluation of cost versus practicality. Many commercially available meters are handheld, battery-powered, and well suited for field applications, while others are primarily suitable for bench tests or laboratory work. Voltmeters can be purchased with an array of features such as large data-storage capacity, oscilloscope display, recording capability, and computer interface connections. A voltmeter should be chosen for its applicability to the work at hand, with the proper balance of cost, durability, and accuracy.

4.1.2 Ammeters

Ammeters are used to measure current. Most ammeters measure only small amounts of current, typically in the range of microamperes, or mA (0.000001 ampere), or in milliamperes, mA (0.001 ampere). Few ammeters can measure more than 10mA, so to increase the range to the ampere level, a shunt is used. A shunt is a resistor that is mounted separately and is used for extending measurement of currents beyond the limits of most self-contained DC ammeters.

To measure above 10 mA, a shunt with an extremely low resistance is placed in series with the load, and the meter is connected across the shunt to measure the resulting voltage drop proportional to current flow.

Shunts located inside the meter case (internal shunts) are generally used to measure values up to 30 amps; shunts located away from the meter case (external shunts) with leads going to the meter are generally used to measure values greater than 30 amps. Above 30 amps of current, the heat generated could damage the meter if an internal shunt were used. The use of a shunt allows the ammeter to derive current in amps by actually measuring the voltage drop across the shunt.

Instructor's Notes:

> **WARNING!**
> Never connect an ammeter in parallel with a load. This will cause a short circuit because of the low resistance in the ammeter, probable damage to the meter and/or circuit, and potential personal injury.

When connecting an ammeter in a DC circuit, you must observe proper polarity. In a DC circuit, you must connect the negative (–) terminal of the meter to the negative or low potential point in the circuit, and connect the positive (+) terminal of the meter to the positive or high potential point in the circuit. The current flow through the ammeter must flow from negative to positive terminal. If you connect the meter with the polarities reversed, the meter coil will move in the opposite direction, and the pointer might strike the left retaining pin. You will not obtain a current reading, and you might bend the pointer of the meter.

A multi-range ammeter is one containing a basic meter movement and several shunts that can be connected across the meter movement. A range switch is normally used to select the particular shunt for the desired current range. Some multi-range ammeters have only one set of values on the scale, even though they measure several different current ranges. For example, if the scale is calibrated in values from 0 to 1 milliamp (mA), and the range switch is in the 1mA position, read the current directly. If the range switch is in the 10mA position, multiply the scale reading by 10 to find the amount of current flowing through the circuit.

In an AC circuit, when measuring AC current at levels greater than one ampere, a clamp-on ammeter is often used. This type of ammeter also measures DC. The clamp-on ammeter may have a mechanical movement or a digital readout. These meters clamp over the wire and do not break the insulation. They sense current flow by measuring the magnetic field surrounding the conductor. When using this type of ammeter, be sure to measure only one conductor at a time. This type of ammeter will often measure up to 1,000 amperes at 600 volts.

Portable, noncontact ammeters are useful instruments on a pipeline, where they are used to measure current magnitude and direction of flow. However, the clamp must fit around the piping or cable. Among the functions they perform are interference testing, finding defective insulation in joints, location of shorts, and the detection of faults. *Figure 8* shows a typical handheld ammeter.

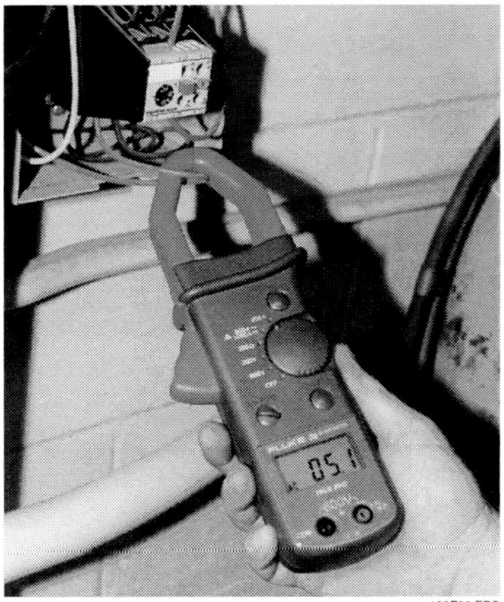
108F08.EPS

Figure 8 ◆ Handheld ammeter.

4.1.3 Ohmmeters

An ohmmeter (*Figure 9*) is an instrument that measures the resistance of a circuit or component. The ohmmeter is also used to locate **open circuits** or shorted circuits. An ohmmeter consists of a DC current meter movement, a low-voltage DC power source (usually a battery), and current-limiting resistors, connected in series.

Before measuring the resistance of an unknown resistor or electrical circuit, the **test leads** are shorted together. This causes the current to flow through the meter movement and the pointer deflects to the right. This indicates that there is zero resistance present across the input terminals of the ohmmeter. The pointer will deflect full scale when there is zero resistance present. The majority of the ohmmeters have a zero adjustment knob on the face of the meter. This knob is used to correct for the fact that, as the batteries of the meter weaken, their output voltage decreases. The meter will no longer deflect to full scale as the voltage of the batteries decreases. By correcting for this change before each use of the meter, the internal resistance of the meter, along with the resistance of the leads, is lowered to deflect the pointer full scale.

Safety

Stress the danger of connecting an ammeter in parallel with a load. Describe the possible consequences.

Classroom

Provide various types of ammeters, shunts, and circuits to measure. Ensure that trainees know when to use shunts.

Safety

Emphasize the importance of verifying and de-energizing the circuit before taking a reading with an ohmmeter.

Laboratory

Discuss the procedure for measuring resistance with an ammeter. Then have trainees practice taking measurements using an ammeter.

Safety

Review proper usage procedures for multimeters. Discuss possible consequences of improper usage.

Laboratory

Have trainees practice using multimeters. Also have them demonstrate proper care of multimeters.

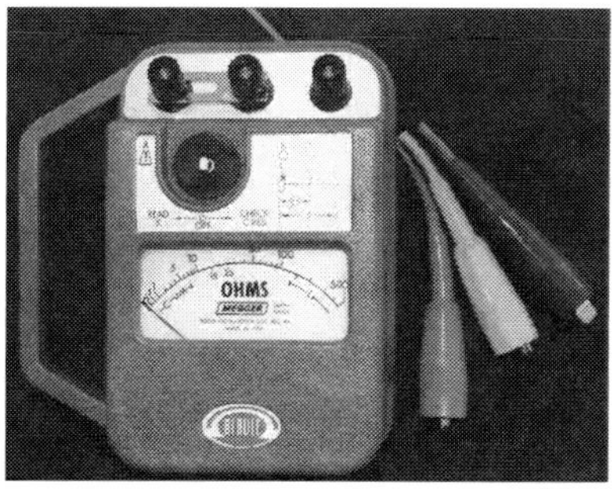

Figure 9 ◆ Ohmmeter.

> **WARNING!**
> The circuit must be verified as de-energized by using a voltmeter prior to taking a reading with the ohmmeter. Failure to de-energize the circuit could cause a damaging current to flow through the meter. This can cause damage to the meter and/or circuit, as well as personal injury.

Each component in the circuit should be tested individually by removing the component from the circuit and connecting the ohmmeter leads across it when making resistance measurement. If the component is not removed from the circuit, the component can be effectively isolated by disconnecting one of its leads from the circuit. However, be careful to ensure that other components are not in parallel with the component being tested or this will lead to an incorrect reading.

4.1.4 Multimeters

The **multimeter** (*Figure 10*) is a volt-ohm-milliammeter (VOM) and is a multipurpose instrument. It is a combination of a voltmeter, milliammeter, and an ohmmeter. There are many different models of the basic multimeter, and they can be analog or digital. For specific instruction on the use of your multimeter, refer to the applicable owner's manual.

A **fuse** is installed in a multimeter to protect the meter circuits on the ohmmeter ranges. It also protects the milliampere ranges from excessive overloads. In addition to the fuse, a varistor protects the indicating instrument circuit. The varistor limits the current through the moving coil in case of overload.

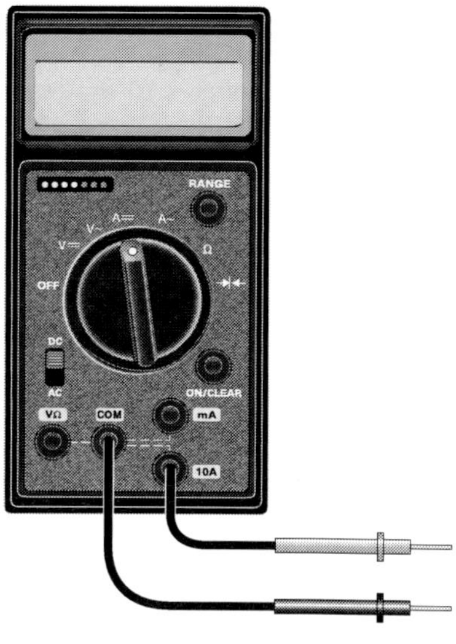

Figure 10 ◆ Multimeter.

The fuse and varistor will prevent serious damage to the meter in most cases of accidental overload. Nevertheless, no overload protection system is foolproof, and improper use of the meter on high-voltage circuits can damage the instrument.

> **CAUTION**
> Always follow the proper usage procedures to avoid meter damage or personal injury.

Some points to remember when using a multimeter are as follows:

- Keep the instrument in a horizontal position when storing and away from the edge of a workbench, shelf, or other area where it may be knocked off and damaged.
- Avoid rapid or extreme temperature changes. Such changes will advance the aging of the meter components and adversely affect the meter life and accuracy.
- Avoid overloading the measuring circuits of the instrument. Always check the range position before connecting the test leads to a circuit.

Instructor's Notes:

- Place the range switch in the TRANSIT position when the instrument is not in use or when it is being moved. Not all meters have a TRANSIT position, but if the meter that you are using has one, this position should be used to avoid damage to the meter.
- Rotate the function and range switches in both directions to wipe the switch contacts, if the meter has not been used for a long period. This will remove most of the tarnish that may have formed on the contacts.

4.1.5 Resistance Test Instruments

Among the first steps to be taken when planning a ground bed is performing a **soil resistivity** test. Soil resistivity is measured in **ohm-cm** and can be tested using a voltmeter, an ammeter, and a battery, although specialized instruments are available. The resistance is the reading on the dial after it has been multiplied by the range switch position. The multiplier (scale) determines the resistance values, which can be anywhere between 0.01 ohm and 11 times 105 ohms. Problems with polarization of the pins is eliminated because the resistance meter uses AC power. This problem usually occurs with DC tests.

In some instances, the corrosion tester is able to estimate the average soil resistivity from an existing ground bed rectifier output or from older soil surveys, but the history of the area should be checked carefully for any changes that may have been made. It is safer to take new resistivity measurements.

One of the most common means of testing the resistivity of soil is the **Wenner four-pin method**. In this method, four metal pins are driven into the earth. The pins must be equidistant from one another and must form a straight line. The pins are spaced at a distance equal to the soil depth to be tested. Current flows between the two outside pins, so the voltage drop between the two inner pins provides the resistance. This method is described in detail in *American Society for Testing and Materials G57* and *Institute of Electrical and Electronics Engineers Standard 81*.

A resistance test meter can be used to measure the voltage drop at the same time that it supplies the necessary current for the outer pins. *Figure 11* shows a typical soil resistance meter.

Resistance meters can be used for determining resistance when installing an anode bed or a grounding system. However, the meter is likely to be damaged if used across energized circuits.

108F11.EPS

Figure 11 ◆ Soil resistance meter.

In addition, accessories such as long, stainless steel soil pins (see *Figure 12*) with T-handles, selector boxes for multiple depth testing, and soil boxes (*Figure 13*) are available with resistance test instruments.

> **CAUTION**
> Soil boxes may allow improper sample handling, compaction, and moisture preservation.

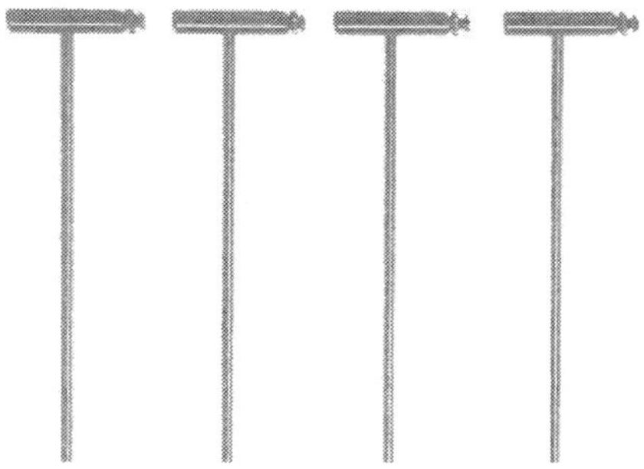
108F12.EPS

Figure 12 ◆ Soil resistance pins.

Safety

Explain proper use of soil resistance boxes. Ask trainees to name possible consequences of improper sample handling, compaction, and moisture preservation.

Laboratory

Have trainees practice measuring soil resistivity by using the Wenner 4-pin method and a soil resistance meter. Also have them practice using soil resistance pins, selector boxes, and soil boxes.

Audiovisual

Show Transparency 9 (Figure 14). Ask a trainee to explain the concept of half cells.

Classroom

Provide a variety of copper sulfate electrodes. Discuss the process of making potential measurements with such electrodes.

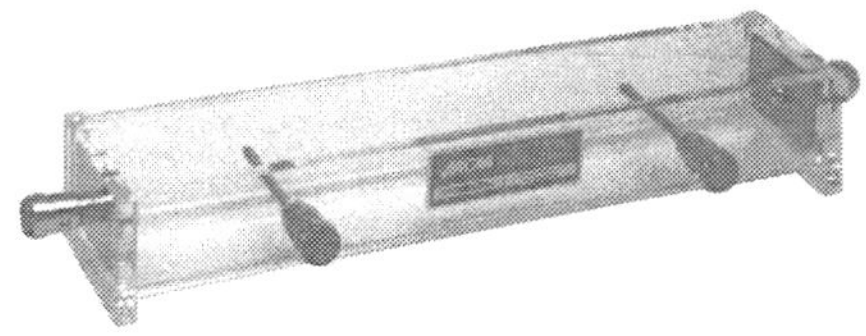

108F13.EPS

Figure 13 ◆ Soil resistance box.

4.1.6 Half Cells (Permanent Buried/Portable)

The copper-copper sulfate electrode is the reference electrode used most commonly in pipeline corrosion work. It may be referred to as a copper sulfate electrode, copper sulfate reference, copper sulfate half cell, or CuSO4 of a copper sulfate electrode.

Electrically, the voltmeter connection is to a pure copper rod. The rod makes electrical contact with the saturated copper sulfate solution within the body of the electrode. Then, actual contact with the earth is made with the solution-saturated porous plug.

The concept of half cells is illustrated in the equivalent circuit, as shown in *Figure 14*. When a voltmeter is connected between the copper sulfate reference electrode and the structure, the observed reading is actually a combination of two values: the potential between the copper rod and earth through the copper sulfate solution, and the potential between the structure and the earth. Each of these values represents a portion of the total cell potential between the piping and the copper rod of the copper sulfate electrode. The structure-to-earth half-cell potential is the variable that must be evaluated. The copper sulfate reference electrode-to-earth half-cell potential is a constant for practical purposes.

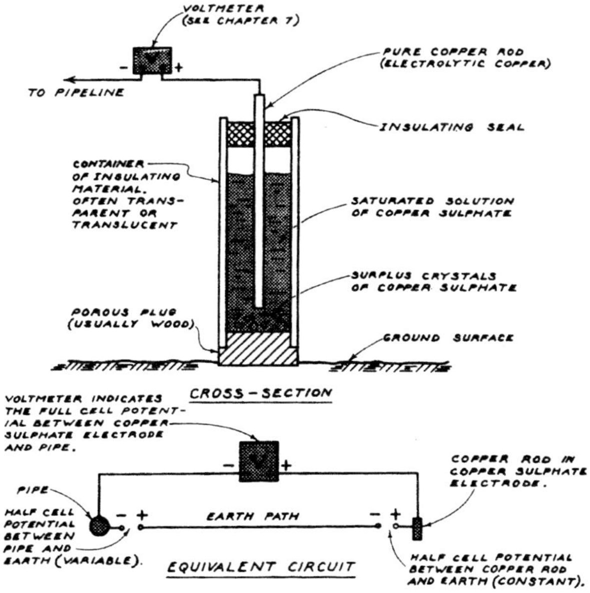

108F14.EPS

Figure 14 ◆ Copper sulfate electrode.

Instructor's Notes:

To make **potential measurements**, the electrode should be placed near the line at the same depth as the pipe; however, this is not feasible in most situations. The usual method for these pipes is to take the measurement at surface level directly above the pipeline; allowing for the potential drop between the pipe and the point above it at the surface, a satisfactory level of protection would be a value of –0.85 volt. *Figure 15* shows a typical copper sulfate electrode.

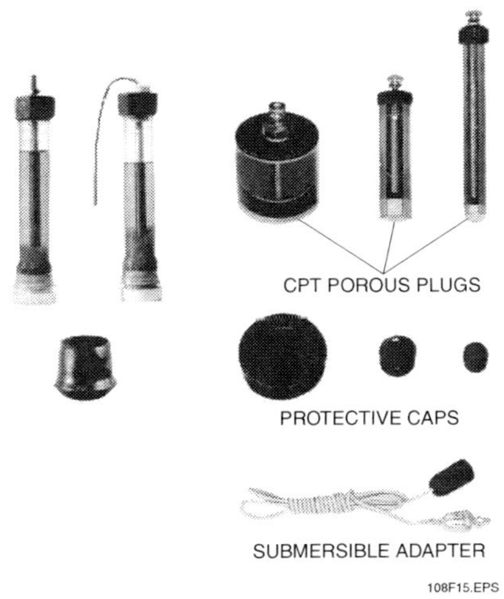

Figure 15 ◆ Copper sulfate electrode.

4.1.7 Silver Chloride (Salt Water)

Permanent portable salt water reference electrodes may be made with zinc/silver-silver chloride. These electrodes are used for sea water installations, such as off-shore facilities, harbor structures, and soils with high chloride content. Portable electrodes are available that are rugged enough for deep water use and can be used in salt water and saturated saline water conditions such as offshore platforms, offshore pipelines, and seawater docks.

The natural potential between zinc and steel is much lower than that between magnesium and steel. When other considerations are equal, the current output of a zinc anode is lower than that of a magnesium anode of the same size. In lower-resistance soil this may not be a disadvantage; but it may make it possible to install anodes for a longer projected life than would otherwise be possible.

At low current output, the efficiency of zinc does not fall off as does that of magnesium. It is possible in many cases to install zinc anodes with a projected life of 20, 30, or more years. Magnesium anodes are seldom practical beyond 10 years.

In some circumstances, the low driving potential of zinc with respect to steel can be an advantage. A generous set of zinc anodes will raise a steel structure to about 1.1 volts with respect to a copper sulfate electrode, but cannot be raised higher because this is about the potential of zinc. If the coating fails in part, or if the resistance of the medium changes, as with tidal water, the current output of the zinc anodes will vary over a wide range with very little shift in the potential of the protected structure. It can function as a constant potential system. Magnesium can do the same, but at the much higher potential of 1.6 volts, which may require four times as much current. *Figure 16* shows a silver chloride reference electrode.

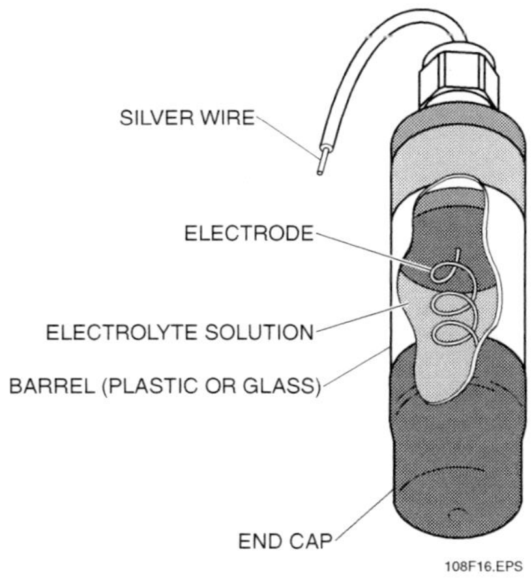

Figure 16 ◆ Silver chloride reference electrode.

4.2.0 Recognizing Electrical Discontinuity

It is important to properly use and understand the data obtained from meter readings. This section describes the following:

- Continuity test
- Meter check and verification of an ohmmeter
- Meter test leads
- Out-of-range requirements

Classroom

Discuss the pros and cons of zinc anodes and magnesium anodes. Ask trainees if they have any questions.

Safety

Stress the importance of turning off the main circuit breaker before attempting a continuity test.

Laboratory

Set up stations with multimeters and audio and visual continuity testers with access to circuits that can be tested. Have trainees practice testing for continuity using each type of device.

Provide multimeters with damaged and undamaged test leads, probes, and plugs. Have trainees practice inspecting these components for damage.

Set up stations with a nonmetallic abrasive pad, distilled water, and a few copper sulfate crystals. Have trainees practice maintaining copper sulfate electrodes.

4.2.1 Continuity Testing

A continuity test is done to determine whether a circuit is open or closed. Testing the continuity of a circuit requires the use of a multimeter or a continuity tester (a continuity tester is a simple device that lights up to indicate continuity). Set the multimeter to the ohm setting. The symbol for ohm is the Greek letter omega (Ω). If there is more than one ohm setting, choose X1. While the probes are not touching anything, the multimeter will indicate a reading of infinity. A reading of infinity means that the circuit is open. When you touch the two probes together, the reading changes to zero. A reading of zero indicates that the circuit is closed or complete. A complete circuit is one that can conduct electricity; an **open circuit** cannot.

> **WARNING!**
> Continuity testing should be performed when current is not present. Always unplug the device or turn off the main circuit breaker before attempting a continuity test.

Touch each probe to one of the terminals (or poles) of the device. If the reading changes to zero, the device has continuity. To test a switch, place a probe on each pole of the switch. When you move the switch from the OFF to the ON position, the meter reading should change from infinity to zero, which implies that the switch is working. To test a component such as a motor, touch a probe to each pole. A reading of zero indicates that motor has continuity and current can pass through it.

An ohmmeter, as part of a multimeter, can be used to test continuity, but this requires carrying an expensive and sensitive test instrument. Pocket-type continuity testers are just as reliable and much more compact and portable.

There are two types of continuity testers: audio and visual. These are specialized instruments for identifying conductors in a conduit by checking continuity.

The audio continuity tester is used to ring out wires in conduit runs. At one end of the conduit run, select one wire, strip off a little insulation, and connect that wire to the conduit. At the other end of the conduit run, clip one lead of the tester to the conduit. Touch the other lead to one wire at a time until the audible alarm sounds, which indicates continuity, where continuity is a closed circuit. Then place matching tags on the wire to indicate that this is a completed circuit. Continue this procedure, one wire at a time, to identify all the other wires in the conduit.

The visual continuity tester is useful if you are working in an area where background noise might make it hard to hear the audio tester. The procedure for the proper operation of this tester is the same as for the audio continuity tester. The light will come on when the proper wire is tested.

4.2.2 Meter Test Leads

Prior to using any meter test lead, inspect the test leads, test probes, and test plugs for any signs of damage, broken wires, or signs of electrical arcs. The test leads are key to performing an accurate test of the equipment. When you use a defective test lead, the equipment test results will not be reliable. Replace any test lead with a test lead designed for the equipment that is being used. Most test equipment specifies a specific model and part number for the leads that are supported by that equipment.

4.2.3 Out-of-Range Requirements

Companies publish standards for the equipment to be tested. When the tests results are out of range of these standards, refer to the company policy for the appropriate corrective action. Company policies vary with application of the equipment.

5.0.0 ◆ VISUAL HALF-CELL INSPECTION

Visually inspect the soil to determine the type of soil present. The soil needs to be a little damp. This may require digging down a little or adding some water to the area.

Copper sulfate electrodes should be kept clean. The copper rod may be cleaned when it becomes encrusted, using a non-metallic abrasive pad. Rinse the rod thoroughly with distilled water before reassembling it. Ensure that the electrode is always full of saturated copper sulfate solution; use a few excess copper sulfate crystals or powder to ensure that the solution remains concentrated. Replace the solution with fresh solution periodically to avoid contamination. To avoid having the electrode dry out, keep the porous plug covered when the rod is not in use. If the electrode does dry out, re-wet the porous plug. Porous plugs seldom need to be replaced. Each test kit should contain at least two copper sulfate electrodes that are matched in potential. The potentials should be checked periodically by measuring the potential difference between the two when placed side by side in soil or water. For best performance, the potential difference should not exceed the value indicated on the cut sheet, or about 10 millivolts.

Instructor's Notes:

When using reference electrodes other than copper-copper sulfate, refer to the comparisons shown in *Table 2* for required adjustments.

6.0.0 ◆ ELECTRICALLY INSULATED FITTINGS

Usually, cathodically protected piping systems require that all fittings be insulated. Insulated fittings are used on transmission lines to electrically isolate the pipeline from terminal facilities or pumping stations. They are also used to divide the pipeline system into sections.

Electrically insulated fittings are used on cathodically protected structures to isolate the structure from terminal facilities, pumping stations, or other structures, and also to divide piping systems into isolated sections. In areas where there are numerous sources of stray currents, it may be helpful to use more closely spaced insulated fittings to control stray current discharge and pickup. The location and spacing of insulated fittings is based on experience with areas in which stray currents are likely and on the information collected during testing.

Field-assembled or preassembled insulated flanges or mechanical fittings may be used. These include insulating bushings, unions, meter bars, and meters with insulating swivels.

Insulated fittings of a cathodic protection system are discussed in the following sections.

6.1.0 Insulated Flanges

Many different types of flange-insulating kits are available from corrosion control equipment suppliers. *Figure 17* shows a typical insulated flange assembly.

Bolt holes on standard flanges are 0.125" larger than the standard bolts. This allows full-length insulating sleeves to be used. As long as the sleeves are properly aligned, they will provide adequate insulation.

Existing flanges can be insulated without taking the pipeline out of service. This can only be done if the existing gasket is made of a good insulating material. It is impossible to insulate an in-service flange if the existing gasket material is metal or graphite-impregnated. To insulate a fitting that has the appropriate gasket material, remove one bolt at a time from the flange, insulate it, and then reinstall it before removing the next bolt.

Sometimes existing flanges are not aligned closely enough for insulating standard bolts. When this occurs, use insulated steel washers and a half-length insulating sleeve on one half of the flange. A half-length sleeve goes through one side of the flange only, so it does not have to be aligned with the other side of the flange. Full bolt insulation can be used if a smaller bolt with a higher strength is used (refer to *Figure 17*).

For new construction, sometimes insulated fittings are assembled and then tested before they are welded into the line. Prefabricated insulated flanges can be tested using an ohmmeter. The resistance values should be measured in the megohm range to indicate proper insulation. *Figure 18* shows shop testing of an insulation flange.

Preassembled flanges also allow for perfect alignment of the flange. This reduces the stresses placed on the insulating material when it is installed in the pipeline.

Table 2 Comparison of Other Reference Electrode Potentials With That of Copper-Copper Sulfate

Type of Comparative Reference Electrode	Structure-to-Comparative Reference Electrode Reading Equivalent to −0.85 Volt with Respect to Copper Sulfate Reference Electrode	To Correct Readings Between Structure and Comparative Reference Electrode to Equivalent Reading with Respect to Copper Sulfate Reference Electrode
Calomel (saturated)	−0.776 volt	Add −0.074 volt
Silver-Silver Chloride (0.1 N KCl Solution)	−0.822	Add −0.028
Silver-Silver Chloride (silver screen with deposited silver chloride)	−0.78	Add −0.07
Pure Zinc (special high grade)	+0.25*	Add −1.10

*Based on zinc having an open circuit potential of −1.10 volt with respect to copper sulfate reference electrode. (From A. W. Peabody, Chapter 5, p. 59, *NACE Basic Corrosion Course*, Houston, TX, 1973.)

108T02.EPS

Classroom

Ask a trainee to name the purposes of electronically insulated fittings. Have another trainee explain where they are used.

Demonstration

Demonstrate how to insulate existing flanges in various installations.

Laboratory

Have trainees practice measuring the resistance across insulated flanges. Then have them practice identifying the cause of a defective insulated flange.

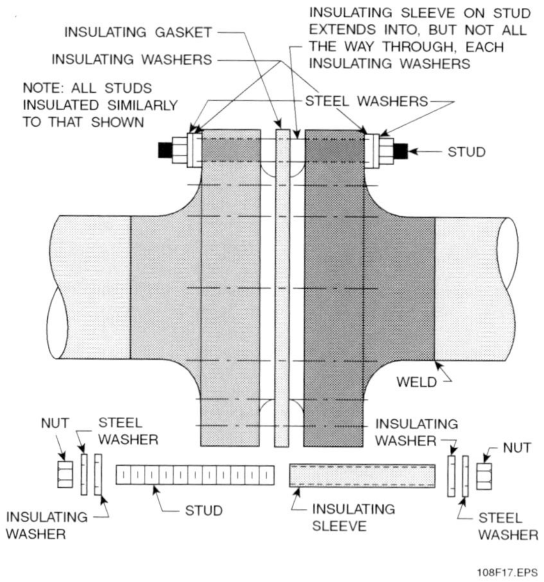

Figure 17 ◈ Insulated flange assembly.

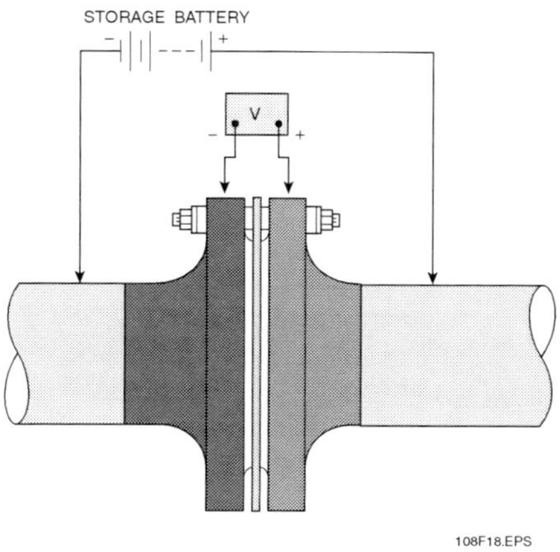

Figure 18 ◈ Shop testing of insulating flanges.

6.2.0 Testing Insulated Conductive Fittings

After an insulated fitting is installed in the pipeline system, the resistance measured across the flange may be only a few ohms.

> **NOTE**
> Measurements of pipe-to-soil potential on either side of the fitting may be used to identify possible shorts. Always use an insulator checker to identify whether the bolts are shorted or the flanges are shorted.

The installed fitting has a substantial length of pipe on each side of the fitting. This reduces the resistance of several million ohms measured before installation to a resistance of only a few ohms after installation. The longer the length of the pipe on each side of the fitting, the lower the resistance will be. The resistance measured is also affected by the quality of the coating and the soil resistivity in the area. Remember that when mea-

Instructor's Notes:

suring the resistance of an installed fitting, you are measuring the fitting resistance as well as the resistance to earth on each side of the fitting.

If a defective insulated flange is found during testing, it may be possible to repair it without taking the pipeline out of service. To determine if the pipeline must be taken out of service, identify the cause of the short.

If the short is a defective insulated bolt, the short may be fixed without taking the pipeline out of service. If the flange is shorted across the gasket, the pipeline must be taken out of service and the gasket replaced. To determine the cause of the short, check the resistance between the flanges and each bolt. Bolts that are shorted will have a low or zero resistance. The insulation can then be replaced on the shorted bolts individually. If all of the bolts have a high resistance and any possible bypassing routes have been isolated, the cause is probably a shorted gasket.

6.3.0 Grounding Cells

In some cases, pipelines may require protection devices if the insulated fittings are subject to lightning damage.

> **CAUTION**
> Lightning damage protection is often required when a pipeline closely parallels high-voltage electrical transmission lines. These utilities are often located along the same right-of-way when possible.

Lightning can cause a very high voltage in the pipeline that can burn through the insulating material of a fitting if it passes through the fitting. Grounding cells can be used to provide adequate protection. Grounding cells provide voltage surge protection.

The value of resistance between the two anodes will depend on the resistivity of the backfill that is used. Properly installed ground cells will have a resistance between 0.2 and 0.5 ohm. When a high voltage surge occurs, the ground cell conducts the voltage across the anodes and lowers the voltage across the insulated fitting. If very high-voltage surges are anticipated, a four-anode ground bed can be used. These ground beds lower the voltage across the insulated flange even further. Also, the zinc anode attached to the unprotected pipeline provides some cathodic protection to the pipe in the flange area.

6.4.0 Isolation Surge Protectors and Polarization Cells

Other devices that provide grounding and protection for insulated fittings are isolation surge protectors (ISPs) and **polarization** cells. ISPs are electronic, and block DC current flow while allowing AC current flow to protect the insulated fitting. These devices can automatically reset after a fault condition. ISPs provide better protection for the system than grounding cells.

Summary

CP is a technique employed to reduce the corrosion rate of a metal surface by making it the cathode of an electrochemical cell. In anodic areas, current flows from the pipeline steel into the surrounding electrolyte and the pipeline will corrode. At the cathodic areas, current flows from the electrolyte onto the pipe surface and the rate of corrosion is reduced. In a CP area, the rate of corrosion is reduced by exposing every bit of exposed metal on the surface of a pipeline to collect this current.

A key element in effective corrosion control of any pipeline is the ability of the corrosion worker to properly use instruments and meters. The use of these meters will be required to perform most all inspections for a pipeline. A voltmeter is used to check voltage at many different levels; an ohmmeter is used to check the resistance in a circuit. There are a few special meters that the corrosion worker uses to increase the accuracy and speed of their inspection.

It is important for corrosion workers to properly inspect and care for the meters and test leads that they use. If the test leads are not serviceable, then the readings that are taken may not be accurate.

The potential difference between a buried pipe and the soil is of considerable importance, both in investigating the corrosive conditions and in evaluating the extent of CP being applied. This quantity actually is measured by connecting an instrument between the pipe and a special electrode placed in contact with the soil. This electrode is known as a half cell and can use different solutions depending on the location and conditions of the test. The most common type of electrode in use is the metal-to-electrolyte junction consisting of copper in contact with a saturated solution of copper sulfate.

There are many types of structures where the structure-to-soil potential will need to be made;

Safety

Ask a trainee to identify some of the common circumstances requiring lightning damage protection. Ask another trainee to explain the consequences of a lightning strike on a pipeline system.

Demonstration

Bring in isolation surge protectors and polarization cells. Ask trainees to explain how each of these devices works.

Classroom

Go over the Review Questions. Answer any questions trainees may have on materials covered to this point.

Examination

Administer the Module Examination. Be sure to record the results of the Exam on Craft Training Report Form 200 and submit the results to the Training Program Sponsor.

Performance Testing

Ensure that all Performance Tasks have been completed and Performance Profile Sheets for each trainee are filled out. Be sure to record the results of the testing on Craft Training Report Form 200 and submit the results to the Training Program Sponsor.

for example, casings, tank bottoms, well casings, and underground storage tanks. These potentials are important in determining the effectiveness of the cathodic corrosion protection for the structure. When a structure is to be measured, the corrosion worker must ensure that the electrolyte is in contact with the half cell, and that company policy is followed. After soil potential readings are taken, ensure that the readings are recorded according to current company policies.

Rectifiers are used to change AC to DC volts, which are used to supply the power for CP systems. The rectifying elements typically used in units are silicon diodes. A silicon diode has high resistance to current flow in one direction and low resistance in the other. When a rectifier is in operation, the corrosion tester needs to verify that the rectifier is operational through visual checks and by taking readings from the unit.

Review Questions

1. The most common form of corrosion is _____.
   a. erosion
   b. metal fatigue
   c. rust
   d. flaking

2. Structures that contain a carrier pipeline are called _____.
   a. casings
   b. super pipelines
   c. tanks
   d. foreign structures

3. Foreign structures include _____.
   a. company-owned pipeline
   b. metallic structures only
   c. structures owned by another company
   d. all structures on the ROW

4. A typical rectifier unit has _____.
   a. frequent arcing
   b. an AC ammeter and an AC voltmeter
   c. two-way current flow in the secondary winding of the transformer
   d. high resistance to current flow in one direction; low in the other

5. If rectifiers do not have a voltmeter or ammeter on the control panel, the corrosion tester must take and record measurements with _____.
   a. a voltmeter or ammeter
   b. test strips
   c. shunts present
   d. all shunts disconnected

6. In anodic areas, current flows from the _____.
   a. electrolyte to the anode
   b. structure into the electrolyte
   c. protective coating to the anode
   d. soil onto the pipeline

7. A(n) _____ is used to attract corrosive action.
   a. anode
   b. cathode
   c. sacrificial metal
   d. test station

8. Impressed current corrosion protection circuits _____.
   a. use an AC power supply for greater efficiency
   b. can use only copper conductors
   c. require an external DC source
   d. provide a limited amount of protection compared to a galvanic anode

9. Impressed current CP systems _____.
   a. can handle a wide range of voltage and current outputs
   b. are less complex than galvanic anode systems
   c. require less maintenance than galvanic anode systems
   d. cost less than galvanic anode systems

10. When planning a CP system, include all of the following *except* _____.
   a. the length of the pipeline
   b. soil conditions
   c. access for CP testing
   d. foreign structures in the area

Instructor's Notes:

11. Remote ground beds are used to provide CP to _____.

a. small sections of piping
b. aboveground structures
c. larger sections of piping
d. pipelines in saltwater

12. Stray current corrosion damage to an underground structure caused by a CP system on another structure is commonly referred to as _____.

a. cross-over damage
b. interference
c. proximity damage
d. interception

13. Resistance test instruments measure _____.

a. soil resistance
b. resistance in CP systems
c. energized circuits
d. resistance of standard steel soil pins

14. Permanent portable salt water reference electrodes are _____.

a. made of silver/silver oxide
b. used in soils with high sulfide content
c. used for sea water installations
d. not used for offshore facilities

15. When you, as a corrosion tester, obtain test results outside your company standards, _____.

a. repeat the test and record the results
b. refer to your company policies for appropriate action
c. calibrate the test instrument
d. double check the results

Instructor's Notes:

GLOSSARY

Glossary

This glossary will help you learn some terms specific to cathodic protection measurement. Trade Terms introduced in this module appear in italic text.

AC meter creep: Faulty operation or connection of a disconnect, which causes power ON.

Ammeter: A meter that measures amperes. Generally available as the direct-connect type and clamp-on type.

**Anode:* The electrode of an electrochemical cell at which oxidation occurs. Electrons flow away from the anode in the external circuit. Corrosion usually occurs and metal ions enter the solution at the anode.

**Cathode:* The electrode of an electrochemical cell at which reduction is the principal reaction. Electrons flow toward the cathode in the external circuit.

Bond: The electrical connection of two structures that usually have two different cathodic protection levels.

Breaker: Switch in the circuit that allows for interruption.

Casing-to-soil potential: The voltage potential of the casing in the soil as related to a reference electrode.

Casing vent: An aboveground metal pipe connected to the buried casing, which provides access to and venting of the atmosphere within the casing.

Copper-copper sulfate electrode (CSE): A commonly used non-polarizing electrode used in corrosion control to measure the electrical potential of a metal structure to a surrounding electrolyte to determine the potential for corrosion damage or to monitor the effectiveness of existing control measures.

**Current:* 1) A flow of electric charge. 2) The amount of electric charge flowing past a specified circuit point per unit time, measured in the direction of net transport of positive charges. (In a metallic conductor, this is the opposite direction of the electron flow.)

**Disbondment:* The loss of adhesion between a coating and the substrate.

Electrical isolation: The electrical separation of two structures.

Electrical potential: The potential energy of a unit charge at any point in an electric circuit, measured with respect to a specified reference point in the circuit or to ground.

**Electrolyte:* A chemical substance containing ions that migrate in an electric field.

**Foreign structure:* Any metallic structure that is not intended as a part of a system under cathodic protection.

Fuse: A safety device used to protect circuits from surges and lightning. It is installed in the circuit to protect other electrical devices.

Galvanic: Pertaining to electricity that flows as a result of a chemical action.

High input impedance voltage meter: An electronic device used to measure structure-to-soil potential.

Interference: The effect upon cathodic protection levels of a structure due to the interaction between its cathodic protection systems and the cathodic protection system(s) of other structures.

**Ion:* An electrically charged atom or group of atoms.

Megger: A test instrument that can generate required current and measure results of induced current at specified points.

Multimeter: A meter that allows measurement of different quantities through the use of a selector switch.

Ohm-cm (Ohms per centimeter): Resistance of a cube of material (earth/soil) with the sides of the cube being measured in centimeters.

Ohm's law: A statement of the relationships between current, voltage, and resistance in an electrical circuit. Current (I) equals voltage (E) divided by resistance (R), generally expressed as a mathematical formula: $I = E \div R$.

Open circuit: A discontinuous circuit.

Output terminal: A terminal associated with the output of a circuit or device.

**Pipe-to-soil potential (also structure-to-electrolyte potential):* The potential difference between the surface of a buried or submerged metallic structure and the electrolyte that is measured with reference to an electrode in contact with the electrolyte.

Polarity: The distinction between positive and negative potential difference at the terminals of a meter.

Potential measurement: The activity of measuring the structure-to-soil potential.

Rectifier: A device used to convert alternating current into direct current.

**Reference electrode:* An electrode whose open-circuit potential is constant under similar conditions of measurement, which is used for measuring the relative potentials of other electrodes.

**Resistivity:* 1) The resistance per unit length of a substance with uniform cross section. 2) A measure of the ability of an electrolyte (e.g., soil) to resist the flow of electric charge (e.g., cathodic protection current). Resistivity data are used to design a groundbed for a cathodic protection system.

Shorted casing: A metallic short is a metal casing that is in contact with the metal carrier pipe. It is usually caused by improper installation or settling of the casing or pipeline. An electrolytic short occurs when the metal casing is not in metallic contact with the carrier pipe; however, there is current flow between the casing and the carrier pipe due to an electrolyte in the casing.

Shunt: A calibrated resistor used for determining current flow in the circuit.

Soil resistivity: Ability of soil to resist the flow of current to or off the pipe.

Structure-to-soil potential: See pipe-to-soil potential.

Tank-to-soil potential: The voltage potential of the tank in the soil as related to a reference electrode.

Test lead: A connection to the structure that is being tested. Usually a wire in a supporting stand with an easy connection point for the meter.

Test station or test point: A specific area or location where a connection to the structure is made to allow a structure-to-soil potential to be measured.

Voltage: The force that makes current flow in a circuit. Voltage (E) is also referred to as electromotive force or potential.

Watts: The basic unit of measurement for electrical power.

Wenner (four-pin) method: Use of four metal probes (pins) placed equal distances apart in a straight line. A current in induced and current and resistance readings are taken at the appropriate points and recorded. Results are applied into the proper formula and charted to meet company specifications.

Instructor's Notes:

REFERENCES

Additional Resources

This module is intended to present thorough resources for task training. The following reference works are suggested for further study. These are optional materials for continued education rather than for task training.

API Recommended Practice 651, *CP of Aboveground Petroleum Storage Tanks*, Second Edition (December 1997). Global Engineering Product Code API CERT 651.

Cathodic Protection Level I Training Manual, 2000. NACE International.

Peabody's Control of Pipeline Corrosion, 2nd Edition, 2001. A.W. Peabody. Houston, TX; NACE International.

Corrosion Prevention by Protective Coatings, 1984. Charles G. Munger. Houston, TX; National Association of Corrosion Engineers (NACE).

ACKNOWLEDGMENTS

Figure Credits

Walter Johnson 108F08

Veronica Westfall 108F09

Figure 2 courtesy of NACE International

(Source: *Peabody's Control of Pipeline Corrosion, Second Edition*, Pg 158, Fig 8.1 Rectifier schematic diagrams) © NACE International – The Corrosion Society, 1440 South Creek Drive, Houston, Texas 77084. Additional publications on corrosion control technology are available from www.nace.org or phone (281) 228-6200

Figure 6 courtesy of NACE International

(Source: *Peabody's Control of Pipeline Corrosion, Second Edition*, Pg 38, Fig 3.10 Electrical shielding in congested areas) © NACE International – The Corrosion Society, 1440 South Creek Drive, Houston, Texas 77084. Additional publications on corrosion control technology are available from www.nace.org or phone (281) 228-6200

Figure 14 courtesy of NACE International

(Source: *Peabody's Control of Pipeline Corrosion* [1967 edition], Pg 34, Fig 5-2 Copper sulfate electrode) © NACE International – The Corrosion Society, 1440 South Creek Drive, Houston, Texas 77084. Additional publications on corrosion control technology are available from www.nace.org or phone (281) 228-6200

Figure 17 courtesy of NACE International

(Source: *Peabody's Control of Pipeline Corrosion, Second Edition*, Pg 250, Fig 12.9 Insulated flange assembly) © NACE International – The Corrosion Society, 1440 South Creek Drive, Houston, Texas 77084. Additional publications on corrosion control technology are available from www.nace.org or phone (281) 228-6200

Figure 18 courtesy of NACE International

(Source: *Peabody's Control of Pipeline Corrosion, Second Edition*, Pg 251, Fig 12.10 Shop testing of insulating flanges) © NACE International – The Corrosion Society, 1440 South Creek Drive, Houston, Texas 77084. Additional publications on corrosion control technology are available from www.nace.org or phone (281) 228-6200

Table 2 courtesy of NACE International

(Source: *Comparison of Other Reference Electrode Potentials With That of Copper-Copper Sulfate*) © NACE International – The Corrosion Society, 1440 South Creek Drive, Houston, Texas 77084. Additional publications on corrosion control technology are available from www.nace.org or phone (281) 228-6200

Instructor's Notes:

MODULE 61108-02 — TEACHING TIPS

The following are suggested activities or instructional methods to help you teach the material in this AIG.

General

When you call on someone to answer a question, the rest of the class relaxes or even tunes out because they expect that the question and answer will take place only between you and the trainee you called on. Instead, use this technique to involve more trainees in answering questions and to keep them on their toes.

1. Ask trainees to define a term or explain a concept.
2. After one trainee has answered, ask a trainee seated nearby if the answer is right. Then ask whether a trainee in the back of the room agrees.
3. Ask trainees to explain why they think an answer is right or wrong.
4. Use the session to clear up incorrect ideas and encourage trainees to learn from their mistakes.

Section 2.2.0 *Measuring Structure-to-Soil Electrolyte Potential*

This exercise gives trainees practice in measuring the potential between a pipeline or structure and a reference electrode. For this exercise, trainees will need test equipment for measuring structure-to-soil electrolyte potential, pencils, their Trainee Guides, and appropriate personal protective equipment. You will need to provide appropriate forms for recording the measurements and access to a pipeline or structure. Allow 30 to 45 minutes for this exercise, not including travel time to and from the site.

1. At the measurement site, divide the class into small groups. Each group needs one set of test equipment. Hand a form to each trainee.
2. Demonstrate the proper way to take the measurements. Point out the importance of proper polarity connections.
3. Assign each group to a particular location. Have trainees take turns using the test equipment and recording the readings. They may refer to their Trainee Guides if necessary. Circulate among the groups and ensure that trainees are following proper procedure.
4. After testing is complete, discuss the results and any problems trainees may have had. Ask trainees if they have any questions.

Sections 2.3.0–2.4.2 *Reading Rectifiers With and Without Meters Installed*

This exercise gives trainees an opportunity to verify the operation of a rectifier. Trainees will need a digital voltmeter, pencils, their Trainee Guides, and appropriate personal protective equipment. You will need to provide appropriate forms for recording the readings and access to a rectifier with and without a meter installed. Allow 30 to 45 minutes for this exercise, not including travel time to and from the site.

1. At the rectifier site, divide the class into small groups. Each group should have a digital voltmeter and ammeter. Hand a form to each trainee.
2. Briefly discuss the differences between reading rectifiers with meters installed and those without meters installed. Then discuss the differences between reading current with and without a shunt. Review the procedure for verifying the operation of a rectifier and safety concerns.
3. Have trainees take turns verifying the operation of each type of rectifier and recording the voltage and amperage. If possible, have them practice taking readings on rectifiers with and without a shunt. If a shunt is present, refer trainees to *Table 1* in the Trainee Guide to help calculate amperage. If the rectifier is not operating correctly, have trainees troubleshoot possible causes of malfunction. If the rectifier is operating correctly, have trainees verify this by using a digital voltmeter. Then ask trainees what they would do if the rectifier were not operating correctly.
4. Review the recorded measurements. Answer any questions trainees may have.

Sections 3.4.0–3.4.3 *Criteria for Cathodic Protection*

This exercise gives trainees the opportunity to learn about criteria for cathodic protection from an experienced professional. For this presentation, invite an expert on cathodic protection and related standards to talk to the class. For this exercise, trainees will need copies of NACE *Standard RP-01-69,* their Trainee Guides, pencils, and paper to take notes for discussion. Allow between 30 and 45 minutes for this exercise.

1. Tell the trainees that a guest speaker will be presenting information on cathodic protection standards. Distribute copies of the standard and have trainees review them before the speaker arrives.
2. Also before the speaker arrives, have trainees brainstorm questions for the guest speaker. Questions could address specific standards and how and why they were developed.
3. Introduce the guest speaker. Remember to include the speaker's name, title, company, and relevant credentials.
4. Following the presentation, allow 5 to 10 minutes for trainees to share their questions with the guest speaker.

MODULE 61108-02 — ANSWERS TO REVIEW QUESTIONS

Answer	Section
1. c	1.0.0
2. a	2.1.2
3. c	2.1.4
4. d	2.3.0
5. a	2.4.0
6. b	3.0.0
7. c	3.2.0
8. c	3.3.0
9. a	3.3.2
10. a	3.5.0
11. c	3.5.2
12. b	3.6.1
13. a	4.1.5
14. c	4.1.7
15. b	4.2.3

CONTREN™ LEARNING SERIES — USER UPDATES

The NCCER makes every effort to keep these textbooks up-to-date and free of technical errors. We appreciate your help in this process. If you have an idea for improving this textbook, or if you find an error, a typographical mistake, or an inaccuracy in NCCER's Contren™ textbooks, please write us, using this form or a photocopy. Be sure to include the exact module number, page number, a detailed description, and the correction, if applicable. Your input will be brought to the attention of the Technical Review Committee. Thank you for your assistance.

Instructors – If you found that additional materials were necessary in order to teach this module effectively, please let us know so that we may include them in the Equipment/Materials list in the Instructor's Guide.

Write: Curriculum Revision and Development Department
National Center for Construction Education and Research
P.O. Box 141104, Gainesville, FL 32614-1104

Fax: 352-334-0932

E-mail: curriculum@nccer.org

Craft ____________________ Module Name ____________________

Copyright Date __________ Module Number __________ Page Number(s) __________

Description

____

____

____

(Optional) Correction

____

____

(Optional) Your Name and Address

____

____

Module 61109-02

Test Station Repair
(CT 2.1, 2.2, 2.3, and 2.4)

Test Station Repair
Annotated Instructor's Guide

Module 61109-02

MODULE OVERVIEW

This module describes some of the activities associated with the repair of corrosion protection test stations.

PREREQUISITES

Please refer to the Course Map in the Trainee Module. Prior to training with this module, it is recommended that the trainee shall have successfully completed the following:

Core Curriculum; Pipeline Core, Modules 66101-02 and 66102-02; Pipeline Corrosion Control Level One, Modules 61103-02 through 61108-02

OBJECTIVES

Upon completion of this module, the trainee will be able to:

1. Identify test lead damage (CT 2.1 and 2.2).
2. Repair a test lead and verify that the repair was performed effectively (CT 2.2).
3. Install test leads by non-exothermic welding methods (CT 2.3).
4. Install test leads by exothermic welding methods (CT 2.4).

PERFORMANCE TASKS

Under the supervision of the instructor, the trainee should be able to:

1. Install test leads by exothermic welding methods (CT 2.4).
2. Install test leads by non-exothermic welding methods (CT 2.3).
3. Inspect and verify test lead continuity (CT 2.1).
4. Repair damaged test leads (CT 2.2).

NCCER STANDARDIZED CRAFT TRAINING PROGRAM

The National Center for Construction Education and Research (NCCER) provides a standardized national program of accredited craft training. Key features of the program include instructor certification, competency-based training, and performance testing. The program provides trainees, instructors, and companies with a standard form of recognition through a National Craft Training Registry. The program is described in full in the *Guidelines for Accreditation*, published by the NCCER. For more information on standardized craft training, contact the NCCER by writing us at P.O. Box 141104, Gainesville, FL 32614-1104; calling 352-334-0911; or e-mailing info@nccer.org. More information may be found at our Web site, www.nccer.org.

HOW TO USE THIS ANNOTATED INSTRUCTOR'S GUIDE

Each page presents two sections of information. The larger section displays each page exactly as it appears in the Trainee Module. The narrow column ties suggested trainee and instructor actions to each page and provides icons to call your attention to material, safety, audiovisual, or testing requirements. The bottom of each page includes space for your notes.

Teaching Tip

If you see the Teaching Tip icon, that means there is a teaching tip associated with this section. Also refer to any suggested teaching tips at the end of the module.

SAFETY CONSIDERATIONS

Ensure that the trainees are equipped with appropriate personal protective equipment. Emphasize the importance of following all safety precautions and procedures when working with power tools.

PREPARATION

Before teaching this module, you should review the Module Outline, Objectives, Performance Tasks, and the Materials and Equipment List. Be sure to allow ample time to prepare your own training or lesson plan and gather all required equipment and materials.

MATERIALS AND EQUIPMENT LIST

Overhead projector and screen

Transparencies

Blank acetate sheets

Transparency pens

Whiteboard/chalkboard

Markers/chalk

Pencils and scratch paper

Appropriate personal protective equipment

Copies of your company's policy and procedures manual

Copies of *49 CFR 192.469*

Copper wire, soldering iron, sponge, water, tinning paste, solder

Appropriate splicing equipment and materials

Appropriate equipment and materials for metallurgical bonding

Appropriate equipment and materials for the four types of mechanical bonding

Multimeters or continuity testers

Copies of sample company and OSHA procedures regarding exposing underground pipeline

Segments or photos of damaged pipe

Module Examinations*

Performance Profile Sheets*

Copies of the Performance Verifications for Covered Tasks 2.1, 2.2, 2.3, and 2.4**

* Located in the Test Booklet.

** Available for download at www.nccer.org.

TEACHING TIME FOR THIS MODULE

An outline for use in developing your lesson plan is presented below. Note that each Roman numeral in the outline equates to one session of instruction. Each session has a suggested time period of 2½ hours. This includes 10 minutes at the beginning of each session for administrative tasks and one 10-minute break during the session. Approximately 5 hours are suggested to cover *Test Station Repair (CT 2.1, 2.2, 2.3, and 2.4)*. You will need to adjust the time required for hands-on activity and testing based on your class size and resources.

Topic	Planned Time
Session I. Construction of Test Stations	
A. Introduction	______
B. Types of Test Stations	______
C. Mounting Test Stations	______
D. Wire-to-Wire Connection	______
1. Laboratory	______
E. Wire-to-Structure Connection	______
1. Laboratory	______
2. Performance Testing	______
Session II. Test Station Repair and AOCs	
A. Aboveground Test Station Repair	______
1. Replace or Repair Defective Parts	______
2. Continuity Testing	______
3. Laboratory	______
4. Verify Effective Repair Using Structure-to-Soil Potential	______
5. Performance Testing	______
B. Belowground Test Station Repair	______
1. Following Excavation Regulations	______
2. Locating Pipeline	______
3. Exposing Pipeline	______
4. Inspecting Damage	______
5. Backfilling	______
6. Laboratory	______
7. Performance Testing	______
C. Abnormal Operating Conditions	______
D. Summary	
1. Summarize module	______
2. Answer questions	______
E. Module Examination	______
1. Trainees must score 70% or higher to receive recognition from the NCCER.	
2. Record the testing results on Craft Training Report Form 200 and submit the results to the Training Program Sponsor.	
F. Performance Testing	______
1. Trainees must perform each task to the satisfaction of the instructor to receive recognition from the NCCER.	
2. Record the testing results on Craft Training Report Form 200 and submit the results to the Training Program Sponsor.	

Module 61109-02

Test Station Repair

(CT 2.1, 2.2, 2.3, and 2.4)

Instructor's Notes:

Instructor's Notes:

Introduction to the Pipeline Operator Qualification Rule, API, and NCCER

It could be said that petroleum products keep America running. Petroleum products like oil and natural gas are the principal sources of energy in the United States and are used in many manufacturing and chemical processes. Petroleum products are primarily transported by pipeline. There are more than a million miles of liquid and natural gas pipeline in this country, ranging in diameter from two inches to four feet. These pipelines bring crude oil and natural gas from production fields to processing plants and refineries—and ultimately to the consumer.

Pipeline Operator Qualification

Petroleum products are considered hazardous, so pipeline personnel must take extra precautions and follow specific procedures when working with them. To help ensure the safe handling and transportation of petroleum products, the United States Department of Transportation (DOT) mandates safe practices as federal regulations.

One such regulation, the Operator Qualification (OQ) rule, requires pipeline operators to develop a written qualification plan to evaluate their personnel and their contract personnel (see *49 CFR Part 192 Subpart N—Gas*, and *49 CFR Part 195 Subpart G—Liquid*). The evaluation hinges on the operator's ability to perform certain covered tasks and to be able to recognize and respond to abnormal operating conditions that may be encountered while performing these covered tasks.

Covered Tasks

A pipeline performance task is deemed *covered* if the activity passes the following four-part test:

1. Is it performed on a pipeline facility?
2. Is it an operations or maintenance task?
3. Is it performed as a requirement of *49 CFR Part 195 (Liquid)* or *49 CFR Part 192 (Gas)*?
4. Does it affect the operation or integrity of the pipeline facility?

If the answer to all four questions is *yes*, then the activity is considered *covered* and must figure into the evaluation process of pipeline personnel.

American Petroleum Institute

The OQ rule and its related covered tasks provided the catalyst to develop a unified pipeline training and assessment program—so that pipeline operators and owners have ready access to the assessment and training tools they need. This effort was the direct result of the vision and leadership provided by The American Petroleum Institute (API) and API members.

Efforts to unify the industry are nothing new to API. Indeed, the impetus for forming the Institute in 1919 was the need to standardize engineering specifications for drilling and production equipment. Today, the most pressing issues revolve around public perceptions and government policies toward the industry. Speaking with one voice on these issues has become as essential as having interchangeable parts in the field. API's members determine the petroleum industry's positions on public policy as well as the standards that govern its day-to-day operations worldwide.

Consortium on Operator Qualification

API assembled a focus group whose primary purpose was to identify covered tasks and provide compliance recommendations to pipeline operators.

The Consortium on Operator Qualification (COOQ) became the guiding force for the identification and development of covered tasks for the liquid pipeline industry. Their work product, API publication 1161, *Guidance Document for the Qualification of Liquid Pipeline Personnel*, formed the basis for the resulting Pipeliner Training and Assessment Program.

The covered task list created in API publication 1161 has now evolved to include gas covered tasks and is maintained by NCCER. The most current covered task list can be downloaded at www.nccer.org.

National Center for Construction Education and Research

With the guidance and support of API, its members, and the Consortium on Operator Qualification, the National Center for Construction Education and Research (NCCER) was able to design and implement the Pipeliner Training and Assessment Program.

NCCER is a not-for-profit 501(c)(3) education foundation established in 1995 by the world's largest and most progressive construction and maintenance contractors and national contractor associations. NCCER was founded to address the

severe workforce shortage facing the construction and maintenance industries and to develop a standardized training process and curricula. Today, NCCER is supported by hundreds of leading construction and maintenance companies, manufacturers, and national associations and developed more than 35 construction and maintenance titles.

NCCER also maintains a National Registry, which contains the records of trainees and instructors who have received NCCER training. All completion information for every trainee is sent to the NCCER and filed within the Registry. The Registry can then confirm training and skills for workers as they move from company to company, state to state, or even within their own company.

Standardized Curriculum (NCCER's Contren™ Learning Series)

NCCER curricula are developed using the expertise of industry subject matter experts (SMEs). The use of SMEs as an integral part of every new development or revision ensures that the curriculum is current, industry-driven, and indicative of industry practices. The development and/or revision process is facilitated from start to finish by NCCER project managers, who work closely with the SMEs and a select team of technical writer(s) on the development and presentation of the curriculum.

Pipeline: Five Training Paths

The pipeline curriculum focuses on essential skills and competencies within five skill areas:

- Corrosion Control
- Electrical and Instrumentation
- Maintenance
- Mechanical
- Operations (Field and Control Center)

SMEs from the pipeline industry have determined the skills and competencies needed based on API's covered tasks and the consensus of the group/industry. Specific company information is generally not incorporated into the curriculum unless the SMEs determine that the information is the industry standard.

Training Modules and Covered Tasks

The smallest unit of instruction in the curriculum is a *module*. Each curriculum is composed of modules (the equivalent of stand-alone chapters) that are logically sequenced into complete *levels* of instruction. Every module is task-oriented and requires the successful completion of a written (knowledge-based) examination and a performance (hands-on) test. Modules may be individually selected and/or combined for efficient task training. Modules in the pipeline curricula, to the extent possible, will align with API-developed covered tasks. While not all modules will relate to covered tasks, all covered tasks will relate to at least one module.

Benefits of Training to the Pipeline Marketplace

Applying NCCER's standardized training process allows pipeline owners and contractors in the United States to establish the benchmark for acceptable skill and competency levels within the pipeline industry. This process will assist training organizations in developing and/or improving standardized pipeline training programs that effectively match their resources and operating conditions while responding to regulatory demands and to the needs of the industry. The process will also recognize training organizations that meet the minimum standards of competence, warranting regulatory, public, and professional confidence. This in turn will foster national unity within the pipeline industry for the mutual benefit of individual technicians, owners, contractors, and ultimately pipeline-users and society.

Additional Resources

To review the Pipeliner Training and Assessment Program Implementation and Resource Guide and the latest version of pipeline covered tasks, visit NCCER's Web site at www.nccer.org, or contact:

National Center for Construction Education and Research
PO Box 141104
Gainesville, FL 32614-1104
(888) 622-3720
Web site: www.nccer.org

For more information on API, please contact:

American Petroleum Institute
1220 L Street, Northwest
Washington, D.C. 20005
(202) 682-8000
Web site: www.api.org

To order *API Pub 1161*, contact Global Engineering Documents by calling 1-800-854-7149 or visiting their Web site at www.global.his.com.

Additional sources of information on the pipeline operator qualification regulation include:

DOT – Office of Pipeline Safety
www.ops.dot.gov

Transportation Safety Institute
www.tsi.dot.gov

Instructor's Notes:

COURSE MAP

This course map shows all of the modules in the first level of the Pipeline Corrosion Control curriculum. The suggested training order begins at the bottom and proceeds up. Skill levels increase as you advance on the course map. The local Training Program Sponsor may adjust the training order.

PIPELINE CORROSION CONTROL LEVEL ONE

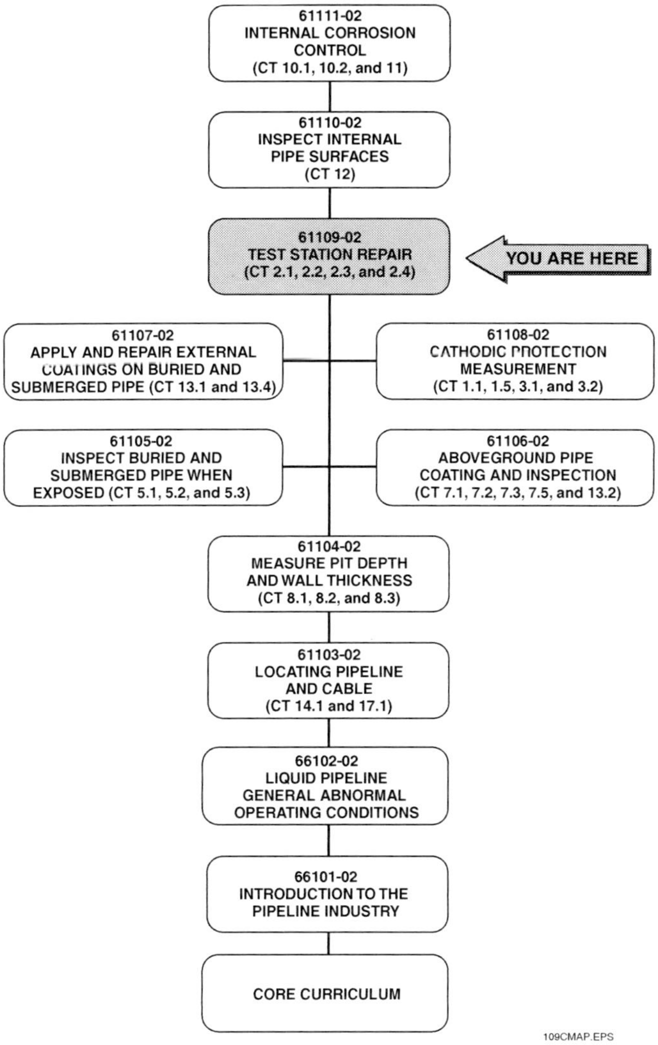

Homework

Assign reading of Module 61109-02.

Instructor's Notes:

MODULE 61109-02 CONTENTS

Figures

Table

Instructor's Notes:

MODULE 61109-02

Test Station Repair (CT 2.1, 2.2, 2.3, and 2.4)

Objectives

When you have completed this module, you will be able to do the following:

1. Identify test lead damage (CT 2.1 and 2.2).
2. Repair a test lead and verify that the repair was performed effectively (CT 2.2).
3. Install test leads by non-exothermic welding methods (CT 2.3).
4. Install test leads by exothermic welding methods (CT 2.4).

Prerequisites

Before you begin this module, it is recommended that you successfully complete the following: Core Curriculum; Pipeline Core, Modules 66101-02 and 66102-02; Pipeline Corrosion Control Level One, Modules 61103-02 through 61108-02.

Required Trainee Materials

1. Pencil and paper
2. Appropriate personal protective equipment

1.0.0 ◈ INTRODUCTION

Each pipeline under cathodic protection (CP) is required by *49 CFR 192.469, Test Stations,* to have sufficient **test stations** or other contact points for electrical measurement to determine the adequacy of CP. Test stations are the best means of electrically reaching a buried pipeline to determine whether it is cathodically protected, and also to perform other tests associated with corrosion control work. Test stations are convenient to use and are constructed so that failures will be minimal. This module describes some of the activities associated with the repair of test stations.

2.0.0 ◈ CONSTRUCTION OF TEST STATIONS

The construction of test stations is discussed in the following sections. Also discussed are types of test stations, mounting of test stations, and test wires and pipeline connections.

2.1.0 Types of Test Stations

Test stations are used to monitor the effectiveness of CP programs. Permanent test stations use electronic leads connected to the pipeline to measure the potential from pipe to soil (see *Figure 1*). Only a very small section of the potential from the piping is actually measured through the test station.

Test stations are designed for specific applications. Aboveground test stations are provided for access to a pipeline or structure for the purpose of corrosion testing. Most consist of a conduit and some sort of terminal box, usually made of steel or tough plastic and sealed against extremes of weather.

In the most common test setup, the conduit contains two wires that are inserted into the ground near the pipeline, cable, or other metallic structure for checking current flow and pipe-to-soil potential (see *Figure 2*). To help identify the leads, the wiring is color coded. Each company has its own color-coding system, which should be followed throughout the line.

Materials

Ensure that you have everything required to teach the course. Check the Materials and Equipment List at the front of this module.

Classroom

Explain that terms shown in bold are defined in the Glossary at the back of this module.

Audiovisual

Show Transparency 1, Course Objectives.

Show Transparency 2, Performance Tasks.

Teaching Tip

See the general Teaching Tip at the end of this module.

Classroom

Distribute copies of *49 CFR 192.469*. Discuss the importance of test stations.

Audiovisual

Show Transparency 3 (Figure 1). Discuss typical applications of each type of test station.

Audiovisual

Show Transparency 4 (Figure 3). Have a trainee distinguish between the four-wire and six-wire test stations.

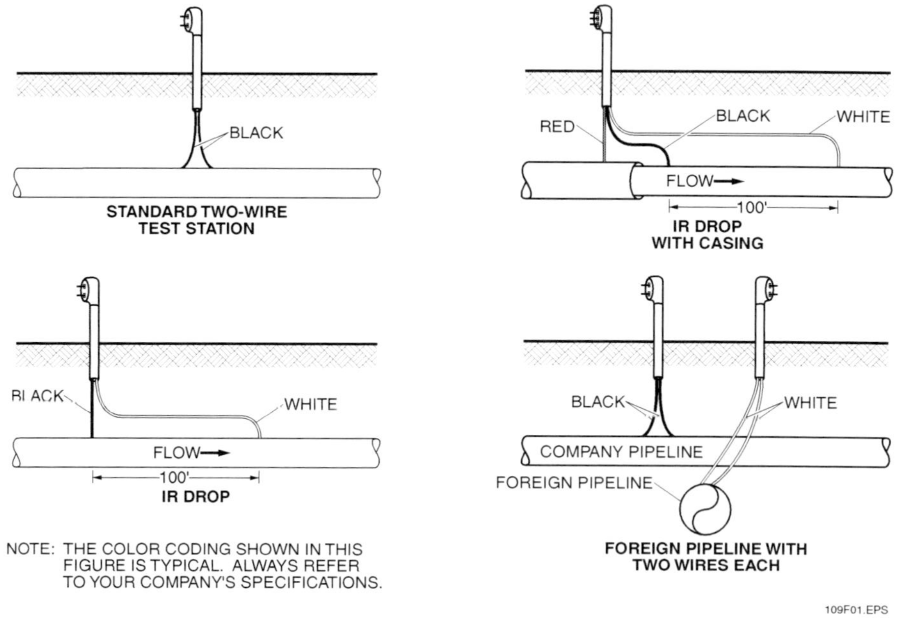

Figure 1 ◆ Types of test stations.

Figure 2 ◆ Test point.

Sometimes there are four or six wires. Four wires are used in the insulated joint setup for more accurate measurement of pipeline current flow; one pair of wires is of a heavier gauge than the other, to allow the insertion of a resistance or solid **bond** across the insulated joint. The second pair of wires provides for dual measurement of pipe-to-soil potentials, one on each side of the insulated joint. *Figure 3* shows typical types of test stations.

In a six-wire combination insulated joint and line current setup (see *Figure 3*), positive measurement of current flow through an insulated flange is permitted. Each side of the insulated joint is provided with one heavier gauge wire for bonding purposes. In situations with terminal insulated flanges, this type of test point is particularly useful. If the flange becomes totally or partially shorted for any reason, it will allow for positive measurement or for measurement of the current through a solid or resistant bond.

Instructor's Notes:

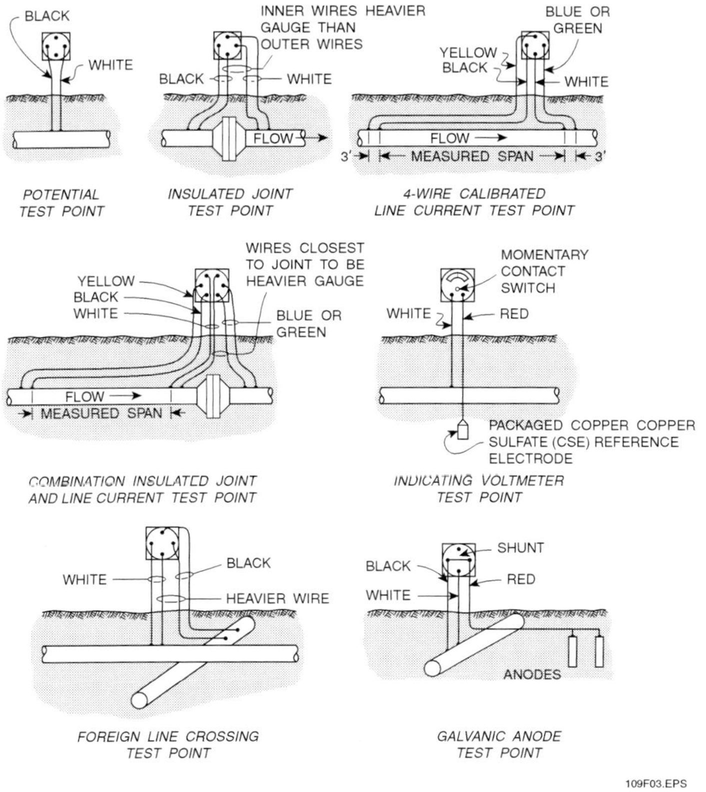

Figure 3 ◈ Typical types of test station.

Some systems have an indicating voltmeter test point (see *Figure 3*) which can be connected between a **reference electrode** and the pipe; scheduled readings may be taken by pipeline corrosion testers. All record-keeping should be done in accordance with company policy.

For testing anodes at only a single location, the galvanic anode test point is useful (see *Figure 4*).

There is also a foreign-line-crossing test point (see *Figure 5*), but use of this type of test point is limited. Many companies do not allow them unless the attachments are made by their own technicians or if company representatives are present. The foreign-line-crossing test point provides two wires to each pipeline. The heavier gauge wire of the two wires is used for bond installation, and the lighter gauge wire is used for **potential measurements**.

The last type of test point described in this text is one containing a **shunt**. A standard shunt does not have the current capacity to handle much power; however, a shunt can be used to control a relay and make it selectable by the user. The presence of a shunt in the terminal box can allow the individual with a **multimeter** or a voltmeter to measure the current from the anodes without disturbing the circuit. The anode current output will not be affected by the common 0.01-ohm shunt, because it does not have sufficient resistance.

Audiovisual

Show Transparencies 5 and 6 (Figures 4 and 5). Discuss the types of test stations illustrated in each figure.

Audiovisual

Show Transparency 7 (Figure 6). Review the requirements for each type of station.

Teaching Tip

See the Teaching Tip at the end of this module.

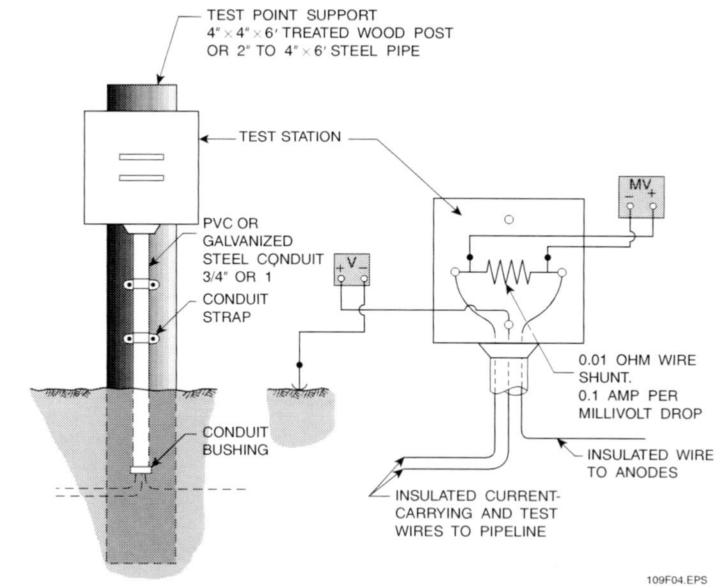

Figure 4 ◈ Galvanic anode test station installation.

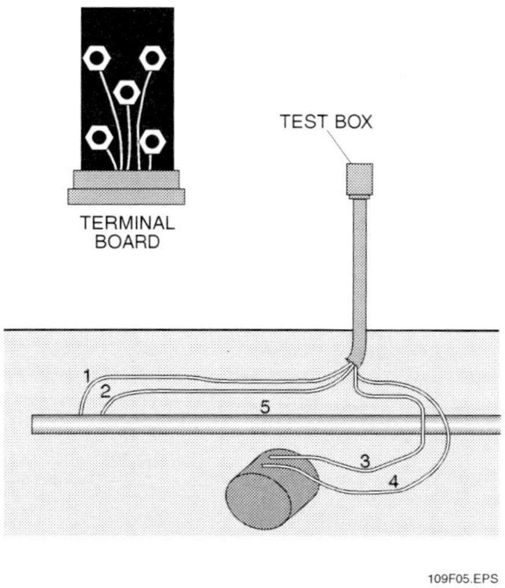

Figure 5 ◈ Foreign-line-crossing test point.

Peabody, in *Control of Pipeline Corrosion*, recommends using an **ammeter** circuit or a clamp-on ammeter to measure the anode current when using a solid link in place of the shunt. However, if a solid link is used instead of the shunt to measure the pipeline potential to a copper sulfate reference electrode, use a separate test wire from the pipeline, as shown in *Figure 5*. No. 8 AWG insulated copper wire can be used for direct burial installation with pipe and anode leads for multiple anode installations. Do not use anything smaller than No. 12 AWG for the separate potential test wire.

Protection is usually provided for the wiring connected to the anode in a galvanic anode installation, making the corrosion of any exposed copper unlikely. Any break in the insulation will allow corrosion to begin, which will eventually sever the wire to the positive terminal of the DC power source.

In a galvanic anode system, from an insulation standpoint, the connection between the copper anode lead and the pipeline is the most important one. If the wire to the positive terminal is severed, the power source will be protected from damage, so the insulation of underground connections on

Instructor's Notes:

galvanic anode installation is not as critical. However, severing the wire will still cause current loss, so the insulation should be done carefully. Also, connections such as crimp or compression should be waterproof. Welded, brazed, or soldered connections should be protected from resistance within the joint.

If there is any possibility of the CP system not covering the piping next to a connection, the dissimilar steel and copper metals should be completely and permanently waterproofed to prevent any possibility of corrosion. The two dissimilar metals can form an electrical shield that prevents the protective current from reaching the steel, thereby allowing it to corrode.

2.2.0 Mounting Test Stations

The purpose of test stations is to provide a convenient testing point for corrosion testers. The location of each station should be carefully determined with this in mind. Post-mounted stations are preferable but may not always be feasible. Grade level or buried test stations may be necessary in those locations.

Many manufacturers offer several types of compatible terminal boxes and blocks. First, make sure the box is sturdy enough to withstand conditions in the environment where the box will be placed. The blocks should have a minimum of 0.24", solid brass studs to fit the boxes. The studs may be plated with cadmium or nickel and fitted with hex nuts and washers. Aboveground test stations may be mounted in a variety of ways (see *Figure 6*).

At least 18" of slack should be allowed to remain below the terminal panel in the housing for grade level test points. The slack must be adequate for backfill settlement and for withdrawing the terminal panel. *Figure 7* shows a typical arrangement for installing a grade level test station.

The terminal boxes should be installed in line with specified reference points and shown on maps or records of the pipeline; this is especially important with grade-level boxes that may be harder to see as dirt and plants encroach on the surface.

Buried test stations are sometimes necessary, but the excavation is time-consuming and costly, so they should only be used when there is no other choice. Keep in mind the possibility of extending the leads for several hundred feet to a suitable aboveground location. It is essential that permanent landmarks and reference points be used to locate buried test stations; their position must be marked on pipeline maps and records.

Classroom

Ask a trainee to name the primary and secondary considerations for selecting test wiring. Review the guidelines for choosing wiring, conduit, and splices.

Audiovisual

Show Transparency 8 (Figure 7). Ask a trainee to identify the function(s) of the components shown in the figure.

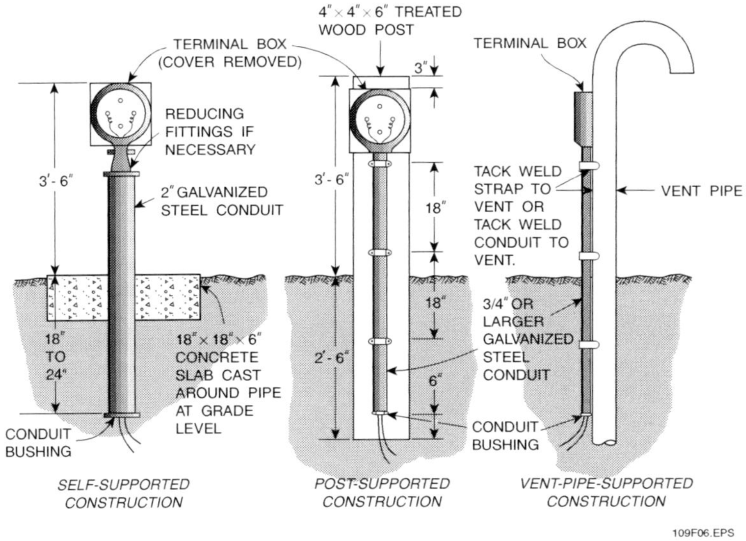

Figure 6 ◆ Typical types of above grade test stations.

Show Transparency 9 (Figure 8).

Set up stations at which trainees can practice making solder connections. Answer any questions trainees may have.

If the test station must be buried, a watertight box is recommended (see *Figure 8*). Inside the box, locate the wires so as to prevent damage when the box must be excavated for use. Place a durable board or cover on the surface directly above the test station.

One basic form of grade-level test station includes test wires, with the ends taped to prevent accidental contact, coiled in a common street valve box with a cover. Each box must have its own fitted terminal board.

2.3.0 Wire-to-Wire Connection

Mechanical strength is the primary consideration in selecting test wiring for potential measurements; electrical requirements are secondary. In a standard environment and circumstances, No. 12 AWG single-conductor stranded wire with 600V insulation, which is available in suitable colors for color coding, is recommended. Unless the test points will carry very large currents or be subjected to severe transit or stray current interference, a heavier wire gauge such as No. 6 or No. 4 AWG is recommended. Under extreme conditions, wire should be selected by experienced technicians who are familiar with such circumstances.

No welding should be attempted after the wires are pulled in; in cases in which a steel structure is used as a support for the test point, requiring it to be tack-welded to the pipe, it must be done before the wire is pulled. Otherwise, the heat from the welding operation will damage the insulation. Remove only enough insulation to allow the weld to be made, and connect only one with each weld. On completion, coat each weld with the required compound.

The conduit carrying the wires may be rigid steel, but it must be carefully reamed after being cut, and the lower end should have a conduit bushing, as shown earlier in *Figure 6*. Allow a minimum of 12" (or follow company standards) of slack for each wire that terminates in the box. Use a cutter to avoid damaging the wire ends. Coil the slack in the manner shown in *Figure 6*. Leaving enough slack is especially important when the wires could be pulled away from the terminals by tension from settling backfill.

Twisted or bolted splices are not recommended for splicing test wires; over time these splices can develop high resistance. Instead, use a good grade

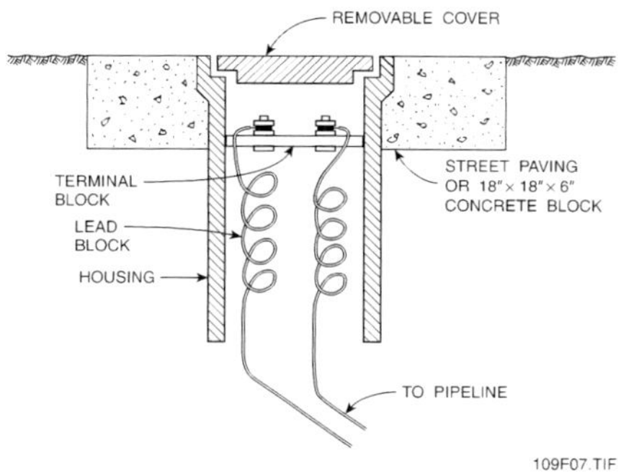

Figure 7 ◆ Typical at-grade types of test stations.

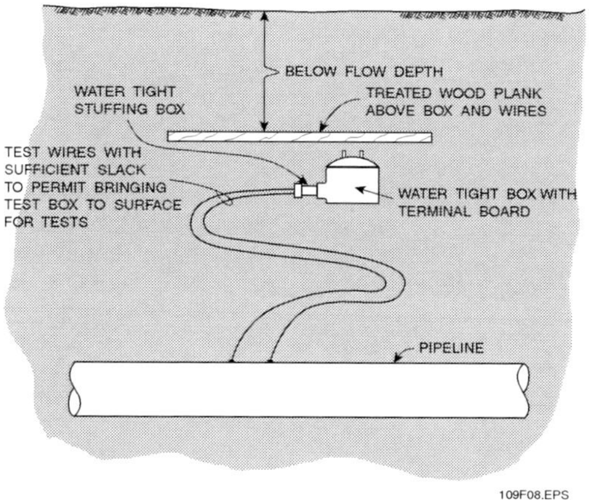

Figure 8 ◆ Typical buried type of test stations.

Instructor's Notes:

of 60-40 lead-tin solder with a noncorrosive flux and make a soldered splice. Then insulate the splice with rubber and plastic tape. To make it watertight, allow the wrapping to overlap the wire insulation enough to seal it.

Use care when installing and repairing wire connections. Broken wire means expensive excavation for repairs. Damage frequently occurs when there is not enough wire for an easy connection to the test point, and the wires have to be pulled taut to make the connection. Before beginning the backfill operation, place the wire as far as it will go under the pipe and ensure that the test point wires have enough slack at the terminal to avoid being pulled out. The wires can also be run along the top of the piping and then taped in place; the tape adds another layer of protection.

Wire-to-wire connections are discussed in the following sections.

2.3.1 Soldering

The most common method for making a wire-to-wire connection is **soldering**. Follow these steps to make a **solder connection**:

Step 1 Clean the portion of the copper wire that will be soldered.

Step 2 Clean the soldering iron tip with a damp sponge after it is heated (see *Figure 9*).

Step 3 Tin the soldering iron tip after it is cleaned (see *Figure 10*).

> **NOTE**
> Always use an approved tinning paste to tin the soldering iron tip.

Step 4 Hold the clean iron tip to the copper and the lead. This will heat both the copper and the lead at the same time.

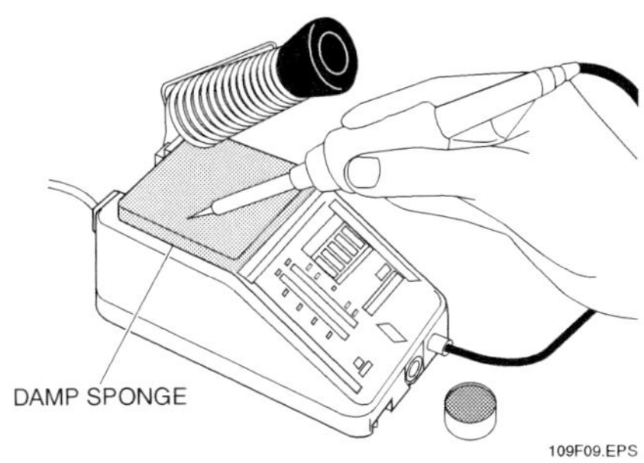

Figure 9 ◈ Cleaning a soldering iron tip.

Step 5 Continue to heat the joint and apply a small amount of solder.

Step 6 Remove the soldering iron from the connection, and let the solder joint cool in the ambient air.

Step 7 Check the soldered connection to make sure there are no dry joints and that proper electrical connection has been made.

Step 8 Recoat the wires according to company specifications.

2.3.2 In-Line Splice Connection

In-line butt splice connections are the easiest connections to insulate. This is because the wire diameters are uniform. In-line butt splice connections can be made of individual WCSF-N type sleeving, or they can be made with the NPK series. *Figure 11* shows insulation of a crimped connection.

Follow these steps to insulate a butt splice crimped connection:

Step 1 Select the proper tubing and verify its maximum use range.

> **NOTE**
> The largest diameter of the splice assembly, including the splice and the wires, must be at or below the maximum application diameter.

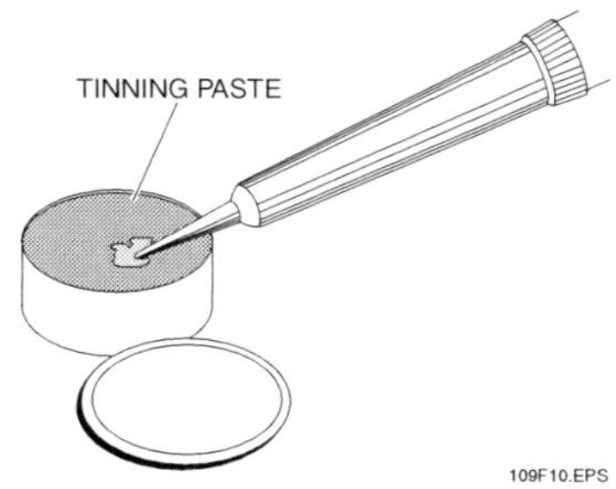

Figure 10 ◈ Tinning a soldering iron tip.

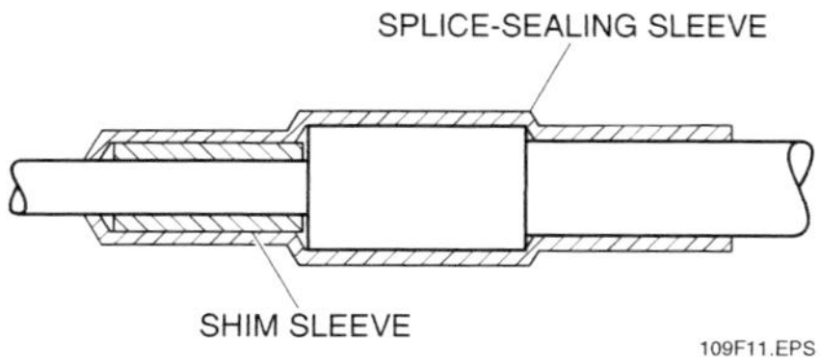

Figure 11 ◈ Insulation of a crimped connection.

Audiovisual

Show Transparencies 10 and 11 (Figures 9 and 10). Ask trainees to explain the procedures shown.

Show Transparencies 12 through 16 (Figures 11 through 15). Familiarize trainees with the insulation procedures detailed in each figure.

Laboratory

Set up stations at which trainees can practice in-line butt splice connections. Have trainees follow the steps in their Trainee Guides.

Safety

Remind trainees not to flex the spliced connection until it cools down sufficiently. Ask a trainee to explain possible consequences of flexing hot connections.

Classroom

Present several hypothetical scenarios in which trainees must choose the proper connection hardware. Have them practice calculating the diameter of the proper tubing.

Step 2 Compare both wire diameters to the minimum application diameter of the tubing selected to determine if shims are required. If either wire diameter is below the minimum, a shim is required to build up the outside diameter of that wire to within the splice-insulating sleeve's use range.

Step 3 Determine the length of the tubing required.

Step 4 Install the proper shims on the wire if they are required.

> **NOTE**
> *Connection length* refers to the total length measured from the insulation cutback of the assembled connection. If shims are used, the length of the splice-insulating sleeve may have to be lengthened to ensure that it overlaps the shims when installed (overlap of ¾" required).

Step 5 Align the shims on the cable jacket within ¼" of the installation cutback.

Step 6 Shrink the cable jacket in place (see *Figure 12*).

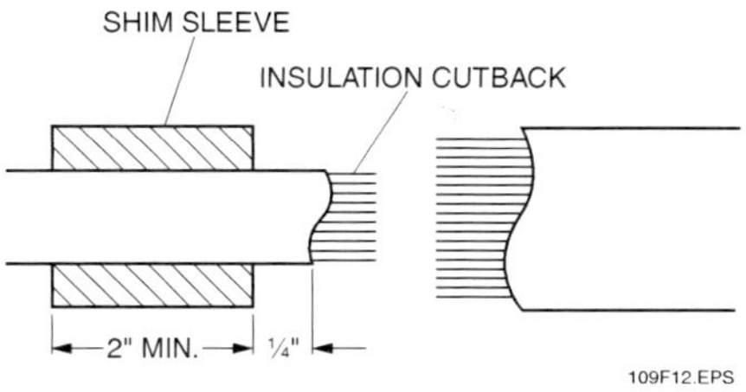

Figure 12 ◈ Applying shim sleeves.

> **NOTE**
> Do not shrink the tubing at this time.

Step 7 Slide the splice-sealing sleeve over either wire to be spliced after the shim is installed.

Step 8 Make the proper conductor connections (see *Figure 13*).

Step 9 Inspect the connection for sharp edges or protruding wire strands.

Step 10 Remove any sharp edges or protruding wire strands with an abrasive cloth or file.

Step 11 Center the splice-sealing sleeve over the connection area.

> **NOTE**
> Where shims are used, the sleeve must overlap the outer shim by ¼".

Step 12 Shrink the sleeve tubing in place (see *Figure 14*).

> **CAUTION**
> Do not flex the connection until it is comfortable to touch. This could cause voids in the connection or possibly result in a crack later.

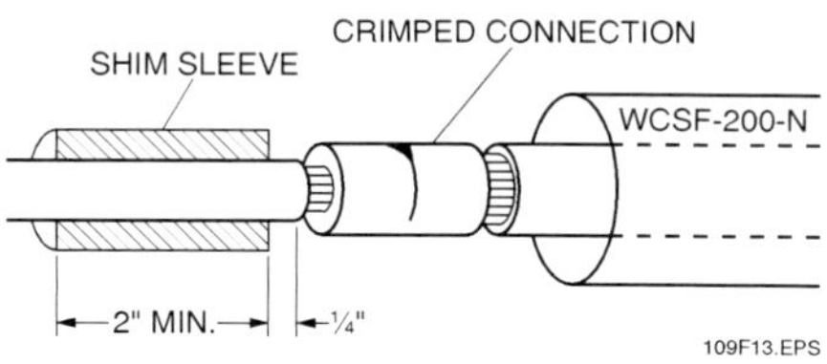

Figure 13 ◈ Making the crimped connection.

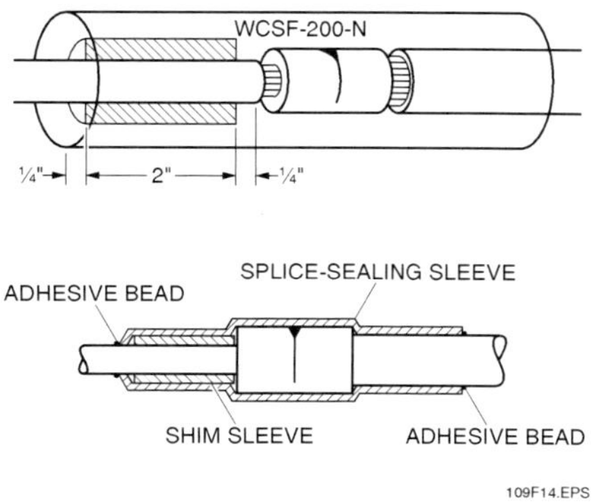

Figure 14 ◈ Centering shrink tube.

Instructor's Notes:

2.3.3 Bolted In-Line Connections

Bolted in-line connections (see *Figure 15*) are usually difficult to insulate. There are normally large differences in diameters over the wires and over the bolt.

Bolted in-line connections and procedures for in-line splice connections follow the same design guidelines.

Connection hardware varies significantly in dimension and fits within the usage range of a given size tubing, based on the diameter of the bolt and lug hardware. Bolt lengths must be selected to be as short as possible. The effective diameter of the bolted connection must never exceed the tubing range.

After the hardware is chosen, calculate the effective diameter over the bolted connection by measuring it with a diameter tape or by using the effective diameter formula (*Figure 16*).

The result is the diameter of the proper tubing size. This diameter must be less than or equal to the maximum use range of the selected tubing.

$D_{eff} + 0.36(W) + 0.64(L)$

Where:

W = Width of largest tang

L = Overall bolt length (including head)

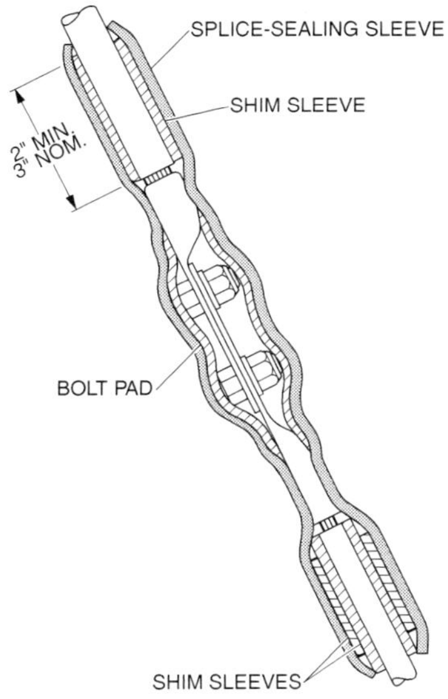

Figure 15 ◆ Typical insulation of a bolted connection.

> **NOTE**
> Reverse L and W in this equation if the lug tang width is greater than the overall bolt length.

A length of uncoated tubing, called a bolt pad, must be applied over the bolts when the overall bolt length or tang width is greater than 0.87". Below 0.87", the use of a bolt pad is optional. The uncoated bolt pad keeps the bolt area free of adhesive, if reentry is desirable. The hardware diameter must be reviewed to include the bolt pad if a bolt pad is used. *Table 1* shows an example of the maximum hardware size combination that can be used with various sizes of tubing.

$D_{eff} = 0.36(W) + 0.64(L)$

D_{eff} = EFFECTIVE DIAMETER
W = WIDTH OF LARGEST TANG (LAG TONGUE)
L = OVERALL BOLT LENGTH (INCLUDING HEAD)

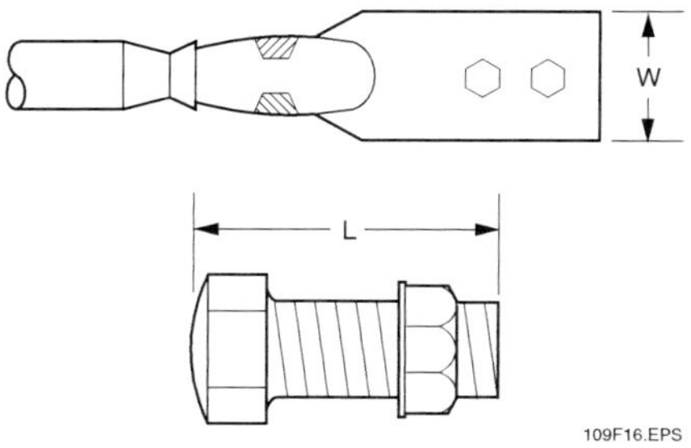

Figure 16 ◆ Effective diameter formula.

Table 1 Example of the Maximum Hardware Size Combination That Can Be Used With Various Sizes of Tubing

Tubing Size	Maximum Bolt Size (Diameter × Length)	Maximum Connector Tang Width
WCSF-200	#10 × ⁵⁄₁₆"	0.40"
WCSF-300	¼" × ⅜"	0.60"
WCSF-500	⅜" × ⅝"	0.87"
WCSF-650 (w/ WCSF 650 Bolt Pad)	½" × ⅝"	1.00"
WCSF-1000 (w/ WCSF 1000 Bolt Pad)	½" × 1¼"	1.90"

109T01.EPS

Classroom

Ask a trainee to explain the two major differences between soldering and welding. Have other trainees name types of solders and fluxes.

Audiovisual

Show Transparency 17 (Figure 16). Give trainees examples for using the formula shown.

Audiovisual

Show Transparency 18 (Figure 17). Familiarize trainees with the proper procedure for exothermic welding.

Laboratory

Prepare stations at which trainees can practice metallurgical bonding techniques. Emphasize safety.

Performance Testing

Have each trainee perform Performance Task 1 to your satisfaction. Fill out a Performance Profile Sheet for each trainee.

Classroom

Have trainees compare and contrast the four types of mechanical bonding. Discuss typical applications for each type.

2.4.0 Wire-to-Structure Connections

Wires can be attached to the pipeline using several different procedures. Company policies should be checked to define the method preferred. The two most common bonding methods are metallurgical and mechanical.

2.4.1 Metallurgical Bonding

There are several types of metallurgical bonding that can be used for attaching wires. The pipe surface must always be properly prepared for bonding and the coating must be restored when attachment is complete. Before using **thermite welding**, company procedures should be checked to ensure that it is allowed.

Always follow company policy while performing these steps for metallurgical bonding:

Step 1 Prepare the pipe surface properly.

Step 2 Perform **ultrasonic testing** (UT) inspection at the weld site to check wall thickness.

Step 3 Determine if the wall thickness is sufficient for the internal pressure.

Step 4 Perform metallurgical bonding.

Step 5 Recoat the weld area on the pipe.

Step 6 Perform a structure-to-soil test to verify that the repair has been properly made.

Metallurgical bonding techniques include:

- Soldering
- **Brazing**
- **Exothermic welding**

Soldering is one of the most common methods of attaching wires to pipes. Soldering is quick, easy, and requires less skill than most other methods of attaching wires. Soldering does have limited use, depending on the location of the wire. This technique does not produce a connection that is as strong as a welded connection. Soldering should only be used in areas that are not stressed.

The two major differences between soldering and welding are:

- The base metals to be joined are not melted or fused together. Only the filler material is melted. When welding, both the filler metal and the base metal are melted.
- Soldering is performed at temperatures below 800°F. Welding is performed at much higher temperatures.

Soldering is used to join metals by causing a lower-melting-point metal to adhere to the joint surfaces and then to freeze in place. Solder is a material that melts below 285°F. Brazing fill melts above that point.

The most common solders are alloys of tin and lead that melt below the melting point of tin. For surfaces to accept the solder, they and the solder must be free from oxide or any other films. When necessary, parts can be cleaned with chemicals or with abrasives.

Fluxes can range from very mild substances to those with extreme chemical activity. Rosin is an effective flux used for electrical connections where the most reliability, freedom from corrosion, and absence of electrical leakage are required. Fluxes leave a residue that may cause corrosion and must be removed from the surface.

To apply solders, the joints are heated using soldering irons, torches, or induction heaters. When using irons or torches, solder is fed to the joint by hand. With induction heaters, preformed shapes are placed close to the joints before fluxing and heating.

Brazing is a process that joins metals by heating to temperatures over 285°F and using a filler metal with a liquid temperature above 285°F. Brazing provides advantages over other connection methods because it permits the joining of dissimilar metals that cannot be joined otherwise. Since the base metals are not melted to be joined, different melting points do not matter. Brazing usually produces less thermally induced distortion than fusion welding.

The important considerations of the brazing process are:

- Filler-metal flow
- Base-metal and filler metal characteristics
- Surface preparation
- Joint design and clearance
- Temperature and time
- Rate and source of heating

Brazing processes can be used on a variety of joints, and are usually modifications of either the butt joint or the lap joint. A butt joint produces a neat single layer but provides less bonding area than a lap joint. The strength of a joint is directly related to the bonding area.

The purpose of brazing fluxes is to wet the base metal with the brazing filler metal. Brazing flux dissolves any oxide left on the base metal after cleaning. It also protects the metal surfaces until the correct temperature is reached.

Cleaning of all surfaces that are involved in the formation of the desired brazed joint is required to achieve successful and repeatable brazed joining. Contaminants on one or both surfaces may result in formation of voids, and restriction or misdirection of filler metal flow. It can also cause inclusion of contaminants within the solidified braze area.

Instructor's Notes:

Exothermic welding is a fusion welding process. This type of joint is produced by heating with superheated liquid metal and **slag**. The heating results from a chemical reaction between a metal oxide and aluminum. Filler metal is obtained from the liquid metal.

Before beginning a powder-welding operation, always refer to the manufacturer's literature for the correct powder charge for the type and size of the pipeline. Welding heat causes localized stress in piping steel, so the correct charge is critical to avoiding later costly repairs. As a general guideline when working on high-pressure steel lines, powder charges should be limited to 15 grams for smaller wire sizes, to reduce any heat-induced damage. When working with wire larger than No. 4, it is advisable to divide the strands into at least two smaller bundles before welding begins; then each wire bundle can be individually powder-welded. After making the connections, be careful to ensure that the coating is properly repaired. *Figure 17* shows wire connections to pipe and the essential features of exothermic weld connections and coating repair.

Figure 18 shows a completed exothermic weld to a CP system.

2.4.2 Mechanical Bonding

Wires may also be attached by mechanical means. Follow these steps to perform mechanical bonding:

Step 1 Prepare the pipe surface properly.

Step 2 Make the attachment according to company and manufacturer's specifications.

Step 3 Coat the pipe attachment surface after repairs are completed to protect from corrosion.

Step 4 Inspect the pipe coatings.

Types of mechanical bonding that may be used include:

- Magnetic
- **Conductive epoxy**
- Clamps
- Split bolts

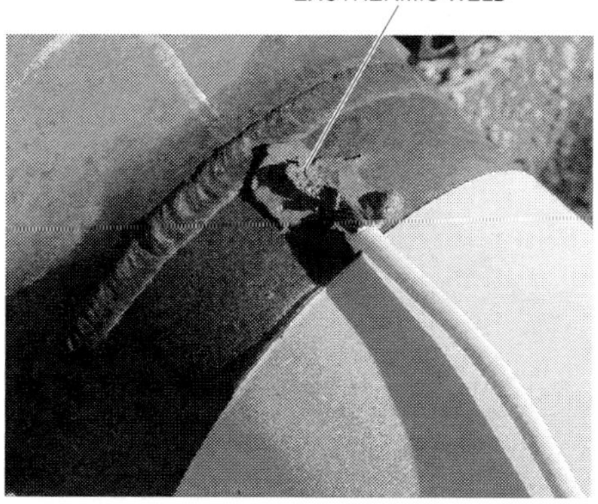

Figure 18 ◆ Completed exothermic weld to a CP system.

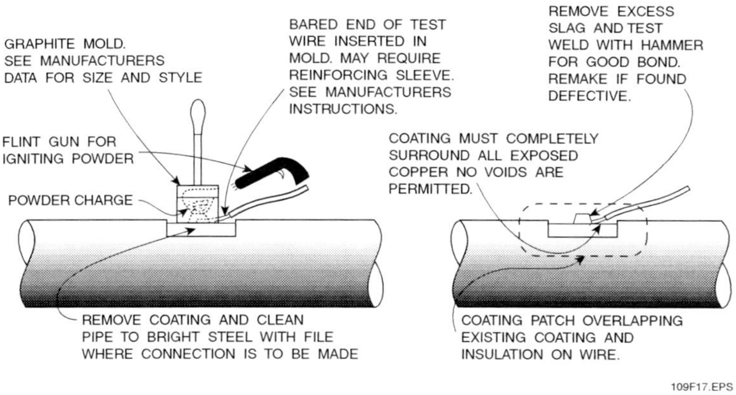

Figure 17 ◆ Wire connections to pipe.

Performance Testing

Have each trainee perform Performance Task 2 to your satisfaction. Fill out a Performance Profile Sheet for each trainee.

Homework

Assign the review of Sections 3.0.0–5.0.0 and completion of the Review Questions for the next session.

Materials

Ensure that you have everything required for teaching and testing during this session.

Teaching Tip

See the Teaching Tip at the end of this module.

Safety

Remind trainees that current must not be present during continuity testing. Ask trainees to explain possible consequences of not following this rule.

Laboratory

Prepare stations at which trainees can practice testing the continuity of a circuit, switch, and motor.

Magnetic attachment is one type of mechanical bonding that is used on pipelines. A **test lead** is attached to the magnet or magnetic bond. Before installing the magnet to the pipeline, the surface of the pipe must be cleaned. This is very important for ensuring that there is good contact between the metal surface and the pipe. If the surface is not clean, this will alter the readings that are observed at the test station. An approved protective coating is applied to the pipe prior to the installation of the magnet. This helps reduce the potential for corrosion. The magnet is then attached to the pipe and may be held on to the pipe with clamps. *Figure 19* shows a typical magnet installed on the pipe.

109F19.EPS

Figure 19 ◆ Magnet installed on a pipe.

After the magnet is installed on the pipe, the installation should be checked to ensure that the proper results are achieved and the test station is providing the correct readings. The magnet installation is then covered with an approved noncorrosive epoxy coating that protects the magnet installation and provides protection from corrosion. *Figure 20* shows a completed magnet installation.

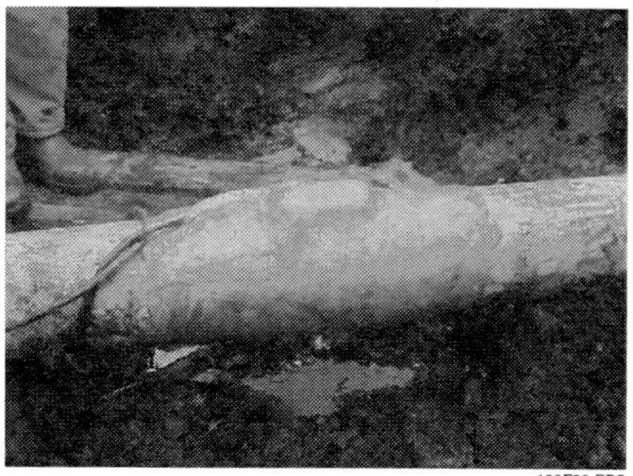

109F20.EPS

Figure 20 ◆ Completed magnet installation.

Conductive epoxy is a polyester resin used to reduce surface resistance and to prevent the accumulation of a static electric charge. Conductive epoxy has a high strength and is low in shrinkage during curing. This makes it an excellent protective coating to use on test stations.

Using a split bolt (see *Figure 21*) is an efficient and cost-effective method of mechanical bonding. Split bolts are manufactured using different high tensile materials that provide a good electrical connection and reduce corrosion. Pressure pads on the split bolts are normally constructed from extruded bars that prevent the pads from cracking and provide for long life of the bolt. Split bolts are normally used for aboveground connections when a flammable or explosive environment prevents the use of thermite welding. The split bolt is inserted into the structure or pipe and the wires are inserted through the middle of the split bolt. The nut is tightened to hold the wires in place.

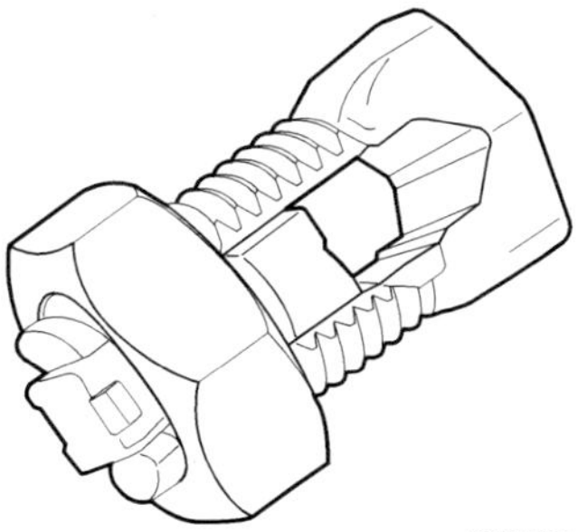

109F21.EPS

Figure 21 ◆ Split bolt.

3.0.0 ◆ ABOVEGROUND TEST STATION REPAIR

The repair of aboveground test stations is discussed in the following sections. The topics covered include replacement and repairing of defective parts, continuity testing, and verifying effective repair using **structure-to-soil potential**.

Instructor's Notes:

3.1.0 Replace or Repair Defective Parts

Test station parts must be repaired or replaced as long as they are in service. Common replacement activities include replacing the wires to the test station. The wires must be replaced with the same gauge and color as the original wiring. The wiring can become defective due to corrosion or physical damage.

Some terminal boxes can be repaired, depending on the damage. It is the employee's responsibility to determine whether the defect should be repaired or replaced. This depends on the cost of the part, availability of the replacement part, and the time and effort required to repair the part.

When repairing or replacing a part, ensure that all company policies and procedures are followed, including the safety requirements for the task and the documentation of the task after it is complete.

3.2.0 Continuity Test

A continuity test is performed to determine if a circuit is open or closed. Testing the continuity of a circuit requires the use of a multimeter or a continuity tester. Set the multimeter to the ohm setting. The symbol for ohm is the Greek letter omega (Ω). If there is more than one ohm setting, select X1. While the probes are not touching anything, the multimeter will indicate a reading of infinity. A reading of infinity means that the circuit is open. When you touch the two probes together, the reading changes to zero. A reading of zero indicates that the circuit is closed or complete. A complete circuit is one that can conduct electricity; an open circuit cannot.

> **NOTE**
> A continuity tester is a simple device that lights up to indicate continuity.

Touch each probe to one of the terminals (or poles) of the device. If the reading changes to zero the device has continuity. To test a switch, place a probe on each pole of the switch. When you move the switch from the off to the on position, the meter reading should change from infinity to zero, which indicates that the switch is working. To test a component such as a motor, touch a probe to each pole. A reading of zero indicates that the motor has continuity and current can pass through it.

> **WARNING!**
> Continuity testing should be performed when current is not present. Always unplug the device or turn off the main circuit breaker before attempting a continuity test.

The ohmmeter part of a multimeter can be used to test continuity, but this requires transporting an expensive and sensitive test instrument. Pocket-type continuity testers are just as reliable and much more compact and portable.

The two types of continuity testers are audio and visual. They are specialized instruments for identifying conductors in a conduit by checking continuity.

The audio continuity tester is used to identify wires in conduit runs. At one end of the conduit run, select one wire, strip off a little insulation, and connect that wire to the conduit. At the other end of the conduit run, clip one lead of the tester to the conduit. Touch the other lead to one wire at a time until the audible alarm sounds, which indicates continuity. Place matching tags on the wire to indicate that this is a completed circuit. Continue this procedure, one wire at a time, to identify all the other wires in the conduit.

The visual continuity tester is useful when working in areas where background noise makes it hard to hear the audio tester. The procedure for operating this tester is the same as for the audio continuity tester. The light will come on when the proper wire is tested.

3.3.0 Verify Effective Repair Using Structure-to-Soil Potential

After a repair is completed, a potential reading should be taken to ensure that the repairs are complete and the CP system is working properly. To measure the effectiveness of the repair, the leads on a high-resistance voltmeter (or multimeter) are connected to the repaired test station and the appropriate half cell.

If a stable reading is obtained (without regard to the positive or negative polarity of the reading), the repairs have been effective. If no reading is obtained, basic troubleshooting should be performed on the meter and the half cell, or the repairs to the test station should be checked.

If a structure-to-soil measurement is required to satisfy Department of Transportation (DOT) regulations, the individual performing the measurement must be Operator Qualification (OQ)-qualified or directly observed by an individual OQ-qualified in this covered task.

Classroom

Ask a trainee to explain the proper response to a failure to obtain a potential reading after a repair.

Performance Testing

Have each trainee perform Performance Task 3 to your satisfaction. Fill out a Performance Profile Sheet for each trainee.

Classroom

Ask a trainee to explain the purpose of excavating. Have another explain the purpose of backfilling.

Discuss possible consequences of not adhering to safety regulations when working on excavation sites.

Ask a trainee to explain the purpose of alerting the system before excavating. Have another trainee name three ways to accurately locate a pipeline.

Classroom

Distribute copies of sample company and OSHA procedures regarding exposing underground pipeline. Review and discuss the different procedures.

Provide sections or photos of damaged pipe. Have trainees identify the types of damage.

Have trainees list at least five guidelines regarding backfilling procedures. Answer any questions trainees may have.

Performance Testing

Have each trainee perform Performance Task 4 to your satisfaction. Fill out a Performance Profile Sheet for each trainee.

Classroom

Ask trainees to name possible causes of an AOC at corrosion protection test stations. Ask a trainee whether all AOCs must be reported.

4.0.0 ◈ BELOWGROUND TEST STATION REPAIR

Excavating and **backfilling** are necessary when parts of the test station belowground must be exposed for repair or replacement. The excavation should be performed so that it does not damage the pipe and cause further problems. The backfilling must provide the protection that is required for the pipeline. Excavation and backfilling can be dangerous work; doing it safely requires planning and precautions.

Repair of belowground test stations is covered in the following sections.

4.1.0 Following Excavation Regulations

Excavation sites are always potentially dangerous, especially those near petroleum pipelines. Workers performing excavation operations or working in an excavation should strictly adhere to all safety regulations. Failure to follow these rules can result in hazards to the worker, the pipeline, and the environment. The following must be considered before beginning any excavation work near a petroleum pipeline:

- OSHA requirements
- One-Call system
- Permits

Refer to the Pipeline Corrosion Control Module, *Locating Pipeline and Cable*, for additional information.

4.2.0 Locating Pipeline

Precise locating of the pipeline before any excavation begins is a critical part of the process. Not only will digging in the wrong place be a waste of time and expense, but it may also damage the pipeline. Company alignment maps should show the approximate location of the pipeline and test stations.

In most cases, you must alert the One-Call system before excavating because other lines may be in the area. When this is done, most companies with buried lines in the area of planned excavation will be alerted and asked to provide temporary markings and information such as depths and line descriptions. Locating the pipeline accurately may involve the following:

- Alignment maps
- **Pipe locators**
- Pipeline marking

Refer to *Locating Pipeline and Cable* for additional information.

4.3.0 Exposing Pipeline

Exposing an underground structure requires special care. The company's procedures will provide specific details on how this should be performed. The exposure process usually includes:

- Probing
- Stripping
- Trenching and shoring
- Excavation

Follow OSHA and company procedures for trenching, shoring, and excavation safety.

4.4.0 Inspecting Damage

Whenever pipe is exposed, and before it is reburied, it must be inspected for damage. Damage could be pre-existing or it could have been caused by excavation. Some types of damage usually encountered include the following:

- **Dents** and **gouges**
- Corrosion
- Loose welds
- Coating damage
- Pipeline instability
- Environmental damage

NOTE
Always report all defects per company policy.

If damage inspection is to be done for DOT compliance, that inspector must be OQ-qualified in this covered task.

4.5.0 Backfilling

A pipeline frequently requires use of a variety of backfilling procedures that must comply with specifications regarding pipe and coating protection.

Adequate fill material must be provided underneath the pipe as well as over it. The backfill material around the pipe must not damage the pipe or coating, and additional material should meet the same specifications as the original construction.

In addition, the backfill must be laid around the pipe in such a manner that it supports the pipe evenly, reducing stresses that may later cause pipe damage. Backfilling a DOT-regulated pipeline may require that the individual be OQ-qualified in that covered task also.

When filling is completed, the surface should be restored as closely as possible to its original condition, including the replacement of topsoil.

Instructor's Notes:

Backfilling correctly and safely involves the following tasks:

- Following DOT requirements
- Preparing the fill
- Positioning skid piles and sandbags
- Placing bedding (sand) or slurry
- Backfilling
- Performing soil compaction

5.0.0 ◈ ABNORMAL OPERATING CONDITIONS (AOCS)

Abnormal operating conditions (AOCs) are those conditions in which operating design limits are exceeded or when an operating condition or event occurs that could cause a hazard to persons, property, or the environment if not corrected. It is important that all personnel be able to recognize and respond to AOCs in their work areas.

The response to each of the specific conditions listed is governed by company procedures. The purpose of each procedure is to control the abnormal condition in a safe and effective manner and to return to normal operating conditions as soon as possible. Personnel are trained by the company on how to respond.

5.1.0 Test Stations

In the specific area of corrosion protection test stations, an AOC might be caused by one of the following:

- Pipe defect in area to be welded
- Failure of an isolation device
- **Arc burn**
- Ignition of hydrocarbons
- Leaks
- Corrosion
- Broken leads
- Burn-through during thermite welding

All such situations must be reported to the appropriate company personnel, but knowing how to immediately respond to each one of these conditions may prevent the situation from becoming a general pipeline emergency.

Summary

Aboveground test station repair involves replacing or repairing defective parts and verifying that the repair is effective and meets company specifications. Structure-to-soil potential measurements are used to verify that the test station is operating correctly.

Belowground test station repair involves excavating to reach the problem. Excavation procedures include using a backhoe to remove the bulk of the soil, a vacuum to clean soil from the vicinity of the station, and hand excavation to expose the pipe. Wires are then repaired, replaced, and spliced as necessary and according to company specifications. Station operation is again verified with structure-to-soil potential readings.

Wires can be attached to pipe in several ways. Metallurgical bonding methods include soldering, brazing, and thermite welding. Mechanical bonding methods include magnetic, conductive epoxy, clamps, and split bolts.

In all cases, the pipe surface must be properly prepared, attachments made according to company specifications, and the area coating restored. The pipe wall thickness must be checked before metallurgical bonding.

Once a pipeline is exposed, the DOT requires that the pipe be examined or inspected. Specific procedures for preparing the trench with stabilization material, bedding, and initial backfill must be followed to give support and protection for the pipeline.

Classroom

Go over the Review Questions. Answer any questions trainees may have.

Examination

Administer the Module Examination. Be sure to record the results of the Exam on Craft Training Report Form 200 and submit the results to the Training Program Sponsor.

Performance Testing

Ensure that all Performance Tasks have been completed and Performance Profile Sheets for each trainee are filled out. Be sure to record the results of the testing on Craft Training Report Form 200 and submit the results to the Training Program Sponsor.

Review Questions

1. The most common potential test setup is the _____ type.
   a. cable
   b. one-wire
   c. two-wire
   d. three-wire

2. For bonding purposes with a foreign-line-crossing test point, a _____ wire is used.
   a. heavier gauge
   b. lighter gauge
   c. shorter
   d. longer

3. Each wire that is terminated in a box should have at least _____ inches of slack.
   a. 8
   b. 10
   c. 12
   d. 14

4. Soldering of wires should be performed with _____ flux.
   a. noncorrosive
   b. corrosive
   c. no
   d. high-acid

5. The last step in a soldering activity is to _____ the wires.
   a. inspect
   b. clean
   c. flux
   d. recoat

6. The simplest type of connection to insulate is the _____ splice.
   a. twisted
   b. in-line butt
   c. overlapping
   d. gap

7. Bolts for bolted connections must be as _____ as possible.
   a. long
   b. short
   c. thick
   d. thin

8. There are _____ metallurgical bonding techniques.
   a. two
   b. three
   c. four
   d. five

9. Soldering is performed at temperatures below _____ °F.
   a. 200
   b. 400
   c. 600
   d. 800

10. Dissimilar metals can be joined using _____.
    a. welding
    b. gluing
    c. brazing
    d. soldering

11. Conductive epoxy has a high strength and low _____.
    a. flexibility
    b. shrinkage
    c. hardness
    d. conductivity

12. Test station wires must be replaced with the same _____ wire.
    a. length
    b. bolted
    c. color
    d. welded

13. After any repair is complete, a(n) _____ reading should be taken.
    a. voltage
    b. potential
    c. amperage
    d. insulation

14. The first critical step of excavating a pipeline is _____.
    a. digging up the line
    b. locating the line
    c. troubleshooting the problem
    d. contacting foreign line operators

15. The last step in backfilling an excavation is to _____.
    a. inspect the pipe
    b. repair the pipe
    c. recoat the pipe
    d. replace the topsoil

Instructor's Notes:

Glossary

This glossary will help you learn some terms specific to test station repair. Trade Terms introduced in this module appear in italic text.

Ammeter: A meter that measures amperes. Generally available as the direct-connect type and clamp-on type.

Abnormal operating conditions (AOCs): Those conditions in which operating design limits are exceeded, or when an operating condition or event occurs that could cause a hazard to persons, property, or the environment if not corrected.

Arc burn: The damage resulting from touching the pipe surface with a welding rod or holder outside the welding area.

Backfilling: The technique for covering a completed pipeline so that adequate fill material is underneath and over the pipe to protect it from loose rocks, abrasion, shifting, and washouts.

Bond: The electrical connection of two structures, for the purpose of regulating current flow, which usually have two different cathodic protection levels.

Brazing: A welding process in which coalescence is produced by heating to suitable temperatures above 842°C and by using a filler metal that must have a liquid temperature above 842°C.

Cadweld: The trade name of a type of thermite welding with filler metal (primary elements are iron oxide and aluminum). When ignited, it gives off large amounts of heat.

Conductive epoxy: A polyester resin used to reduce surface resistance and to prevent the accumulation of a static electric charge.

Dent: A depression that produces a reversal in the curvature of the pipe wall without reducing the wall thickness.

Exothermic welding: Welding with molten iron, which is obtained by igniting aluminum and an iron oxide in a mold, whereby the aluminum floats to the top of the molten metal and is poured off.

Gouge: Surface damage that reduces the pipe wall thickness by displacement or removal of metal.

High input impedance voltage meter: An electronic device used to measure structure-to-soil potential.

Magnetic attachment: An attachment whereby the test lead is attached to the magnet or magnetic bond.

Multimeter: A meter that allows measurement of different AC and DC electrical units.

Pipe locator: A device used in leakage surveys to locate affected pipe. It sends out radio signals from underground and traces the signal the length of the pipe. It will work on plastic or steel pipe, since the plastic pipe has a metal wire running its entire length.

Potential measurement: The activity of measuring structure-to-soil potential.

Rectifier: A device used to convert AC current to DC current. In an impressed cathodic protection system it supplies current to the structure in the soil.

**Reference electrode:* An electrode whose open-circuit potential is constant under similar conditions of measurement, which is used for measuring the relative potentials of other electrodes.

Shunt: A conductor joining two points in an electrical circuit to form a parallel or alternative path through which a portion of the current may pass.

Slag: The residue of aluminum oxide on the surface after the aluminum reduces the iron oxide to molten iron.

Solder connection: Connection made to a structure using soldering equipment and techniques, usually through the application of controlled heat to a metallic material that is susceptible to melting.

Soldering: The joining of metals by causing a lower-melting point metal to wet or alloy with the joint surfaces and then to freeze in place. Solder is a joining material that melts below 800.6°F.

Spark gun: The device used to ignite the crucible containing the thermite.

**Structure-to-soil potential (also structure to electrolyte potential):* The potential difference between the surface of a buried or submerged metallic structure and the electrolyte that is measured with reference to an electrode in contact with the electrolyte.

Test lead: A connection to the structure being tested, usually a wire in a supporting stand with an easy connection point for the meter.

Test station: Specific area or location where structure-to-soil potential is measured.

Thermite welding: Welding with a fire-hazardous mixture of ferric oxide and powdered aluminum; upon ignition by a magnesium ribbon, it reaches a temperature of 4,000°F, sufficient to soften steel.

Ultrasonic (UT) tester: An instrument that determines the thickness of a structure.

* Term published in NACE *Glossary of Corrosion-Related Terms.* For a copy of the complete *Glossary of Corrosion-Related Terms*, contact NACE International, 1440 South Creek Dr., Houston, TX 77084. The Glossary is revised periodically, so information in an outdated version may not be accurate.

Instructor's Notes:

REFERENCES

Additional Resources

This module is intended to present thorough resources for task training. The following reference works are suggested for further study. These are optional materials for continued education rather than for task training:

"Close Interval Potential Survey (CIPS) Technique," 2000, Dr. M. Tullmin. Ontario, Canada; Corrosion Club.com.

Control of Pipeline Corrosion, Second Edition (2001), A.W. Peabody. Houston, TX; NACE International.

ACKNOWLEDGMENTS

Figure Credits

Figure 3 courtesy of NACE International

(Source: *Peabody's Control of Pipeline Corrosion, Second Edition* Pg 240, Fig 12.1 Typical types of test station) © NACE International – The Corrosion Society, 1440 South Creek Drive, Houston, Texas 77084. Additional publications on corrosion control technology are available from www.nace.org or phone (281) 228-6200.

Figure 4 courtesy of NACE International

(Source: *Peabody's Control of Pipeline Corrosion, Second Edition* Pg 199, Fig 9.9 Galvanic anode test station installation) © NACE International – The Corrosion Society, 1440 South Creek Drive, Houston, Texas 77084. Additional publications on corrosion control technology are available from www.nace.org or phone (281) 228-6200.

Figure 6 courtesy of NACE International

(Source: *Peabody's Control of Pipeline Corrosion, Second Edition* Pg 241, Fig 12.2 Typical types of above grade test stations) © NACE International – The Corrosion Society, 1440 South Creek Drive, Houston, Texas 77084. Additional publications on corrosion control technology are available from www.nace.org or phone (281) 228-6200.

Figure 7 courtesy of NACE International

(Source: *Peabody's Control of Pipeline Corrosion, Second Edition* Pg 242, Fig 12.3 Typical at-grade types of test stations) © NACE International – The Corrosion Society, 1440 South Creek Drive, Houston, Texas 77084. Additional publications on corrosion control technology are available from www.nace.org or phone (281) 228-6200.

Figure 8 courtesy of NACE International

(Source: *Peabody's Control of Pipeline Corrosion, Second Edition* Pg 243, Fig 12.4 Typical buried type of test stations) © NACE International – The Corrosion Society, 1440 South Creek Drive, Houston, Texas 77084. Additional publications on corrosion control technology are available from www.nace.org or phone (281) 228-6200.

Figure 17 courtesy of NACE International

(Source: *Peabody's Control of Pipeline Corrosion, Second Edition* Pg 244, Fig 12.5 Wire connections to pipe) © NACE International – The Corrosion Society, 1440 South Creek Drive, Houston, Texas 77084. Additional publications on corrosion control technology are available from www.nace.org or phone (281) 228-6200.

Instructor's Notes:

MODULE 61109-02—TEACHING TIPS

The following are suggested activities or instructional methods to help you teach the material in this AIG.

General

When you call on someone to answer a question, the rest of the class relaxes or even tunes out because they expect that the question and answer will take place only between you and the trainee you called on. Instead, use this technique to involve more trainees in answering questions and to keep them on their toes.

1. Ask trainees to define a term or explain a concept.
2. After one trainee has answered, ask a trainee seated nearby if the answer is right. Then ask whether a trainee in the back of the room agrees.
3. Ask trainees to explain why they think an answer is right or wrong.
4. Use the session to clear up incorrect ideas and encourage trainees to learn from their mistakes.

Section 2.2.0 *Mounting Test Stations*

This exercise gives trainees the opportunity to apply what they have learned about mounting test stations. For this exercise, trainees will need paper, pencils, and their Trainee Guides. Allow 30 to 45 minutes for this exercise.

1. Divide trainees into pairs. Describe several hypothetical situations in which test stations are needed. Write the important factors for each situation on the board or an overhead.
2. Have each pair of trainees work together to choose the best type of station for each location. Have them explain their choices in writing.
3. Ask for volunteers to share their responses with the class. Answer any questions trainees may have.

Section 2.4.2 *Mechanical Bonding*

This exercise gives trainees the opportunity to practice the four types of mechanical bonding. Trainees will need personal protective equipment and their Trainee Guides. You will need to prepare at least one station for each type of mechanical bonding. Allow 1 to 2 hours for this exercise.

1. Review the steps for each type of mechanical bonding. Have trainees familiarize themselves with the relevant specifications for making each type of attachment. Also review safety procedures.
2. Divide trainees into as many groups as there are stations. Have each trainee take turns making mechanical bonds. Have each group rotate through each of the stations until trainees have practiced all four types of mechanical bonding.
3. Inspect trainee work. Ask trainees if they have any questions.

Section 3.1.0 *Replace or Repair Defective Parts*

This exercise will allow trainees to implement what they have learned about replacing or repairing defective test station parts. Trainees will need paper, pencils, and their Trainee Guides. Allow 30 to 45 minutes for this exercise.

1. Present several hypothetical scenarios regarding defective test station parts.
2. Have trainees work in pairs to determine, for each scenario, whether the parts can be repaired or whether they must be replaced. Have them write down the considerations that helped determine their response.
3. Ask for volunteers to share their answers with the class. Answer any questions trainees may have.

MODULE 61109-02—ANSWERS TO REVIEW QUESTIONS

Answer	Section
1. c	2.1.0
2. a	2.1.0
3. c	2.2.0
4. a	2.3.0
5. d	2.3.1
6. b	2.3.2
7. b	2.3.3
8. b	2.4.1
9. d	2.4.1
10. c	2.4.1
11. b	2.4.2
12. c	3.1.0
13. b	3.3.0
14. b	4.2.0
15. d	4.5.0

CONTREN™ LEARNING SERIES — USER UPDATES

The NCCER makes every effort to keep these textbooks up-to-date and free of technical errors. We appreciate your help in this process. If you have an idea for improving this textbook, or if you find an error, a typographical mistake, or an inaccuracy in NCCER's Contren™ textbooks, please write us, using this form or a photocopy. Be sure to include the exact module number, page number, a detailed description, and the correction, if applicable. Your input will be brought to the attention of the Technical Review Committee. Thank you for your assistance.

Instructors – If you found that additional materials were necessary in order to teach this module effectively, please let us know so that we may include them in the Equipment/Materials list in the Instructor's Guide.

Write: Curriculum Revision and Development Department
National Center for Construction Education and Research
P.O. Box 141104, Gainesville, FL 32614-1104

Fax: 352-334-0932

E-mail: curriculum@nccer.org

Craft ______________________ Module Name ______________________

Copyright Date __________ Module Number __________ Page Number(s) __________

Description

____

____

____

____

(Optional) Correction

____

____

____

(Optional) Your Name and Address

____

____

____

Module 61110-02

Inspect Internal Pipe Surfaces

(CT 12)

Inspect Internal Pipe Surfaces
Annotated Instructor's Guide

Module 61110-02

MODULE OVERVIEW

This module familiarizes trainees with the procedures for inspecting a pipeline for signs of corrosion or other damage. Trainees will learn how to use instruments to measure the extent of corrosion.

PREREQUISITES

Please refer to the Course Map in the Trainee Module. Prior to training with this module, it is recommended that the trainee shall have successfully completed the following:

Core Curriculum; Pipeline Core, Modules 66101-02 and 66102-02; Pipeline Corrosion Control Level One, Modules 61103-02 through 61109-02

OBJECTIVES

Upon completion of this module, the trainee will be able to:

1. Properly orient the pipe to be removed by identifying the top, bottom, left side, right side, upstream direction, and downstream direction (CT 12).
2. Recognize internal corrosion (CT 12).
3. Properly operate wall thickness and pit depth measurement equipment (CT 12).

PERFORMANCE TASK

Under the supervision of the instructor, the trainee should be able to:

1. Inspect internal pipe surface (CT 12).

NCCER STANDARDIZED CRAFT TRAINING PROGRAM

The National Center for Construction Education and Research (NCCER) provides a standardized national program of accredited craft training. Key features of the program include instructor certification, competency-based training, and performance testing. The program provides trainees, instructors, and companies with a standard form of recognition through a National Craft Training Registry. The program is described in full in the *Guidelines for Accreditation*, published by the NCCER. For more information on standardized craft training, contact the NCCER by writing us at P.O. Box 141104, Gainesville, FL 32614-1104; calling 352-334-0911; or e-mailing info@nccer.org. More information may be found at our Web site, www.nccer.org.

HOW TO USE THIS ANNOTATED INSTRUCTOR'S GUIDE

Each page presents two sections of information. The larger section displays each page exactly as it appears in the Trainee Module. The narrow column ties suggested trainee and instructor actions to each page and provides icons to call your attention to material, safety, audiovisual, or testing requirements. The bottom of each page includes space for your notes.

If you see the Teaching Tip icon, that means there is a teaching tip associated with this section. Also refer to any suggested teaching tips at the end of the module.

SAFETY CONSIDERATIONS

Ensure that the trainees are equipped with appropriate personal protective equipment. Emphasize the importance of following all safety precautions and procedures when working with power tools.

PREPARATION

Before teaching this module, you should review the Module Outline, Objectives, Performance Tasks, and the Materials and Equipment List. Be sure to allow ample time to prepare your own training or lesson plan and gather all required equipment and materials.

MATERIALS AND EQUIPMENT LIST

Overhead projector and screen

Transparencies

Blank acetate sheets

Transparency pens

Whiteboard/chalkboard

Markers/chalk

Pencils and scratch paper

Appropriate personal protective equipment

Copies of your company's policy and procedures manual

Compasses

Pipeline inspection documents, two sets

Compasses, pit gauges, pocket knives, and plastic for wrapping the samples

Field corrosion report forms

Examples of coating failures

Corroded pipe sections

Company forms for recording pit length and depth measurements

Ultrasonic thickness gauges and calibration standards

Pipeline inspection forms

Access to pipeline that needs to be inspected and related equipment

Solvent or mechanical means of removing hydrocarbon residue

Copies of Quick Quiz**

Module Examinations*

Performance Profile Sheets*

Copies of the Performance Verification for Covered Task 12†

* Located in the Test Booklet.

** Located at the end of this module.

† Available for download at www.nccer.org.

TEACHING TIME FOR THIS MODULE

An outline for use in developing your lesson plan is presented below. Note that each Roman numeral in the outline equates to one session of instruction. Each session has a suggested time period of 2½ hours. This includes 10 minutes at the beginning of each session for administrative tasks and one 10-minute break during the session. Approximately 7½ hours are suggested to cover *Inspect Internal Pipe Surfaces (CT 12)*. You will need to adjust the time required for hands-on activity and testing based on your class size and resources.

Topic	Planned Time
Session I. Pipe Orientation, Damage, and Defects	
A. Introduction	________
B. Determine Pipe Orientation	________
1. Based on Flow Direction	________
2. From Compass Readings	________
C. Inspecting for Damage and Defects	________
1. In-Depth Inspection	________
2. Coating Failure and Corrosion	________
Session II. Visual Inspections and Ultrasonic Meters	
A. Types of Pit Gauges	________
B. Calibrate Gauge and Check for Worn Point	________
1. Laboratory	________
C. Use Pit Gauge to Measure Pit Depth	________
D. Use Pit Gauge to Measure Pit Length	________
1. Laboratory	________
E. Take Multiple Readings to Gather Remaining Strength Data	________
F. Basic Ultrasonic Meter Operation	________
Session III. Ultrasonic Thickness Measurements, Documentation, and Inspection of Internal Pipe Surfaces	
A. Meter Calibration	________
B. Completing a Pipeline Inspection Report Form	________
1. Laboratory	________
C. Procedure for Inspecting Internal Pipe Surfaces	________
1. Laboratory	________
D. Summary	
1. Summarize module	________
2. Answer questions	________
E. Module Examination	________
1. Trainees must score 70% or higher to receive recognition from the NCCER.	
2. Record the testing results on Craft Training Report Form 200 and submit the results to the Training Program Sponsor.	
F. Performance Testing	________
1. Trainees must perform each task to the satisfaction of the instructor to receive recognition from the NCCER.	
2. Record the testing results on Craft Training Report Form 200 and submit the results to the Training Program Sponsor.	

Module 61110-02

Inspect Internal Pipe Surfaces
(CT 12)

Instructor's Notes:

Instructor's Notes:

Introduction to the Pipeline Operator Qualification Rule, API, and NCCER

It could be said that petroleum products keep America running. Petroleum products like oil and natural gas are the principal sources of energy in the United States and are used in many manufacturing and chemical processes. Petroleum products are primarily transported by pipeline. There are more than a million miles of liquid and natural gas pipeline in this country, ranging in diameter from two inches to four feet. These pipelines bring crude oil and natural gas from production fields to processing plants and refineries—and ultimately to the consumer.

Pipeline Operator Qualification

Petroleum products are considered hazardous, so pipeline personnel must take extra precautions and follow specific procedures when working with them. To help ensure the safe handling and transportation of petroleum products, the United States Department of Transportation (DOT) mandates safe practices as federal regulations.

One such regulation, the Operator Qualification (OQ) rule, requires pipeline operators to develop a written qualification plan to evaluate their personnel and their contract personnel (see *49 CFR Part 192 Subpart N—Gas,* and *49 CFR Part 195 Subpart G—Liquid*). The evaluation hinges on the operator's ability to perform certain covered tasks and to be able to recognize and respond to abnormal operating conditions that may be encountered while performing these covered tasks.

Covered Tasks

A pipeline performance task is deemed *covered* if the activity passes the following four-part test:

1. Is it performed on a pipeline facility?
2. Is it an operations or maintenance task?
3. Is it performed as a requirement of *49 CFR Part 195 (Liquid)* or *49 CFR Part 192 (Gas)*?
4. Does it affect the operation or integrity of the pipeline facility?

If the answer to all four questions is *yes*, then the activity is considered *covered* and must figure into the evaluation process of pipeline personnel.

American Petroleum Institute

The OQ rule and its related covered tasks provided the catalyst to develop a unified pipeline training and assessment program—so that pipeline operators and owners have ready access to the assessment and training tools they need. This effort was the direct result of the vision and leadership provided by The American Petroleum Institute (API) and API members.

Efforts to unify the industry are nothing new to API. Indeed, the impetus for forming the Institute in 1919 was the need to standardize engineering specifications for drilling and production equipment. Today, the most pressing issues revolve around public perceptions and government policies toward the industry. Speaking with one voice on these issues has become as essential as having interchangeable parts in the field. API's members determine the petroleum industry's positions on public policy as well as the standards that govern its day-to-day operations worldwide.

Consortium on Operator Qualification

API assembled a focus group whose primary purpose was to identify covered tasks and provide compliance recommendations to pipeline operators.

The Consortium on Operator Qualification (COOQ) became the guiding force for the identification and development of covered tasks for the liquid pipeline industry. Their work product, API publication 1161, *Guidance Document for the Qualification of Liquid Pipeline Personnel*, formed the basis for the resulting Pipeliner Training and Assessment Program.

The covered task list created in API publication 1161 has now evolved to include gas covered tasks and is maintained by NCCER. The most current covered task list can be downloaded at www.nccer.org.

National Center for Construction Education and Research

With the guidance and support of API, its members, and the Consortium on Operator Qualification, the National Center for Construction Education and Research (NCCER) was able to design and implement the Pipeliner Training and Assessment Program.

NCCER is a not-for-profit 501(c)(3) education foundation established in 1995 by the world's largest and most progressive construction and maintenance contractors and national contractor associations. NCCER was founded to address the

severe workforce shortage facing the construction and maintenance industries and to develop a standardized training process and curricula. Today, NCCER is supported by hundreds of leading construction and maintenance companies, manufacturers, and national associations and developed more than 35 construction and maintenance titles.

NCCER also maintains a National Registry, which contains the records of trainees and instructors who have received NCCER training. All completion information for every trainee is sent to the NCCER and filed within the Registry. The Registry can then confirm training and skills for workers as they move from company to company, state to state, or even within their own company.

Standardized Curriculum (NCCER's Contren™ Learning Series)

NCCER curricula are developed using the expertise of industry subject matter experts (SMEs). The use of SMEs as an integral part of every new development or revision ensures that the curriculum is current, industry-driven, and indicative of industry practices. The development and/or revision process is facilitated from start to finish by NCCER project managers, who work closely with the SMEs and a select team of technical writer(s) on the development and presentation of the curriculum.

Pipeline: Five Training Paths

The pipeline curriculum focuses on essential skills and competencies within five skill areas:

- Corrosion Control
- Electrical and Instrumentation
- Maintenance
- Mechanical
- Operations (Field and Control Center)

SMEs from the pipeline industry have determined the skills and competencies needed based on API's covered tasks and the consensus of the group/industry. Specific company information is generally not incorporated into the curriculum unless the SMEs determine that the information is the industry standard.

Training Modules and Covered Tasks

The smallest unit of instruction in the curriculum is a *module*. Each curriculum is composed of modules (the equivalent of stand-alone chapters) that are logically sequenced into complete *levels* of instruction. Every module is task-oriented and requires the successful completion of a written (knowledge-based) examination and a performance (hands-on) test. Modules may be individually selected and/or combined for efficient task training. Modules in the pipeline curricula, to the extent possible, will align with API-developed covered tasks. While not all modules will relate to covered tasks, all covered tasks will relate to at least one module.

Benefits of Training to the Pipeline Marketplace

Applying NCCER's standardized training process allows pipeline owners and contractors in the United States to establish the benchmark for acceptable skill and competency levels within the pipeline industry. This process will assist training organizations in developing and/or improving standardized pipeline training programs that effectively match their resources and operating conditions while responding to regulatory demands and to the needs of the industry. The process will also recognize training organizations that meet the minimum standards of competence, warranting regulatory, public, and professional confidence. This in turn will foster national unity within the pipeline industry for the mutual benefit of individual technicians, owners, contractors, and ultimately pipeline-users and society.

Additional Resources

To review the Pipeliner Training and Assessment Program Implementation and Resource Guide and the latest version of pipeline covered tasks, visit NCCER's Web site at www.nccer.org, or contact:

National Center for Construction Education and Research
PO Box 141104
Gainesville, FL 32614-1104
(888) 622-3720
Web site: www.nccer.org

For more information on API, please contact:

American Petroleum Institute
1220 L Street, Northwest
Washington, D.C. 20005
(202) 682-8000
Web site: www.api.org

To order *API Pub 1161*, contact Global Engineering Documents by calling 1-800-854-7149 or visiting their Web site at www.global.his.com.

Additional sources of information on the pipeline operator qualification regulation include:

DOT – Office of Pipeline Safety
www.ops.dot.gov

Transportation Safety Institute
www.tsi.dot.gov

Instructor's Notes:

COURSE MAP

This course map shows all of the modules in the first level of the Pipeline Corrosion Control curriculum. The suggested training order begins at the bottom and proceeds up. Skill levels increase as you advance on the course map. The local Training Program Sponsor may adjust the training order.

PIPELINE CORROSION CONTROL LEVEL ONE

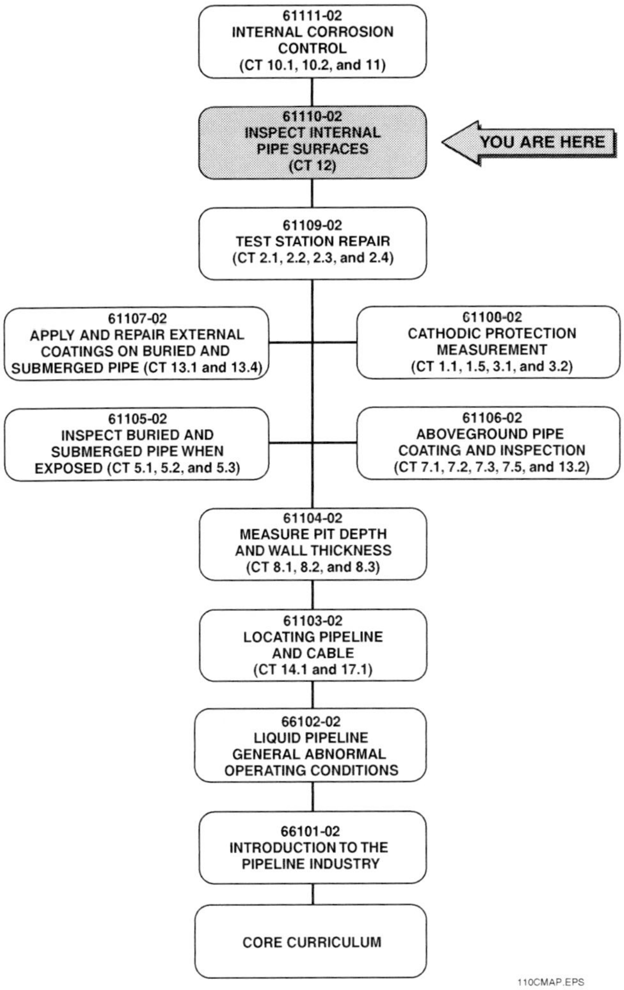

Homework

Assign reading of Module 61110-02.

Instructor's Notes:

MODULE 61110-02 CONTENTS

MODULE 61110-02 CONTENTS (Continued)

Instructor's Notes:

Figures

Table

Instructor's Notes:

MODULE 61110-02

Inspect Internal Pipe Surfaces (CT 12)

Objectives

When you have completed this module, you will be able to do the following:

1. Properly orient the pipe to be removed by identifying the top, bottom, left side, right side, upstream direction, and downstream direction (CT 12).
2. Recognize internal corrosion (CT 12).
3. Properly operate wall thickness and pit depth measurement equipment (CT 12).

Prerequisites

Before you begin this module, it is recommended that you successfully complete the following: Core Curriculum; Pipeline Core, Modules 66101-02 and 66102-02; Pipeline Corrosion Control Level One, Modules 61103-02 through 61109-02.

Required Trainee Materials

1. Pencil and paper
2. Appropriate personal protective equipment

1.0.0 ◆ INTRODUCTION

Each year, **corrosion** costs the pipeline industry millions of dollars in material damage and downtime. External corrosion on a pipe is visible when the pipe is exposed; internal corrosion is not as easily detected.

Underground corrosion is caused by a natural phenomenon of metal returning to its original lowest energy state, such as steel returning to its original state of iron oxide or rust. A general definition of corrosion is the deterioration of a material because of a reaction with its environment. Metallic corrosion is considered an electrochemical process, in that it consists of a series of chemical reactions with a measurable electrical current flow.

Maintenance and repair begins with an in-depth inspection of the pipeline to determine the extent of corrosion or damage that is present. This procedure should include a visual inspection for corrosion such as pitting, metal loss, and general corrosion on pipe, as well as mechanical damage such as gouges, scrapes, and dents.

2.0.0 ◆ DETERMINE PIPE ORIENTATION

Documenting the correct orientation of the pipeline serves two purposes. When an individual is sent into the field to conduct an inspection, the inspection should be performed the same way each time. And, when a section of pipe is sent to the laboratory for testing, lab personnel must know the orientation of the pipe in order to correctly interpret the data.

2.1.0 Pipeline Orientation Based on Flow Direction

To establish the orientation of the pipeline based on flow direction, follow these steps:

Step 1 Determine the direction of product flow in the pipeline.

> **NOTE**
> If this is a bi-directional line with product flowing in either direction at times, orientation should be done using compass readings instead of product flow direction. Bi-directional lines will be addressed in a later section.

Materials

Ensure that you have everything required to teach the course. Check the Materials and Equipment List at the front of this module.

Classroom

Explain that terms shown in bold are defined in the Glossary at the back of this module.

Audiovisual

Show Transparency 1, Course Objectives.

Show Transparency 2, Performance Task.

Teaching Tip

See the general Teaching Tip at the end of this module.

Classroom

Have a trainee explain the cause of underground corrosion. Ask trainees to name types of corrosion and mechanical damage on pipes.

Audiovisual

Show Transparency 3 (Figure 1). Ask a trainee to interpret the drawing.

Audiovisual

Show Transparencies 4 and 5 (Figures 2 and 3). Ask a trainee to explain how the drawing in Figure 3 was created from the pipeline drawn in Figure 2.

Teaching Tip

See the Teaching Tip at the end of this module.

Step 2 On the pipeline inspection document provided by your company, draw a circle representing a cross-section of the pipeline from your viewpoint facing upstream (against product flow).

> **NOTE**
> The basic layout of a section of pipe may already be printed on the form your company uses.

Step 3 Using the hours as they appear on a clock face, label the four major points on the clockface, or 12:00 (top of pipe), 3:00 (right side), 6:00 (bottom of pipe) and 9:00 (left side of pipe). *Figure 1* shows a typical inspection form orientation drawing.

> **NOTE**
> The clock face orientation corresponds to compass headings. The top, or 12:00 corresponds to 360 degrees; the right side, or 3:00, 90 degrees. Continuing around the circumference of the pipe, the bottom, or 6:00, corresponds to 180 degrees and the left side, or 9:00, 270 degrees.

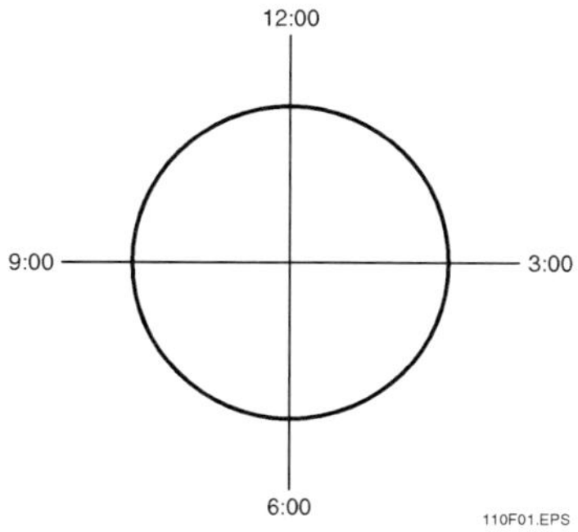

Figure 1 ◆ Inspection form orientation drawing.

Step 4 Visually examine the pipe for pitting and/or deposits.

2.2.0 Pipeline Orientation From Compass Readings

To establish the orientation of the pipeline based on compass headings, follow these steps:

Step 1 Using a compass, determine the heading of the pipeline.

Step 2 On the pipeline inspection document provided by your company, draw a circle representing a cross-section of the pipeline from your viewpoint facing the compass direction of your choice. To ensure uniformity of performing this task, it is recommended that you always choose the orientation facing the higher of the two compass readings. *Figure 3* shows a typical inspection form showing pipeline orientation of the pipe shown in *Figure 2* using the recommended approach.

> **NOTE**
> A straight section of pipe will have two headings 180 degrees apart from each other. If one end of the pipe is pointed to the west (270 degrees), the other end will be pointed to the east (90 degrees) or 180 degrees difference on the compass. Figure 2 shows a pipeline situated along a northwest-southeast line. The compass reading for this line is 315 degrees (NW) and 135 degrees (SE). NW (315 degrees) minus SE (135 degrees).

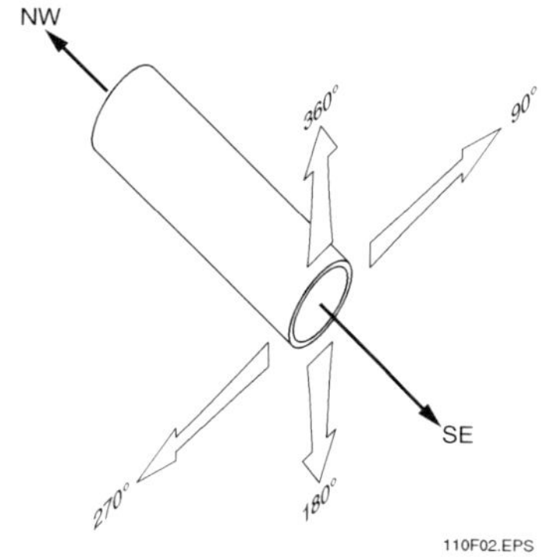

Figure 2 ◆ Pipeline situated along a northwest-southeast line.

Step 3 Using the hours as they appear on a clock face, label the four major points on the clockface, or 12:00 (top of pipe), 3:00 (right side), 6:00 (bottom of pipe) and 9:00 (left side of pipe). As an option, you may choose to label the pipe using compass headings for reference. In this case, the top of the pipe would be labeled 360 degrees and the right side as you face it, 90 degrees. The bottom is represented as

Instructor's Notes:

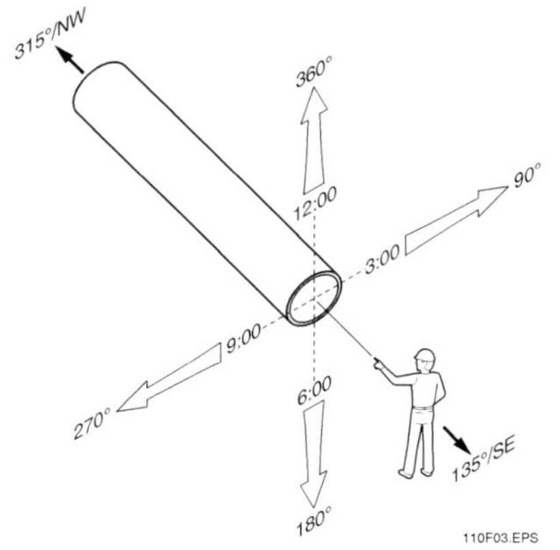

Figure 3 ◈ Inspection form showing pipeline orientation.

180 degrees and the left side as you face it is the 270 degrees point. *Figure 4* shows a pipe labeled with compass readings.

Step 4 Visually examine the pipe for pitting and/or deposits.

> **NOTE**
> The basic layout of a section of pipe may already be printed on the form your company uses.

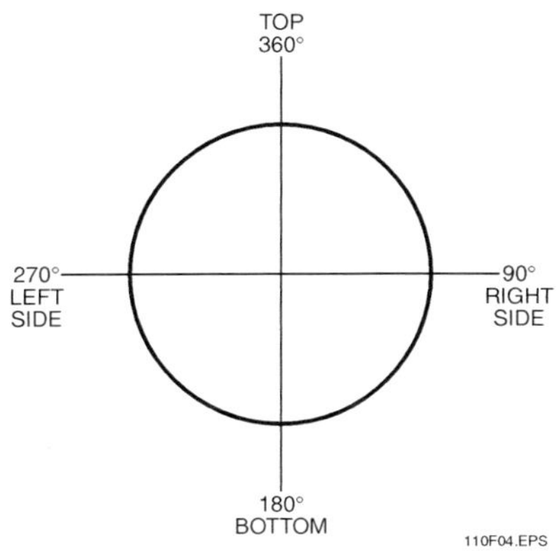

Figure 4 ◈ Pipe labeled with compass readings.

3.0.0 ◈ INSPECTING FOR DAMAGE AND DEFECTS

Generally, coatings that are suitable for use underground or underwater are not suitable for exposure to sunlight and weather.

In recent years, the petroleum industry has become more aware of the threat of leaks from underground fuels and hazardous chemicals. Studies conducted by the federal government, in preparation for the implementation of regulations on underground liquid storage systems, have pinpointed corrosion as the major cause of leakage in underground liquid storage systems. Inspecting for damage and defects is discussed in the following sections.

3.1.0 In-Depth Inspection

An in-depth inspection of a pipeline for corrosion and/or damage includes looking for all of the following:

- *Pitting* – Sometimes referred to as localized corrosion, pitting refers to corrosion that only affects a defined area on the metal surface. Pitting corrosion is the formation of cavities or pits. *Figure 5* shows a profile view of typical pitting corrosion.
- *General corrosion* – Used to describe corrosion that involves metal loss distributed over a wide area with no discernable pitting.
- *Mechanical damage* – Typically refers to dents, gouges, and scrapes on the internal surface of the pipeline. Mechanical damage to internal pipe surfaces usually occurs during pigging operations. *Figure 6* shows a scrape along the outer surface of a pipe.
- *Deposits* – Refers to an accumulation of debris that may occur at any point along the run of the pipeline. Deposits include **scale** and parafin buildup. *Figure 7* shows deposits in a line.

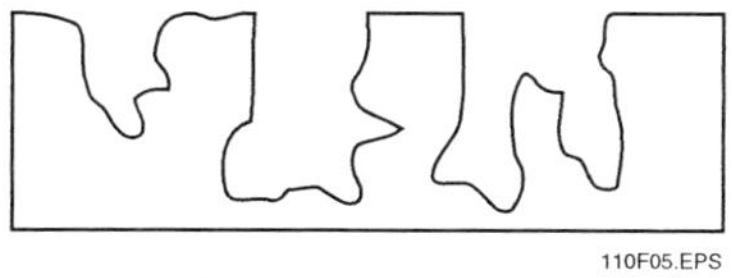

Figure 5 ◈ Typical pitting corrosion.

3.1.1 Address Wall Loss and Deposits

If pitting or wall loss is discovered after a visual examination of the pipe, capture the following data in order to address the damage sufficiently:

Audiovisual

Show Transparency 6 (Figure 5). Ask a trainee to distinguish between pitting and general corrosion.

Classroom

Set up stations by placing pipe segments that exhibit the four types of corrosion. Have trainees identify each type of corrosion.

Safety

Stress that iron sulfide must be wiped off any samples. Explain that iron sulfide is extremely flammable.

Teaching Tip

See the Teaching Tip at the end of this module.

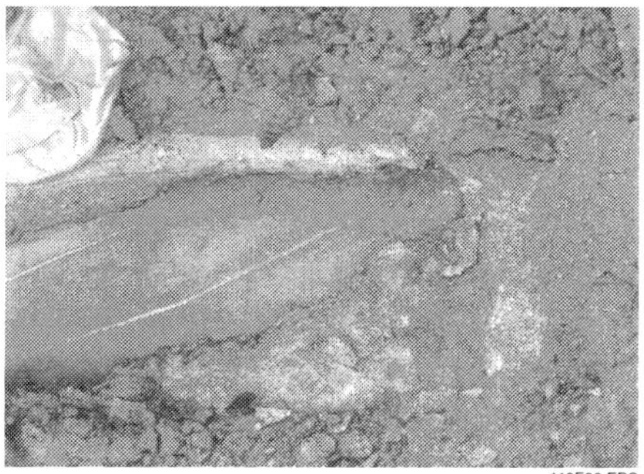

110F06.EPS

Figure 6 ◆ Scrape on outer surface of a pipe.

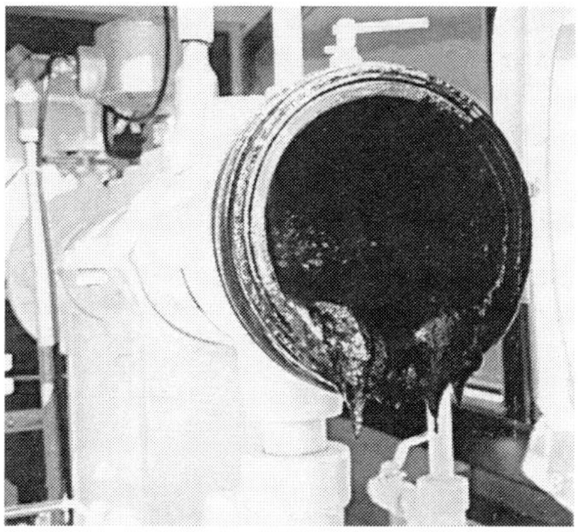

110F07.EPS

Figure 7 ◆ Deposits in a line.

- Determine the orientation of the pits or wall loss (for example, on the bottom at 6:00).
- Determine the depth and breadth of the pits or wall loss.
- Determine whether the pits are scattered discretely or are all in one location, or if the wall loss occurred only on the bottom (at both 5:00 and 7:00, etc.).
- Determine if there is material in the bottom of the pits or wall loss.
- Determine the shape of the pits or wall loss (pit morphology).

Record the observations with sketches and photographs. If possible, place something easily recognizable next to the pit or wall loss for a size reference.

If there is material at the bottom of the pits or wall loss, scrape some of it out with a sterile swab or tongue depressor and inoculate sessile serial dilution vials (run a sessile bacteria test).

NOTE
A sessile bacteria test can only be performed with fresh samples, preferably no more than a couple of hours old.

If pitting or general wall loss is discovered in the area immediately adjacent to welds, determine whether this is the only place the metal loss is occurring, and apply the pitting and wall loss steps discussed earlier.

If there are visual buildups or deposits on the pipe, obtain the following information:

- Determine the thickness of the deposits.
- Determine the type of material (for example, iron sulfide, scale, iron oxide, etc.).
- Determine the adherence of the deposit by prying it up with a pocket knife.
- Determine if there is corrosion underneath the deposit. If there is, apply the pitting and wall loss steps discussed earlier.
- Photograph and/or sketch the deposit.
- Collect a sample of the deposit for laboratory analysis.
- Determine whether the deposit is magnetic or non-magnetic.

If possible, determine if the deposit is water soluble, hydrocarbon soluble, or acid soluble. While checking for acid solubility, check for a rotten egg odor or a bright yellow color.

If the sample is to be sent to the laboratory, do not clean it in any way unless directed to do so. Mark the orientation of the pipe, state the location, date, and type of product on the pipe, and indicate any tests that may be required. Typically, your company will have a form for filing internal corrosion date. (*Figure 8* shows a sample Field Corrosion Report Form.) Wrap the sample in plastic, making it as airtight as possible, or cap the ends. Be sure to get the sample to the lab as quickly as possible.

WARNING!
If a significant amount of iron sulfide is discovered on the pipe, wipe most of it off of the sample. Iron sulfide can be extremely flammable when dry and can auto combust in direct sunlight.

Instructor's Notes:

Field Corrosion Report

Company: ______ **Date:** ______ **Time:** ______
Pipeline Name: ______
Pipeline System: ______ **Pipeline Segment:** ______
Location:
Milepost: ______ Station: ______ GIS Information: ______
Latitude: ______ Longitude: ______ Section: ______
Block: ______ Town: ______ County: ______ State: ______
Directions to location: ______
Type or types of product: ______ **Pipe diameter:** ______
Pipe wall thickness as per drawings: ______
Pipe wall thickness as per ultrasonic thickness tester at location: ______
Nominal allowable pipe thickness (allowable thickness range as per pressure code): ______
Was the pipe exposed? Y N **Soil Type:** ______ **Soil pH:** ______
What was the cathodic protection reading prior to digging? ______
What were the cathodic protections readings after the exposure was repaired? ______

External Metal Loss
Was there any obvious external metal loss? Y N
Describe the location: ______
Describe the shape of the metal loss: ______
What was the depth of the metal loss? ______
How was this measured? ______
Draw a diagram of the metal loss or make a rubbing of the affected area. Label position according to the clock, upstream and downstream if possible, and all pertinent dimensions and depths. Attach photo.

Long Seam Corrosion
Does the pipe have seams running longitudinally on the pipe (along the length of the pipe)? Y N
Is there any metal loss along these weld seams or within 2" of the seams? Y N
Draw a diagram of the metal loss or make a rubbing of the affected area. Label position according to the clock, upstream and downstream if possible, and all pertinent dimensions and depths. Attach photo.

External Deposits
Was there any deposit on the pipe? Y N
Was a sample collected? Y N
Were field bacteria tests run on the deposits? Y N
Name and telephone number of contact for laboratory: ______

External Coating
Was the pipe coated or bare? Coated Bare
What type of coating? ______
Did the coating require repair? Y N
Describe the size of the repair: ______
Describe the type of repair performed: ______
Attach photos of before and after the repair.

Internal Metal Loss
Was the pipe opened up and inspected internally? Y N
Were there any obvious signs of internal metal loss? Y N
Describe the location: ______
Describe the shape of the metal loss: ______
What was the depth of the metal loss? ______
How was this measured? ______
Draw a diagram of the metal loss or make a rubbing of the affected area. Label position according to the clock, upstream and downstream if possible, and all pertinent dimensions and depths. Attach photo.

Liquids
Were there free liquids? Y N
Describe the liquids: ______
Were field water chemistry tests done on the liquids? Y N
Were field bacteria tests run on the liquids? Y N
Were samples sent to a laboratory? Y N
Describe the results: ______
Name and telephone number of contact for laboratory: ______
Attach photo.

Deposits
Were there deposits ? Y N
Describe the location of the deposits: ______
Describe the deposits: ______
What was the pH of the deposits? ______
Were field bacteria tests run on the deposits? Y N
Where were they shipped? ______
Name and telephone number of contact for laboratory: ______
Attach photo.

Remaining Wall Strength Data
- Number of pits = total number of pits
- Length of pit is measured along distance of pipe, not circumference; this number is reported in inches.
- Depth of pit is the distance from the original pipe wall to the deepest portion of the pit; this number is reported in inches.
- Pit depth to wall thickness ratio (pit depth ÷ actual wall thickness)
- Percent remaining wall thickness (1.0 –[pit depth ÷ wall thickness])

Pit Number	Length	Depth	Pit Depth to Wall Thickness Ratio	Percent Remaining Wall Thickness	80% Remaining Wall Thickness
					Yes No
					Yes No
					Yes No

Name of Inspector ______ *Signature* ______ *Date* ______

108F08.EPS

![Classroom]

Ask trainees to name the functions of a coating. Ask a trainee to identify the cause of most coating failures.

Ask trainees to distinguish between sagging and cracking failures. Have a trainee explain how to minimize cracking.

3.2.0 Coating Failure and Corrosion

Corrosion can be caused by a number of conditions that may exist on a site. These conditions may include one or more of the following:

- Dissimilar metal corrosion
- Highly conductive and poor soils
- Differential soils
- Differential aeration
- Stray electrical currents from other structures

For protection, coatings can be applied to the piping to increase the resistance between the pipe surface and the soil environment. Cathodic protection can be employed to create a net increase in electrical potential, thus inducing a current flow that will overcome and control the currents associated with corrosion.

Widely accepted estimates are that 70% of all coating failures are the result of poor surface preparation. Correctly following surface preparation, application, and inspection procedures offers the best protection against coating failure. While this may seem like a costly approach, it is much less costly than performing repeated repairs over the life of the pipeline.

A coating is a complex mix of interacting ingredients such as resins, plasticizers, pigments, extenders, catalysts, fungicides, and solvents. Coatings are usually applied as thin films so any solvents, or carriers, are allowed to evaporate. The remaining hardened coating must adhere to the surface, forming an insoluble surface and providing an acceptable appearance as well. The atmospheric coating must protect the surface from weather, atmosphere, chemicals, and other forces. Underground coatings must create an insoluble barrier between the pipe and the surrounding environment.

Protective coatings available today are expected to resist corrosive environments and many other potentially damaging variables. With the complex materials involved, and the wide range of forces in play, there can be many causes for coating failures. The following sections describe some common failures associated with coatings applied to **buried** or **submerged pipe**.

3.2.1 Sagging

Sagging failures are usually the result of coating thickness problems. Many coatings continue to cure over time and in areas where the coating application was too heavy, they often check or crack, leading to rapid coating failure. Runs and sags in pipe coatings are most often the result of incorrect or poor application practices.

Examples of improperly applied coatings include material that was not brushed out sufficiently to eliminate thick areas, or sprayed materials applied with incorrect spray gun adjustments. Holding the spray gun too close to the surface or not moving it quickly enough over the surface can result in thick applications and/or runs.

Coating sag can be caused by soil stress. Moisture and/or temperature changes may cause the soil to shift, causing damage to the coating. Sag may result from the direct effect of those same changes on the coating material.

3.2.2 Cracking

Cracking failures are often related to the coating formula and result in premature aging and weathering of the coating material. Cracking is defined as a surface problem but one in which the cracks at the point of failure extend completely through the coating material to the surface of the pipe.

Cracking occurs in varying weather and moisture conditions. Coatings can also experience cracking failure as a result of the contraction and expansion of the pipe or structure being protected. This is more likely to occur when the coating has been applied in a thicker film. The thicker the coating film, the greater the stress within the material as curing takes place.

It is possible to minimize cracking through the correct use of resins, plasticizers, and pigments. Reinforcing pigments, fibrous in nature, are effective for reinforcing a coating against cracking failure. The addition of resins or plasticizers also helps to soften the coating, making it more elastic and less susceptible to the stresses caused by temperature changes, weather, and curing. *Figure 9* shows general cracking of a coating.

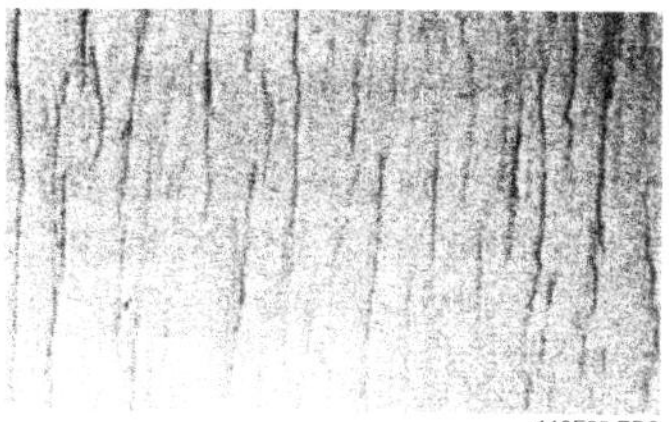
110F09.EPS

Figure 9 ◆ General cracking of a coating.

Instructor's Notes:

3.2.3 Wrinkling

Wrinkling, like sagging, is usually associated with coatings that are applied too thickly. As opposed to **shrinkage**, as is the case with checking and cracking, wrinkling is the result of the expansion of a coating. In this case, the coating surface expands more rapidly than the body of the coating during the curing process. Wrinkling is often a problem with oil-based coatings, alkyds, and similar materials.

Many oil-based coating materials contain drying agents. If a coating material contains too much surface drier, wrinkling may occur in spots where the coating layer is thicker. Cobalt driers are commonly used as surface driers. Zinc and lead compounds are classified as through-drying driers.

Temperature can also impact wrinkling. Coatings that cure without wrinkling at standard application temperatures may wrinkle if the curing is accelerated by heat. An increase in temperature tends to cure the surface too rapidly, often leading to wrinkles in the coating surface.

Wrinkling is not only detrimental to appearance, but may also reduce the corrosion resistance of a coating. If, however, the coating dries thoroughly and retains its adhesion, the coating should still be effective.

If a coating material seems to have a tendency to wrinkle, apply it in several thin layers so there are no overly thick areas. Eliminating thick areas during the application process will minimize the potential for wrinkling. *Figure 10* shows wrinkling of a coating.

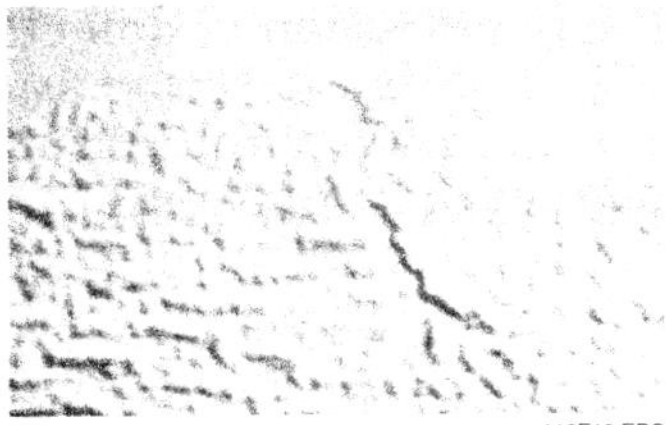
110F10.EPS

Figure 10 ◆ Wrinkling of a coating.

3.2.4 Blistering

Blistering is a common type of coating failure. Blisters, whether large or small, tend to be round in appearance and raised from the surface. Blisters might be liquid-filled or dry. *Figure 11* shows a typical hemispherical blistered surface.

Blistering is usually caused by moisture working its way under the surface of the coating in areas where adhesion may break down. This moisture, once it gains an entrance into or under the coating material, can condense and form a liquid-filled blister, or it may form a dry blister from the pressure of the moisture vapor. Blisters typically form in two locations: underneath the coating, separating it from the protected surface; or between coating layers.

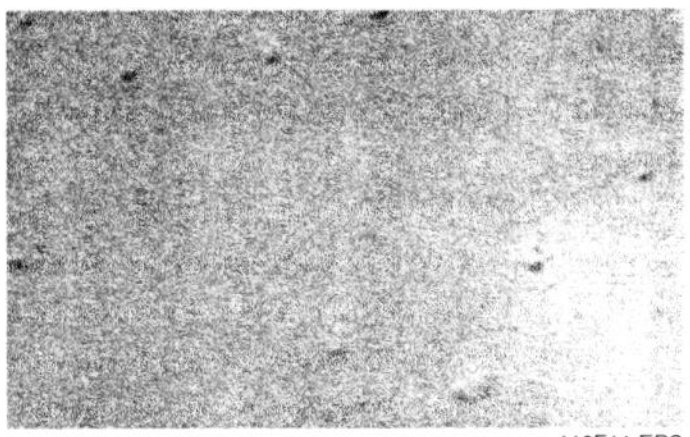
110F11.EPS

Figure 11 ◆ Hemispherical blistered surface.

Blistering can be caused by hydrogen gas that forms on the pipeline surface under the coating. Gas pressure separates the coating from the pipe surface. It might also occur where surfaces have been corroded and some remaining salts or oxides provide the conditions that support blistering.

Blistering is a serious problem and should be corrected before it can lead to further damage. Localized spots can be treated by removing and re-applying the coating. In the case of general, or widespread blistering, the maintenance crew must remove the coating, clean and prepare the surface to accept a new coating, and replace the coating with a suitable material for the location and environment.

3.2.5 Peeling

When a coating peels (*Figure 12*) , it is usually the result of an adhesion problem rather than contamination by invading chemicals attacking the coating surface or from the use of an unsuitable coating material. Peeling is often related to expansion or contraction of the coating material to the extent that it loses its **surface adhesion**.

Peeling can occur when improper surface preparation techniques are used or when of some form of contamination is introduced between coating layers as they are applied, causing one or more layers of the coating to peel away. In addition to these failures, coatings that are applied too thickly often develop into a film with surface tension that is strong enough to overcome the adhesion of the coating to the pipe surface, resulting in a loss of adhesion and eventual widespread peeling.

Classroom

Ask trainees to explain possible consequences of moisture and heat on coatings.

Classroom

Bring in examples of each type of coating failure. Have trainees identify the type of failure and possible causes.

Teaching Tip

See the Teaching Tip at the end of this module.

Homework

Assign the review of Sections 4.0.0–5.1.3 for the next session.

110F12.EPS

Figure 12 ◆ Peeling of a coating.

3.2.6 Flaking

Flaking (*Figure 13*) is an adhesion-related failure similar in appearance to peeling. In the case of flaking, the coating material usually becomes hard and brittle from curing too quickly or beyond the desired level of flexibility and, in reaching this inflexible state, shrinks and pulls away from the surface of the pipe. As the coating material shrinks and cracks begin to appear in the surface, further tension caused by the shrinking causes the cracked pieces to curl at the edges and continue to peel, or pull back, away from the pipeline surface.

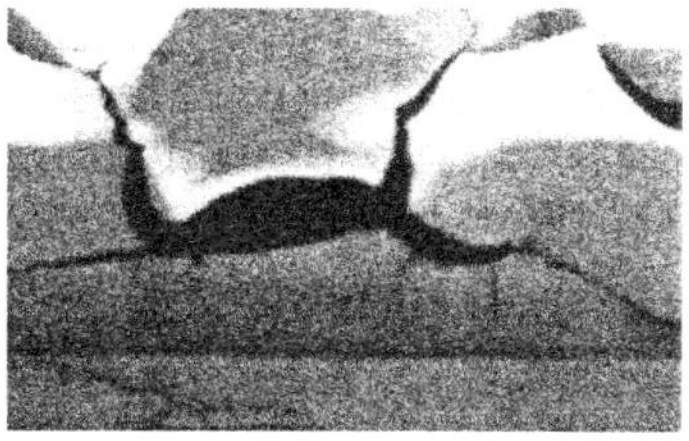

110F13.EPS

Figure 13 ◆ Flaking of a heavy organic coating.

3.2.7 Intercoat Delamination

Delamination is the term used to describe that condition where the adhesion between coating layers is lost (for example, from blistering) and the layers begin to separate from each other. Coatings often fail in this manner when new coating material is applied over an existing coating, after maintenance or repairs have been completed. The old coating material, which may have been in place for an extended period of time, may have started to deteriorate or may not have been properly prepared to receive the new material. If the new coating fails to adhere completely to the old, delamination may result. *Figure 14* shows delamination of a coal tar epoxy coating on the interior of a large pipe.

Other delamination causes include: incompatibility of materials where a like material is not used to recover a repair; coating application too thick as described in the earlier section on flaking; chemical invasion from acids; or separation of coating layers caused by strong solvents.

Localized or spotty delamination can be repaired in the same way as blistering or peeling. Small areas can be treated by removing and reapplying the coating material. When repairing widespread delamination, the pipeline maintenance crew must remove the coating, clean and prepare the surface to accept a new coating, and replace the coating with a suitable material for the location and environment.

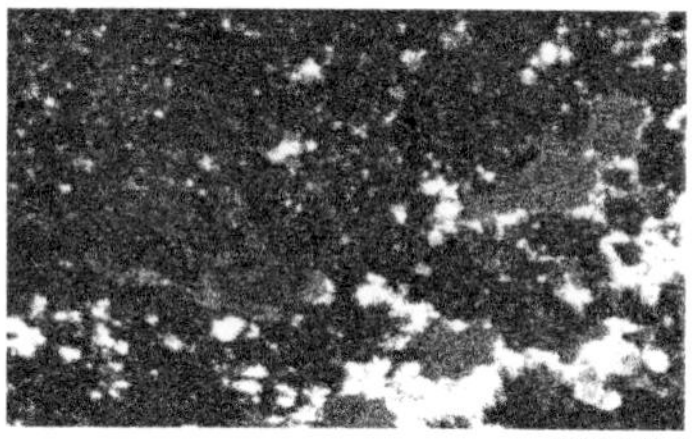

110F14.EPS

Figure 14 ◆ Delamination of a coal tar epoxy coating.

3.2.8 Undercutting

When corrosion, or rust, forms at a break in the coating material, often caused by damage or other deterioration, it can spread under the coating over the surface of the pipe. This form of corrosion, referred to as undercutting, can also occur without this breakdown in coating material integrity if the coating has been applied over a corroded or contaminated surface.

This form of corrosion makes it critical that pipeline inspections are careful and thorough, as it is easy to overlook minor damage to the coating surface that may lead to undercutting corrosion. *Figure 15* shows an example of undercutting corrosion that can occur at the location of a minor cut or abrasion on the coated surface.

Undercutting corrosion can quickly lead to more widespread corrosion once the protective coating surface has been breached. Corrosion builds up and spreads between the coating material and the pipe surface where typical conditions exist for the formation of an ideal corrosion cell. *Figure 16* shows undercutting beginning at a break in the coating. The coating is actually being turned

Instructor's Notes:

110F15.EPS

Figure 15 ◈ Undercutting from minor abrasion.

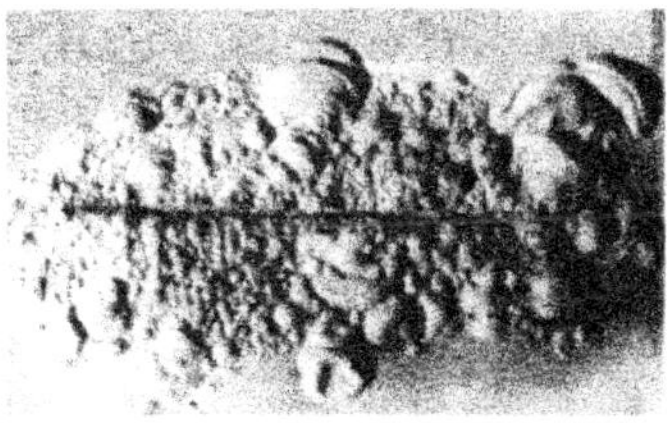

110F16.EPS

Figure 16 ◈ Undercutting at a coating break.

up at the sides of the break as the corrosion product works underneath.

Damaging levels of pitting can result from undercutting. In areas susceptible to serious corrosion, detailed planning and execution are needed to reduce or eliminate undercutting corrosion.

4.0.0 ◈ VISUAL INSPECTIONS

When corrosion or damage is discovered, individuals must determine the remaining wall thickness of the pipe in order to set the priority for repair or replacement of the corroded section of pipe.

It is important to identify any anomalies and to document their relative positions. One of the most common tools used to check the extent of corrosion to a pipe is the **pit gauge**. These gauges provide an accurate measurement of pit depth and other anomalies. Pit gauges are discussed in the following sections.

4.1.0 Types of Pit Gauges

There are several different types of pit gauges that are common to the petroleum industry. The common types of pit gauges, mechanical and micrometer, are discussed in the following sections.

4.1.1 Mechanical

The common mechanical pit gauge *(Figure 17)* is a small, sturdy, easy-to-use tool for making a quick determination of the extent of corrosion damage present. These gauges are usually made of polished stainless steel and can easily be carried in a tool kit. Care must be taken to use these gauges properly, and to recognize and understand their limitations.

Inaccurate readings can result if the gauge is not aligned properly and resting on an even surface. Where severely pitted or partially coated pipe and associated structures are involved, these handheld pit gauges provide the user with only an approximation of pit depth. The depth measurement scale on the mechanical pit gauge is marked in sixteenths of an inch and in mils (thousands of an

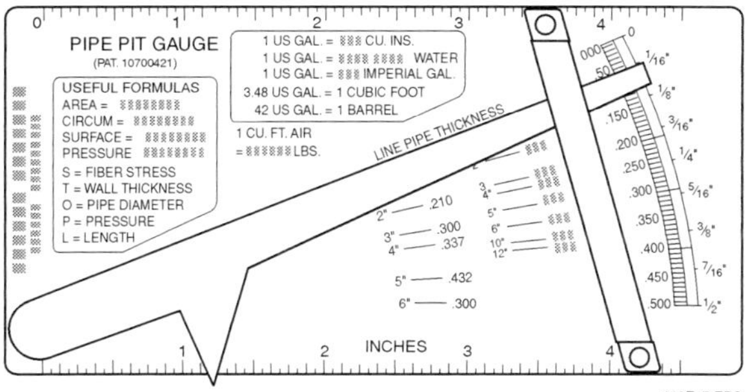

110F17.EPS

Figure 17 ◈ Mechanical pit gauge.

Materials

Ensure that you have everything required for demonstrations and teaching during this session.

Demonstration

Bring in a mechanical pit gauge and a micrometer pit gauge. Discuss and demonstrate how to use and care for each type of gauge.

Laboratory

Have trainees practice calibrating each type of pit gauge. Also show them how to check for damage or worn points.

Safety

Discuss typical safety procedures for measuring pipe that is still connected to the system.

inch). Most mechanical pit gauges include useful formulas, tables, and data giving outside diameter (OD), wall thickness, and weight per foot for various sizes of pipe.

4.1.2 Micrometer Pit Gauge

When pit depth measurements are critical and accuracy is a priority, micrometer or dial pit gauges, designed for a more accurate measurement of corrosion pit depth, may be used. Most micrometers are capable of measuring depth to extremely close tolerances. On the down side, these indicators are more fragile than mechanical gauges and should be handled with care to ensure their repeated accuracy.

A micrometer pit gauge (see *Figures 18 and 19*) will provide the most accurate measurement of corrosion pit depth. Some models of manually operated micrometer pit depth indicators measure to an accuracy of +/–0.0002". These measuring tools are somewhat fragile and quite expensive compared to other indicators.

Micrometer pit gauges should be handled with care to ensure their repeated accuracy and to safeguard the company's investment in these costly tools. Before using a micrometer pit depth indicator, check to see if it has been calibrated and certified as accurate according to your company's tool calibration procedure.

4.2.0 Calibrate Gauge and Check for Worn Point

Before a pit gauge is used for measurement, it must be inspected for any damage and for worn points. During this inspection, always closely check the point of the instrument for any worn spots. If a pit gauge does have a worn point, the pit gauge should be replaced. Only use an instrument that is in excellent condition, otherwise your inspection may not be accurate. Prior to using the pit gauge, clean the pit in accordance with your company's approved policy.

4.3.0 Use Pit Gauge to Measure Pit Depth

To use a standard pit gauge, align the gauge lengthwise with the pipe to be inspected. The gauge must rest squarely on the pipe surface to provide an accurate reading of the pits. Move the arm on the gauge so that the pointer is at the bottom of the pit. Read and record the corresponding number of the scale of the gauge. Continue this process for a general sampling of the pits to be measured. *Figure 20* shows the proper technique for measuring the pit depth.

110F18.EPS

Figure 18 ◆ Fowler Ultra-Digit Mark V electronic indicator.

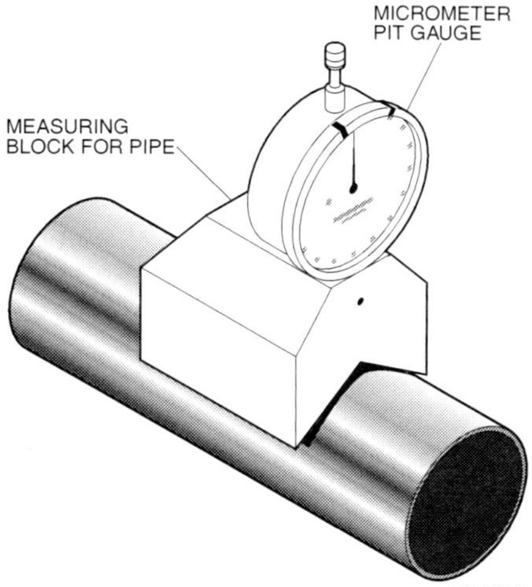

110F19.EPS

Figure 19 ◆ Micrometer pit gauge positioned on pipe.

WARNING!
The pipe is still connected to the system. Make sure you follow your company policies and procedures for work in a hazardous environment.

Instructor's Notes:

When using a dial type pit gauge, the gauge must first be zeroed. This will calibrate the gauge. Then place the gauge squarely on the pipe surface. Push the pointer into the pit to be measured with the control button on the side of the gauge. This should register a reading on the gauge. Record this measurement. Continue this process until an adequate sample of readings has been recorded.

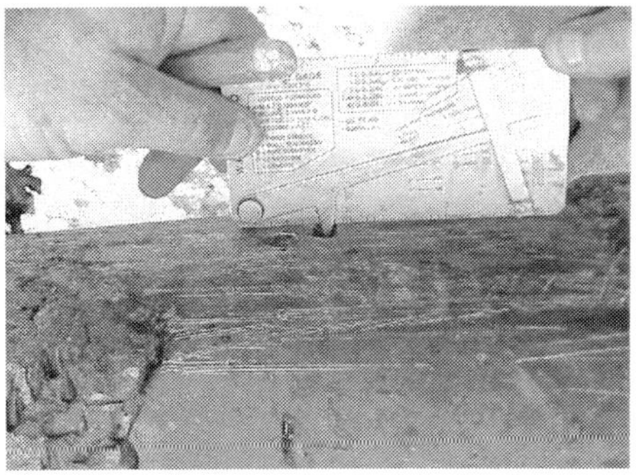

110F20.EPS

Figure 20 ◆ Measuring pit depth.

4.4.0 Use Pit Gauge to Measure Pit Length

With the pointer of the pit gauge in the pit, observe the length of the pit on the bottom scale of the mechanical pit gauge. After this measurement is taken, make sure that the measurement is recorded on the appropriate company forms.

4.5.0 Take Multiple Readings to Gather Remaining Strength (RSTRENG) Data

The B31G and RSTRENG (remaining strength) corrosion assessment methods are used by pipeline individuals to evaluate the structural integrity of pipelines with corrosion-based metal loss. The basis for these two calculation approaches has been in existence for more than 25 years. Both methodologies have been validated through burst testing of pipe with corrosion and failures of pipe in service.

While acceptable, the B31G method is recognized to be very conservative in nature. The RSTRENG criterion was developed to offset this conservative characteristic. Pipeline individuals are required by *49 CFR 192* and *195* to use either ASME/ANSI B31G or RSTRENG to evaluate the remaining strength in any corroded pipeline. The B31G method calculates predicted failure points based on the maximum length and maximum depth of the corrosion defect on the pipe.

The RSTRENG approach combines this approach with a more detailed profile of the defect by evaluating a greater number of depth measurements (minimum of three with one being the maximum depth in the defect) taken along the horizontal length of the defect. RSTRENG works equally well with single defects or with several corrosion arrays that interact together. *Figure 21* shows the typical river bottom profile of the corrosion defect as determined through the RSTRENG method.

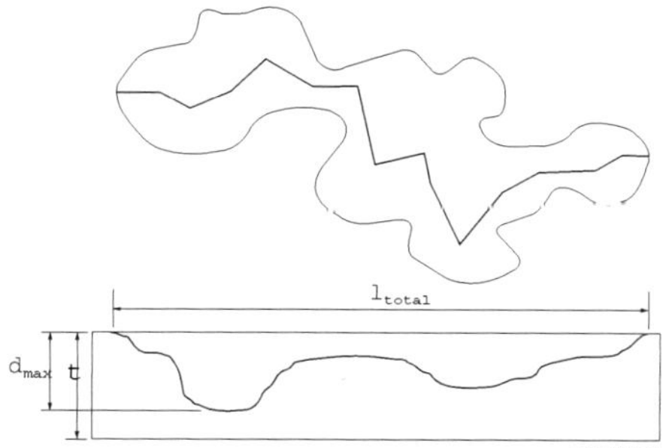

110F21.EPS

Figure 21 ◆ River bottom profile.

5.0.0 ◆ ULTRASONIC THICKNESS (UT) MEASUREMENTS

Testing on-line equipment is a common occurrence in most pipeline operations. A common testing method is digital **ultrasonic thickness** (UT) **testing** for the measurement of remaining pipe wall thickness. Ultrasonic thickness gauging is widely used to determine remaining wall thickness, one factor in overall equipment integrity. This is especially useful when internal access to an equipment item is impractical.

Using a digital ultrasonic thickness gauge and available **transducers** (probes), the pipeline maintenance individual can quickly and easily measure and record the wall thickness on systems and equipment subject to corrosion damage. To avoid unnecessary downtime and loss of production, these inspections can be made on in-service lines. *Figure 22* shows a typical ultrasonic wall thickness meter and transducer probe in use. The use and calibration of general ultrasonic meters are discussed in the following sections.

Laboratory

Have trainees practice measuring pit depth and length using a pit gauge. Then have them record their measurements on the appropriate forms.

Classroom

Have trainees compare and contrast the B31G and RSTRENG corrosion assessment methods. Answer any questions trainees may have.

Classroom

Have a trainee explain how digital ultrasonic thickness gauges work.

Ask trainees to name the problems that arise when measuring the thickness of pipes protected by coating. Then have them explain how to overcome the problems.

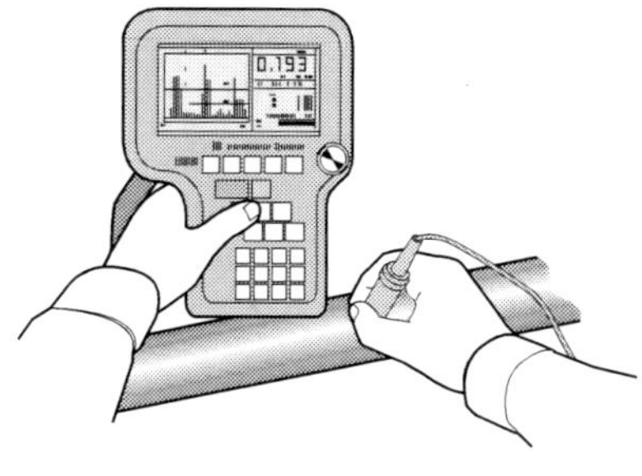

110F22.EPS

Figure 22 ◆ Operating an ultrasonic thickness gauge.

5.1.0 Basic Ultrasonic Meter Operation

Ultrasonic thickness gauges, their basic operation, and management of gathered data are discussed in the following sections, which cover digital ultrasonic thickness gauge function, thru-coating measurement, data loggers/recorders, and data analysis.

5.1.1 Digital Ultrasonic Thickness Gauge Function

The basic operation of the ultrasonic thickness gauge begins with the generation of an electrical energy pulse. The transducer probe converts this pulse into a high-frequency sound wave. The sound wave, emitted by the transducer probe as it rests on the surface of the pipe being measured, travels through the pipe wall and is reflected from the interior surface (backwall) of the pipe.

The reflected wave is picked up by the transducer probe upon its return to the surface side of the pipe. The wave is converted back into electrical energy and passed on to the gauge. Charts are available to show the pipeline maintenance individual the velocity of sound through a given material. Using that value for the pipe material of construction, the time required for the emitted sound wave to travel to the interior wall of the pipe and return to the probe can be converted into a thickness value. Using the correct velocity of sound for the pipe material being measured is critical for accurate results.

Ultrasonic gauges are reliable tools for measuring equipment subject to internal corrosion. Where corrosion damage exists, the damaged area has a different molecular structure and is not a true part of the pipe or equipment material. The transmitted signal reflects off the line between the good and damaged materials and can be used to determine the thickness of the good material.

5.1.2 Thru-Coating Measurement

Pipeline systems and equipment can also fall victim to environmental corrosion damage. Typically, aboveground equipment is protected by a thick layer of paint or one of many coating materials. This poses a problem for ultrasonic measurement because the individual does not know the exact thickness of the coating.

Another problem is caused by the extremely slow velocity of sound values for most coating materials. Because of this difference in velocity of sound values between the coating and the underlying metal, the ultrasonic thickness gauge measures the typical coating as being two to three times thicker than it actually is. This error in thickness could give the maintenance individual false data.

In the past, accurate thickness readings could only be obtained by removing the coating and measuring only the pipe material itself. The paint-stripped area was then cleaned and re-coated, a costly process in terms of material and labor hours.

An improved method for measuring coating thickness is available with the use of dual-element transducer probes. These probes are designed to emit a more focused signal beam that is capable of more accurate readings in areas of varying thicknesses than single beam probes. *Figure 23* shows how the transducer probe is placed on the pipe to generate a readout of wall thickness.

Ultrasonic thickness gauges can be equipped with a multiple echo readout capability. This type of thickness gauge is used to identify and measure metal thickness apart from coating thickness and display the results in a single readout. *Figure 24* shows in detail how the transducer signals are converted into digital values for both the coating and the pipe material.

Individuals may need to adjust the gauge for starting point and gate width for accurate measurement of pipe or equipment material thickness.

5.1.3 Data Analysis

Depending on the size of the equipment being tested, the number of thickness readings could be hundreds or thousands. The use of software programs to analyze large amounts of thickness data, schedule future inspections, and predict the service life of the equipment is vital to any mechanical integrity program.

The data management software program can interpret massive amounts of data to make decision-making easy. By viewing the data in color

Instructor's Notes:

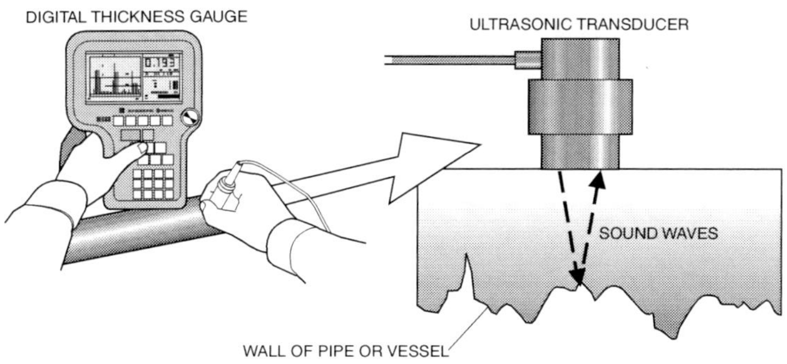

Figure 23 ◆ Transducer probe used to generate wall-thickness readout.

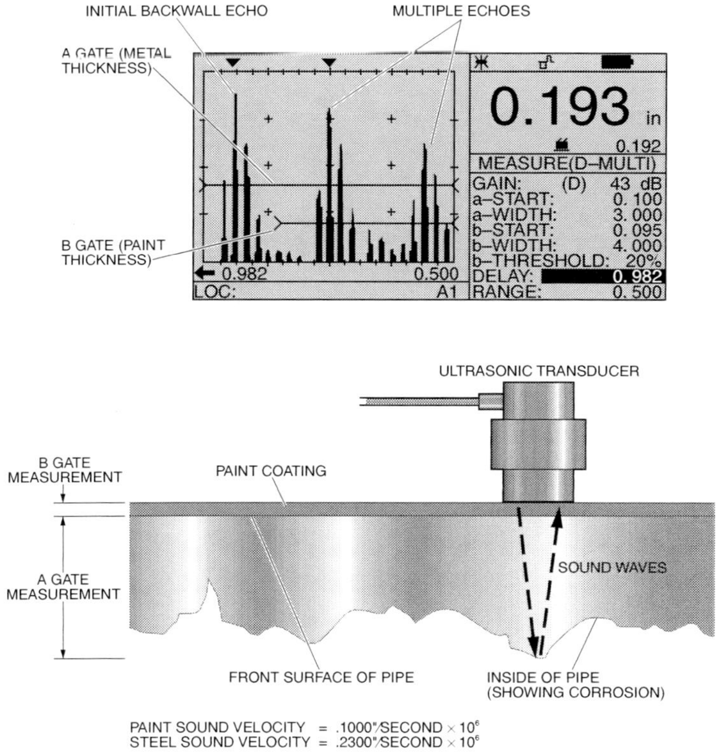

Figure 24 ◆ DMS digital A-scan thickness readout.

Classroom

Discuss how to read the screen displays in Figures 24 through 26. Ask trainees to list the benefits of using software programs to analyze wall thickness data.

Homework

Assign review of Sections 5.2.0–7.0.0 and completion of the Review Questions for the next session.

Materials

Ensure that you have everything required for teaching and testing during this session.

Classroom

Ask trainees to explain why it is important to calibrate the ultrasonic thickness gauge.

levels related to thickness, it is easy to see that part of a component is below the minimum thickness value. *Figure 25* shows a typical thickness level as displayed in grid block view.

High-level software programs are typically used in large refineries, chemical processing plants, or power generation plants. Databases, history and complex analysis calculations within the program can predict the remaining service life of an in-service equipment item or component, and calculate when the item needs re-inspection. *Figure 26* shows a typical thickness measurement location line graph showing corrosion rates and remaining life.

Documentation of testing results, record retention and comparison to historical data are other added benefits of these programs. Individuals can store historic thickness data, make notes of the inspection, and identify their work. Storage of this history and analysis of equipment performance is an essential part of some code performance.

The high-level software programs that aid in data analysis provide the user with many benefits, such as:

- Predict service life of an equipment or component item
- Predict and establish inspection intervals
- Review plant wide inspection intervals and schedule work accordingly
- Track all equipment that is overdue for inspection
- Alert operations to equipment that is potentially unsafe
- Document all inspections on equipment
- Document recommended repairs on equipment
- Document the inspector that performed the inspection
- Report minimum required thickness and maximum allowable operating pressure of the equipment

5.2.0 Meter Calibration

Before any measurement is taken, the ultrasonic thickness gauge must be calibrated in order to provide accurate data. When calibrating the ultrasonic thickness gauge, always follow the proper procedures as recommended by the manufacturer. These procedures will vary depending on the manufacturer of the test instrument. The general calibration requirements are discussed in the following sections.

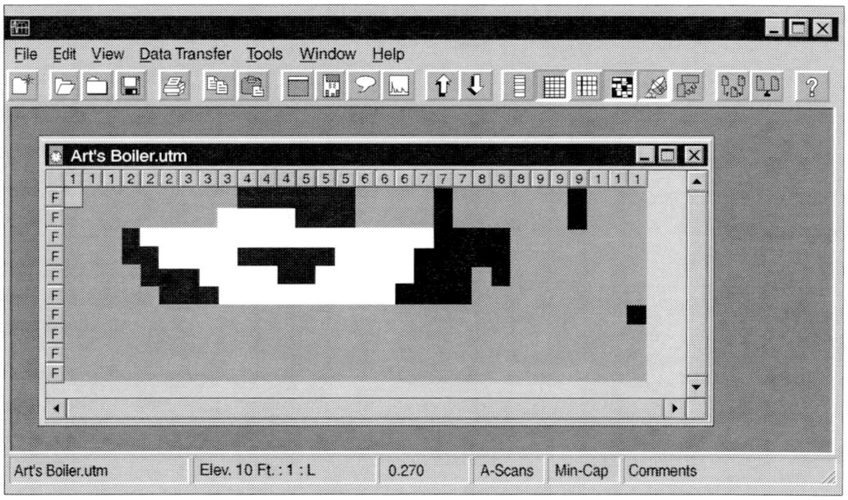

110F25.EPS

Figure 25 ◆ Typical thickness level grid block readout.

Instructor's Notes:

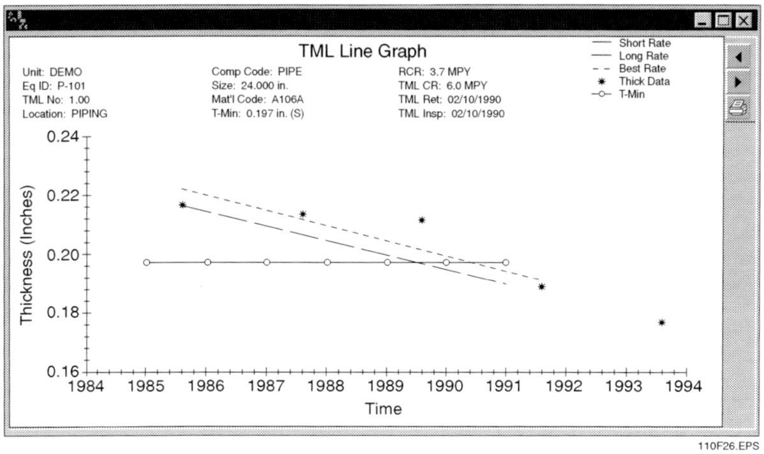

Figure 26 ◈ Typical thickness measurement line graph.

5.2.1 Calibration of Sound Velocity

The principle of operation of an ultrasonic thickness gauge is that the instrument measures the time of flight of an ultrasonic pulse through the test piece and multiplies this time by the **velocity of sound** in the material. Thickness measuring error is minimized by ensuring that the sound velocity that the instrument is calibrated to is the same as the sound velocity of the material being tested. Actual sound velocities in materials often vary significantly from the values found in published tables. Best results are always obtained if the instrument is calibrated on a velocity reference block made from the same material as the test piece. This block should be flat, smooth, and as thick as the maximum thickness of the test piece.

Individuals should be aware that the sound velocity may not be constant in the material being tested. Heat treating, for example, can cause significant changes in sound velocity. This must be considered when evaluating the accuracy of the thickness provided by this instrument. Instruments should always be calibrated before testing, and the calibration should be checked after testing, to minimize testing errors.

5.2.2 Effects of Temperature on Calibration

Temperature variations will change the sound velocity of materials and transducer delay lines. This also changes calibration requirements. All calibrations should be performed on-site, and with test blocks at the same temperature as the test piece, to minimize errors due to temperature variations.

5.2.3 Transducer Selection

The transducer used in testing must be in good condition without noticeable wear of the front surface. Badly worn transducers will have a reduced effective measuring range. The specified range of the transducer must include the complete range of thickness to be tested. The temperature of the material to be tested must be within the transducer's temperature range.

5.2.4 Use of Couplants

Individuals must be familiar with the use of ultrasonic **couplants**. Testing skills must be developed so that couplants are used and applied in a consistent manner to minimize variations in couplant layer thickness and errors in test results. This effect is known as doubling. Coupling can occur below the minimum specified range of the transducer. If the transducer being used is worn, doubling is possible at a thickness greater than the minimum of the specified range.

When using a new transducer, any reading that less than twice the minimum specified range of the transducer may be a doubled reading. The thickness of the material being tested should be verified by the use of other methods. If the transducer shows any sign of wear, doubling may occur at a

Classroom

Ask a trainee to explain how to obtain the best results when calibrating an ultrasonic thickness gauge.

Have a trainee explain doubling. Ask another trainee to explain how to verify that doubling has not occurred.

Teaching Tip

See the Teaching Tip at the end of this module.

Audiovisual

Show Transparencies 7 and 8 (Figures 27 and 28).

thickness greater than twice the minimum of the specified range. This thickness should be determined by calibrating the instrument/transducer combination on reference blocks that represent the complete range of possible thickness that may be encountered in testing. This is particularly important when the test piece is being ultrasonically measured for the first time or in any case where the history of thickness of the test specimen is unknown.

5.2.5 Calibration to a Known Thickness, 1-Point

Follow this procedure to calibrate the ultrasonic thickness gauge to a known thickness. These are the general steps required for this calibration; always refer to your test instrument operating manual for the procedure required for your specific instrument.

A calibration standard of the same material and velocity of sound as the material to be measured is required. For best results, the thickness of the calibration standard should be equal to, or slightly greater than the thickest area to be measured.

Step 1 Press CAL/ON button.

Step 2 With the CAL flashing, couple the probe to the calibration standard. *Figure 27* shows a stable CAL reading.

> **NOTE**
> Be sure the coupling indicator lights and that the reading is stable. The displayed value may not match the known thickness of the calibration standard at this point. Keep the probe coupled or uncouple the probe and wipe the excess couplants from the probe face.

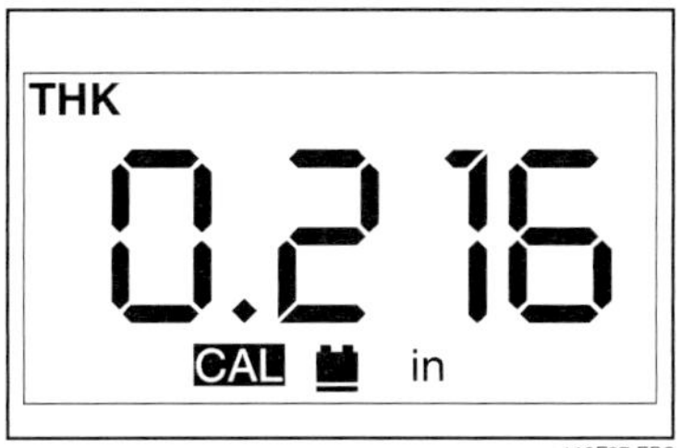

Figure 27 ◆ Stable CAL reading.

Step 3 Use the UP and DOWN buttons to adjust the displayed value to match the thickness of the calibration standard. *Figure 28* shows how the display should appear with a 0.200" calibration standard.

> **NOTE**
> Normally the scrolling speed accelerates as the button is held. If a button is released for less than 1 second, scrolling will resume at the same speed.

Figure 28 ◆ Display with a 0.200" calibration standard.

Step 4 Press CAL/ON button again to end the calibration procedure.

5.2.6 Calibration to a Known Thickness, 2-Point

Follow this procedure to calibrate the ultrasonic thickness gauge to a known thickness. These are the general steps required for this calibration; always refer to your test instrument-operating manual for the procedure required for your instrument.

This procedure requires two calibration standards (or one standard with two thickness) of the same material, velocity, and curvature as the material under test.

For best results, the thicker standard should be equal to, or slightly greater than, the maximum thickness to be measured. The thickness of the thin standard should be as close as possible to the lower end of the expected measurement range.

Step 1 Press the CAL/ON button to display the current status.

Step 2 Use the UP and DOWN button to enable 2-PT (see *Figure 29*).

Instructor's Notes:

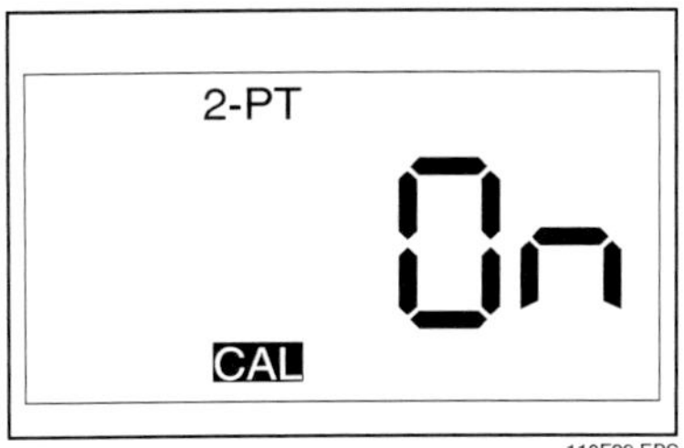

110F29.EPS

Figure 29 ◈ Enable 2-PT.

Step 3 Press CAL/ON button again. The 2-PT indicator now flashes and the prompt to couple the LO (thin) calibration standard should appear (see *Figure 30*).

Step 4 Couple the probe to the thin calibration standard. Ensure that the coupling indicator light comes on and that the reading is stable. Keep the probe coupled or uncouple the probe and wipe the excess couplants from the probe face.

Step 5 Use the UP and DOWN button to adjust the displayed value to match the thickness of the thin calibration standard.

Step 6 Press CAL/ON button again. The prompt to couple the HI (thick) calibration standard should be displayed.

Step 7 Couple the probe to the thick calibration standard. Ensure that the coupling indicator light comes on and that the reading is stable.

Step 8 Use the UP and DOWN button to adjust the displayed value to match the thick calibration standard.

110F30.EPS

Figure 30 ◈ LO display.

Step 9 Press the CAL/ON button to exit 2-PT calibration mode (see *Figure 31*). The 2-PT indicator remains illuminated to show that the calibration mode is 2-PT and that the automatic probe is set at zero.

The 2-PT calibration procedure is complete and the instrument is ready for thickness measurement.

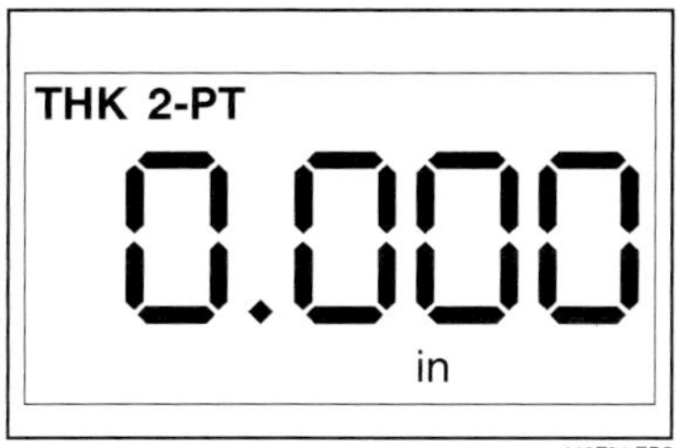

110F31.EPS

Figure 31 ◈ 2-PT calibration readout.

5.2.7 Calibration to a Known Velocity

Follow this procedure to calibrate the ultrasonic thickness gauge to a known velocity. These are the general steps required for this calibration. Always refer to your test instrument operating manual for the procedure required for your instrument.

When performing this procedure for velocity, the 2-PT calibration mode must be set to the OFF position.

If you know the velocity of sound of the material to be measured, the instrument can be calibrated to the known velocity without using calibration standard. However, the calibration should be always be verified by measuring a known thickness (or calibration standard) of the same material and velocity as the material to be measured.

Step 1 Press the MODE button until VEL illuminates. *Figure 32* shows the current velocity setting.

Step 2 Press CAL/ON button to enable calibration mode.

Step 3 With the CAL indicator flashing, use the UP and DOWN button to set the displayed velocity to that of the material under test. *Figure 33* shows the velocity setting of 0.2328" per microsecond.

Step 4 Press CAL/ON button again to exit velocity calibration mode and return to the THK mode.

Audiovisual

Show Transparencies 9 through 13 (Figures 29 through 33).

Laboratory

Have trainees practice measuring the thickness of pipe with an ultrasonic thickness gauge. Have them record their findings on a pipeline inspection report form.

Provide access to a pipeline that is going to be inspected. Have trainees practice inspecting the internal pipe surfaces, following the steps outlined in the Trainee Guide.

Safety

Ask a trainee to explain the proper response to a failure of an isolation device. Have another trainee explain the proper response to fire or explosion. Ask another trainee what to do if he or she notices excessive internal corrosion.

110F32.EPS

Figure 32 ◆ Current velocity setting.

110F33.EPS

Figure 33 ◆ Velocity setting of 0.2328" per microsecond.

Table 1 shows typical sound velocities for some common materials. The values may not precisely match the velocity in the material to be measured, because processing, exact composition, and temperature affect velocity.

6.0.0 ◆ DOCUMENT FINDINGS

When determining the thickness of pipe, obtain the alignment sheet of the pipeline that you are inspecting. Determine your exact location on the pipeline alignment sheet. Obtain the required information from the alignment sheet.

Obtain the required company pipeline inspection report form. The form for each company will differ but most will have the same information to be completed. As a minimum, the form must contain the exact location of the inspection and the results that you find during your inspection, along with your contact information.

Always follow your company procedures when completing an inspection report. This will ensure that you provide your company with the information needed to evaluate the extent of the corrosion. In addition, historic reports can be used to compare to your report to see the past rate and current rate of any corrosion that may be visible. *Figure 34* shows a typical pipeline inspection form.

Table 1 Sound Velocities

Material	in/µs	m/s
Aluminum	0.2500	6300
Brass	0.1700	4300
Cadmium	0.1100	2800
Cast Iron	0.1800	4600
Copper	0.1800	4600
Epoxy Resin	0.1100	2800
Glass (Crown)	0.2200	5600
Glass (Window)	0.2700	6800
Gold	0.1300	3300
Inconel	0.2200	5600
Lead	0.0850	2200
Magnesium	0.2300	5800
Manganese	0.1800	4600
Molybdenum	0.2500	6300
Monel	0.2100	5300
Neoprene	0.0630	1600
Nickel	0.2200	5600
Nylon	0.1000	2500
Phenolic	0.0560	1400
Platinum	0.1600	4100
Plexiglas	0.1100	2800
Polyethylene	0.0700	1800
Polystyrene	0.0930	2400
Polyurethane	0.0700	1800
Porcelain	0.2200	5600
Rubber (Butyl)	0.0730	1900
Rubber (Vulc.)	0.0900	2300
Silver	0.1400	3600
Steel	0.2300	5800
Tin	0.1300	3300
Titanium	0.2400	6100
Tungsten	0.2100	5300
Zinc	0.1600	4100
Zircaloy 2	0.1900	4700

110T01.EPS

7.0.0 ◆ INSPECTING INTERNAL PIPE SURFACES

The following steps detail a basic procedure of how to inspect internal pipe surfaces. Please note that these steps are very general. Your company may have more specific procedures in place, which you should always follow. If you have any questions, always contact your supervisor. Remember to obtain and properly use the appropriate personal protective equipment for each task.

Step 1 First, record all data from your initial visual inspection of the inside of the pipe as required by the Department of Transportation (DOT) and your company's specific policies and procedures.

Instructor's Notes:

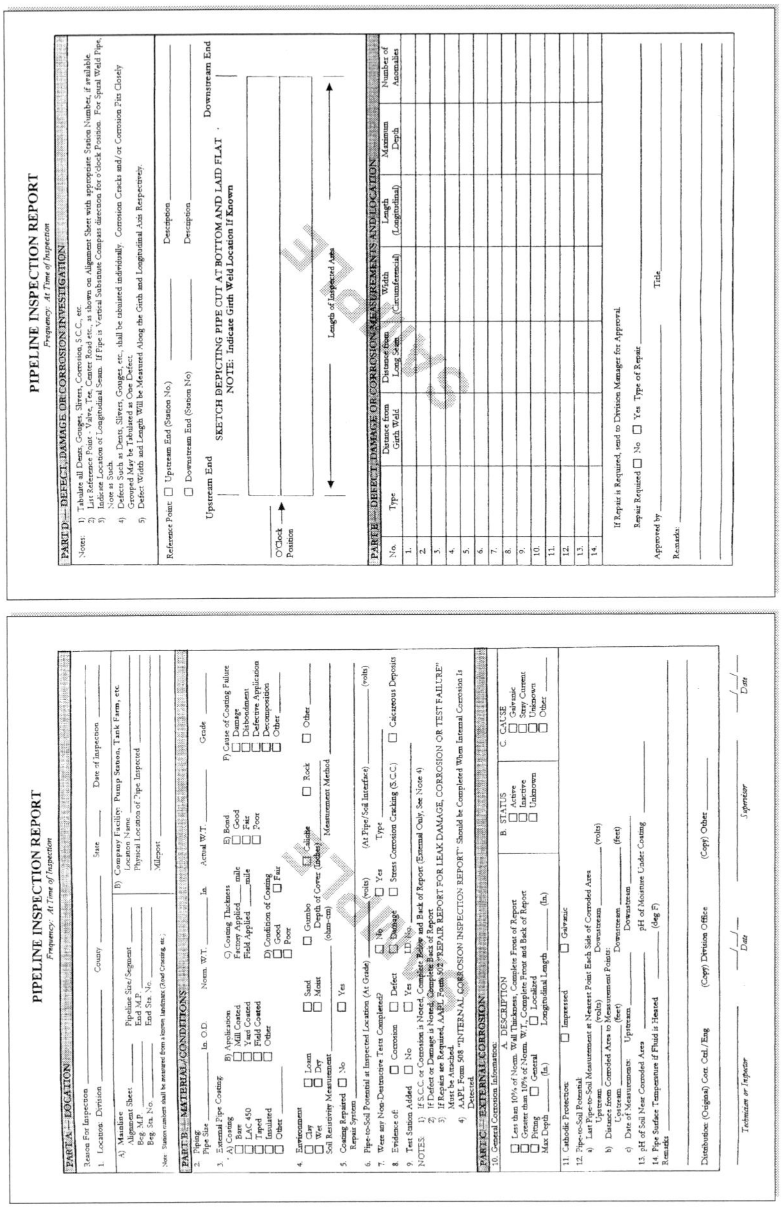

110F34.EPS

Figure 34 ◈ Typical pipeline inspection form.

Classroom

Go over the Review Questions. Answer any questions trainees may have on materials covered to this point.

Examination

Administer the Module Examination. Be sure to record the results of the Exam on Craft Training Report Form 200 and submit the results to the Training Program Sponsor.

Performance Testing

Have each trainee perform Performance Task 1 to your satisfaction. Fill out a Performance Profile Sheet for each trainee.

Ensure the Performance Task has been completed and Performance Profile Sheets for each trainee are filled out. Be sure to record the results of the testing on Craft Training Report Form 200 and submit the results to the Training Program Sponsor.

> **WARNING!**
> Failure of isolation devices may result in an unexpected hazardous liquid or gas leak. If safe to do so, isolate the leak and clean and/or ventilate the area. Contact appropriate company personnel for assistance.

Step 2 Hydrocarbon must be removed to accurately view the effects of corrosive agents on the pipe. Your next step is to remove any hydrocarbon residue with an approved solvent or by appropriate mechanical means. Dispose of waste in an approved container. Do not disturb the scale that has built up inside the pipe.

> **WARNING!**
> In the case of fire, explosion, or ignition of hydrocarbon, evacuate the area immediately and notify the local emergency personnel.

Step 3 Now record all data from your visual inspection of the pipe interior after cleaning as required by DOT and your company's policies and procedures. Carefully record the patterns and locations of pitting or scale buildup.

> **WARNING!**
> In the event that you notice excessive internal corrosion, notify your supervisor immediately.

Step 4 Next, measure and record internal wall loss using the appropriate equipment.

Step 5 Take care during transport and storage of the sample pipe so that you do not cause any additional mechanical damage to it. Secure storage is necessary for future reference and analysis.

Step 6 All data pertaining to the removed sample must be filed and distributed according to company policies and DOT requirements.

Step 7 Thoroughly document your findings and make company notifications as applicable. Proper documentation is critical to future analysis and identification of problem areas.

Step 8 Always be alert for potential abnormal operating conditions. Identify them and determine the appropriate action(s) to be taken.

Summary

Keeping critical pipelines and system components on-line and operating is a huge challenge for the industry. Constant inspection and testing helps to ensure that pipeline individuals and maintainers know the condition of their system and discover any deterioration before it becomes a catastrophic failure.

Pipeline tests for pits are done by using visual observations and a variety of mechanical instruments. Ultrasonic thickness tests are used to predict failure by measuring wall thickness on pipes and other equipment. Test data is recorded and tracked to monitor the amount of corrosion that has occurred and determine the remaining wall thickness for the tested items. Informed decisions can then be made for planning and scheduling repair actions.

The most important part of this process is the timely and correct completion of all inspection reports. Always follow your company policies and procedures to ensure that you have reliable documentation to work with in making these decisions.

Instructor's Notes:

Review Questions

1. The general definition of corrosion is _____.
   a. an electro-mechanical process
   b. deterioration of material caused by a reaction with its environment
   c. external damage to a structure
   d. internal damage to a structure

2. Coatings suited for buried pipe may not be suitable for _____.
   a. exposure to sun and weather
   b. high-pressure lines
   c. cold temperatures
   d. moist soil environments

3. General corrosion can be described as _____.
   a. a defined area with pits and cavities
   b. a wearing away of material by abrasive and corrosive product flow
   c. corrosion spread over a wide area with no visible pitting
   d. damage to the internal surface of the pipeline

4. Corrosion can be caused by all of the following *except* _____.
   a. electrically conductive soil conditions
   b. differential aeration
   c. dissimilar metals interacting
   d. high operating pressures and temperatures

5. Individuals involved in coating a pipeline for burial can minimize cracking failures by _____.
   a. applying thicker layers of coating
   b. using minimal brushing
   c. using more drying agents to speed the process
   d. correctly using resins, plasticizers, and fibrous pigment

6. Pipeline corrosion is commonly checked using all of the following *except* for the _____.
   a. outside micrometer
   b. mechanical pit gauge
   c. micrometer pit gauge
   d. ultrasonic thickness gauge

7. If a mechanical pit gauge has a worn and rounded point, you should _____.
   a. file the point to obtain a sharp contact area
   b. replace the pit gauge
   c. limit measurements to larger pits on the surface
   d. ensure the gauge is adjusted to the zero point before using

8. Individuals can quickly and easily measure and record the thickness of a pipe wall by using _____.
   a. a mechanical pit gauge
   b. a micrometer pit gauge
   c. a digital ultrasonic thickness gauge with transducer
   d. the RSTRENG or B31G methods for evaluating integrity

9. Measuring remaining pipe wall thickness through a coating on the pipeline is possible with a digital ultrasonic thickness gauge equipped with a _____.
   a. recording circuit
   b. reflected wave translator
   c. dual-element transducer probe
   d. velocity of sound calibration

10. Velocity of sound is a term used to describe the _____.
    a. constant speed of sound
    b. speed at which an ultrasonic pulse travels through a material
    c. transducer calibration point
    d. gauge measurement error

11. Temperature variations can affect thickness readings by _____.
    a. reducing the velocity of sound for a material
    b. requiring an increase in couplant layer thickness
    c. changing the transducer transmission frequency
    d. changing the velocity of sound for the material being tested

12. Transducer selection involves all of the following *except* that _____.
    a. dual-element transducers are preferred for 2-point thickness tasks
    b. a worn transducer will have a reduced effective measurement range
    c. transducer range must match the thickness of the material to be tested
    d. temperature of the material to be tested must fall within the transducer temperature range

13. Calibration of the ultrasonic thickness gauge to a known thickness is done by _____.
    a. use of a calibration standard of the same material and velocity of sound
    b. use of a test block of the same thickness as the material to be tested
    c. zeroing out the thickness function on the gauge
    d. placing the 2-point calibration mode in the OFF position

14. Calibration of the ultrasonic thickness gauge to a known velocity can be accomplished with each of the following *except* _____.
    a. known velocity of sound information
    b. standard zero calibration for the gauge
    c. calibration standard of the same material as that to be tested
    d. settings from the velocity of sound table

15. Documentation of findings from pipe thickness measurements is used for all the following *except* _____.
    a. evaluating the rate of corrosion
    b. evaluating the extent of corrosion
    c. determining the location of the pipe
    d. recording existing conditions

Instructor's Notes:

GLOSSARY

Trade Terms Introduced in This Module

Buried pipe: Pipeline run beneath the ground surface.

Caliper: A measurement device used to determine changes in pipe wall thickness.

**Corrosion:* The deterioration of a material, usually a metal, that results from a reaction with its environment.

Couplant: Used to facilitate ultrasonic testing, such as testing wall thickness in a pipeline. A couplant is a liquid, grease, or paste, or a pliable solid placed between the transducer of the ultrasonic tester and the surface of the test piece to permit transmission of ultrasonic energy between them. Couplants, often referred to as coupling medium or coupling film, retard sound transmission by air between the transducer and the work surface.

**Erosion:* The progressive loss of material from a solid surface due to mechanical interaction between that surface and a fluid, a multicomponent fluid, or solid particles carried with the fluid.

Mechanical Damage: Dents, gouges or other surface damage.

Pit Depth Gauge: Measures the depth of the depressions caused by corrosion and the remaining wall thickness.

**Pitting:* Localized corrosion of a metal surface that is confined to a small area and takes the form of cavities called pits.

**Scaling:* (1) The formation at high temperatures of thick corrosion-product layers on a metal surface. (2) The deposition of water-insoluble constituents on a metal surface.

Shrinkage: Coating defect caused by rapid contraction of coating material.

Submerged pipe: Pipeline run beneath the underwater ground surface.

Surface adhesion: The tendency for surface molecules in a coating material to cling together.

Ultrasonic Testing: Test methods utilizing ultrasonic frequencies to determine thickness of material.

Wrinkling: (1) Distortion occurring in a film of paint that gives the appearance of ripples. (2) Development of a roughened surface on a film of sealant.

ACKNOWLEDGMENTS

Figure Credits

Fred V. Fowler Co., Inc.	110F17
Allen Carlille	110F19
Art Leach, Krautkramer	110F21, 110F22, 110F23, 110F24
EnhanceCo., Inc.	110F08

Instructor's Notes:

MODULE 61110-02—TEACHING TIPS

The following are suggested activities or instructional methods to help you teach the material in this AIG.

General

When you call on someone to answer a question, the rest of the class relaxes or even tunes out because they expect that the question and answer will take place only between you and the trainee you called on. Instead, use this technique to involve more trainees in answering questions and to keep them on their toes.

1. Ask trainees to define a term or explain a concept.
2. After one trainee has answered, ask a trainee seated nearby if the answer is right. Then ask whether a trainee in the back of the room agrees.
3. Ask trainees to explain why they think an answer is right or wrong.
4. Use the session to clear up incorrect ideas and encourage trainees to learn from their mistakes.

Sections 2.0.0–2.2.0

Determine Pipe Orientation

This exercise gives trainees the opportunity to practice determining pipe orientation. For this exercise, trainees will need pencils, paper, and their Trainee Guides. You will need to provide pipeline inspection documents and a compass and to coordinate a trip to a pipeline. Allow 30 to 45 minutes for this exercise, not including travel time to and from the site.

1. Review the steps for determining pipeline orientation based on flow direction and compass readings.
2. Pair up trainees. Have them determine pipeline orientation by following the steps in their Trainee Guides.
3. Review the completed inspection documents. Ask trainees if they have any questions.

Section 3.1.1

Address Wall Loss and Deposits

This exercise will allow trainees to practice collecting the data necessary to address pitting or wall loss. For this exercise, trainees will need pencil, paper, and their Trainee Guides. You will need to provide Field Corrosion Report Forms and coordinate access to pipelines that show pitting or wall loss. You may also need to provide compasses, pit gauges, pocket knives, and plastic for wrapping the sample(s). Allow 45 to 60 minutes for this exercise, not including travel time to and from the site.

1. Review the guidelines for collecting the necessary data.
2. Divide class into small groups. Assign each group to a damaged pipeline and have them collect the necessary data. Then have each group complete a field corrosion report form.
3. Review the completed forms. Answer any questions trainees may have.

Sections 3.2.0–3.2.8

Coating Failure and Corrosion

This Quick Quiz will familiarize trainees with the types of coating failures. You will need photocopies of the quiz provided on the following page. Trainees will need pencils. If you allow trainees to use the Trainee Guide, decrease the amount of time you give them to complete the quiz.

1. Make as many photocopies as you will need of the quiz.
2. Give trainees between 5 and 10 minutes to complete the quiz.
3. Go over the answers to the quiz.
4. Ask trainees whether they have questions about the types of coating failures.

Quick Quiz *Coating Failure and Corrosion*

For each description listed, identify the type of coating failure the text best describes. Write the corresponding letter in the blank provided.

______ **1.** Rust forms at a break in the coating material and spreads under the coating over the pipe surface.

______ **2.** Moisture works its way under the surface of the coating and adhesion breaks down.

______ **3.** The coating material becomes brittle from curing too quickly and becomes inflexible.

______ **4.** The coating surface expands more rapidly than the body of the coating during the curing process.

______ **5.** Coatings are applied too thickly.

a. blistering
b. sagging
c. wrinkling
d. undercutting
e. flaking

Sections 3.2.0–3.2.8

Answers to Quick Quiz

1. d. undercutting
2. a. blistering
3. e. flaking
4. c. wrinkling
5. b. sagging

Sections 5.2.5–5.2.7

Calibration

This exercise gives trainees practice in calibrating ultrasonic thickness gauges to 1 point, 2 point, and a known velocity. For this exercise, trainees will need pencils, paper, and their Trainee Guides. You will need to provide ultrasonic thickness gauges and calibration standards; you will also need Transparencies 7 through 13 (*Figures 27* through *33*). Allow 60 to 75 minutes for this exercise.

1. Set up calibration stations. Review the procedure for each type of calibration, using the transparencies as appropriate.
2. Divide the class into small groups and assign each group to a station. Have trainees take turns calibrating the gauge to 1 point, 2 point, and a known velocity. Ask them to take notes on the procedure and any problems they encountered.
3. Discuss the procedure with the class. Ask trainees if they have any questions.

MODULE 61110-02—ANSWERS TO REVIEW QUESTIONS

Answer	Section
1. b	1.0.0
2. a	3.0.0
3. c	3.1.0
4. d	3.2.0
5. d	3.2.2
6. a	4.0.0
7. b	4.3.0
8. c	5.0.0
9. c	5.1.2
10. b	5.2.1
11. d	5.2.2
12. a	5.2.3
13. a	5.2.5
14. b	5.2.7
15. c	6.0.0

CONTREN™ LEARNING SERIES — USER UPDATES

The NCCER makes every effort to keep these textbooks up-to-date and free of technical errors. We appreciate your help in this process. If you have an idea for improving this textbook, or if you find an error, a typographical mistake, or an inaccuracy in NCCER's Contren™ textbooks, please write us, using this form or a photocopy. Be sure to include the exact module number, page number, a detailed description, and the correction, if applicable. Your input will be brought to the attention of the Technical Review Committee. Thank you for your assistance.

Instructors – If you found that additional materials were necessary in order to teach this module effectively, please let us know so that we may include them in the Equipment/Materials list in the Instructor's Guide.

Write: Curriculum Revision and Development Department
National Center for Construction Education and Research
P.O. Box 141104, Gainesville, FL 32614-1104

Fax: 352-334-0932

E-mail: curriculum@nccer.org

Craft ______________________ Module Name ______________________

Copyright Date ____________ Module Number ____________ Page Number(s) ____________

Description

____

____

____

____

(Optional) Correction

____

____

____

(Optional) Your Name and Address

____

____

____

Module 61111-02

Internal Corrosion Control
(CT 10.1, 10.2, and 11)

MODULE OVERVIEW

This module describes the various types of internal corrosion. Trainees will learn methods of controlling corrosion, including mechanical cleaning and the use of corrosion treatment chemicals.

PREREQUISITES

Please refer to the Course Map in the Trainee Module. Prior to training with this module, it is recommended that the trainee shall have successfully completed the following:

Core Curriculum; Pipeline Core, Modules 66101-02 and 66102-02; Pipeline Corrosion Control Level One, Modules 61103-02 through 61110-02

OBJECTIVES

Upon completion of this module, the trainee will be able to:

1. Safely insert and retract corrosion coupons and document vital information (CT 10.1).
2. Accurately record readings of probes with corrosion measurement tools (CT 10.2).
3. Adjust inhibitor injection rates to achieve an acceptable company standard (CT 11).

PERFORMANCE TASKS

Under the instructor's supervision, the trainee should be able to:

1. Insert and remove coupons according to manufacturer and company specifications (CT 10.1).
2. Properly monitor on-line probes (CT 10.2).
3. Perform internal corrosion remediation (CT 11).

NCCER STANDARDIZED CRAFT TRAINING PROGRAM

The National Center for Construction Education and Research (NCCER) provides a standardized national program of accredited craft training. Key features of the program include instructor certification, competency-based training, and performance testing. The program provides trainees, instructors, and companies with a standard form of recognition through a National Craft Training Registry. The program is described in full in the *Guidelines for Accreditation*, published by the NCCER. For more information on standardized craft training, contact the NCCER by writing us at P.O. Box 141104, Gainesville, FL 32614-1104; calling 352-334-0911; or e-mailing info@nccer.org. More information may be found at our Web site, www.nccer.org.

HOW TO USE THIS ANNOTATED INSTRUCTOR'S GUIDE

Each page presents two sections of information. The larger section displays each page exactly as it appears in the Trainee Module. The narrow column ties suggested trainee and instructor actions to each page and provides icons to call your attention to material, safety, audiovisual, or testing requirements. The bottom of each page includes space for your notes.

Teaching Tip

If you see the Teaching Tip icon, that means there is a teaching tip associated with this section. Also refer to any suggested teaching tips at the end of the module.

SAFETY CONSIDERATIONS

Ensure that the trainees are equipped with appropriate personal protective equipment. Emphasize the importance of following all safety precautions and procedures when working with power tools.

PREPARATION

Before teaching this module, you should review the Module Outline, Objectives, Performance Tasks, and the Materials and Equipment List. Be sure to allow ample time to prepare your own training or lesson plan and gather all required equipment and materials.

MATERIALS AND EQUIPMENT LIST

Transparencies

Markers/chalk

Blank acetate sheets

Transparency pens

Pencils and scratch paper

Module Examinations*

Performance Profile Sheets*

Overhead projector and screen

Copies of Performance Verifications for Covered Tasks 10.1, 10.2, and 11**

Copies of *49 CFR Part 195* (Liquid) and *49 CFR 192* (Gas)

Whiteboard/chalkboard

Appropriate personal protective equipment

Copies of your company's policies and procedures manual

Equipment and tools necessary for monitoring and adjusting chemical injection systems

Equipment and tools necessary for inserting and retracting coupon holders

A variety of probes, manufacturers' instructions, and company forms

* Located in the Test Booklet.

** Available for download at www.nccer.org.

TEACHING TIME FOR THIS MODULE

An outline for use in developing your lesson plan is presented below. Note that each Roman numeral in the outline equates to one session of instruction. Each session has a suggested time period of 2½ hours. This includes 10 minutes at the beginning of each session for administrative tasks and one 10-minute break during the session. Approximately 7½ hours are suggested to cover *Internal Corrosion Control* (CT 10.1, 10.2, and 11). You will need to adjust the time required for hands-on activity and testing based on your class size and resources.

Topic	Planned Time
Session I. Internal Corrosion	
A. Types of Corrosion	________
B. Factors Affecting Internal Corrosion	________
1. Acid Gases	________
2. pH Level	________
3. Water Content	________
4. Temperature	________
5. Partial Pressure	________
6. Velocity	________
C. Pigging for Corrosion	________
1. Utility Pigs	________
2. Gel Pigs	________
Session II. Corrosion Inhibitor Injection and Pipe Corrosion Coupons	
A. Identifying Injection Locations	________
1. Identifying Injection Rate	________
2. Monitoring and Adjusting	________
3. Laboratory	________
B. Pipe Corrosion Coupons	________
1. Weight Loss and Scale Coupons	________
2. Coupon Mounting Devices	________
3. Coupon Orientation	________
4. Inserting Retractable Coupon Holders	________
5. Retracting Coupon Holders	________
6. Performance Testing	________
Session III. Probes	
A. Electrical Resistance (ER) Probes	________
B. Linear Polarization Resistance (LPR) Probes	________
C. Galvanic Probes	________
D. Hydrogen Probes	________
E. Reading a Probe	________
F. Summary	________
1. Summarize module	________
2. Answer questions	________
G. Module Examination	________
1. Trainees must score 70% or higher to receive recognition from the NCCER.	
2. Record the testing results on Craft Training Report Form 200 and submit the results to the Training Program Sponsor.	
H. Performance Testing	________
1. Trainees must perform each task to the satisfaction of the instructor to receive recognition from the NCCER.	
2. Record the testing results on Craft Training Report Form 200 and submit the results to the Training Program Sponsor.	

Module 61111-02

Internal Corrosion Control

(CT 10.1, 10.2, and 11)

Instructor's Notes:

Instructor's Notes:

Introduction to the Pipeline Operator Qualification Rule, API, and NCCER

It could be said that petroleum products keep America running. Petroleum products like oil and natural gas are the principal sources of energy in the United States and are used in many manufacturing and chemical processes. Petroleum products are primarily transported by pipeline. There are more than a million miles of liquid and natural gas pipeline in this country, ranging in diameter from two inches to four feet. These pipelines bring crude oil and natural gas from production fields to processing plants and refineries—and ultimately to the consumer.

Pipeline Operator Qualification

Petroleum products are considered hazardous, so pipeline personnel must take extra precautions and follow specific procedures when working with them. To help ensure the safe handling and transportation of petroleum products, the United States Department of Transportation (DOT) mandates safe practices as federal regulations.

One such regulation, the Operator Qualification (OQ) rule, requires pipeline operators to develop a written qualification plan to evaluate their personnel and their contract personnel (see *49 CFR Part 192 Subpart N—Gas*, and *49 CFR Part 195 Subpart G—Liquid*). The evaluation hinges on the operator's ability to perform certain covered tasks and to be able to recognize and respond to abnormal operating conditions that may be encountered while performing these covered tasks.

Covered Tasks

A pipeline performance task is deemed *covered* if the activity passes the following four-part test:

1. Is it performed on a pipeline facility?
2. Is it an operations or maintenance task?
3. Is it performed as a requirement of *49 CFR Part 195 (Liquid)* or *49 CFR Part 192 (Gas)*?
4. Does it affect the operation or integrity of the pipeline facility?

If the answer to all four questions is *yes*, then the activity is considered *covered* and must figure into the evaluation process of pipeline personnel.

American Petroleum Institute

The OQ rule and its related covered tasks provided the catalyst to develop a unified pipeline training and assessment program—so that pipeline operators and owners have ready access to the assessment and training tools they need. This effort was the direct result of the vision and leadership provided by The American Petroleum Institute (API) and API members.

Efforts to unify the industry are nothing new to API. Indeed, the impetus for forming the Institute in 1919 was the need to standardize engineering specifications for drilling and production equipment. Today, the most pressing issues revolve around public perceptions and government policies toward the industry. Speaking with one voice on these issues has become as essential as having interchangeable parts in the field. API's members determine the petroleum industry's positions on public policy as well as the standards that govern its day-to-day operations worldwide.

Consortium on Operator Qualification

API assembled a focus group whose primary purpose was to identify covered tasks and provide compliance recommendations to pipeline operators.

The Consortium on Operator Qualification (COOQ) became the guiding force for the identification and development of covered tasks for the liquid pipeline industry. Their work product, API publication 1161, *Guidance Document for the Qualification of Liquid Pipeline Personnel*, formed the basis for the resulting Pipeliner Training and Assessment Program.

The covered task list created in API publication 1161 has now evolved to include gas covered tasks and is maintained by NCCER. The most current covered task list can be downloaded at www.nccer.org.

National Center for Construction Education and Research

With the guidance and support of API, its members, and the Consortium on Operator Qualification, the National Center for Construction Education and Research (NCCER) was able to design and implement the Pipeliner Training and Assessment Program.

NCCER is a not-for-profit 501(c)(3) education foundation established in 1995 by the world's largest and most progressive construction and maintenance contractors and national contractor associations. NCCER was founded to address the

severe workforce shortage facing the construction and maintenance industries and to develop a standardized training process and curricula. Today, NCCER is supported by hundreds of leading construction and maintenance companies, manufacturers, and national associations and developed more than 35 construction and maintenance titles.

NCCER also maintains a National Registry, which contains the records of trainees and instructors who have received NCCER training. All completion information for every trainee is sent to the NCCER and filed within the Registry. The Registry can then confirm training and skills for workers as they move from company to company, state to state, or even within their own company.

Standardized Curriculum (NCCER's Contren™ Learning Series)

NCCER curricula are developed using the expertise of industry subject matter experts (SMEs). The use of SMEs as an integral part of every new development or revision ensures that the curriculum is current, industry-driven, and indicative of industry practices. The development and/or revision process is facilitated from start to finish by NCCER project managers, who work closely with the SMEs and a select team of technical writer(s) on the development and presentation of the curriculum.

Pipeline: Five Training Paths

The pipeline curriculum focuses on essential skills and competencies within five skill areas:

- Corrosion Control
- Electrical and Instrumentation
- Maintenance
- Mechanical
- Operations (Field and Control Center)

SMEs from the pipeline industry have determined the skills and competencies needed based on API's covered tasks and the consensus of the group/industry. Specific company information is generally not incorporated into the curriculum unless the SMEs determine that the information is the industry standard.

Training Modules and Covered Tasks

The smallest unit of instruction in the curriculum is a *module*. Each curriculum is composed of modules (the equivalent of stand-alone chapters) that are logically sequenced into complete *levels* of instruction. Every module is task-oriented and requires the successful completion of a written (knowledge-based) examination and a performance (hands-on) test. Modules may be individually selected and/or combined for efficient task training. Modules in the pipeline curricula, to the extent possible, will align with API-developed covered tasks. While not all modules will relate to covered tasks, all covered tasks will relate to at least one module.

Benefits of Training to the Pipeline Marketplace

Applying NCCER's standardized training process allows pipeline owners and contractors in the United States to establish the benchmark for acceptable skill and competency levels within the pipeline industry. This process will assist training organizations in developing and/or improving standardized pipeline training programs that effectively match their resources and operating conditions while responding to regulatory demands and to the needs of the industry. The process will also recognize training organizations that meet the minimum standards of competence, warranting regulatory, public, and professional confidence. This in turn will foster national unity within the pipeline industry for the mutual benefit of individual technicians, owners, contractors, and ultimately pipeline-users and society.

Additional Resources

To review the Pipeliner Training and Assessment Program Implementation and Resource Guide and the latest version of pipeline covered tasks, visit NCCER's Web site at www.nccer.org, or contact:

National Center for Construction Education and Research
PO Box 141104
Gainesville, FL 32614-1104
(888) 622-3720
Web site: www.nccer.org

For more information on API, please contact:

American Petroleum Institute
1220 L Street, Northwest
Washington, D.C. 20005
(202) 682-8000
Web site: www.api.org

To order *API Pub 1161*, contact Global Engineering Documents by calling 1-800-854-7149 or visiting their Web site at www.global.his.com.

Additional sources of information on the pipeline operator qualification regulation include:

DOT – Office of Pipeline Safety
www.ops.dot.gov

Transportation Safety Institute
www.tsi.dot.gov

Instructor's Notes:

COURSE MAP

This course map shows all of the modules in the first level of the Pipeline Corrosion Control curriculum. The suggested training order begins at the bottom and proceeds up. Skill levels increase as you advance on the course map. The local Training Program Sponsor may adjust the training order.

PIPELINE CORROSION CONTROL LEVEL ONE

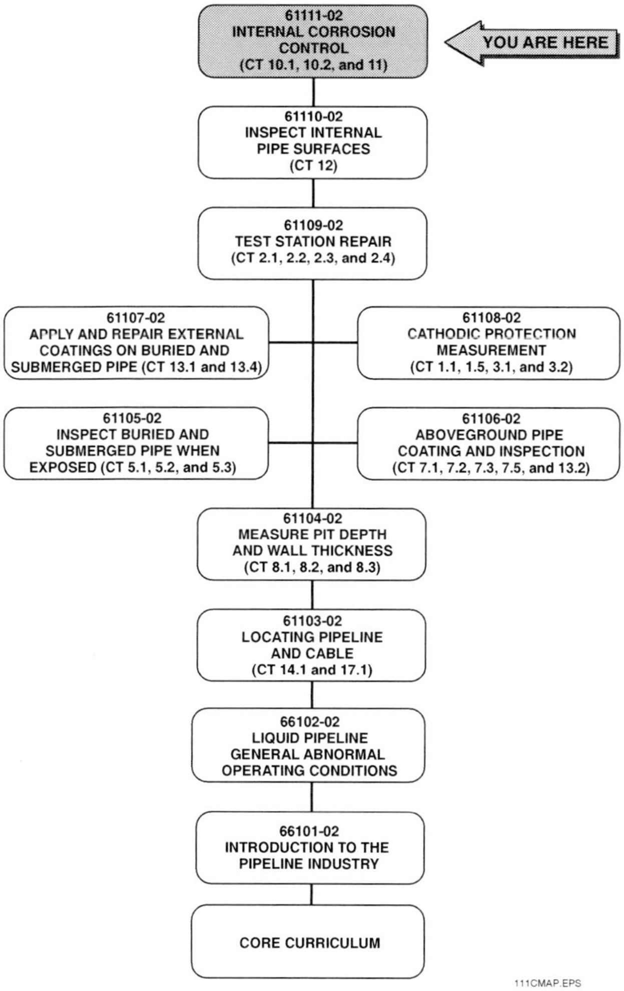

111CMAP.EPS

Homework

Assign reading of Module 61111-02.

Instructor's Notes:

MODULE 61111-02 CONTENTS

Figures

Instructor's Notes:

MODULE 61111-02

Internal Corrosion Control (CT 10.1, 10.2, and 11)

Objectives

When you have completed this module, you will be able to do the following:

1. Safely insert and retract corrosion coupons and document vital information (CT 10.1).
2. Accurately record readings of probes with corrosion measurement tools (CT 10.2).
3. Adjust inhibitor injection rates to achieve an acceptable company standard (CT 11).

Prerequisites

Before you begin this module, it is recommended that you successfully complete the following: Core Curriculum; Pipeline Core, Modules 66101-02 and 66102-02; Pipeline Corrosion Control Level One, Modules 61103-02 through 61110-02.

Required Trainee Materials

1. Pencil and paper
2. Appropriate personal protective equipment

1.0.0 ◆ INTRODUCTION

Corrosion is the destruction of a substance, usually a metal or an alloy, caused by a reaction with its environment. Each year, corrosion costs the pipeline industry millions of dollars in material damage and downtime. External corrosion on a pipe is visible when the pipe is exposed; internal corrosion is not as easily detected.

There are several approaches that can be used to control internal corrosion. Mechanical cleaning, operational changes, physical changes, and the use of corrosion treatment chemicals all help control internal corrosion.

2.0.0 ◆ INTERNAL CORROSION

Corrosion in a pipeline system can lead to expensive maintenance, de-rating or abandonment of assets, as well as environmental and safety-related issues. Corrosion can occur anywhere there is an electrolyte in the pipeline, such as water. Corrosion can also occur under deposits, in heat-affected zones around welds, and at locations where there are significant changes in temperature or pressure. Deposits in a pipeline system can lead to operational problems, including decreased flow, loss of heat transfer capacity and corrosion problems. Free water in a system can lead to internal corrosion and bacterial growth. Analysis of fluids, solids, and gases will help define the possible corrosion mechanisms.

2.1.0 Types of Corrosion

Corrosion is the gradual destruction, dissolving, or wearing away of a substance, usually a metal or an alloy, caused by a chemical or electrochemical reaction with its environment. Corrosion can be a problem in different locations along the pipeline, both externally and internally.

- *General corrosion* – Used to describe corrosion that is distributed more or less uniformly over the surface of a material.
- *Localized or pitting corrosion* – Refers to corrosion that only affects a defined area on the metal surface. Pitting corrosion refers to the formation of cavities or pits. *Figure 1* shows typical pitting corrosion.
- *Under-deposit corrosion* – Defines an area of degradation occurring beneath a deposit due to a difference in potential between the area under the deposit and the surrounding area.

Materials

Ensure that you have everything required to teach the course. Check the Materials and Equipment List at the front of this module.

Classroom

Explain that terms shown in bold are defined in the Glossary at the back of this module.

Audiovisual

Show Transparency 1, Course Objectives.

Show Transparency 2, Performance Tasks.

Teaching Tip

See the general Teaching Tip at the end of this module.

Classroom

Ask a trainee to define corrosion. Have trainees discuss ways to control internal corrosion.

Discuss the differences among general, localized, underdeposit, and erosion corrosion. Have a trainee identify the type of corrosion shown in Figure 1.

Audiovisual

Show Transparencies 3 through 5 (Figures 2 through 4). Ask trainees to identify the acid gas that usually causes each type of corrosion depicted in the figures.

Show Transparency 6 (Figure 5). Read the following pairs of numbers and have a trainee identify which number reflects a more acidic pH value: 8.0 and 4.3; 5.0 and 5.7; 6.9 and 9.6; and 2.0 and 3.0.

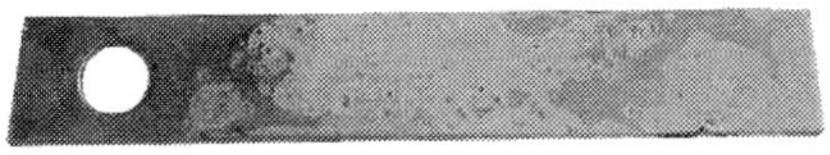

111F01.EPS

Figure 1 ◈ Typical pitting corrosion.

- *Erosion corrosion* – Defined as a dual action involving corrosion and erosion in the presence of a moving corrosive fluid or a material moving through the fluid, leading to an accelerated loss of material.

2.2.0 Factors Affecting Internal Corrosion

Some of the factors influencing internal corrosion are acid gases, pH, chlorides, temperature, water content (including **water of condensation**), partial pressure, velocity, and bacterial growth. These factors are discussed in the following sections.

2.2.1 Acid Gases

The major acid gases are oxygen, carbon dioxide, and hydrogen sulfide. These gases react with water to form acids.

Carbon dioxide, or CO_2, creates carbonic acid, a weak acid that attacks metal when in contact with water. This attack often forms what is described as Mesa pitting (see *Figure 2*).

Hydrogen sulfide (H_2S), due to its sulfur content, can create a weak acid that has a characteristic rotten-egg smell by which it can be identified. Iron sulfide creates a black deposit, FeS. Hydrogen sulfide (H_2S) corrosion is usually identifiable by the etching in the bottom of the pits. *Figure 3* shows pitting with etching in pits.

Oxygen, or O_2, is a strong depolarizer and can greatly accelerate the action of other corrosive agents. Corrosion caused by oxygen is usually found on surface equipment where air has an entry point. O_2 corrosion typically forms deep pits (see *Figure 4*).

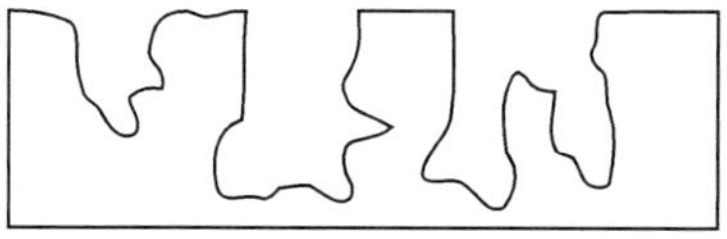

111F02.EPS

Figure 2 ◈ Mesa pitting.

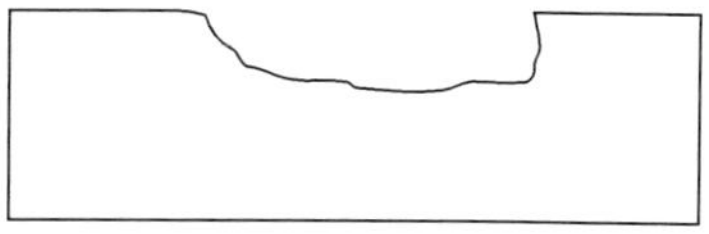

111F03.EPS

Figure 3 ◈ Pitting with etching in pits.

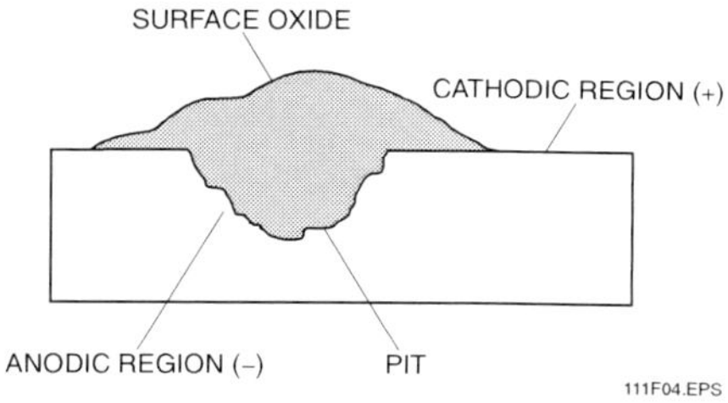

111F04.EPS

Figure 4 ◈ Deep pits created during the oxygen corrosion process.

2.2.2 pH Level

The normal range for pH is from 0 to 12 with 0 pH being most acidic, 7 pH being neutral, and 12 pH being very basic. The pH of water indicates the tendency of the water to be corrosive or to form scales. Refer to the pH scale in *Figure 5*.

The lower the pH reading, the more acidic the water, and the greater the tendency of the water to be corrosive. The higher the pH value, the more likely it is to form scale.

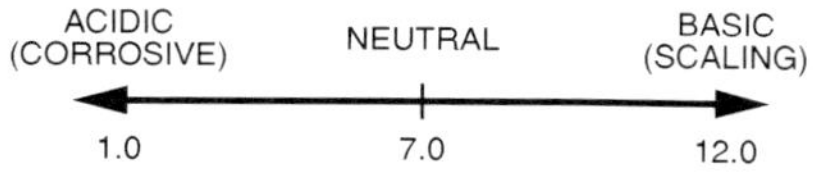

111F05.EPS

Figure 5 ◈ pH scale.

2.2.3 Water Content

Water content of the liquid or gas sample is the amount of water available in the sample. This water serves as the electrolyte needed to complete the corrosion circuit. Any vapor that can condense will, and any water that is available will condense on the pipe wall. Water of condensation forms when the temperature of the exterior of the pipeline drops significantly below the temperature of the product.

Instructor's Notes:

There are four requirements for the makeup of a four-part corrosion cell. Their names and functions are:

- *Anode* – The electrode of a cell where oxidation occurs. Electrons flow away from the anode.
- *Cathode* – The cell electrode at which reduction occurs. Electrons flow toward the cathode.
- *Metallic path* – Conductor that electrically connects the anode with the cathode. Typically, this will be the pipe itself.
- *Electrolyte* – The anode and cathode must be immersed in an electrically conductive electrolyte (water).

Figure 6 shows a typical corrosion circuit with anode, cathode, electrolyte, and metallic path for current flow.

2.2.4 Temperature

Temperature becomes a factor for internal corrosion above approximately 110°F. Above this temperature there is a significant increase in corrosion with an increase in temperature while all other corrosion factors remain the same.

2.2.5 Partial Pressure

Partial pressure is the percentage of the gas of interest in the total amount of all of the gas present. For example, if there are 100 parts of CO_2 in a total of 1000 parts of gas, the partial pressure is 100/1000, or 10%.

2.2.6 Velocity

Velocity is the speed of fluids or gases as they move through the pipeline. Erosion occurs as these fluids or gases flow over the surface and wear away minute amounts of material. Liquids and gases passing through a pipeline at high velocities can cause turbulence and increase corrosion rates by stripping any protective layer or inhibitor from the metal surface, allowing fresh, unprotected surfaces to corrode.

Individuals responsible for controlling corrosion should be familiar with the system, and should review it regularly to help identify the locations of possible trouble spots.

3.0.0 ◈ PIGGING FOR CORROSION

Pipeline pigs are devices that are inserted into and travel through the pipeline, driven by product flow. They were originally developed to remove any deposits that could impede product flow through the pipeline. They are used today to remove construction debris, mill scale, and welding icicles, as well as for cleaning and checking the geometry of the pipeline. After a pipeline has been constructed and cleaned, it is a standard practice to pig the line periodically to maintain line efficiency and control corrosion. Although each pipeline has its own set of characteristics, there are basically three reasons for sending a pig through a pipeline:

- To separate dissimilar products
- For displacement purposes
- To perform an inspection and/or cleaning

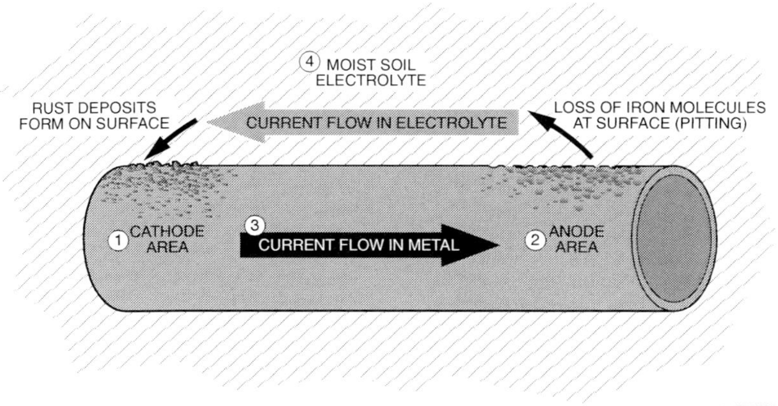

Figure 6 ◈ Typical corrosion circuit with anode, cathode, electrolyte, and metallic path for current flow.

Audiovisual

Show Transparency 7 (Figure 6). Have trainees take turns using a transparency pen to label and explain the functions of the parts of a four-part corrosion cell.

Classroom

Have a trainee describe how temperature, partial pressure, and velocity affect corrosion rates.

Classroom

Have trainees name five uses of utility pigs. Ask a trainee to identify the types of debris that are commonly removed from pipeline walls.

Pigs used to accomplish these tasks can be divided into two general categories:

- Utility pigs are used for cleaning, separating, and dewatering.
- Gel pigs are used in conjunction with utility pigs to optimize cleaning, dewatering, and drying of the pipeline after hydrostatic testing.

3.1.0 Utility Pigs

After construction and cleanout have been completed, utility pigs are used for hydrostatic testing and acceptance of dry pipelines. If it is a liquid line, a pig will also be used to fill the line for commissioning and startup. Utility pigs are also used to sweep liquids from the pipeline or to provide an interface between two dissimilar products in the line. *Figure 7* shows a pig being launched into the pipeline.

111F07.EPS

Figure 7 ◆ Pig being launched into the pipeline.

Utility pigs are used to remove accumulated solids and debris from the walls of the pipeline. In crude oil pipelines, this is usually paraffin; in gas pipelines it is most often the substance formed by solvents used along with corrosion inhibitors. These utility pigs are also used to scour corrosion deposits and to remove water, microbes, corrosion products, and food for microbes. There are four commonly used types of utility pigs: mandrel pigs, spheres, solid cast urethane pigs, and foam pigs (referred to as polly-pigs). *Figure 8* shows a mandrel pig and *Figure 9* shows a variety of foam pigs.

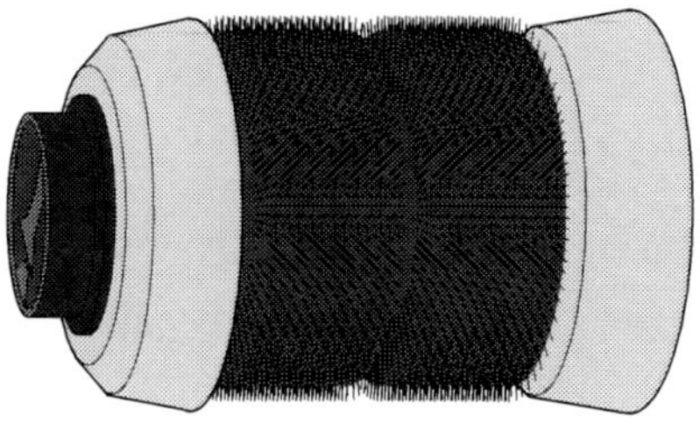

111F08.EPS

Figure 8 ◆ Mandrel pig.

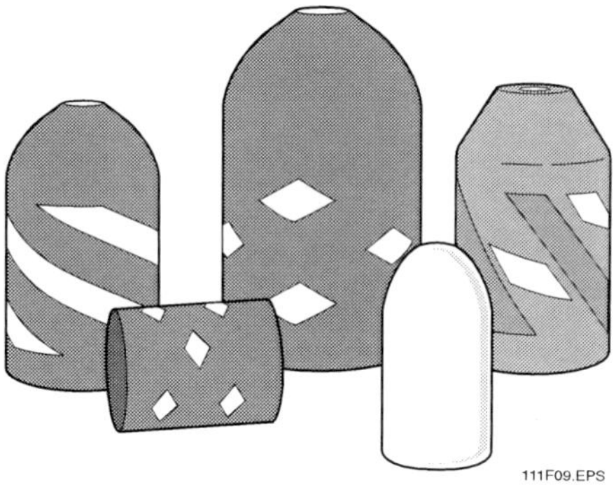

111F09.EPS

Figure 9 ◆ Foam pigs.

3.2.0 Gel Pigs

Gel pigs consist of a series of gelled liquid systems developed specifically for pipeline operations. Most pipeline gels are water-based, but a range of chemicals, solvents, and acids can be gelled.

Four main types of gel are used in pipeline applications. They are:

- Batching, or separator, gels
- Debris pickup gels
- Hydrocarbon gels
- Dehydrating gels

Gel can be pumped through any line that will accept liquids. Gel pigs can be used alone (in liquid lines) or in conjunction with other conventional pigs. When used with conventional pigs, gell pigs can improve the overall performance with very little risk of sticking a pig. Gel pigs do not wear out like conventional pigs, but are susceptible to dilution or gas cutting. Placing a conventional pig at

Instructor's Notes:

the back of a gel pig train will minimize the fluid bypass. *Figure 10* shows a utility pig/gel pig combination used for cleaning.

Special gels are available to seal gate valves during hydrostatic testing. Gels with controlled gellation times and controlled viscosity are often used for temporary pipeline isolation purposes.

4.0.0 ◆ CORROSION INHIBITOR INJECTION

Chemical inhibition can be used to control internal corrosion. Inhibitors are added to a liquid to change the characteristics of the liquid or to form a protective coat on the inner surface of the pipeline. Some inhibitors, instead of forming a film on the entire surface, deposit a film only on anodic or cathodic locations along the interior surface. Others work by solubilizing in the electrolyte (water) phase, and preventing the water from becoming an effective electrolyte.

4.1.0 Identifying Injection Locations

Knowledgeable company personnel should determine injection locations and ensure that the following steps are accomplished:

Step 1 Verify that the injection location is appropriate.

Step 2 Verify that the correct inhibitor is selected.

Step 3 Check the safety placard for health, flammability, explosiveness, protective gear, and reactivity.

Step 4 Confirm that the injection rate is set per company procedure.

4.1.1 Identifying Injection Rate

Injection rate is the amount of chemical to be injected within a specified time and is defined in units of measure in time, such as gallons per hour (gph) or gallons per day (GPD). Chemical injection rates are normally determined by knowledgeable company personnel and verified by using sight glass and tank level changes for reference.

4.1.2 Monitoring and Adjusting

All chemical injection systems should be closely monitored for correct injection rates. As a prerequisite, individuals authorized to make adjustments to the chemical injection rate should have a thorough knowledge of the injection system and of the permissible range for injection rates and volumes over time.

All chemical injection systems should be closely monitored for correct injection rates. As a prerequisite, individuals authorized to make adjustments to the chemical injection rate should have a thorough knowledge of the injection system and of the permissible range for injection rates and volumes over time.

Figures 11 and *12* show typical chemical injection systems. It is important to defer to company policies and procedures to make adjustments to the injection rates. However, sample injection calibration checklists follow:

Step 1 Ensure that there is power to the pump, either by electricity or gas.

> **NOTE**
> If using the pump to calibrate, the pump must be started and stopped.

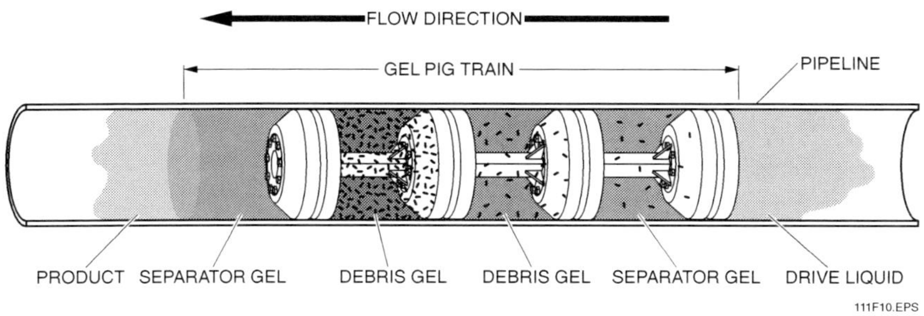

Figure 10 ◆ Utility pig/gel pig combination used for cleaning.

Audiovisual

Show Transparency 8 (Figure 10). Have a trainee explain how a utility pig/gel pig combination cleans the pipeline.

Homework

Have trainees review Sections 4.0.0–5.4.0 for the next session.

Materials

Ensure that you have everything required for the laboratory and testing during this session.

Laboratory

Have trainees simulate the procedure for identifying injection locations.

Classroom

Ask a trainee to define injection rate.

Teaching Tip

See the Teaching Tip for Section 4.1.2 at the end of this module.

Audiovisual

Show Transparency 9 (Figure 12). Ask a trainee to explain how the injection system operates.

Safety

Stress the importance of making sure the pump is not running when allowing the sight glasses to fill.

Figure 11 ◈ Typical chemical tanks injection system.

Step 2 Allow the sight glass to fill to a convenient level by using either the built-in fill valve or by shutting the pump discharge valve.

> **CAUTION**
> Be sure the pump is not running when allowing the sight glass to fill.

Step 3 Close the tank discharge valve between the tank and the sight glass.

Step 4 Mark the sight glass at the existing level.

Step 5 Open the pump discharge valve to the pipeline.

Step 6 Turn on the pump and allow it to run for one minute, then turn it off again.

Step 7 Read the change in the fluid level of the sight glass. On most sight glasses there will be an inside and an outside set of marks. The inside marks on the left are usually graduated in quarts per twenty-four hours. The inside marks on the right are in liters per twenty-four hours. The outside marks usually indicate the total volume of chemical available. The outside marks on the left indicate gallons. The outside marks on the right indicate liters.

Step 8 Re-calibrate two more times, average the results, and this will be your rate.

Step 9 Adjust as necessary to conform to required injection rates.

Step 10 Make sure to leave the tank valve in the OPEN position when you leave.

Instructor's Notes:

The following are general steps as outlined by Kenco Engineering. For more specific procedures, contact your supervisor and/or consult company literature.

> **NOTE**
> If using the ball valve to calibrate, the pump remains running.

Step 1 Depress the valve handle (or close the ball valve), closing the gauge glass off from the drum.

Step 2 Hold the valve closed for one minute and observe the change in level on the scale cast inside the gauge frame. The scale on the left-hand side of the gauge is calibrated in quarts per twenty-four hours and the right is calibrated in liters per twenty-four hours.

Possible items to document during inspections include:

- Amount of chemical in tank
- Injection rate
- Physical condition of system components
- Change in injection rate
- Follow-up items for next inspection
- Repairs affected/spare parts used

Always follow company policies and procedures when performing inspection tasks. *Figure 12* shows an example of a chemical injection system attached to a pipeline.

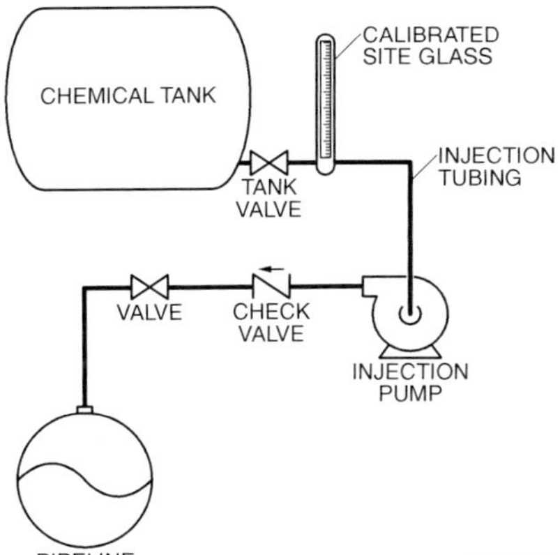

111F12.EPS

Figure 12 ◆ Chemical injection system attached to a pipeline.

5.0.0 ◆ PIPE CORROSION COUPONS

Pipeline corrosion **coupons** are the most widely used corrosion monitoring devices. The coupons are individual homogeneous specimens of steel, pre-weighed and cleaned, and are designed to represent the interior of the pipe wall. These test specimens are inserted into the system, exposed to fluids or gas in the pipeline for a specific period of time, then removed, cleaned, and re-weighed. The change in overall mass is recorded as weight loss in milligrams and the coupons are visually inspected for obvious signs of corrosion. The weight loss during the exposure time is extrapolated into weight loss for one year and normally expressed in **mils per year**. One mil equals 1/1000" and is expressed as (0.001).

Corrosion coupon documentation typically includes:

- Coupon number
- Coupon location
- Coupon initial weight
- Date installed
- Date removed
- Visual observations

Always follow company policies and procedures when removing and replacing corrosion coupons. *Figure 13* shows a corrosion coupon mailing pouch.

5.1.0 Weight Loss and Scale Coupons

Coupons are usually either weight loss coupons or scale coupons. Weight loss coupons can be any shape or size and can be made of any material. These coupons are usually homogeneous, although specialty coupons may be made up of more than one material. Results gained from the use of weight loss coupons are reported as weight loss over time. *Figure 14* shows different coupon types.

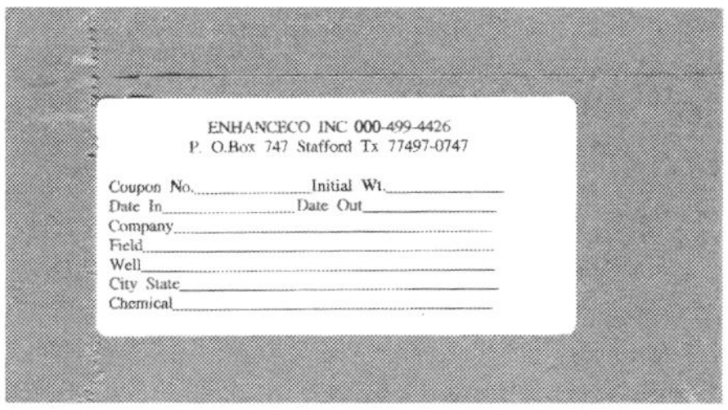

111F13.EPS

Figure 13 ◆ Corrosion coupon mailing pouch.

Classroom

Ask trainees to identify the information that may be documented during chemical injection system inspections.

Ask a trainee to explain the purpose of coupons. Have a trainee identify the unit usually used to express weight loss. Answer any questions trainees may have.

Audiovisual

Show Transparency 10 (Figure 14). Ask a trainee to explain how weight loss coupons work. Have a trainee explain how the results for the use of scale coupons are reported.

Classroom

Ask a trainee to distinguish between fixed and retractable mounting devices.

Audiovisual

Show Transparency 11 (Figure 16). Ask trainees to indicate which orientations are acceptable or not, and explain why.

Safety

Stress the importance of standing to the side of the coupon holder when opening the valve system.

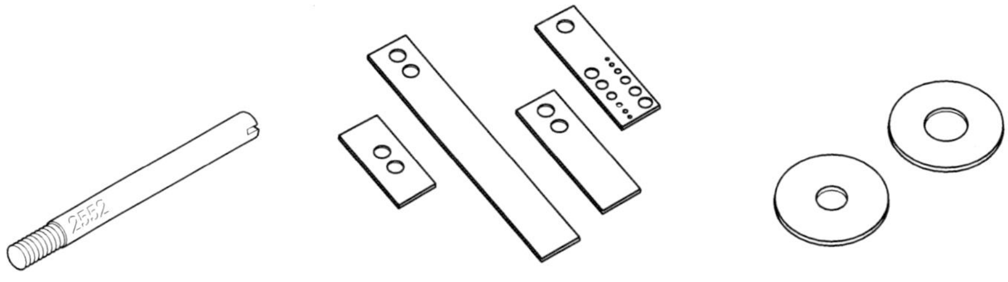

Figure 14 ◆ Different coupon types.

Scale coupons are designed to create differential pressure between the upstream and downstream sides of the coupon. This encourages the formation of scale on the coupon. Usually this is accomplished by drilling multiple holes of increasing diameter in a flat coupon. The results gained from the use of scale coupons are reported as weight gained and the type of scale formed.

5.2.0 Coupon Mounting Devices

Coupons are held in place by a variety of mounting devices. These fall into two categories: fixed and retractable. A fixed mounting device (*Figure 15*) typically can only be removed with the system de-pressurized. A retractable device can be removed and replaced while the pipeline remains in service. All mounting devices will have a non-conductive insulator to electrically isolate the coupon from the mounting device.

5.2.1 Coupon Orientation

It is important to properly orient flat coupons when installed in the path of product flow. Flat coupons should be fixed so product flows over their narrow profile. Because of their shape, the placement of rod and disc coupons is not as critical. *Figure 16* shows both correct and incorrect orientation and location of the coupon within the pipe.

5.3.0 Inserting Retractable Coupon Holders

The instructions contained in this section are written using an EnhanceCo, Inc. retractable coupon holder as an example and are intended to serve as general guidelines only. If you are using another brand of coupon holder, refer to the instructions provided by the manufacturer for detailed step-by-step procedures. This procedure applies to both threaded and flanged coupon holder models. *Figure 17* shows a retractable coupon holder.

> **NOTE**
> Do not touch the coupon with your bare hands. External contaminants such as acid present on human hands can affect weight loss of the coupon.

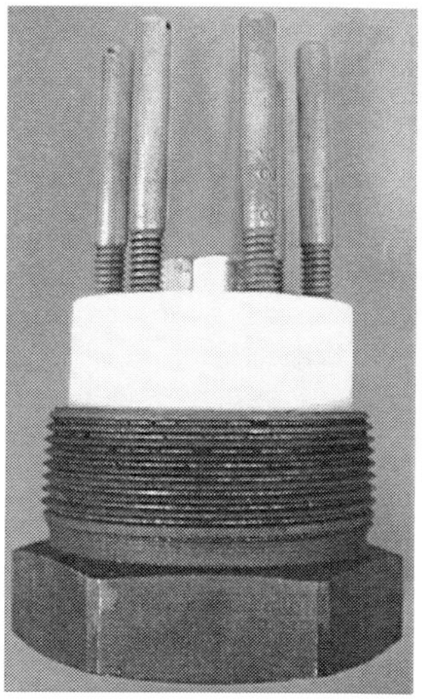

Figure 15 ◆ Fixed coupon mounting device.

Instructor's Notes:

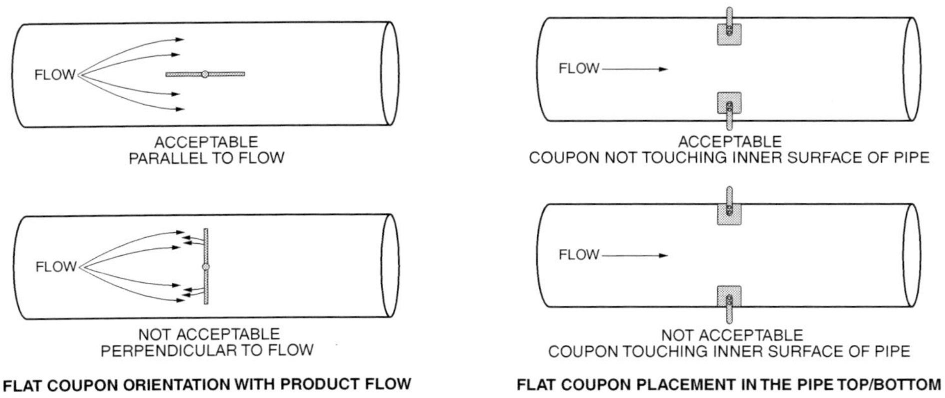

Figure 16 ◈ Coupon orientation

Step 1 Select the type of coupon holder to install.

Step 2 Install corrosion coupons on the insulator portion of the retractable holder.

Step 3 Fully retract the rod so the coupon holder housing protects the coupon.

Step 4 Tighten the packing gland until the probe rod will not move.

> **NOTE**
> Apply appropriate amount of thread sealant according to company specifications.

> **NOTE**
> Using a new flange gasket, install the coupon holder on the existing system valve. To tighten, follow company policy.

Step 5 Make sure the bleed valve is closed and the plug removed.

Step 6 Confirm that the probe rod is completely retracted.

Step 7 Open the system valve slowly and carefully.

> **WARNING!**
> Stand well to the side of the coupon holder when performing this step.

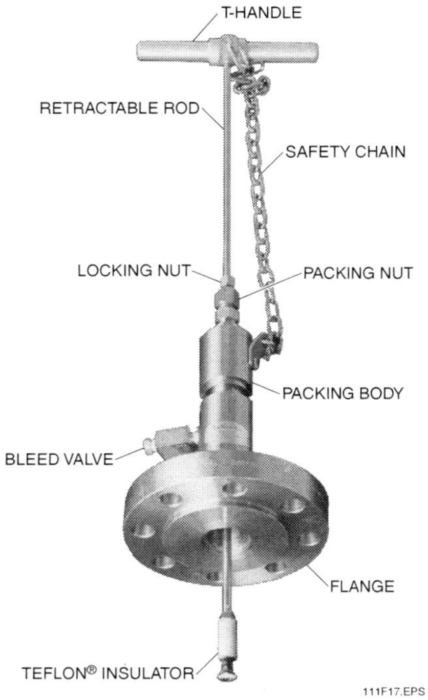

Figure 17 ◈ Retractable coupon holder.

Safety

Remind trainees that the coupon assembly is still pressurized even when it is isolated from the main line. Ask a trainee to explain the proper procedure to follow if the pressure does not bleed off.

Step 8 Check for leaks when you can hear that the holder has reached line pressure.

> **NOTE**
> Experienced individuals will be able to distinguish the sound of line pressure equalizing.

Step 9 Take the following measures to troubleshoot leaks that do occur:

- If a leak occurs at the threaded connection, shut the system valve, bleed pressure through the bleed valve (needle valve), remove the probe, re-tape the probe with thread tape and dope, then repeat Steps 4 through 8.
- If a leak occurs at the bleed valve, close the valve.
- If a leak occurs at the packing gland, tighten the packing slightly until the leak stops.
- If a leak occurs at the flange faces, bleed the pressure down, remove the tool and clean the gasket and flange faces.
- Replace the retractable coupon holder and re-torque the bolts according to the flange manufacturer's recommendations.

> **NOTE**
> Ensure there is no trash or other foreign matter on either side of the gasket.

Step 10 Loosen the packing gland slightly.

Step 11 Insert the coupon to the desired depth.

> **NOTE**
> Coupon should be in the proper location in accordance with company specification. Do not allow metal-to-metal contact. Care should be taken to position the coupon properly in the flow.

Step 12 Using the bleed valve, bleed out air through the valve, then close it.

Step 13 Reinstall the plug in the bleed valve.

Step 14 Follow company procedure to tag the probe and valve.

Step 15 On lines that can be pigged, if the probe could be damaged during the pigging process, tag both the upstream and downstream pig launching traps with warnings about devices in the lines.

Step 16 Mark the shaft of the retractable coupon holder just above the packing gland with an indelible marker. This will provide an indication if the instrument has moved.

Step 17 Tighten the safety chain.

5.4.0 Retracting Coupon Holders

These instructions refer to an EnhanceCo, Inc. retractable coupon holder as an example and are intended to serve as general guidelines only. This procedure applies to both threaded and flanged coupon holder models.

Step 1 Check to see whether the retractable coupon holder has moved. Look at the mark drawn above the packing gland.

Step 2 If the mark has moved, tighten the large packing nut until the retractable coupon holder is held securely.

Step 3 If the mark has not moved, proceed with caution.

Step 4 While holding the handle securely, loosen the packing nut slightly. The packing should not be leaking.

Step 5 Retract the coupon.

Step 6 As soon as the probe is fully retracted, shut the system valve.

Step 7 The retractable coupon holder is now isolated from the main line.

> **WARNING!**
> Do not forget that the coupon assembly is still pressurized.

Step 8 Bleed off pressure using the bleed valve. Leave the valve open after bleeding off pressure.

> **CAUTION**
> If the system valve does leak, you will hear pressure escaping from the bleed valve. If the pressure does not bleed off: shut the bleed valve, open and shut the system valve a couple of times, then shut the system valve and slowly re-bleed the coupon.

Instructor's Notes:

Step 9 Threaded model: Unthread the retractable coupon holder from the valve.

Step 10 Flanged model: Unbolt the retractable coupon holder from the valve.

Step 11 Clean the probe rod thoroughly before reuse.

> **CAUTION**
> Inspect the rod during cleaning. A badly pitted or bent rod may not seal at the packing gland. Contact appropriate personnel if this is the case.

Step 12 Visually inspect the entire retractable coupon holder for problems.

Step 13 Remove the coupon and replace it in its' storage sack. Record the date installed and removed, as well as any other pertinent observations in the space provided. *Figure 18* shows a coupon after the test period has been completed.

111F18.EPS

Figure 18 ◆ Coupon after the test period has been completed.

Step 14 Replace the coupon and reinstall the retractable coupon holder.

> **NOTE**
> If the retractable coupon holder will be left out for any length of time, you should plug the system valve with a bull plug, using thread tape, thread dope, or a blind flange.

Step 15 Send the coupons to the appropriate facility for analysis.

> **NOTE**
> Do not touch the coupon with your bare hands.

6.0.0 ◆ PROBES

Probes are sensing elements that extend into gas or liquid streams to measure variables. In corrosion detection and control, probes are used to measure electrical potential and the corrosivity of certain substances to determine a system's corrosive tendencies.

6.1.0 Electrical Resistance (ER) Probes

Electrical **resistance** probes work as automatic coupons. They function on the principle that as the diameter of a conductor decreases, the resistance to current flow increases. These probes pass a current through an element exposed to the fluid or gas product in a pipeline. As the element corrodes, there is a corresponding increase in resistance to current flow. This change in resistance is measured and recorded over a period of time.

The surface area of the exposed element can be calculated at the time of insertion into the product stream. As the change in resistance occurs, the change in surface area (metal loss) over time can be determined. This metal loss over time is then extrapolated into weight loss in mils per year.

These probes do not have to be immersed in an electrolyte to function properly. The element can be manufactured from almost any conductive metal. These probes are available in either fixed or retractable styles. ER probes require the use of a handheld reader, a transmitter, or a combination transmitter and datalogger to provide corrosion readings. *Figure 19* shows a typical ER probe.

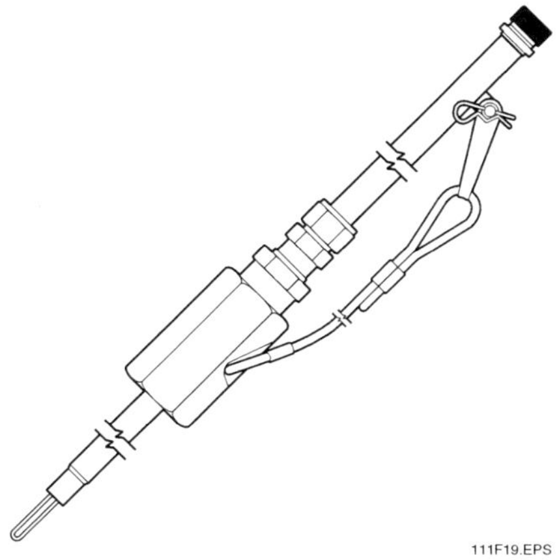

111F19.EPS

Figure 19 ◆ ER probe.

Safety

Stress the importance of inspecting the probe rod during cleaning. Ask a trainee to explain the possible problems resulting from a badly pitted or bent rod.

Teaching Tip

See the Teaching Tip for Sections 5.3.0–5.4.0 at the end of this module.

Performance Testing

Have each trainee perform Performance Task 1 to your satisfaction. Fill out a Performance Profile Sheet for each trainee.

Homework

Have trainees review Sections 6.0.0–6.5.0 and answer the Review Questions for the next session.

Materials

Ensure that you have everything required for teaching and testing during this session.

Audiovisual

Show Transparency 12 (Figure 19). Ask a trainee to explain how ER probes work. Discuss their advantages and disadvantages.

Audiovisual

Show Transparency 13 (Figure 20). Have a trainee explain the principle by which linear polarization probes function. Ask a trainee to explain how LPR probes work.

Show Transparencies 14 and 15 (Figures 21 and 22). Ask a trainee to explain how a galvanic probe works.

Show Transparency 16 (Figure 23). Have a trainee explain how hydrogen probes function. Ask a trainee to distinguish between the two types of hydrogen probes.

6.2.0 Linear Polarization Resistance (LPR) Probes

Linear polarization probes are capable of providing instantaneous corrosion rate information. Linear polarization probes function on the principle of polarization decay. These probes are available in two- and three-probe configurations. LPR probes must be immersed in an electrolyte to function properly. They work best in a very clean electrolyte.

LPR probes *(Figure 20)* require either a handheld reader or a combination transmitter and datalogger to provide readings.

6.3.0 Galvanic Probes

A galvanic probe functions using the inherent differences in potential between two dissimilar metals. The **probe elements** are commonly made of brass and steel. If there is an electrolyte between the two metal elements, the potential difference between the two metals causes a corrosion current to flow. This difference in potential, or current flow, is measured using a meter or datalogger connected to the probe. *Figure 21* shows a schematic diagram of a galvanic probe.

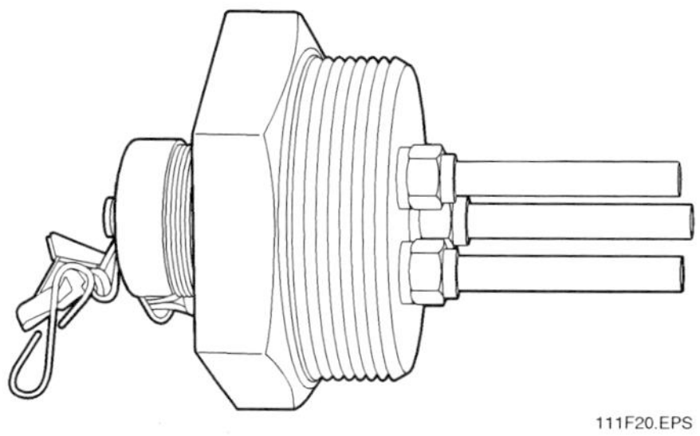

Figure 20 ◆ LPR probe.

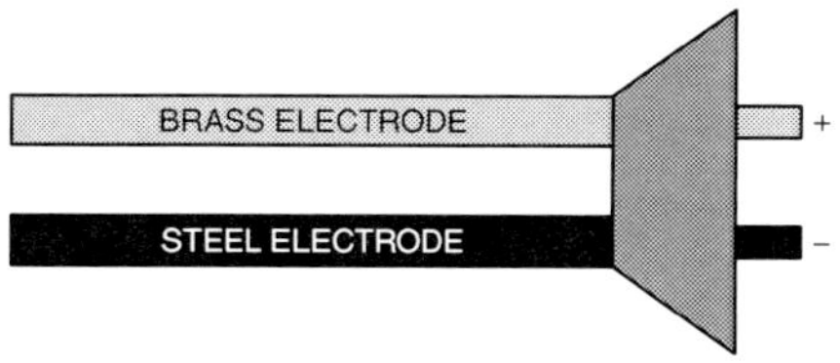

Figure 21 ◆ Schematic diagram of a galvanic probe.

The probe elements must be in contact with the electrolyte to function properly. Any change in the makeup of the electrolyte will result in a change in the reading obtained from the probe. Galvanic probes *(Figure 22)* are often used to detect oxygen ingress into systems and to predict corrosion inhibitor film life.

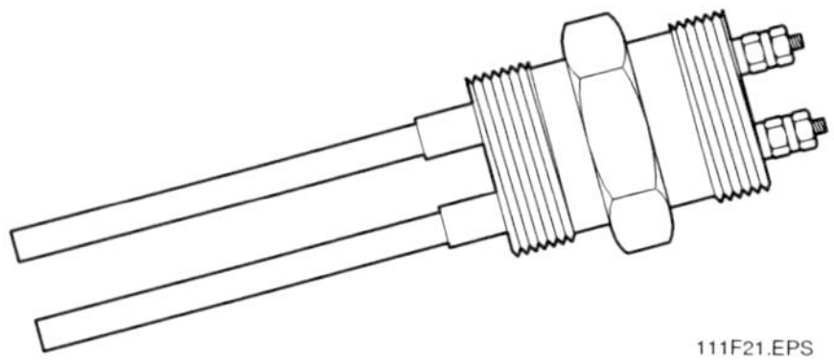

Figure 22 ◆ Galvanic probe.

6.4.0 Hydrogen Probes

Hydrogen probes function by reacting to the hydrogen film formed by the corrosion process. As the hydrogen film is formed, some of the atomic hydrogen migrates through the grain boundaries of the probe element and combine with other atomic hydrogen atoms in the void space inside the probe element to form molecular hydrogen. *Figure 23* shows a schematic diagram of a hydrogen probe.

When two hydrogen atoms combine to form molecular hydrogen, the molecule is too large to migrate back through the probe element grain boundary. As additional hydrogen molecules are formed, the void space starts to fill, causing a corresponding increase in pressure inside the element. This change in pressure is recorded over time. A hydrogen probe does not need to have its elements immersed in electrolyte to function properly.

There are two basic types of hydrogen probes. One type consists of a thin carbon steel tube with a solid rod inside creating a small annular space. This tube is inserted into the flow stream. Hydrogen atoms migrate from the product into the flow stream. As hydrogen molecules collect, the pres-

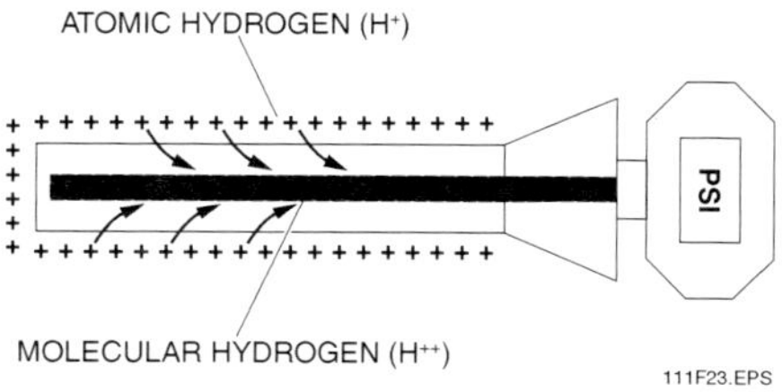

Figure 23 ◆ Schematic diagram of a hydrogen probe.

Instructor's Notes:

sure increases and is displayed by a pressure gauge. *Figure 24* shows a high-pressure hydrogen probe.

The second type, the patch probe, operates on the same principle. In this case, the patch is sealed to the exterior surface of the pipeline and collects hydrogen atoms that penetrate the wall. Hydrogen probes are more suited for identifying trends than for determining exact rates of corrosion. *Figure 25* shows a patch probe.

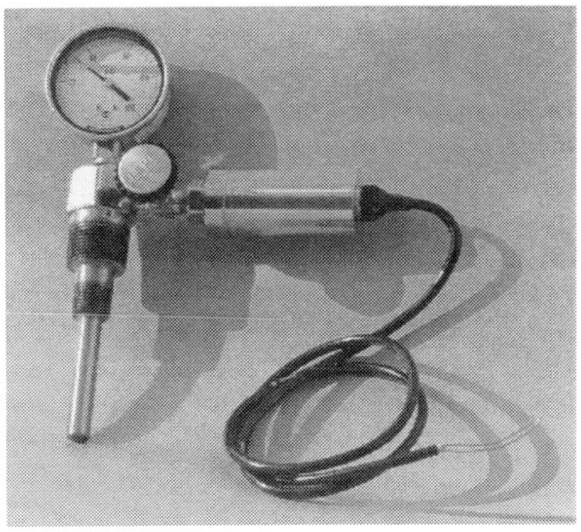

111F24.EPS

Figure 24 ◈ High-pressure hydrogen probe.

111F25.EPS

Figure 25 ◈ Patch probe.

6.5.0 Reading a Probe

The following steps are intended to serve as a generic guide to reading probes. Always follow your own company's policies and procedures and adhere to the specific step-by-step instructions provided with the equipment you are using, keeping in mind that different probe types require the use of different types of test equipment for accurate, reliable readings.

Step 1 Make sure the equipment you are using is in proper working order.

Step 2 Make sure the leads are in good condition.

Step 3 Remove the probe cover to gain access to the probe elements.

> **NOTE**
> Visually inspect the probe installation for leaks, frayed wires and overall condition. Document any unsatisfactory condition and notify the appropriate person of your findings.

Step 4 Properly connect the leads to the reading instrument and to the probe head.

Step 5 Document the reading on appropriate company forms.

Step 6 Securely replace the cover on the probe.

Summary

Controlling internal corrosion is an integral part of safe and profitable pipeline operations. The ability to correctly read probes, remove and replace coupons, and control chemical injection rates, facilitates the collection of meaningful data that is used to make operations decisions.

The ability of a qualified person to recognize and respond to abnormal operating conditions leads to reduced operating costs, decreased down time, and less environmental impact.

Teaching Tip

See the Teaching Tip for Section 6.5.0 at the end of this module.

Classroom

Go over the Review Questions. Answer any questions trainees may have.

Examination

Administer the Module Examination. Be sure to record the results of the Exam on Craft Training Report Form 200 and submit the results to the Training Program Sponsor.

Performance Testing

Have each trainee perform Performance Tasks 2 and 3 to your satisfaction. Fill out a Performance Profile Sheet for each trainee.

Ensure that all Performance Tests have been completed and Performance Profile Sheets for each trainee are filled out. Be sure to record the results of the testing on Craft Training Report Form 200 and submit the results to the Training Program Sponsor.

Review Questions

1. The term used to describe a limited area of corrosion characterized by cavities and pits is _____.
   a. general corrosion
   b. erosion corrosion
   c. localized corrosion
   d. under-deposit corrosion

2. The major gases that contribute to corrosive damage in a pipeline include the following *except* for _____.
   a. carbon monoxide
   b. oxygen
   c. carbon dioxide
   d. hydrogen sulfide

3. The normal range used to describe the tendency of water to be corrosive is typically shown on a scale with the minimum and maximum values of _____.
   a. –1 to +6
   b. 0 to 12
   c. 1 to 16
   d. 5 to 20

4. The four-part corrosion circuit must contain all of the following *except* _____.
   a. cathodic protection
   b. an anode
   c. a metallic path for current flow
   d. an electrolyte

5. Pipeline tools used for cleaning and dewatering tasks are commonly referred to as _____.
   a. smart pigs
   b. utility pigs
   c. mapping tools
   d. geometry tools

6. Some pipeline cleaning tasks are performed by utility pigs in combination with _____ pigs.
   a. metal-loss
   b. scanner
   c. smart
   d. gel

7. Corrosion inhibitors do all of in the following *except* _____.
   a. chemically removing corrosion damage
   b. forming a protective coat on the inner surface of the pipe
   c. depositing a film on the anodic and cathodic areas only
   d. preventing water from acting as an electrolyte

8. Chemical inhibitor injection rates are normally set by _____.
   a. the inhibitor vendor
   b. management procedure
   c. OSHA standard
   d. knowledgeable individuals in the company

9. The most widely used corrosion monitoring device is the _____.
   a. ultrasonic gauge
   b. visual inspection
   c. corrosion coupon
   d. mechanical pit gauge

10. All corrosion coupon mounting devices are _____.
    a. electrically isolated from the coupon
    b. fixed in place
    c. mounted on the top half of the pipe
    d. retractable to prevent pig damage

11. Retractable coupon holders come in _____ models.
    a. insulated and non-insulated
    b. automatic and manual
    c. metric and ASME-standard
    d. threaded and flanged

12. When inserting a retractable coupon holder, refer to the detailed procedure contained in _____.
    a. this document
    b. DOT standards for corrosion testing
    c. the manufacturer's instructions
    d. company safety procedures

13. Electrical resistance probes provide corrosion readings with all the following *except* for a(n) _____.
    a. ultrasonic transmitter and transducer
    b. handheld reader
    c. transmitter
    d. transmitter/datalogger combination

14. Galvanic probes function by detecting the difference in potential between _____.
    a. corrosion damaged areas and the pipe wall
    b. dissimilar metals
    c. the probe elements
    d. anodic and cathodic areas on the pipe

15. Hydrogen atoms migrate through the outer skin of the hydrogen probe element and combine with other hydrogen atoms to form _____.
    a. corrosion
    b. a film
    c. molecular hydrogen
    d. a barrier to corrosion

Instructor's Notes:

GLOSSARY

Glossary

This glossary will help you learn some terms specific to internal corrosion control. Trade Terms introduced in this module appear in italic text.

Coupon: A metal strip inserted into a system to monitor corrosion rate and inhibitor effectiveness.

**Corrosion:* The deterioration of a material, usually a metal, that results from a reaction with its environment.

**Erosion-corrosion:* A conjoint action involving corrosion and erosion in the presence of a moving corrosive fluid or a material moving through the fluid, leading to accelerated loss of material.

General corrosion: Used to describe corrosion that involves metal loss distributed over a wide area with no discernible pitting.

Insulating assembly: A device utilized to hold and insulate coupon from surrounding metals.

Isolation or process valve: A device utilized to isolate the coupon holder from the process.

Localized or pitting corrosion: Refers to corrosion that only affects a defined area on the metal surface. Pitting corrosion refers to the formation of cavities or pits.

Mils per year: The rate of corrosion measured in 1/1000 of an inch/year.

National Electrical Code® (NEC): Published by the National Fire Protection Association, the NEC establishes minimum requirements for the installation of electrical services.

Parts per million (PPM): Concentration of chemical/inhibitor per volume treated as calculated to control internal corrosion.

Pressure rating: Maximum allowable working pressure (MAOP).

Probe element: Device utilized to monitor corrosion through the changes in electrical resistance of the element material in an electrolyte. Actual, exposed working portion of the probe.

Probe span: The life of the probe element in a unit of time based on the expected corrosion rate.

Resistance: An electrical property that opposes the flow of current through a circuit. Resistance (R) is measured in ohms.

Retrieval tool/extractor tool: A device used to remove and replace coupons.

Under-deposit corrosion: Defines an area of degradation occurring beneath a deposit due to a difference in potential between the area under the deposit and the surrounding area.

Water of condensation: Condensation that forms when water is present in the product and the temperature of the exterior of the pipeline drops significantly below the temperature of the product.

REFERENCES & ACKNOWLEDGMENTS

Additional Resources

This module is intended to present thorough resources for task training. The following reference works are suggested for further study. These are optional materials for continued education rather than for task training.

The Art of Pigging, 1998. Girard Industries. Houston, Texas.

Electrical Resistance E/R™ Monitoring. Metal Samples Company. Munford, AL.

Linear Polarization Resistance (LPR). Metal Samples Company. Munford, AL.

Figures

EnhanceCo, Inc.	111F01, 111F13, 111F15, 111F17, 111F24, 111F25
Metal Samples	111F18
PII Group Limited	111F07

Instructor's Notes:

MODULE 61111-02—TEACHING TIPS

The following are suggested activities or instructional methods to help you teach the material in this AIG.

General

When you call on someone to answer a question, the rest of the class relaxes or even tunes out because they expect that the question and answer will take place only between you and the trainee you called on. Instead, use this technique to involve more trainees in answering questions and to keep them on their toes.

1. Ask trainees to define a term or explain a concept.
2. After one trainee has answered, ask a trainee seated nearby if the answer is right. Then ask whether a trainee in the back of the room agrees.
3. Ask trainees to explain why they think an answer is right or wrong.
4. Use the session to clear up incorrect ideas and encourage trainees to learn from their mistakes.

Section 4.1.2 *Monitoring and Adjusting Chemical Injection Systems*

This exercise will give trainees practice in monitoring and adjusting chemical injection systems. Trainees will need pencils, paper, appropriate personal protective equipment, and their Trainee Guides. You will need to arrange access to a chemical injection system and provide all necessary equipment and tools, including company guidelines if applicable. Alternatively, have trainees simulate or observe the process. Allow 1 to 2 hours for this exercise.

1. Have trainees familiarize themselves with the process of monitoring and adjusting chemical injection systems. Ensure that they understand the difference between inside marks and outside marks on the sight glass.
2. Supervise trainees as they follow the procedure in their Trainee Guide or company guidelines, record their findings, and calculate the average rate.
3. Check trainees' final averaged rate before they adjust the injection system.
4. Ask trainees if they have any questions.

Sections 5.3.0–5.4.0 *Inserting and Retracting Coupon Holders*

This exercise will give trainees an opportunity to practice inserting and retracting coupon holders. Trainees will need pencils, paper, appropriate personal protective equipment, and their Trainee Guides. You will need to arrange access to a coupon installation and provide all necessary equipment and tools, including manufacturer's instructions. Alternatively, have trainees simulate or observe the process. Allow 1 to 2 hours for this exercise.

1. Have trainees identify the type of coupon holder to be installed.
2. Supervise trainees as they install the coupon holder. If leaks occur, have trainees troubleshoot them. If no leaks occur, have trainees simulate troubleshooting measures.
3. After the successful installation of the coupon holder, have trainees retract it. Then have them perform the following final steps:
   - clean the probe rod
   - visually inspect the coupon holder
   - record the date the coupon was installed and removed
4. Discuss the process with trainees. Answer any questions they may have.

Section 6.5.0 *Reading a Probe*

In this exercise, trainees will practice reading a variety of probes. Trainees will need pencils, paper, appropriate personal protective equipment, and their Trainee Guides. You will need to provide a variety of probes with manufacturer's instructions and company forms and arrange access to equipment for testing. Alternatively, have trainees simulate or observe the process. Allow 1 to 2 hours for this exercise.

1. Discuss methods of ensuring that equipment is in proper working order and that the leads are in good condition.
2. Divide the class into small groups. Set up stations for the use of each probe. If time permits, have trainees practice using each type of probe. Remind trainees to follow the manufacturer's instructions for each type of probe and complete company forms.
3. Alternatively, assign one group to each station. After all groups have finished, have each group make a presentation to the class, explaining how to use the probe and discussing their results.
4. Ask trainees if they have any questions.

MODULE 61111-02 — ANSWERS TO REVIEW QUESTIONS

Answer	Section
1. c	2.1.0
2. a	2.2.1
3. b	2.2.2
4. a	2.2.3
5. b	3.0.0
6. d	3.0.0
7. a	4.0.0
8. d	4.1.1
9. c	5.0.0
10. a	5.2.0
11. d	5.3.0
12. c	5.3.0
13. a	6.1.0
14. b	6.3.0
15. c	6.4.0

CONTREN™ LEARNING SERIES — USER UPDATES

The NCCER makes every effort to keep these textbooks up-to-date and free of technical errors. We appreciate your help in this process. If you have an idea for improving this textbook, or if you find an error, a typographical mistake, or an inaccuracy in NCCER's Contren™ textbooks, please write us, using this form or a photocopy. Be sure to include the exact module number, page number, a detailed description, and the correction, if applicable. Your input will be brought to the attention of the Technical Review Committee. Thank you for your assistance.

Instructors – If you found that additional materials were necessary in order to teach this module effectively, please let us know so that we may include them in the Equipment/Materials list in the Instructor's Guide.

Write: Curriculum Revision and Development Department
National Center for Construction Education and Research
P.O. Box 141104, Gainesville, FL 32614-1104

Fax: 352-334-0932

E-mail: curriculum@nccer.org

Craft ______________________ Module Name ______________________

Copyright Date ____________ Module Number ____________ Page Number(s) ____________

Description

____

____

____

____

(Optional) Correction

____

____

____

(Optional) Your Name and Address

____

____

____

Index

Index

Index

Index